ANNUAL REVIEW OF
GENOMICS AND HUMAN GENETICS

ANNUAL REVIEW OF GENOMICS AND HUMAN GENETICS

VOLUME 2, 2001

ERIC LANDER, *Editor*
Whitehead Institute for Biomedical Research,
Massachusetts Institute of Technology

DAVID PAGE, *Associate Editor*
Howard Hughes Medical Institute,
Massachusetts Institute of Technology

RICHARD LIFTON, *Associate Editor*
Yale University

www.AnnualReviews.org   science@AnnualReviews.org   650-493-4400

ANNUAL REVIEWS
4139 El Camino Way • P.O. BOX 10139 • Palo Alto, California 94303-0139

ANNUAL REVIEWS
Palo Alto, California, USA

International Standard Serial Number: 1527-8204
International Standard Book Number: 0-8243-3702-6

TYPESET BY TECHBOOKS, FAIRFAX, VA
PRINTED AND BOUND IN THE UNITED STATES OF AMERICA

Annual Review of Genomics and Human Genetics,
Volume 2

CONTENTS

ERRATA
An online log of corrections (if any) to the *Annual Review of Genomics and
Human Genetics* chapters may be found at http://genom.AnnualReviews.org

RELATED ARTICLES

Errata

Kalim U. Mir and Edwin M. Southern, SEQUENCE VARIATION
IN GENES AND GENOMIC DNA: Methods for Large-Scale Analysis
(*Annu. Rev. Genomics Hum. Genet.* **2000. 1:329–60**)

Erratum: *ANALYSIS OF SINGLE-BASE CHANGES/Discovery of
Single-Nucleotide Polymorphisms and Mutations/Efficient Scanning Methods:
Analysis of Conformational Change*

Please note, the following sentence should be included after the third sentence of
paragraph 2.

"Therefore the sites of cleavage will be different in variants."

Erratum: ANALYSIS OF SINGLE-BASE CHANGES/Discovery of Single-
Nucleotide Polymorphisms and Mutations/Scanning and Resequencing on
Oligonucleotide Arrays

Please note, the second line of paragraph 2 erroneously refers the reader to
"Figure 2." This line should refer the reader to Figure 3.

Erratum: Figure 2

Please note revised version of Figure 2 (see online Errata, http://genom.
annualreviews.org).

Erratum: Figure 3

Please note the following revisions to Figure 3 (see online Errata, http://genom.
annualreviews.org) and Figure 3 Legend.

"Multiplex resequencing on scanning arrays. Scanning arrays are fabricated by
a flood method. Clockwise: A lozenge-shaped reaction cell produces stripes of
oligonucleotides that form a tiling-path of complements to the target of known
sequence. If oligonucleotides are tethered by their 5′ ends, they can act as primers
in a polymerase chain extension reaction. The target is resequenced when applied to
a scanning array split into channels with a different base applied to each. When all
four fluorescent bases are applied in a single channel, analysis can be multiplexed,
each channel resequencing a separate target."

Annu. Rev. Genomics Hum. Genet. 2001. 2:1–8

Hundred-Year Search
for the Human Genome

Frank Ruddle

*Department of Molecular, Cellular, and Developmental Biology, Yale University,
New Haven, Connecticut 06520; e-mail: frank.ruddle@yale.edu*

Key Words   genome, human genome initiative, human gene map

■ **Abstract**   The human genome has been an article of interest since the rediscovery of Mendel's laws at the turn of the century (1900–1901). Much progress was made during the first decade (1900–1910) with respect to our understanding of fundamental aspects of human genetics, such as the chromosomal basis of heredity, biochemical genetics, and population genetics. The development of these fields of inquiry languished for several decades but then advanced rapidly. However, human gene mapping stalled until 1970 when somatic cell genetic methods were introduced. The contributions of hybrid cell mapping to physical methods of genome analysis are described, and its legacy as an antecedent to the human genome initiative is discussed. Lastly, some properties of the 2000–2001 version of the human genome are briefly outlined.

> GENOME: the haploid chromosome set together with its inclusive genes. A term coined by Winkler in 1920 as a conjunction between GENe and chromosOME.

The year 2001 has seen the realization of a major accomplishment: a sequence map of the human genome. The mouse genome will be close behind, and sequence maps for major model organisms are already in hand: *Saccharomyces*, *Caenorabditis*, *Drosophila*, *Arabidopsis*, plus a number of protostome species. Certainly, things will not be the same. Much the same was said a century ago when Mendel's laws of genetic transmission were rediscovered. The search for the human genome had its beginnings then without the searchers knowing exactly what was being sought.

We have to give our forebears credit; they accomplished a great deal in a few years following the rediscovery. Mendel's laws were thoroughly confirmed and exceptions shown to prove the rule. In addition to plants, fowl (Bateson in 1902) and mice (Cuenot in 1902) were shown to be no different than peas with respect to genetic transmission. Chromosomes were demonstrated to be the fundamental vehicles of heredity by Sutton & Boveri in 1902. Castle introduced *Drosophila* as a model organism. Morgan and his associates then used it to show that genes are the fundamental units of heredity, strung in linkage alignment along the chromosome. The introduction of cytogenetics bridged the gap between chromosome cytology and genetic transmission. Population genetics (Weinberg 1908; Hardy 1908) was born only to mature some 30 years later. Genetics was established as a vital discipline, with major consequences for our ways of thinking about all of biology, but especially evolution and development (8, 33).

But what of the human genome? Here also, surprising advances were made. As early as 1902, Garrod showed that *Alcaptonuria* mendelized and attributed it to a gene mutation that affected tyrosine metabolism. Biochemical genetics was thus anticipated to flourish only decades later. X-chromosome linkage was first demonstrated in *Drosophila*, based on earlier findings on sex determination that involved heteromorphic sex chromosomes. X linkage was demonstrated in humans soon afterward, and from then, there has been a steady assignment of genes to the X chromosome as the first human linkage group. However, the acquisition of genes assignable to human autosomal linkage groups would make little progress for at least 50 years (8, 33).

Why was it so difficult to assign genes to autosomal linkage groups in humans? Primarily, it is the large size of the human genome, measuring roughly 3 billion base pairs or 3000 cM. The probability of any two segregating genes being located within 10 cM or less is extremely low. The problem is exacerbated by the difficulty of finding suitable genetic variants that coexist within families amenable to genetic analysis. A large number of serological markers were accumulated following the work of Landsteiner at the turn of the twentieth century (15), but evidence for their linkage was virtually nonexistent as late as 1970 (10). Some exceptions did exist, relating to closely linked genetic clusters, such as the globins and the major histocompatibility complex. Autosomal linkage in humans was still an unsolved problem seventy years after the rediscovery of Mendel's laws.

The intellectual environment in the 1960s was vastly different than that of the early 1900s. DNA was shown to be the chemical basis of heredity by Avery and associates in 1944 (1), and an elegant structural model that predicted its functional properties was put forward by Watson & Crick in 1953 (30). One could now think of the linkage problem in terms of a single linear DNA duplex molecule that extends the length of the chromatid, the genes encoded along its length. This model served as the basis of a purely physical approach to gene mapping. Previous years had seen the formulation of parasexual approaches to genetic analysis based on analogs to Mendelian transmission genetics. These were, notably, somatic recombination in *Drosophila* by Stern in 1936 (26), sexuality in bacteria by Lederberg & Tatum in 1946 (16), and chromosome segregation in fungi by Pontecorvo in 1958 (19). The question in the minds of some in the early 1960s was whether it would be possible to fashion a human somatic cell genetics system that could solve the linkage problem by a purely parasexual approach.

The requirements for a functional somatic cell genetics turned out to be the following: (*a*) a tissue culture system to permit the propagation of somatic cells in vitro (12), (*b*) a cell hybridization system that allowed the combination of genetically diverse parental cells, for example, mouse and human (2), (*c*) an enrichment protocol that served to select for interspecific hybrid cells (17, 29), (*d*) segregation of human chromosomes from the hybrid cells to allow the assignment of genes to specific human chromosomes (31), (*e*) genetic marker systems that allowed the facile discrimination of parental orthologous genes or their products (21), and (*f*) cytological techniques that permitted the specific identification of the individual human chromosomes (6). These desiderata were realized by the late 1960s. In addition to the cell hybridization approach, methods were devised to enable the transfer of partial donor parental genomes into recipient somatic cells by a variety of methods, including single chromosomes (9), chromosome fragments (14), and DNA (32), thus allowing the ordering of linked genes over short distances and at high resolution. Approximately 100 genes comprising 20 linkage groups were cited in the report of the first Human Gene Mapping Conference in 1973 (3), and this number, which was achieved largely by somatic cell genetic methodologies, grew to more than 2000 entries in 24 linkage groups by the Tenth Human Gene Mapping Workshop in 1989 (22).

Somatic cell genetics contributed significantly to our ability to analyze and understand the human genome. First, by cloning a human chromosome or a part of a chromosome in a hybrid cell, one could begin to deconstruct a large complex genome into smaller elements that could then be reconstructed into a fully characterized (mapped) entity. This approach was further elaborated by the application of microbial cloning systems such as yeast artificial chromosomes (YACs), bacterial artificial chromosomes (BACs), cosmids, and plasmids. Second, somatic cell genetic systems were introduced that made use of DNA methodologies to map genes. These were principally of two types: (*a*) in situ hybridization (18) and (*b*) restriction fragment length polymorphism (RFLP) discrimination between interspecific donor and recipient orthologous gene sequences in hybrid cells (28). Concomitantly, investigators realized that RFLPs were sufficiently frequent within the human population so that this approach could be used profitably to map genes that segregated within families (4). Third, the gene transfer methodologies pioneered by somatic cell genetics were adapted to the genetic transformation of complex organisms. Nuclear injection of one-cell mouse embryos allowed the stable incorporation of foreign genes into the murine germ line, the so-called transgenic mouse (11). Homologous recombination was demonstrated in murine embryonal stem cells, allowing the targeting of genes for the purpose of producing loss of function (knockout) or gain of function (knockin) mutations (25). Transgenesis now provided a means to modify the genome at will, allowing its functional analysis in ways not previously possible. It also ushered in a new era where humans could produce new strains of organisms with novel genetic properties. Fourth, the success of somatic cell genetics in producing a human gene map and the demonstration of its benefits in biomedicine paved the way for the acceptance of a national program, The Human Genome Project, to fully map and sequence the human genome.

The Human Genome Project, conceived in 1984 and implemented in 1990, had as its principal goals the complete mapping and sequencing of the human genome by 2003. The research program divided into three principal parts. One was concerned with Mendelian analysis of the genome, using DNA technologies. The second employed physical mapping approaches to deconstruct and reconstruct the genome, using gene-cloning methodologies. The third sought to sequence the whole genome in combination with approaches one and two. All three efforts were to be interconnected informatically.

The Mendelian mapping effort was greatly enhanced by the discovery of microsatellite elements that are scattered throughout the genome in abundance. In conjunction with the newly developed polymerase chain reaction (PCR) technology, these simple sequence length polymorphisms (SSLPs) serve as genetic markers throughout the genome, providing a framework for the mapping of unmapped genes. Currently, thousands of SSLP markers have been mapped at a high density, greatly enhancing the expeditious mapping of genes in families. It must be emphasized that only the Mendelian mapping approach can be used to map genes that encode complex traits that are expressed on the organismal level, for example, eye color, behavior, and obesity. One of the significant accomplishments of human genetics in recent years has been the mapping of genes that encode complex traits and their characterization at a molecular level by positional cloning (5).

Physical mapping has provided a means to disassemble the genome into overlapping subgenomic fragments that can be characterized individually and then reassembled into the whole. Reassembly is made possible by the identification of unique DNA markers termed sequence tagged sites (STSs) that allow the joining of individual overlapping fragments into contiguous fragment arrays (contigs). A particularly useful development has been the sequencing of mRNAs to produce expressed sequence tags (ESTs) that provide probes to localize expressed genes. EST probe libraries have a number of useful properties, such as correlating gene expression with particular cells or tissue types, use as STSs, and the isolation of genes by positional cloning, among others (5).

The physical sequencing of the genome proceeded using standard sequencing methodologies. Present day sequencing machines can produce approximately 400 kb of sequence per day, and a newer model under development is designed to produce one megabase of sequence daily. Thus, a battery of only 100 machines can produce a rough sequence of the human genome in about one year, taking into account 10-fold redundancy. The current world sequencing capacity of academic, governmental, and commercial laboratories is sufficiently large to allow the sequencing of the complex genomes of additional organisms, such as the laboratory mouse (largely completed), selected primates, representatives of the orders of mammals, important crop plants, and a variety of additional life forms of commercial and biomedical interest. The information gained will have a major impact on our understanding of evolution, development, and a variety of biomedical concerns. Already, comparative genomics has established an unusually high degree of genetic relatedness between diverse organisms, to a degree entirely unsuspected even a few years ago.

Comparative genetics can be expected to take on new importance. Already, sequence comparisons of the genomes of species with various degrees of evolutionary relatedness are being used to discover noncoding DNA motifs of functional significance, such as promoters, enhancers, and silencers, among others (24).

Multiple alignments that use sequence information from multiple species promise to increase both the resolution of motif definition and the probability of motif discovery (27). Genome comparisons are beginning to yield interesting information. Gene number can now be precisely determined and compared between species: *Hemophilus* with 1709, *Saccharomyces* with 6241, *Caenorhabditis* with 18,424, *Drosophila* with 13,601, and humans with an estimate between 30,000 (14a) and 80,000 (20). It is interesting that as genome size increases, gene number increases correspondingly less. This suggests that higher organisms may engage in combinatorial interactions between genes to achieve developmental complexity. Alternatively, noncoding DNA in higher forms may have taken on functional roles not present or less developed in lower organisms. It comes as a definite surprise that the lowly worm should have more genes than the high-flying fly (20).

Comparative genetic analysis within organisms will be equally important. The analysis of single nucleotide polymorphisms (SNPs) shows that genes, including coding and noncoding regions, are highly variable within populations. SNPs correlate with the functional properties of particular genes that segregate within kinships. In some instances, the SNP may be causal to the functional variant, but in most instances it is neutral, having a close linkage relationship with a DNA polymorphism that is causal. In this latter case, the SNP may or may not correlate with a functional property. Genes may accumulate SNPs randomly over time; and viewed from a population perspective, the SNP signature may be highly complex, with many thousands of SNP combinations possible for a particular gene. However, when one examines the linkage combinations of SNPs, one finds that chromosomally phased SNPs or haplotypes (HAPs) may occur orders of magnitude less frequently and show a direct relationship with the functional properties of a variant gene. The human *beta2-adrenergic* receptor gene harbors 13 SNPs, allowing a theoretical maximum of 8192 SNP combinations. Haplotype analysis shows that only 12 major HAPs exist in the population at large. Interestingly, major differences exist in the distribution of HAPs in racially distinct populations. Moreover, certain HAP pairs show distinct bronchodialator responsiveness to a particular beta-agonist in asthmatics (7). The implications for haplotype analysis in human populations are immense with respect to diagnosis, prognosis, and selective modes of treatment for particular disease conditions. Genome analysis now makes possible a practical pharmacogenetic approach to the management of human disease. Now that a first pass at sequencing the human genome is drawing to a close, a new frontier is opening, namely, an in depth analysis of genome sequence variability.

Genome sequencing is bringing to light genetic modules of unusual structural design and function. One such is the *Hox* gene cluster that plays an important role in axial patterning. *Hox* gene cluster(s) are broadly distributed in animal metazoans, with the possible exception of the Porifera. Mammals have four Hox clusters. Those in humans and mice are similar in both structure and function, strongly supporting their origination by genome duplication events. Each mammalian cluster is approximately 100 kb in length. Recently, full sequence comparisons between Hox cluster orthologs in various species have been made possible. Such comparisons reveal a high degree of similarity in both coding and noncoding regions over divergent times that span well over 300 million years (23). Alignments can be readily made across the entire cluster, as demonstrated by comparisons between the human A cluster and its shark (*Heterodontus*) ortholog (13). The spacing between coding regions is highly conserved and suggests a special functional relationship between coding and noncoding regions. This feature is further emphasized by an almost total absence of line and sine elements in the *Hox* clusters, although these elements are ubiquitously present elsewhere in the genome. Only a full nucleotide sequence comparison could reveal these interesting and probably significant features. One can predict with some certainty that additional interesting genomic domains will be found, as genome sequencing and sequence comparisons become commonplace.

The molecular analysis and manipulation of the human genome is beginning to branch out in a number of directions. Informatics and EST probes have allowed the identification of virtually all of the expressed genes. Sequence probes specific for particular gene families allow the recovery of these families. The genome-wide coding sequences are being used to generate microarrays to monitor expression patterns of the genes at the cell and tissue level at specific stages of the life cycle and under different physiological conditions. DNA microarrays in conjunction with chromatin-derived probes are being used to establish the web of control interactions between DNA binding proteins and their gene targets, again, under specific developmental and physiological conditions. Now, all the human genes can be expressed as protein products in gene expression systems, giving rise to the new field of proteomics (proteome: the full set of proteins encoded by the genome). Protein and peptide microarrays are being used to screen for protein, DNA, and small molecules, including drugs that interact or bind to particular proteins. We are suddenly on the threshold of understanding the human genome not only as a static linear set of genes, but as a dynamic web of informational interactions that sustain the developmental process and allow its evolutionary modification.

What is the human genome? It will be the sum total of a number of genomic studies, namely, (*a*) the physical genome: the gene map, control motif map, and full DNA sequence; (*b*) the functional genome: an understanding of what the genes do; (*c*) the population genome: variation of genes in the human population; (*d*) the comparative genome: the comparison of the human genome with other genomes; and (*e*) the integrative genome: the functional interaction of genes within the genome, among others to be invented in the future.

And what can we say of the future of the human genome and genomic studies? Thomas Hunt Morgan (8), when reviewing the progress made by *Drosophila* genetics in 1932, concluded his remarks by saying, "Should you ask me how these discoveries [i.e., those of the future] are to be made, I should become vague and resort to generalities." Following his example, I would only conclude that we will continue to study our genome, eventually understand it, and in all likelihood in the not too distant future, for good or bad, modify it.

Visit the Annual Reviews home page at www.AnnualReviews.org

LITERATURE CITED

1. Avery OT, MacLeod CM, McCarty M. 1944. Studies on the chemical nature of the substance inducing transformation of pneumococcal types. *J. Exp. Med.* 79:137–58

2. Barski G, Soriel S, Cornefert F. 1960. Production dans des cultures in vitro de deux souches cellulaires en association, de cellules de caractere "hybride." *C. R. Acad. Sci. Paris* 251:1825–27

3. Bergsma D, ed. 1973. *Human Gene Mapping New Haven Conference*. New York/ London: Intercont. Med. Book Corp.

4. Botstein D, White RL, Skolnick M, Davis RW. 1980. Construction of a genetic linkage map in man using restriction fragment length polymorphisms. *Am. J. Hum. Genet.* 32:314–31

5. Brown TA. 1999. Genomes. New York: Wiley

6. Caspersson T, Lomakka G, Zeck L. 1971. The 24 fluorescence patterns of the human metaphase chromosomes. Distinguishing characters and variability. *Hereditas* 67:89–102

7. Drysdale C, McGraw DW, Stack CB, Stephens JC, Judson FS, et al. 2000. Complex promoter and coding region beta2-adrenergic receptor haplotypes alter receptor expresssion and predict in vivo responsiveness. *Proc. Natl. Acad. Sci. USA* 97:10483–88

8. Dunn LC. 1965. *A Short History of Genetics*. New York: McGraw-Hill

9. Fournier REK, Ruddle FH. 1977. Stable association of the human transgenome and hostmurine chromosomes demonstrated using trispecific microcell hybrids. *Proc. Natl. Acad. Sci. USA* 74:3937–41

10. Giblett ER. 1969. *Genetic Markers in Human Blood*. Oxford, UK: Blackwell Sci.

11. Gordon JW, Scangos GA, Plotkin DJ, Barbosa JA, Ruddle FH. 1980. Genetic transformation of mouse embryos by microinjection of purified DNA. *Proc. Natl. Acad. Sci. USA* 77:7380–84

12. Harris M. 1964. *Cell Culture and Somatic Variation*. New York: Holt, Reinhart, & Winston

13. Kim C-B, Amemiya C, Bailey W, Kawasaki K, Mezey J, et al. 2000. Hox cluster genomics in the horn shark, *Heterodontus francisci. Proc. Natl. Acad. Sci. USA* 97:1655–60

14. Klobutcher LA, Ruddle FH. 1979. Phenotype stabilization and integration of transferred material in chromosome-mediated gene transfer. *Nature* 280:657–60

14a. Lander ES, Rogers J, Waterson RH, Hawkins T, Gibbs RA, et al. 2001. Initial sequencing and analysis of the human genome. *Nature* 409:860–922

15. Landsteiner K. 1901. Über Agglutinationserscheinungen normalen menschlichen Blutes. *Wien. Klin. Wochenschr.* 14:1132–34

16. Lederberg J, Tatum EL. 1946. Gene recombination in *Escherichia coli. Nature* 45:558

17. Littlefield JW. 1964. Selection of hybrids from matings of fibroblasts in vitro and their presumed recombinants. *Science* 145:709–10

18. Pardue ML, Gall JG. 1970. Chromosomal localization of mouse satellite DNA. *Science* 168:1356–58

19. Pontecorvo G. 1958. *Trends in Genetic Analysis*. New York: Columbia Univ. Press

20. Rubin GM, Yandell MD, Wortman JR, Gabor Miklos GL, Nelson CR, et al. 2000. Comparative genomics of the eukaryotes. *Science* 287:2204–15

21. Ruddle FH. 1968. Isozymic variants as genetic markers in somatic cell populations

in vitro. *Natl. Cancer Inst. Monogr.* 29:9–13

22. Ruddle FH, Kidd KK. 1989. Tenth international workshop on human gene mapping. *Cytogenet. Cell Genet.* 51:1–1148

23. Ruddle FH, Amemiya C, Carr JL, Kim C-B, Shashikant CS, Wagner GP. 1999. Evolution of chordate Hox Gene Clusters. In *Molecular Strategies in Biological Evolution*, ed. LH Caporale, 870:238–48. New York: NY Acad. Sci.

24. Shashikant CS, Kim C-B, Borbely MA, Wang WC, Ruddle FH. 1998. Comparative studies on mammalian Hoxc8 early enhancer sequence reveal a baleen whale-specific deletion of a cis-acting element. *Proc. Natl. Acad. Sci. USA* 95:15446–51

25. Smithies O, Gregg SS, Koralewski MA, Kucherlapati R. 1985. Insertion of DNA sequences into the human chromosomal beta-globin locus by homologous recombination. *Nature* 317:230–34

26. Stern C. 1936. Somatic crossing over and segregation in *Drosophila melanogaster*. *Genetics* 21:625–730

27. Sumiyama K, Kim CB, Ruddle FH. 2001. An efficient cis-element discovery method using multiple sequence comparisons based on evolutionary relationships. *Genomics* 71:260–62

28. Swan D, D'Eustachio P, Leinwand L, Seidman J, Keithley D, Ruddle F. 1979. Chromosomal assignment of the mouse light chain genes. *Proc. Natl. Acad. Sci. USA* 76:2735–39

29. Szybalski W, Szybalska EH, Ragni G. 1962. Genetic studies with human cell lines. *Natl. Cancer Inst. Monogr.* 7:75–89

30. Watson JD, Crick FHC. 1953. The Structure of DNA. *Cold Spring Harbor Symp. Quant. Biol.* 23:123–31

31. Weiss MC, Green H. 1967. Human-mouse hybrid cell lines containing partial complements of human chromosomes and functioning human genes. *Proc. Natl. Acad. Sci. USA* 58:1104–11

32. Wigler M, Silverstein S, Lee LS, Pellicer A, Cheng YC, Axel R. 1977. Transfer of purified herpes virus thymidine kinase gene to cultured mouse cells. *Cell* 11:223–32

33. Wilson EB. 1928. The *Cell in Development and Inheritance*. New York: Macmillan. 3rd ed.

Annu. Rev. Genomics Hum. Genet. 2001. 2:9–39

PHARMACOGENOMICS: The Inherited Basis for Interindividual Differences in Drug Response

William E. Evans[1] and Julie A. Johnson[2]

[1]*St. Jude Children's Research Hospital, Department of Pharmaceutical Sciences, Memphis, Tennessee 38105; e-mail: william.evans@stjude.org, and University of Tennessee Health Science Center, Memphis, Tennessee 38163*
[2]*University of Florida, Gainesville, Florida 32610-0486; e-mail: johnson@cop.ufl.edu*

Key Words   pharmacogenetics, polymorphism, drug metabolism, drug therapy

■ **Abstract**   It is well recognized that most medications exhibit wide interpatient variability in their efficacy and toxicity. For many medications, these interindividual differences are due in part to polymorphisms in genes encoding drug metabolizing enzymes, drug transporters, and/or drug targets (e.g., receptors, enzymes). Pharmacogenomics is a burgeoning field aimed at elucidating the genetic basis for differences in drug efficacy and toxicity, and it uses genome-wide approaches to identify the network of genes that govern an individual's response to drug therapy. For some genetic polymorphisms (e.g., thiopurine S-methyltransferase), monogenic traits have a marked effect on pharmacokinetics (e.g., drug metabolism), such that individuals who inherit an enzyme deficiency must be treated with markedly different doses of the affected medications (e.g., 5%–10% of the standard thiopurine dose). Likewise, polymorphisms in drug targets (e.g., beta adrenergic receptor) can alter the sensitivity of patients to treatment (e.g., beta-agonists), changing the pharmacodynamics of drug response. Recognizing that most drug effects are determined by the interplay of several gene products that govern the pharmacokinetics and pharmacodynamics of medications, pharmacogenomics research aims to elucidate these polygenic determinants of drug effects. The ultimate goal is to provide new strategies for optimizing drug therapy based on each patient's genetic determinants of drug efficacy and toxicity. This chapter provides an overview of the current pharmacogenomics literature and offers insights for the potential impact of this field on the safe and effective use of medications.

INTRODUCTION

There are often large differences among individuals in the way they respond to medications, whether the endpoint is host toxicity, treatment efficacy, or both. Potential causes for variability in drug effects include the nature and severity of the disease being treated, the individual's age and race, organ function, concomitant therapy, drug interactions, and concomitant illnesses. Although these factors are often important, inherited differences in the metabolism and disposition of drugs,

1527-8204/01/0728-0009$14.00

9

and genetic polymorphisms in the targets of drug therapy (e.g., receptors), can have an even greater influence on the efficacy and toxicity of medications. Clinical observations of inherited differences in drug effects were first documented in the 1950s (19, 60, 69), giving rise to the field of pharmacogenetics, which has now been rediscovered by the pharmaceutical industry and a broader spectrum of academia, giving birth to pharmacogenomics. Although the two terms are often used interchangeably, pharmacogenomics is used herein to describe a genome-wide approach to identifying the network of genes that govern an individual's response to drug therapy. The ultimate goal of pharmacogenomics is to define the contributions of genetic differences in drug disposition or drug targets to drug response, thereby to improve the safety and efficacy of drug therapy through use of genetically guided, individualized treatment. With more sophisticated molecular tools available for detection of gene polymorphisms, advances in bioinformatics and functional genomics, and the wealth of new data emerging from the human genome projects, the genetic determinants of drug disposition and effects are being rapidly elucidated, and these data are already being translated into more rational drug therapy (40, 102, 104).

The elucidation of the molecular genetic basis for inherited differences in drug metabolism began in the late 1980s, with the initial cloning of a polymorphic human gene encoding the drug metabolizing enzyme debrisoquin hydroxylase (CYP2D6) (52). The human genes involved in many such pharmacogenetic traits have now been isolated, their molecular mechanisms elucidated, and their clinical importance more clearly defined, as highlighted herein. Inherited differences in individual drug metabolizing enzymes are typically monogenic traits, and their influence on the pharmacokinetics and pharmacologic effects of medications is determined by the importance of these polymorphic enzymes for the activation or inactivation of drug substrates. The effects can be profound toxicity for medications that have a narrow therapeutic index and are inactivated by a polymorphic enzyme (e.g., mercaptopurine, azathioprine, fluorouracil) (78) or reduced efficacy of medications that require activation by an enzyme exhibiting genetic polymorphism (e.g., codeine) (30). Conversely, for drugs that have a very wide therapeutic index (e.g., metoprolol), the altered pharmacokinetics in CYP2D6-deficient individuals translates into clinically unimportant changes in drug effects.

However, the overall pharmacologic effects of medications are more often polygenic traits, determined by numerous genes encoding proteins involved in multiple pathways of drug metabolism, disposition, and effects (see Figure 1). Such polygenic traits are more difficult to elucidate in clinical studies, especially when a medication's metabolic fate and mechanism(s) of action are poorly defined. However, as the molecular mechanisms of pharmacologic effects, genetic determinants of disease pathogenesis, and polymorphisms in genes that govern drug metabolism and disposition are clarified, these genetic determinants become more tractable. Furthermore, the human genome project, coupled with functional genomics, bioinformatics, and high-throughput screening methods, is providing powerful tools for elucidating polygenic determinants of disease pathogenesis and drug response.

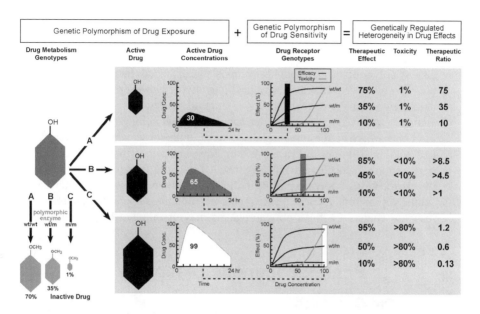

Figure 1 The potential polygenic nature of drug response is illustrated, depicting the hypothetical effects of two polymorphic genes, one determining the extent of drug inactivation and the other, drug receptor sensitivity. The polymorphic drug metabolizing enzyme, which exhibits codominant inheritance in this example, determines drug concentrations in individual patients, whereas the polymorphic receptor determines drug response at any given drug concentration. Thus, in this example, an individual with homozygous wild-type drug metabolism and drug receptors would have a high probability of therapeutic efficacy and a low probability of toxicity (therapeutic ratio = 75), in contrast to an individual with homozygous mutant genotypes for the drug metabolizing enzyme and the drug receptor, in whom the likelihood of efficacy is low and toxicity high (therapeutic ratio = <0.13). [Modified from (40)]

Pharmacogenomics aims to elucidate the network of genes that determine the efficacy and toxicity of specific medications and to capitalize on these insights to discover new therapeutic targets and optimize drug therapy. Such knowledge should make it possible to select drug therapy based on each patient's inherited ability to metabolize, eliminate, and respond to specific medications.

GENETIC POLYMORPHISMS IN DRUG METABOLISM AND DISPOSITION

Initially, inherited differences in drug metabolism were discovered following clinical observations of marked interindividual differences in drug response (e.g., pronounced hypotension following debrisoquin) (94). These obesrvations were then followed by population studies of drug disposition phenotype, then biochemical, and eventually molecular elucidation of the genetic defect responsible for the phenotypic outliners. This clinically based approach made it likely that such genetic polymorphisms would have clinical consequences for drug effect because their discovery was based on a clinical phenotype. However, the framework for discovery of genetic polymorphisms of drug disposition or response is evolving. With recent advances in molecular sequencing technology, gene (DNA) polymorphisms may be the initiating discoveries, with subsequent biochemical and finally clinical studies to assess whether the genomic polymorphisms have phenotypic consequence in patients (Figure 2). This latter framework may allow for the clarification of polymorphisms in drug metabolizing enzymes that have more subtle, yet important, consequences for interindividual variability in human drug response (Figure 3). Such polymorphisms may or may not have clear clinical

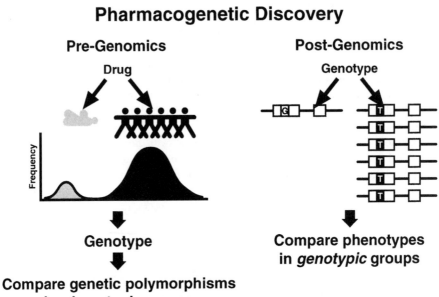

Figure 2   Two strategies for the discovery of pharmacogenetic traits: The "pre-genomics" strategy (before 2000) was first to discover an unusual drug response or drug metabolism phenotype, and then to conduct family studies to elucidate inheritance patterns. These steps were followed by cloning of the involved gene and sequencing to identify genotypes that conferred the inherited phenotype. The "post-genomics" strategy (beginning in ~2000) capitalizes on high-throughput sequencing methods and databases generated from the human genome project, to first identify mutations [e.g., single nucleotide polymorphisms (SNPs)], and then search for associations with drug response phenotypes, without necessarily knowing a priori the mechanisms involved in potential genotype-phenotype associations (from 102).

Character of Population Phenotype Distributions Differs Based on Mutation Type

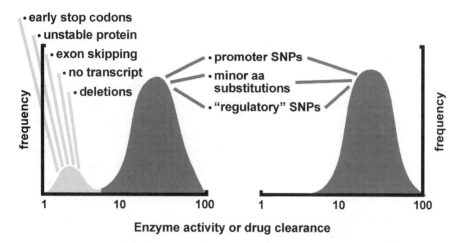

Figure 3   The left panel depicts the population distribution of drug metabolism phenotypes for most of the common genetic polymorphisms identified to date, which largely involve mutations that confer complete or near-complete loss of enzyme activity. The right panel depicts the population distribution of drug metabolism phenotypes for mutations associated with altered enzyme activity but not compete loss of function, reflecting the normal distribution of activity that typifies the metabolism and clearance of most medications. It is anticipated that at least some of the variability associated with the range of activity within a normal distribution will be inherited and due to mutations in the promoter regions of involved genes or amino acid changes that decrease or increase activity, but that do not eliminate activity (Adapted from M. Relling, personal communication).

significance for affected medications, depending on the importance of the enzyme for the overall metabolism of a medication, the expression of other drug metabolizing enzymes in the patient, the therapeutic index of the drug, the presence of concurrent medications or illnesses, and other polygenic factors that impact drug response. It is likely that almost every gene involved in drug metabolism is subject to genetic polymorphisms, although the phenotypic consequences may be subtle, such as placing individuals on one end or the other of a normal distribution of drug metabolism phenotypes, instead of conferring a complete deficiency of the encoded enzyme (Figure 3). Thus, inactivating polymorphisms can be broadly categorized into two groups, those that confer complete or near-complete loss of activity of encoded proteins and those that confer more subtle changes in function via more modest changes in expression, regulation, stability, or catalytic activity, but without complete loss of function.

For drug metabolizing enzymes, the molecular mechanisms of inactivation include splice site mutations resulting in exon skipping (e.g., DPD, CYP2C19), microsatellite nucleotide repeats (e.g., CYP2D6), gene duplication (e.g., CYP2D6), point mutations resulting in early stop codons (e.g., CYP2D6), enhanced proteolysis (e.g., TPMT), altered promoter functions (e.g., CYP2A5, UGTIA1), critical amino acid substitutions (e.g., NAT2, CYP2D6, CYP2C19, CYP2C9), or large gene deletions (e.g., GSTM1, CYP2D6). Conversely, gene duplication can be associated with enhanced activity for some drug metabolizing enzymes (e.g., CYP2D6). For many genes encoding drug metabolizing enzymes, the frequency of single nucleotide polymorphisms (SNPs) and other genetic defects appears to be more common than the "1 per 1000 nucleotide" frequency that is frequently cited for the human genome [recently revised to ~1 SNP per 1900 bases in the human genome, with 1 SNP per ~1080 bases in exons (61a)]. It may be that genetic polymorphisms of drug metabolizing enzymes are quite common because these enzymes are not essential from an evolutionary perspective. However, some essential receptors have more mutations than would be predicted from the 1 in 1000 rate (e.g., B1AR, B2AR), although these mutations do not confer complete loss of receptor function. Nonetheless, these common polymorphisms in drug receptors and drug metabolizing enzymes are often major determinants of interindividual differences in drug response.

No completely inactivating mutations have been reported for the gene encoding CYP3A4, which may be because CYP3A4 is required for the metabolism of critical endogenous glucocorticoid and sex hormones and thus its complete inactivation may be incompatible with life. However, a common polymorphism in the promoter for CYP3A4 (CYP3A4*1B) has been described (122) that may affect the extent to which CYP3A4 is inducible, rather than affecting constitutive levels of the enzyme, or its importance may be that it is in linkage disequilibrium with other functional polymorphisms at the CYP3A locus (80). Allelic variants have also been identified (*CYP3A4*2*, *CYP3A4*3*), one of which (*CYP3A4*2*) has altered catalytic activity for nifedipine but not testosterone (131). Furthermore, a polymorphism recently discovered in intron 3 of the human CYP3A5 gene creates an ectopic splice site, leading to a premature codon in the encoded mRNA. This common allelic variant (*CYP3A5*3*) is the principal genetic basis for polymorphic CYP3A5 expression in humans (80). Because most *CYP3A4* substrates are also substrates for *CYP3A5*, this *CYP3A5* polymorphism influences overall CYP3A activity in humans, which would be expected to shift subjects to the higher end of the population distribution for CYP3A activity (Figure 3, right panel).

Table 1 provides a list of human drug metabolizing enzymes that exhibit functional genetic polymorphisms, their substrates, and clinical consequences of polymorphisms in their genes, when applicable. Essentially all polymorphisms studied to date differ in frequency among ethnic and racial groups. Marked racial diversity in the frequency or type of functional defects in drug metabolizing enzymes means that the optional dose of medications may differ among world populations, an important consideration with the globalization of drug development.

Several adverse drug reactions have been linked to specific drug metabolizer phenotypes. Among the earliest examples are the associations of the slow acetylator phenotype with isoniazid-induced neuropathies, hydralazine or proc-ainamide-induced lupus, dye-associated bladder cancer, and sulfonamide-induced hypersensitivity reactions (53, 142). In each of these cases, acetylation of parent drug or an active metabolite was an inactivating pathway. N-acetyltransferase is a phase II enzyme that conjugates substrates with a more water-soluble small molec-ular moiety. Such conjugation reactions are frequently, but not always, detoxifying by "masking" a reactive functional group and typically enhancing urinary or biliary excretion of substrates.

Thiopurine S-methyltransferase (TPMT) is a polymorphic phase II enzyme that catalyzes the S-methyltransferase of thiopurine medications (mercaptopurine, thioguanine, and azathioprine), which is the inactivation pathway in hematopoi-etic tissues (78). For the TPMT polymorphism, all patients who inherit two non-functional TPMT alleles will develop dose-limiting hematopoietic toxicity that can be fatal if these patients are treated with full doses of thiopurine medica-tions (39, 84, 103, 137). However, TPMT-deficient patients can tolerate thiopurine therapy, without acute toxicity, if they are treated with 5%–10% of the con-ventional dose of these medications. Ten percent of patients who are heterozy-gous at the TPMT locus (with one wild-type allele) are also at greater risk of thiopurine toxicity, but these patients can usually be safely treated with only modest dose reduction (11, 38b, 125). More recently, an adverse interaction was observed among the TPMT polymorphism, thiopurine therapy, and cranial irradi-ation (126). Among patients cured of acute lymphoblastic leukemia (ALL) with therapy that included concurrent thiopurine chemotherapy and cranial irradiation, those with TPMT-deficiency (heterozygous or homozygous deficient) had a signif-icantly higher frequency of developing a malignant brain tumor as a consequence of ALL treatment (see Figure 4). In the absence of cranial irradiation, the inci-dence of brain tumors was essentially zero in patients treated with thiopurines, but in those who received cranial irradiation, the cumulative incidence of brain tumors was 40% in patients who were TPMT-deficient versus 8.3% in those with wild-type TPMT activity. Subsequent studies have begun to elucidate potential mechanisms by which high thioguanine levels perturb DNA repair mechanisms (77). These data illustrate the potential nature of interactions among genetic poly-morphisms in drug metabolism, their drug substrates, and other components of treatment or the environment. The molecular genetic basis of the TPMT poly-morphism has been clarified (79, 149, 150), and molecular diagnostics are now available to prospectively identify TPMT-deficient and heterozygotes patients (161).

Several phase I enzymes exhibit functional genetic polymorphism (Table 1), such that a subset of the population inherits a deficiency of the enzyme activ-ity. These inherited deficiencies can be associated with increased pharmacologic effects for medications that are primarily inactivated by these enzymes, such as several tricyclic antidepressants and *CYP2D6* (27), antipsychotics and CYP2D6 (27), fluoxetine and *CYP2D6* (129), warfarin and *CYP2C9* (3, 47, 146, 153), and 5-fluorouracil and dihydropyrimidine dehydrogenase (31, 51). For example,

TABLE 1 Genetic polymorphisms of human drug metabolizing enzymes and transporters

Enzyme	Substrates	Consequences of polymorphism for drug effects	Reference
Phase I enzymes			
CYP1A1	Benzo(a)pyrene, phenacetin	Not yet elucidated	(20, 48, 71, 93)
CYP1A2	Acetaminophen, amonafide, caffeine, paraxanthine, ethoxyresorufin, propranolol, fluvoxamine	Not yet elucidated	(87)
CYP1B1	Estrogen metabolites	Not yet elucidated	(8)
CYP2A6	Coumarin, nicotine, halothane	Cigarette addiction	(90, 117)
CYP2B6	Cyclophosphamide, aflatoxin, mephenytoin	Not yet elucidated	(24)
CYP2C8	Retinoic acid, paclitaxel	Not yet elucidated	
CYP2C9	Tolbutamide, warfarin, phenytoin, non-steroidal anti-inflammatories	Anticoagulant effect of warfarin	(3, 47, 146, 153)
CYP2C19	Mephenytoin, omeprazole, hexobarbital, mephobarbital, propranolol, proguanil, phenytoin	Peptic ulcer response to omeprazole	(46, 50)
CYP2D6	Beta blockers, antidepressants, antipsychotics, codeine, debrisoquin, dextromethorphan, encainide, flecainide, fluoxetine, guanoxan, methoxy-amphetamine, N-propylajmaline, perhexiline, phenacetin, phenformin, propafenone, sparteine	Tardive dyskinesia from antipsychotics, narcotic side effects, efficacy, and dependence, imipramine dose requirement, beta blocker effect	(15, 17, 27, 30, 66, 70, 82, 86, 94, 118, 129, 140, 154, 163)
CYP2E1	N-nitrosodimethylamine, acetaminophen, ethanol	Possible effect on alcohol consumption	(18, 41, 55, 88, 152)
CYP3A4/3A5/3A7	Macrolides, cyclosporin, tacrolimus, calcium channel blockers, midazolam, terfenadine, lidocaine, dapsone, quinidine, triazolam, etoposide, teniposide, lovastatin, alfentanil, tamoxifen, steroids, benzo(a)pyrene	Not yet elucidated, polymorphic 3A5 expression linked to 3A5 polymorphism	(43, 80, 122, 131)
Aldehyde dehydrogenase (ALDH2)	Cyclophosphamide, vinyl chloride	SCE frequency in lymphocytes	(159)
Alcohol dehydrogenase (ADH3)	Ethanol	Increased alcohol consumption and dependence	(55, 152, 158)
Dihydropyrimidine dehydrogenase	Fluorouracil	5-fluorouracil neurotoxicity	(31, 51)
NQO1 (DT-diaphorase)	Ubiquinones, menadione, mitomycin C	Menadione-associated urolithiasis	(127, 128, 136, 139)

and urine, distribution of drug into "therapeutic sanctuaries," such as the brain and testes, and transport into sites of action, such as cardiovascular tissue, tumor cells, and infectious microorganisms. Many transporters include members of the adenosine triphosphate (ATP)-binding-cassette family, which share many physicochemical characteristics. Transporters include the P-glycoprotein (MDR1), alpha-1-acid glycoprotein, MRP1-6 (multidrug resistance proteins), and SPGP (sister PGP). It has been proposed that p-glycoprotein may not be essential for viability, because knockout mice appear normal until challenged with xenobiotics (134), whereas other transporters play critical roles in transport of endogenous substances, such as bilirubin and glutathione conjugates, and some medications (13). Although some polymorphisms in p-glycoprotein have been reported (106), and such variation may have functional significance for drug absorption and elimination, the clinical relevance of polymorphisms in drug transporters has yet to be fully elucidated. Transporters for neurotransmitters (e.g., serotonin and dopamine, see Table 1) exhibit genetic polymorphism (35, 45), and some of these have been linked to drug response (e.g., clozapine response in schizophrenia has been linked to six genetic polymorphisms in four human genes, including the serotonin transporter) (6).

DRUG TARGET PHARMACOGENETICS AND PHARMACOGENOMICS

In recent years, there has been an increasing focus on genetic polymorphisms in drug targets, with an interest in defining their impact on drug efficacy and/or toxicity. For the purposes of this review, a drug target is defined as the direct protein target of a drug (e.g., a receptor or enzyme), proteins involved in the pharmacologic response (e.g., signal transduction proteins or downstream proteins), or proteins associated with disease risk or pathogenesis that is altered by the drug. The broad objective of drug target pharmacogenomics research is to identify the inherited basis for interindividual variability in drug response and toxicity, particularly when this variability is not explained by differences in drug concentration (pharmacokinetics).

Although studies of drug metabolism pharmacogenetics date back to the 1950s, the literature on drug target pharmacogenetics essentially began in the mid to late 1990s. Moreover, drug target pharmacogenomics is rapidly moving from a monogenic to a polygenic (genomic) focus, largely because most drug effects are polygenic in nature, and tools are now available for high throughput genotyping. Herein, we describe examples of drug target pharmacogenomics from each of the drug target categories defined above, and we discuss the rationale and benefits of moving from a single gene approach to a genomic strategy in order to provide more clinically useful information.

Most of the early drug target pharmacogenetic studies focused on a single polymorphism in a single gene, with the gene most often being the direct target of the drug (i.e., the receptor). Table 2 lists examples of genetic polymorphisms

TABLE 2 Example of single gene drug target pharmacogenomics

Gene/gene product	Medication	Drug effect associated with polymorphism	Reference
ACE	ACE inhibitors (e.g., enalapril)	Renoprotective effects, BP reduction, left ventricular mass reduction, endothelial function improvement, ACE inhibitor induced cough	(64, 76, 111, 112, 115, 116, 119, 130, 143)
Bradykinin B2 receptor	ACE inhibitors	ACE inhibitor induced cough	(107)
β_2-adrenergic receptor	β_2-agonists (e.g., albuterol, terbutaline)	Bronchodilation, susceptibility to agonist-induced desensitization, cardiovascular effects (e.g., increased heart rate, cardiac index, peripheral vasodilation)	(23, 34, 54, 59, 62, 89, 99, 151)
Gs protein α	β-blockers (e.g., metoprolol)	Antihypertensive effect	(65)
ACE	Fluvastatin	Lipid changes (e.g., reductions in total LDL-cholesterol and apolipoprotein B); progression/regression of atherosclerotic lesions	(96)
Platelet FC receptor (FCRII)	Heparin	Heparin-induced thrombocytopenia	(14)
Glycoprotein IIIa subunit of glycoprotein IIb/IIIa receptor	Aspirin/glycoprotein IIb/IIIa inhibitors (e.g., abciximab)	Antiplatelet effect	(105)
ALOX5	Leukotriene biosynthesis inhibitors (e.g., ABT-761-zileuton-derivative)	Improvement in FEV_1	(32)

Estrogen receptor	Conjugated estrogens	Bone mineral density increases	(113)
Sulfonylurea receptor	Sulfonylureas (e.g., tolbutamide)	Sulfonylurea-induced insulin release	(56)
Inositol-p1p	Lithium	Response of manic depressive illness	(145)
Dopamine receptors (D2, D3, D4)	Antipsychotics (e.g., haloperidol, clozapine, thioridazine nemorapride)	Antipsychotic response (D2, D3, D4), antipsychotic-induced tardive dyskinsia (D3), antipsychotic-induced acute akathisia (D3), hyperprolactinemia in females (D2)	(6, 9, 25, 36, 61, 68, 133, 148)
Dopamine receptor	Levodopa & dopamine	Drug induced hallucinations	(95)
5HT2A, 5HT6	Antipsychotics (e.g., clozapine, typical antipsychotics)	Clozapine response (5HT2A, 5HT6), typical antipsychotic response and long term outcomes (5HT2A)	(67, 100, 162)
G protein $\beta 3$	Antidepressants (various)	Response to antidepressant therapy	(164)
Serotonin transporter (5-HTT)	Antidepressants (e.g., clomipramine, fluoxetine, paroxetine, fluvoxamine)	5-HT neurotransmission, antidepressant response	(73, 141, 157)
Ryanodine receptor	Anesthetics (e.g., halothane)	Malignant hyperthermia	(101)

of drug targets that have been investigated for their contribution to drug response variability, and it summarizes the relevant drug (drug class) and the drug response or adverse effect associated with the polymorphism. Review of this body of literature reveals that although single gene/single polymorphism drug target studies have identified numerous associations between polymorphisms and the anticipated response, they have also been somewhat disappointing in terms of the consistency with which these associations have been documented. For example, the insertion/deletion (I/D) polymorphism of the angiotensin converting enzyme (ACE) gene is probably one of the most extensively studied of the drug target polymorphisms. The DD genotype has been consistently associated with increased ACE activity and ACE I/D genotypes have been associated with various clinical effects of ACE inhibitors, including renoprotective effects, blood pressure reduction, left ventricular hypertrophy reduction, and improvements in endothelial function. However, these studies are not concordant, in that some show no association between response and ACE I/D genotype (16), some show the II genotype is associated with greater drug response (64, 76, 111, 112, 115), whereas others show the DD genotype is associated with the best response (57, 116, 119, 130, 143).

The β_2-adrenergic receptor polymorphisms and their influence on β_2-agonist–mediated effects have also been the focus of numerous investigations. Studies of the bronchodilator effects of β_2-agonists have been equivocal, with some showing the β_2-agonist–mediated bronchodilator effects are dependent on a certain genotype or allele, whereas others have reported that the opposite genotype/allele is a predictor of response (34, 62, 89, 99, 151). Studies of the cardiovascular effects of β_2-agonists are also conflicting (23, 54, 59). Like many genes, the β_2-adrenergic receptor gene has multiple polymorphisms, some of which are in linkage disequilibrium. Recent studies suggest that analysis of haplotype, rather than analysis of individual polymorphisms, may markedly enhance the ability to detect important associations between gene polymorphisms and drug response (34).

Numerous studies have also focused on polymorphisms in various genes that are direct targets of psychiatric medications, particularly polymorphisms in the dopamine receptor genes and the serotonin (5HT) receptor genes (6, 25, 36, 61, 67, 68, 73, 100, 133, 144, 148, 157, 162). Consistent with the ACE and β_2-adrenergic receptor gene polymorphisms, the 5HT receptor genetic polymorphisms have been associated in some, but not all, studies with efficacy or toxicity. The serotonin transporter is the only example in Table 2 where multiple studies have consistently shown an association between response and genotype. Such apparent discordance among studies may reflect the endpoints utilized to measure drug effects, the time course over which effects were assessed, the nature of the study population, and other biological (e.g., genetic) or environmental differences among studies.

Although focusing on a polymorphism in a direct protein target of a drug may seem a logical approach, it is also easy to explain why such an approach may lead to equivocal results, and why a genomic approach may provide more reproducible results. Take the ACE inhibitors as an example. ACE is the direct protein target of the ACE inhibitors. Inhibition of ACE decreases production of angiotensin II and decreases degradation of bradykinin and other vasodilator substances, each having their own signal transduction cascade. Angiotensin II, for example, binds to its G protein coupled receptor (AT_1), with the ensuing signal transduction cascade, ultimately involving 20 to 30 proteins to generate the eventual cellular effects. Given estimates that polymorphisms occur about every 1900 bp (61a), it would be anticipated that, on average, the genes for each of these 20–30 proteins could have one or two polymorphisms. Therefore, it is easy to appreciate why a single polymorphism in the ACE gene may not adequately predict the variability in patient response to ACE inhibitors. Given that most drugs act on enzymes or receptors that have similar or more complex signal transduction cascades, it seems apparent that genomic approaches will be needed to more adequately elucidate the genetic basis for drug response variability.

Table 3 summarizes single gene/single polymorphism pharmacogenetic studies focused on polymorphisms that are associated with altered disease risk. Unlike many of the single gene/direct target pharmacogenetic studies, the findings in this category have been more consistent across studies. The first examples in Table 3 highlight how a drug known to cause a certain adverse event, in combination with a genetic polymorphism associated with the same adverse event, can lead to a marked increase in the risk of drug toxicity. For example, oral contraceptive use in patients with Factor V or prothrombin mutations produces a markedly higher risk of a thrombotic event than either the mutation alone or oral contraceptive use alone (97, 98). Similarly, gene mutations in cardiac potassium and sodium channels that are associated with long QT syndrome are also associated with increased risk of drug-induced Torsade de Pointes (2, 33, 109). This may represent a useful paradigm for identifying patients at greatest risk of serious drug-induced adverse effects.

Methylation of cytosines within C_pG-rich regions of human gene promoters is a well-established mechanism of transcriptional inactivation. It was recently shown that interindividual differences in methylation of the promoter region of O6-methylaguanine-DNA methyltransferase (MGMT) are significantly related to the response of grade III–IV gliomas to carmustine therapy (37). Responses were documented in 12 of 19 patients with methylated MGMT promoters in their tumors, compared to responses in only 1 of 28 patients with unmethylated MGMT promoters (37). The putative mechanism is that those with methylated MGMT promoters had lower expression of MGMT, an enzyme that reverses the alkylation of DNA that is critical to the mechanism of carmustine's anticancer effects. Future studies are needed to verify these initial findings, determine the precise mechanism(s) involved, and clarify the molecular basis for interindividual differences in MGMT promoter methylation.

The remaining examples in Table 3 highlight how disease-associated polymorphisms may also impact drug efficacy, even when the proteins of interest are not directly involved in the pharmacologic actions of the drug. The outcome literature with HMG CoA-reductase inhibitors (statins) are particularly interesting in this regard. Studies of the statins that were conducted in fairly large populations with

TABLE 3 Disease pathogenesis polymorphisms associated with altered drug effect

Gene/gene product	Disease or response association	Medication	Impact of polymorphism on drug effect/toxicity	Reference
Prothrombin and factor V	Deep vein thrombosis and cerebral vein thrombosis	Oral contraceptives	Increased deep vein and cerebral vein thrombosis risk with oral contraceptives	(97, 98)
HERG, KvLQT1, Mink, MiRP1	Congenital long QT syndrome	Various such as: erythromycin, terfenadine, cisapride, clarithromycin, antiarrhythmic drugs (e.g., quinidine)	Increased risk of drug-induced Torsade de Pointes	(2, 33, 109)
APOE	Atherosclerosis progression; ischemic cardiovascular events	Statins (e.g., simvastatin)	Enhanced survival prolongation with simvastatin	(29, 49, 110, 114)
APOE	Alzheimer's disease	Tacrine	Clinical improvement with tacrine	(42)
MGMT	Glioma	Carmustine	Response of glioma to carmustine	(37)
CETP	Atherosclerosis progression; HDL–cholesterol levels	Statins (e.g., pravastatin)	Slowing atherosclerosis progression by pravastatin	(81)
Stromelysin-1	Atherosclerosis progression	Statins (e.g., pravastatin)	Reduction in cardiovascular events by pravastatin (e.g., death, myocardial infarction, stroke, angina, etc.), reduction in repeat angioplasty	(28)
Parkin	Parkinson's disease	Levodopa	Clinical improvement and L-dopa induced dyskinesias	(91)

specific clinical endpoints (beyond lipid changes) have focused on the polymorphic genes coding the following proteins: apolipoprotein E, cholesteryl ester transfer protein (CETP), stromelysin-1, and β-fibrinogen (28, 29, 49, 81, 110, 114). These studies and their results have been strikingly consistent. First, investigators studied a genetic polymorphism thought to be or documented as associated with an adverse clinical outcome (e.g., greater atherosclerosis progression, cardiovascular event, or death), but not directly associated with the drug's pharmacologic effect. In all cases, the placebo arm of the study supported this genotype-clinical outcome association. The studies were also remarkably consistent in showing that those who carried the polymorphism associated with increased risk of adverse outcomes were the patients who derived the greatest benefit from statin therapy, whereas the lower risk group derived less or no benefit from the statin. These studies also consistently showed that lipid changes associated with statin therapy did not differ by genotype, suggesting that the clinical benefits are somewhat independent of lipid lowering and thus independent of the presumed direct pharmacological action of the drugs. Similar findings have been documented with apolipoprotein E polymorphisms and tacrine response in Alzheimer's disease (42) as well as *parkin* polymorphisms and efficacy or adverse effects of levodopa therapy in Parkinson's disease (91).

Polymorphisms in the genes of pathogenic agents (notably HIV and *Helicobacter pylori*, among others) are also important determinants of response to antiinfective agents. However, these pharmacogenetic examples are not discussed here, as the focus of this review is human and not pathogen polymorphisms.

In many ways, the single polymorphism/single gene studies described above provide a "proof of concept," specifically that variability in drug response might be attributable to genetic variability. Although these studies have documented that some variability in drug response can be explained by genetic variability, they do not always provide a level of predictability that could be useful clinically. For example, the apolipoprotein E-tacrine response in Alzheimer's is one of the most widely cited pharmacogenomic studies (42). Although the authors provide clear evidence of a difference in response to tacrine based on genotype, the predictive value is not adequate to withhold therapy based on genotype alone, especially in this setting where there are few alternative drug therapy choices. For example, 83% of patients lacking the APO ε4 allele had a positive response to tacrine, whereas only 41% of patients with an APO ε4 allele had a positive response. Thus, even with these clear response differences by genotype, it would not be appropriate to withhold therapy based on the presence of the APO ε4 allele because a substantial portion of patients with this allele do benefit from such therapy. In fact, the greatest improvement with tacrine in this study was observed in a patient carrying an APO ε4 allele. Moreover, the difference in response between genotypes for most of the other examples cited above are much less dramatic than the tacrine-APO ε4 example. Thus, to move drug target pharmacogenomics to the next level, where genetic information can be used to reliably predict drug response (or toxicity), with subsequent therapeutic decisions based on this information, more sophisticated genomic approaches will be necessary in most cases.

Although there are only a few examples of genomic approaches that relate drug target polymorphisms to response or toxicity, the available studies suggest that such approaches will provide more useful information than the single gene/single polymorphism approach. One approach is a "candidate gene" pharmacogenomic approach, where polymorphisms in multiple genes known or suspected to contribute to drug effects or disposition are studied for their association with response or toxicity.

For example, a recent study of clozapine response in schizophrenic patients used a multiple candidate gene approach to gain insight into the genetic contribution to response variability to clozapine (6). This study evaluated the relationship between clozapine response and 19 polymorphisms in 10 genes (including genes for α-adrenergic receptors, dopamine receptors, serotonin receptors, histamine receptors, and the serotonin transporter). A combination of the six polymorphisms showing the strongest association with response provided a positive predictive value of 76%, a negative predictive value of 82%, with a sensitivity of 96% for identifying schizophrenic patients showing improvement with clozapine, and a specificity of 38% for identifying patients with minimal response to clozapine. Additional studies will be needed to validate this model in a larger patient population, to assess their ability to predict response to other antipsychotic agents (or even placebo), and to expand the repertoire of genes and polymorphisms investigated. Although the approaches described above can probably be improved upon, it provides insights into an approach that is likely to prove much more powerful in the quest to move pharmacogenomics into the clinical setting.

Another approach in pharmacogenomics is a genome scanning approach, which does not rely on knowledge of the drug's pharmacologic actions, as does the candidate gene approach. Use of SNP maps to identify polymorphisms in genes involved in drug disposition or effects, or in linkage disequilibrium with these polymorphisms, and thereby associated with efficacy or toxicity, is a pharmacogenomics approach that many large pharmaceutical companies plan to employ in their drug development process. Actual implementation of such an approach may further advance our understanding of the genetic basis for variability in drug efficacy and toxicity, although the mechanism underlying such associations will require further study once the predictive SNPs have been identified. For this to be a cost-effective strategy, the cost of SNP detection will have to decrease markedly, or the number of SNPs tested will have to be constrained well below the estimated 1.42 million SNPs in each human genome (61a).

THE FUTURE

It is clear from the numerous examples reviewed here that genetic polymorphism can be an important determinant of drug disposition and response in humans. It is equally clear that we are at the early stages of defining these pharmacogenomic

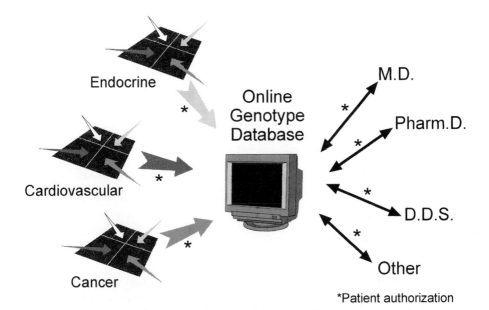

Figure 6 Ultimately, a secured online database should be developed in which each individual's informative genetic profile will be stored and available to authorized clinicians. With current technologies, these informative pharmacogenomic genotypes will likely be determined in panels that are potentially important for their current illness, but with advances in genotyping technologies, it should eventually be possible to perform genome-wide detection of hundreds of thousands of informative mutations and to deposit these data well prior to the need to make treatment decisions.

determinants and that a broader genomics approach will be required to elucidate polygenic determinants for most medications. Once the network of genes that govern drug responses in humans is defined, it will then be possible to more accurately optimize drug therapy based on each patient's ability to metabolize, transport, and respond to medications. The vision is that in the future, authorized clinicians will be able to access a secured database in which their patient's genetic polymorphisms will have been deposited, as they are determined for specific classes of medications, based on their illnesses (Figure 6). Technology will ultimately make it possible to perform a focused genome-wide scan for polymorphisms that are associated with disease risk or drug response, such that these data will be determined a priori and thus will be available to clinicians for preventive health and prospective treatment decisions. The end result will be the optimal selection of medications and their dosages based on the individual patient and not treatment based on the average experience from the entire universe of patients with a similar diagnosis.

ACKNOWLEDGMENTS

The authors gratefully acknowledge the excellent contributions of Rose Phillips-Williams and Teri Smith in the preparation of the manuscript, and Nancy Kornegay for her outstanding computer savvy in assembling the references, tables, and text. This work was supported in part by the following: NIH R37 CA36401, RO1 CA78224, R01 HL64691, Cancer Center Support Grant CA21765, a Center of Excellence grant from the State of Tennessee, and the American Lebanese Syrian Associated Charities.

Visit the Annual Reviews home page at www.AnnualReviews.org

LITERATURE CITED

1. Aarbakke J, Janka-Schaub G, Elion GB. 1997. Thiopurine biology and pharmacology. *Trends Pharmacol. Sci.* 18:3–7

2. Abbott GW, Sesti F, Splawski I, Buck ME, Lehmann MH, et al. 1999. MiRP1 forms IKr potassium channels with HERG and is associated with cardiac arrhythmia. *Cell* 97:175–87

3. Aithal GP, Day CP, Kesteven PJ, Daly AK. 1999. Association of polymorphisms in the cytochrome P450 CYP2C9 with warfarin dose requirement and risk of bleeding complications. *Lancet* 353:717–19

4. Aksoy S, Raftogianis R, Weinshilboum R. 1996. Human histamine N-methyltransferase gene: structural characterization and chromosomal location. *Biochem. Biophys. Res. Commun.* 219:548–54

5. Ando Y, Saka H, Asai G, Sugiura S, Shimokata K, Kamataki T. 1998. UGT1A1 genotypes and glucuronidation of SN-38, the active metabolite of irinotecan. *Ann. Oncol.* 9:845–47

6. Arranz MJ, Munro J, Birkett J, Bollonna A, Mancama D, et al. 2000. Pharmacogenetic prediction of clozapine response. *Lancet* 355:1615–16

7. Baez S, Segura-Aguilar J, Widersten M, Johansson AS, Mannervik B. 1997. Glutathione transferases catalyse the detoxication of oxidized metabolites (o-quinones) of catecholamines and may serve as an antioxidant system preventing degenerative cellular processes. *Biochem. J.* 324:25–28

8. Bailey LR, Roodi N, Dupont WD, Parl FF. 1998. Association of cytochrome P450 1B1 (CYP1B1) polymorphism with steroid receptor status in breast cancer. *Cancer Res.* 58:5038–41

9. Basile VS, Masellis M, Badri F, Paterson AD, Meltzer HY, et al. 1999. Association of the MscI polymorphism of the dopamine D3 receptor gene with tardive dyskinesia in schizophrenia. *Neuropsychopharmacology* 21:17–27

10. Beutler E, Gelbart T, Demina A. 1998. Racial variability in the UDP-glucuronosyltransferase 1 (UGT1A1) promoter: a balanced polymorphism for regulation of bilirubin metabolism? *Proc. Natl. Acad. Sci. USA* 95:8170–74

11. Black AJ, McLeod HL, Capell HA, Powrie RH, Matowe LK, et al. 1998. Thiopurine methyltransferase genotype predicts therapy-limiting severe toxicity from azathioprine. *Ann. Intern. Med.* 129:716–18

12. Blum M, Demierre A, Grant DM, Heim M, Meyer UA. 1991. Molecular mechanism of slow acetylation of drugs and carcinogens in humans. *Proc. Natl. Acad. Sci. USA* 88:5237–41

13. Borst P, Evers R, Kool M, Wijnholds J. 2000. A family of drug transporters: the

multidrug resistance-associated proteins. *J. Natl. Cancer Inst.* 92:1295–302

14. Brandt JT, Isenhart CE, Osborne JM, Ahmed A, Anderson CL. 1995. On the role of platelet Fc gamma RIIa phenotype in heparin-induced thrombocytopenia. *Thromb. Haemost.* 74:1564–72

15. Brosen K. 1989. Clinical significance of the sparteine/debrisoquine oxidation polymorphism. *Eur. J. Clin. Pharmacol.* 36:537–47

16. Cannella G, Paoletti E, Barocci S, Massa-rino F, Delfino R, et al. 1998. Angiotensin-converting enzyme gene polymorphism and reversibility of uremic left ventricular hypertrophy following long-term antihypertensive therapy. *Kidney Int.* 54:618–26

17. Caraco Y, Sheller J, Wood AJ. 1996. Pharmacogenetic determination of the effects of codeine and prediction of drug interactions. *J. Pharmacol. Exp. Ther.* 278:1165–74

18. Carriere V, Berthou F, Baird S, Belloc C, Beaune P, de Waziers I. 1996. Human cytochrome P450 2E1 (CYP2E1): from genotype to phenotype. *Pharmacogenetics* 6:203–11

19. Carson PE, Flanagan CL Ickes CE, Alving AS. 1956. Enzymatic deficiency in primaquine-sensitive erythrocytes. *Science* 124:484–85

20. Cascorbi I, Brockmoller J, Roots I. 1996. A C4887A polymorphism in exon 7 of human CYP1A1: population frequency, mutation linkages, and impact on lung cancer susceptibility. *Cancer Res.* 56:4965–69

21. Chen H, Juchau MR. 1998. Recombinant human glutathione S-transferases catalyse enzymic isomerization of 13–cis-retinoic acid to all-trans-retinoic acid in vitro. *Biochem. J.* 336:223–26

22. Choi KH, Chen CJ, Kriegler M, Roninson IB. 1988. An altered pattern of cross-resistance in multidrug-resistant human cells results from spontaneous mutations in the mdr1 (P-glycoprotein) gene. *Cell* 53:519–29

23. Cockcroft JR, Gazis AG, Cross DJ, Wheatley A, Dewar J, et al. 2000. Beta(2) adrenoceptor polymorphism determines vascular reactivity in humans. *Hypertension* 36:371–75

24. Code EL, Crespi CL, Penman BW, Gonzalez FJ, Chang TK, Waxman DJ. 1997. Human cytochrome P4502B6: interindividual hepatic expression, substrate specificity, and role in procarcinogen activation. *Drug Metab. Dispos.* 25:985–93

25. Cohen BM, Ennulat DJ, Centorrino F, Matthysse S, Konieczna H, et al. 1999. Polymorphisms of the dopamine D4 receptor and response to antipsychotic drugs. *Psychopharmacology* 141:6–10

26. Cole SP, Deeley RG. 1998. Multidrug resistance mediated by the ATP-binding cassette transporter protein MRP. *BioEssays* 20:931–40

27. Dahl ML, Bertilsson L. 1993. Genetically variable metabolism of antidepressants and neuroleptic drugs in man. *Pharmacogenetics* 3:61–70

28. de Maat MP, Jukema JW, Ye S, Zwinderman AH, Moghaddam PH, et al. 1999. Effect of the stromelysin-1 promoter on efficacy of pravastatin in coronary atherosclerosis and restenosis. *Am. J. Cardiol.* 83:852–56

29. de Maat MP, Kastelein JJ, Jukema JW, Zwinderman AH, Jansen H, et al. 1998. -455G/A polymorphism of the beta-fibrinogen gene is associated with the progression of coronary atherosclerosis in symptomatic men: proposed role for an acute-phase reaction pattern f fibrinogen. REGRESS group. *Arterioscler. Thromb. Vasc. Biol.* 18:265–71

30. Desmeules J, Gascon MP, Dayer P, Magistris M. 1991. Impact of environmental and genetic factors on codeine analgesia. *Eur. J. Clin. Pharmacol.* 41:23–26

31. Diasio RB, Beavers TL, Carpenter JT. 1988. Familial deficiency of dihydropyrimidine dehydrogenase. Biochemical basis for familial pyrimidinemia and

severe 5-fluorouracil-induced toxicity. *J. Clin. Invest.* 81:47–51

32. Drazen JM, Yandava CN, Dube L, Szczerback N, Hippensteel R, et al. 1999. Pharmacogenetic association between *ALOX5* promoter genotype and the response to anti-asthma trreatment. *Nat. Genet.* 22:168–70

33. Drici MD, Barhanin J. 2000. Cardiac K+ channels and drug-acquired long QT syndrome. *Therapie* 55:185–93

34. Drysdale CM, McGraw DW, Stack CB, Stephens JC, Judson RS, et al. 2000. Complex promoter and coding region beta 2-adrenergic receptor haplotypes alter receptor expression and predict in vivo responsiveness. *Proc. Natl. Acad. Sci. USA* 97:10483–88

35. Ebstein RP, Levine J, Geller V, Auerbach J, Gritsenko I, Belmaker RH. 1998. Dopamine D4 receptor and serotonin transporter promoter in the determination of neonatal temperament. *Mol. Psychiatry* 3:238–46

36. Eichhammer P, Albus M, Borrmann-Hassenbach M, Schoeler A, Putzhammer A, et al. 2000. Association of dopamine D3-receptor gene variants with neuroleptic induced akathisia in schizophrenic patients: a generalization of Steen's study on DRD3 and tardive dyskinesia. *Am. J. Med. Genet.* 96:187–91

37. Esteller M, Garcia-Foncillas J, Andion E, Goodman SN, Hidalgo OF, et al. 2000. Inactivation of the DNA-repair gene MGMT and the clinical response of gliomas to alkylating agents. *N. Engl. J. Med.* 343:1350–54

38. Evans DAP, Manley KA, McKusick VA. 1960. Genetic control of isoniazid metabolism in man. *Br. Med. J.* 2:485–91

38a. Evans WE, Hon YY, Bamgaars L, Coutre S, Holdsworth M, et al. 2001. Preponderance of thiopurine S-methyltransferase deficiency and heterozygosite among patients intolerant to mercaptoputine or azathioprine. *J. Clin. Oncol.* 19:2293–301

39. Evans WE, Horner M, Chu YQ, Kalwinsky D, Roberts WM. 1991. Altered mercaptopurine metabolism, toxic effects, and dosage requirement in a thiopurine methyltransferase-deficient child with acute lymphocytic leukemia. *J. Pediatr.* 119:985–89

40. Evans WE, Relling MV. 1999. Pharmacogenomics: translating functional genomics into rational therapeutics. *Science* 286:487–91

41. Fairbrother KS, Grove J, de Waziers I, Steimel DT, Day CP, et al. 1998. Detection and characterization of novel polymorphisms in the CYP2E1 gene. *Pharmacogenetics* 8:543–52

42. Farlow MR, Lahiri DK, Poirier J, Davignon J, Schneider L, Hui SL. 1998. Treatment outcome of tacrine therapy depends on apolipoprotein genotype and gender of the subjects with Alzheimer's disease. *Neurology* 50:669–77

43. Felix CA, Walker AH, Lange BJ, Williams TM, Winick NJ, et al. 1998. Association of CYP3A4 genotype with treatment-related leukemia. *Proc. Natl. Acad. Sci. USA* 95:13176–81

44. Fijal BA, Hall JM, Witte JS. 2000. Clinical trials in the genomic era: effects of protective genotypes on sample size and duration of trial. *Control Clin. Trials* 21:7–20

45. Fujiwara Y, Yamaguchi K, Tanaka Y, Tomita H, Shiro Y, et al. 1997. Polymorphism of dopamine receptors and transporter genes in neuropsychiatric diseases. *Eur. Neurol.* 38 (Suppl. 1):6–10

46. Furuta T, Ohashi K, Kamata T, Takashima M, Kosuge K, et al. 1998. Effect of genetic differences in omeprazole metabolism on cure rates for Helicobacter pylori infection and peptic ulcer. *Ann. Intern. Med.* 129:1027–30

47. Furuya H, Fernandez-Salguero P, Gregory W, Taber H, Steward A, et al. 1995. Genetic polymorphism of CYP2C9 and its effect on warfarin maintenance dose requirement in patients undergoing

anticoagulation therapy. *Pharmacogenetics* 5:389–92

48. Garte S. 1998. The role of ethnicity in cancer susceptibility gene polymorphisms: the example of CYP1A1. *Carcinogenesis* 19:1329–32

49. Gerdes LU, Gerdes C, Kervinen K, Savolainen M, Klausen IC, et al. 2000. The apolipoprotein epsilon4 allele determines prognosis and the effect on prognosis of simvastatin in survivors of myocardial infarction: a substudy of the Scandinavian simvastatin survival study. *Circulation* 101:1366–71

50. Goldstein JA, Faletto MB, Romkes-Sparks M, Sullivan T, Kitareewan S, et al. 1994. Evidence that CYP2C19 is the major (S) mephentoin 4′-hydroxylase in humans. *Biochemistry* 33:1743–52

51. Gonzalez FJ, Fernandez-Salguero P. 1995. Diagnostic analysis, clinical importance and molecular basis of dihydropyrimidine dehydrogenase deficiency. *Trends Pharmacol. Sci.* 16:325–27

52. Gonzalez FJ, Skoda RC, Kimura S, Umeno M, Zanger UM, et al. 1988. Characterization of the common genetic defect in humans deficient in debrisoquine metabolism. *Nature* 331:442–46

53. Grant DM, Hughes NC, Janezic SA, Goodfellow GH, Chen HJ, et al. 1997. Human acetyltransferase polymorphisms. *Mutat. Res.* 376:61–70

54. Gratze G, Fortin J, Labugger R, Binder A, Kotanko P, et al. 1999. Beta 2-adrenergic receptor variants affect resting blood pressure and agonist-induced vasodilation in young adult Caucasians. *Hypertension* 33:1425–30

55. Grove J, Brown AS, Daly AK, Bassendine MF, James OF, Day CP. 1998. The RsaI polymorphism of CYP2E1 and susceptibility to alcoholic liver disease in Caucasians: effect on age of presentation and dependence on alcohol dehydrogenase genotype. *Pharmacogenetics* 8:335–42

56. Hansen T, Echwald SM, Hansen L, Møller AM, Almind K, et al. 1998. Decreased tolbutamide-stimulated insulin secretion in healthy subjects with sequence variants in the high-affinity sulfonylurea receptor gene. *Diabetes* 47:598–605

57. Hernandez D, Lacalzada J, Salido E, Linares J, Barragan A, et al. 2000. Regression of left ventricular hypertrophy by lisinopril after renal transplantation: role of ACE gene polymorphism. *Kidney Int.* 58:889–97

58. Hirata N, Takeuchi K, Ukai K, Sakakura Y. 1999. Expression of histidine decarboxylase messenger RNA and histamine N-methyltransferase messenger RNA in nasal allergy. *Clin. Exp. Allergy* 29:76–83

59. Hoit BD, Suresh DP, Craft L, Walsh RA, Liggett SB. 2000. Beta 2-adrenergic receptor polymorphisms at amino acid 16 differentially influence agonist-stimulated blood pressure and peripheral blood flow in normal individuals. *Am. Heart J.* 139:537–42

60. Hughes HB, Biehl JP, Jones AP, Schmidt LH. 1954. Metabolism of isoniazid in man as related to occurrence of peripheral neuritis. *Am. Rev. Tuberc.* 70:266–73

61. Hwu HG, Hong CJ, Lee YL, Lee PC, Lee SF. 1998. Dopamine D4 receptor gene polymorphisms and neuroleptic response in schizophrenia. *Biol. Psychiatry* 44:483–87

61a. Int. SNP Map Work. Group. 2001. A map of the human genome sequence variation containing 1.42 million single nucleotide polymorphisms. *Nature* 409:928–33

62. Israel E, Drazen JM, Liggett SB, Boushey HA, Cherniack RM, et al. 2000. The effect of polymorphisms of the beta(2)-adrenergic receptor on the response to regular use of albuterol in asthma. *Am. J. Respir. Crit. Care Med.* 162:75–80

63. Iyer L, King CD, Whitington PF, Green MD, Roy SK, et al. 1998. Genetic predisposition to the metabolism of irinotecan (CPT-11). Role of uridine diphosphate glucuronosyltransferase isoform 1A1 in

the glucuronidation of its active metabolite (SN-38) in human liver microsomes. *J. Clin. Invest.* 101:847–54

64. Jacobsen P, Rossing K, Rossing P, Tarnow L, Mallet C, et al. 1998. Angiotensin converting enzyme gene polymorphism and ACE inhibition in diabetic nephropathy. *Kidney Int.* 53:1002–6

65. Jia H, Hingorani AD, Sharma P, Hopper R, Dickerson C, et al. 1999. Association of the G(s)alpha gene with essential hypertension and response to beta-blockade. *Hypertension* 34:8–14

66. Johansson I, Lundqvist E, Bertilsson L, Dahl ML, Sjoqvist F, Ingelman-Sundberg M. 1993. Inherited amplification of an active gene in the cytochrome P450 CYP2D locus as a cause of ultrarapid metabolism of debrisoquine. *Proc. Natl. Acad. Sci. USA* 90:11825–29

67. Joober R, Benkelfat C, Brisebois K, Toulouse A, Turecki G, et al. 1999. T102C polymorphism in the 5HT2A gene and schizophrenia: relation to phenotype and drug response variability. *J. Psychiatry Neurosci.* 24:141–46

68. Kaiser R, Konneker M, Henneken M, Dettling M, Muller-Oerlinghausen B, et al. 2000. Dopamine D4 receptor 48-bp repeat polymorphism: no association with response to antipsychotic treatment, but association with catatonic schizophrenia. *Mol. Psychiatry* 5:418–24

69. Kalow W. 1956. Familial incidence of low pseudocholinesterase level. *Lancet* 211:576–77

70. Kapitany T, Meszaros K, Lenzinger E, Schindler SD, Barnas C, et al. 1998. Genetic polymorphisms for drug metabolism (CYP2D6) and tardive dyskinesia in schizophrenia. *Schizophr. Res.* 32:101–6

71. Kawajiri K, Watanabe J, Hayashi S. 1996. Identification of allelic variants of the human CYP1A1 gene. *Methods Enzymol.* 272:226–32

72. Keppler D, Leier I, Jedlitschky G. 1997. Transport of glutathione conjugates and glucuronides by the multidrug resistance proteins MRP1 and MRP2. *Biol. Chem.* 378:787–91

73. Kim DK, Lim SW, Lee S, Sohn SE, Kim S, et al. 2000. Serotonin transporter gene polymorphism and antidepressant response. *NeuroReport* 11:215–19

74. Kinnula K, Linnainmaa K, Raivio KO, Kinnula VL. 1998. Endogenous antioxidant enzymes and glutathione S-transferase in protection of mesothelioma cells against hydrogen peroxide and epirubicin toxicity. *Br. J. Cancer* 77:1097–102

75. Kioka N, Tsubota J, Kakehi Y, Komano T, Gottesman MM, et al. 1989. P-glycoprotein gene (MDR1) cDNA from human adrenal: normal P-glycoprotein carries Gly185 with an altered pattern of multidrug resistance. *Biochem. Biophys. Res. Commun.* 162:224–31

76. Kohno M, Yokokawa K, Minami M, Kano H, Yasunari K, et al. 1999. Association between angiotensin-converting enzyme gene polymorphisms and regression of left ventricular hypertrophy in patients treated with angiotensin-converting enzyme inhibitors. *Am. J. Med.* 106:544–49

77. Krynetskaia NF, Cai X, Nitiss JL, Krynetski EY, Relling MV. 2000. Thioguanine substitution alters DNA cleavage mediated by topoisomerase II. *FASEB J.* 14:2339–44

78. Krynetski EY, Evans WE. 1998. Pharmacogenetics of cancer therapy: getting personal. *Am. J. Hum. Genet.* 63:11–16

79. Krynetski EY, Schuetz JD, Galpin AJ, Pui C-H, Relling MV, Evans WE. 1995. A single point mutation leading to loss of catalytic activity in human thiopurine S-methyltransferase. *Proc. Natl. Acad. Sci. USA* 92:949–53

80. Kuehl P, Zhang J, Lin Y, Watkins P, Maurel P, et al. 2001 Sequence diversity in the *CYP3A* promoters and characterization of the genetic basis of polymorphic CYP3A5 expression. *Nat. Genet.* 27:383–91

81. Kuivenhoven JA. 1998. The role of a common variant of the cholesteryl ester transfer protein gene in the progression of

coronary atherosclerosis. The Regression Growth Evaluation Statin Study Group. *N. Engl. J. Med.* 338:86–93

82. Lee JT, Kroemer HK, Silberstein DJ, Funck-Brentano C, Lineberry MD, et al. 1990. The role of genetically determined polymorphic drug metabolism in the beta-blockade produced by propafenone. *N. Engl. J. Med.* 322:1764–68

83. Lennard L. 1998. Clinical implications of thiopurine methyltransferase-optimization of drug dosage and potential drug interactions. *Ther. Drug Monit.* 20:527–31

84. Lennard L, Gibson BE, Nicole T, Lilleyman JS. 1993. Congenital thiopurine methyltransferase deficiency and 6-mercaptopurine toxicity during treatment for acute lymphoblastic leukaemia. *Arch. Dis. Child.* 69:577–79

85. Lennard L, Van Loon JA, Lilleyman JS, Weinshilboum RM. 1987. Thiopurine pharmacogenetics in leukemia: correlation of erythrocyte thiopurine methyltransferase activity and 6-thioguanine nucleotide concentrations. *Clin. Pharmacol. Ther.* 41:18–25

86. Lennard MS, Tucker GT, Silas JH, Freestone S, Ramsay LE, Woods HF. 1983. Differential stereoselective metabolism of metoprolol in extensive and poor debrisoquin metabolizers. *Clin. Pharmacol. Ther.* 34:732–37

87. Liang HC, Li H, McKinnon RA, Duffy JJ, Potter SS, et al. 1996. Cyp1a2(−/−) null mutant mice develop normally but show deficient drug metabolism. *Proc. Natl. Acad. Sci. USA* 93:1671–76

88. Lieber CS. 1997. Cytochrome P-4502E1: its physiological and pathological role. *Physiol. Rev.* 77:517–44

89. Lima JJ, Thomason DB, Mohamed MHN, Eberle LV, Self TH, Johnson JA. 1999. Impact of genetic polymorphisms of the beta 2-adrenergic receptor on albuterol bronchodilator pharmacodynamics. *Clin. Pharmacol. Ther.* 65:519–25

90. London SJ, Idle JR, Daly AK, Coetzee GA. 1999. Genetic variation of CYP2A6, smoking, and risk of cancer. *Lancet* 353:898–99

91. Lucking CB, Durr A, Bonifati V, Vaughan J, De Michele G, et al. 2000. Association between early-onset Parkinson's disease and mutations in the parkin gene. French Parkinson's Disease Genetics Study Group. *N. Engl. J. Med.* 342:1560–67

92. MacKenzie PI, Owens IS, Burchell B, Bock KW, Bairoch A, et al. 1997. The UDP glycosyltransferase gene superfamily: recommended nomenclature update based on evolutionary divergence. *Pharmacogenetics* 7:255–69

93. MacLeod S, Sinha R, Kadlubar FF, Lang NP. 1997. Polymorphisms of CYP1A1 and GSTM1 influence the in vivo function of CYP1A2. *Mutat. Res.* 376:135–42

94. Mahgoub A, Idle JR, Dring LG, Lancaster R, Smith RL. 1977. Polymorphic hydroxylation of Debrisoquine in man. *Lancet* 2:584–86

95. Makoff AJ, Graham JM, Arranz MJ, Forsyth J, Li T, et al. 2000. Association study of dopamine receptor gene polymorphisms with drug-induced hallucinations in patients with idiopathic Parkinson's disease. *Pharmacogenetics* 10:43–48

96. Marian AJ, Safavi F, Ferlic L, Dunn JK, Gotto AM, Ballantyne CM. 2000. Interactions between angiotensin-I converting enzyme insertion/deletion polymorphism and response of plasma lipids and coronary atherosclerosis to treatment with fluvastatin: the lipoprotein and coronary atherosclerosis study. *J. Am. Coll. Cardiol.* 35:89–95

97. Martinelli I, Sacchi E, Landi G, Taioli E, Duca F, Mannucci PM. 1998. High risk of cerebral-vein thrombosis in carriers of a prothrombin-gene mutation and in users of oral contraceptives. *N. Engl. J. Med.* 338:1793–97

98. Martinelli I, Taioli E, Bucciarelli P, Akhavan S, Mannucci PM. 1999. Interaction between the G20210A mutation of

the prothrombin gene and oral contraceptive use in deep vein thrombosis. *Arterioscler. Thromb. Vasc. Biol.* 19:700–3

99. Martinez FD, Graves PE, Baldini M, Solomon S, Erickson P. 1997. Association between genetic polymorphisms of the beta 2-adrenoceptor and response to albuterol in children with and without a history of wheezing. *J. Clin. Investig.* 100:3184–88

100. Masellis M, Basile V, Meltzer HY, Lieberman JA, Sevy S, et al. 1998. Serotonin subtype 2 receptor genes and clinical response to clozapine in schizophrenia patients. *Neuropsychopharmacology* 19:123–32

101. McCarthy TV, Quane KA, Lynch PJ. 2000. Ryanodine receptor mutations in malignant hyperthermia and central core disease. *Hum. Mutat.* 15:410–17

102. McLeod HL, Evans WE. 2001. Pharmacogenomics: unlocking the human genome for better drug therapy. *Annu. Rev. Pharmacol. Toxicol.* 41:101–21

103. McLeod H, Krynetski E, Relling MV, Evans WE. 2000. Genetic polymorphism of thiopurine methyltransferase and its clinical relevance for childhood acute lymphoblastic leukemia. *Leukemia* 14:567–72

104. Meyer UA. 2000. Pharmacogenetics and adverse drug reactions. *Lancet* 356:1667–71

105. Michelson AD, Furman MI, Goldschmidt-Clermont P, Mascelli MA, Hendrix C, et al. 2000. Platelet GP IIIa Pl(A) polymorphisms display different sensitivities to agonists. *Circulation* 101:1013–18

106. Mickley LA, Lee JS, Weng Z, Zhan Z, Alvarez M, et al. 1998. Genetic polymorphism in MDR-1: a tool for examining allelic expression in normal cells, unselected and drug-selected cell lines, and human tumors. *Blood* 91:1749–56

107. Mukae S, Aoki S, Itoh S, Iwata T Ueda H, Katagiri T. 2000. Bradykinin B(2) receptor gene polymorphism is associated with angiotensin-converting enzyme inhibitor-related cough. *Hypertension* 36:127–31

108. Nakamura H, Uetrecht J, Cribb AE, Miller MA, Zahid N, et al. 1995. In vitro formation, disposition and toxicity of N-acetoxy-sulfamethoxazole, a potential mediator of sulfamethoxazole toxicity. *J. Pharmacol. Exp. Ther.* 274:1099–104

109. Napolitano C, Schwartz PJ, Brown AM, Ronchetti E, Bianchi L, et al. 2000. Evidence for a cardiac ion channel mutation underlying drug-induced QT prolongation and life-threatening arrhythmias. *J. Cardiovasc. Electrophysiol.* 11:691–96

110. Nemeth A, Szakmary K, Kramer J, Dinya E, Pados G, et al. 1995. Apolipoprotein E and complement C3 polymorphism and their role in the response to gemfibrozil and low fat low cholesterol therapy. *Eur. J. Clin. Chem. Clin. Biochem.* 33:799–804

111. Ohmichi N, Iwai N, Uchida Y, Shichiri G, Nakamura Y, Kinoshita M. 1997. Relationship between the response to the angiotensin converting enzyme inhibitor imidapril and the angiotensin converting enzyme genotype. *Am. J. Hypertens.* 10:951–55

112. Okamura A, Ohishi M, Rakugi H, Katsuya T, Yanagitani Y, et al. 1999. Pharmacogenetic analysis of the effect of angiotensin-converting enzyme inhibitor on restenosis after percutaneous transluminal coronary angioplasty. *Angiology* 50:811–22

113. Ongphiphadhanakul B, Chanprasertyothin S, Payatikul P, Tung SS, Piaseu N, et al. 2000. Oestrogen-receptor-alpha gene polymorphism affects response in bone mineral density to oestrogen in postmenopausal women. *Clin. Endocrinol.* 52:581–85

114. Ordovas JM, Lopez-Miranda J, Perez-Jimenez F, Rodriguez C, Park JS, et al. 1995. Effect of apolipoprotein E and A-IV phenotypes on the low density lipoprotein response to HMG CoA reductase inhibitor therapy. *Atherosclerosis* 113:157–66

115. Penno G. 1998. Effect of angiotensin-converting enzyme (ACE) gene polymorphism on progression of renal disease and the influence of ACE inhibition in IDDM patients: findings from the EUCLID Randomized Controlled Trial. EURODIAB Controlled Trial of Lisinopril in IDDM. *Diabetes* 47:1507–11

116. Perna A, Ruggenenti P, Testa A, Spoto B, Benini R, et al. 2000. ACE genotype and ACE inhibitors induced renoprotection in chronic proteinuric nephropathies1. *Kidney Int.* 57:274–81

117. Pianezza ML, Sellers EM, Tyndale RF. 1998. Nicotine metabolism defect reduces smoking. *Nature* 393:750

118. Poulsen L, Brosen K, Arendt-Nielsen L, Gram LF, Elbaek K, Sindrup SH. 1996. Codeine and morphine in extensive and poor metabolizers of sparteine: pharmacokinetics, analgesic effect and side effects. *Eur. J. Clin. Pharmacol.* 51:289–95

119. Prasad A, Narayanan S, Husain S, Padder F, Waclawiw M, et al. 2000. Insertion-deletion polymorphism of the ACE gene modulates reversibility of endothelial dysfunction with ACE inhibition. *Circulation* 102:35–41

120. Preuss CV, Wood TC, Szumlanski CL, Raftogianis RB, Otterness DM, et al. 1998. Human histamine N-methyl-transferase pharmacogenetics: common genetic polymorphisms that alter activity. *Mol. Pharmacol.* 53:708–17

121. Raftogianis RB, Wood TC, Otterness DM, Van Loon JA, Weinshilboum RM. 1997. Phenol sulfotransferase pharmacogenetics in humans: association of common SULT1A1 alleles with TS PST phenotype. *Biochem. Biophys. Res. Commun.* 239:298–304

122. Rebbeck TR, Jaffe JM, Walker AH, Wein AJ, Malkowicz SB. 1998. Modification of clinical presentation of prostate tumors by a novel genetic variant in CYP3A4. *J. Natl. Cancer Inst.* 90:1225–29

123. Reilly DK, Rivera-Calimlim L, Van Dyke D. 1980. Catechol-O-methyltransferase activity: a determinant of levodopa response. *Clin. Pharmacol. Ther.* 28:278–86

124. Relling MV, Hancock ML, Boyett JM, Pui C-H, Evans WE. 1999. Prognostic importance of 6-mercaptopurine dose intensity in acute lymphoblastic leukemia. *Blood* 93:2817–23

125. Relling MV, Hancock ML, Rivera GK, Sandlund JT, Ribeiro RC, et al. 1999. Mercaptopurine therapy intolerance and heterozygosity at the thiopurine S-methyltransferase gene locus. *J. Natl. Cancer Inst.* 91:2001–8

126. Relling MV, Rubnitz JE, Rivera GK, Boyett JM, Hancock ML, et al. 1999. High incidence of secondary brain tumors related to irradiation and antimetabolite therapy. *Lancet* 354:34–39

127. Ross D. 1996. Metabolic basis of benzene toxicity. *Eur. J. Haematol.* 60:111–18

128. Rothman N, Smith MT, Hayes RB, Traver RD, Hoener B, et al. 1997. Benzene poisoning, a risk factor for hematological malignancy, is associated with the NQO1 609C –> T mutation and rapid fractional excretion of chlorzoxazone. *Cancer Res.* 57:2839–42

129. Sallee FR, DeVane CL, Ferrell RE. 2000. Fluoxetine-related death in a child with cytochrome P-450 2D6 genetic deficiency. *J. Child Adolesc. Psychopharmacol.* 10:27–34

130. Sasaki M, Oki T, Iuchi A, Tabata T, Yamada H, et al. 1996. Relationship between the angiotensin converting enzyme gene polymorphism and the effects of enalapril on left ventricular hypertrophy and impaired diastolic filling in essential hypertension: M-mode and pulsed doppler echocardiographic studies. *J. Hypertens.* 14:1403–8

131. Sata F, Sapone A, Elizondo G, Stocker P, Miller VP, et al. 2000. *CYP3A4* allelic variants with amino acid substitutions in exon 7 and 12: evidence for an allelic variant with altered catalytic activity. *Clin. Pharmacol. Ther.* 67:48–56

132. Satoh K, Sato R, Takahata T, Suzuki S,
     Hayakari M, et al. 1999. Quantitative
     differences in the active-site hydropho-
     bicity of five human glutathione S
     -transferase isoenzymes: water-soluble
     carcinogen-selective properties of the
     neoplastic GSTP1-1 species. *Arch. Bio-
     chem. Biophys.* 361:271–76
133. Scharfetter J, Chaudhry HR, Hornik K,
     Fuchs K, Sieghart W, et al. 1999. Dopa-
     mine D3 receptor gene polymorphism and
     response to clozapine in schizophrenic
     Pakastani patients. *Eur. Neuropsycho-
     pharmacol.* 10:17–20
134. Schinkel AH. 1997. The physiological
     function of drug-transporting P-glyco-
     proteins. *Semin. Cancer Biol.* 8:161–70
135. Schuetz JD, Connelly MD, Sun D, Paibir
     SG, Flynn PM, et al. 1999. MRP4: a pre-
     viously unidentified factor in resistance
     to nucleoside-based antiviral drugs. *Nat.
     Med.* 5:1048–51
136. Schulz WA, Krummeck A, Rosinger I,
     Schmitz-Drager BJ, Sies H. 1998. Predis-
     position towards urolithiasis associated
     with the NQO1 null-allele. *Pharmacoge-
     netics* 8:453–54
137. Schutz E, Gummert J, Mohr F, Oellerich
     M. 1993. Azathioprine-induced myelo-
     suppression in thiopurine methyltrans-
     ferase deficient heart transplant recipient.
     Lancet 341:436
138. Seidegard J, Ekstrom G. 1997. The role
     of human glutathione transferases and
     epoxide hydrolases in the metabolism
     of xenobiotics. *Environ. Health Perspect.*
     105(Suppl. 4):791–99
139. Siegel D, McGuinness SM, Winski SL,
     Ross D. 1999. Genotype-phenotype rela-
     tionships in studies of a polymorphism
     in NAD(P)H:quinone oxidoreductase 1.
     Pharmacogenetics 9:113–21
140. Sindrup SH, Poulsen L, Brosen K, Are-
     ndt-Nielsen L, Gram LF. 1993. Are poor
     metabolisers of sparteine/debrisoquine
     less pain tolerant than extensive metabo-
     lisers? *Pain* 53:335–39
141. Smeraldi E, Zanardi R, Benedetti F, Di
     Bella D, Perez J, Catalano M. 1998. Poly-
     morphism within the promoter of the sero-
     tonin transporter gene and antidepressant
     efficacy of fluvoxamine. *Mol. Psychiatry*
     3:508–11
142. Spielberg SP. 1996. N-acetyltransferases:
     pharmacogenetics and clinical conse-
     quences of polymorphic drug metabolism.
     J. Pharmacokinet. Biopharm. 24:509–19
143. Stavroulakis GA, Makris TK, Krespi PG,
     Hatzizacharias AN, Gialeraki AE, et al.
     2000. Predicting response to chronic an-
     tihypertensive treatment with fosinopril:
     the role of angiotensin-converting enzyme
     gene polymorphism. *Cardiovasc. Drugs
     Ther.* 14:427–32
144. Steen VM, Lovlie R, MacEwan T, Mc-
     Creadie RG. 1997. Dopamine D3-re-
     ceptor gene variant and susceptibility to
     tardive dyskinesia in schizophrenic pa-
     tients. *Mol. Psychiatry* 2:139–45
145. Steen VM, Lovlie R, Osher Y, Belmaker
     RH, Berle JO, Gulbrandsen AK. 1998.
     The polymorphic inositol polyphosphate
     1-phosphatase gene as a candidate for
     pharmacogenetic prediction of lithium-
     responsive manic-depressive illness.
     Pharmacogenetics 8:259–68
146. Steward DJ, Haining RL, Henne KR, Da-
     vis G, Rushmore TH, et al. 1997. Genetic
     association between sensitivity to war-
     farin and expression of CYP2C9*3. *Phar-
     macogenetics* 7:361–67
147. Strautnieks SS, Bull LN, Knisely AS,
     Kocoshis SA, Dahl N, et al. 1998. A gene
     encoding a liver-specific ABC transporter
     is mutated in progressive familial intra-
     hepatic cholestasis. *Nat. Genet.* 20:233–
     38
148. Suzuki A, Mihara K, Kondo T, Tanaka
     O, Nagashima U, et al. 2000. The relation-
     ship between dopamine D2-receptor poly-
     morphism at the Taq1 A locus and
     therapeutic response to nemonapride, a
     selective dopamine antagonist, in schi-
     zophrenic patients. *Pharmacogenetics* 10:
     335–41
149. Szumlanski C, Otterness D, Her C, Lee

D, Brandriff B, et al. 1996. Thiopurine methyltransferase pharmacogenetics: human gene cloning and characterization of a common polymorphism. *DNA Cell Biol.* 15:17–30

150. Tai HL, Krynetski EY, Yates CR, Loennechen T, Fessing MY, et al. 1996. Thiopurine S-methyltransferase deficiency: two nucleotide transitions define the most prevalent mutant allele associated with loss of catalytic activity in Caucasians. *Am. J. Hum. Genet.* 58:694–702

151. Tan S. 1997. Association between beta 2-adrenoceptor polymorphism and susceptibility to bronchodilator desensitisation in moderately severe stable asthmatics. *Lancet* 350:995–99

152. Tanaka F, Shiratori Y, Yokosuka O, Imazeki F, Tsukada Y, Omata M. 1997. Polymorphism of alcohol-metabolizing genes affects drinking behavior and alcoholic liver disease in Japanese men. *Alcohol Clin. Exp. Res.* 21:596–601

153. Taube J, Halsall D, Baglin T. 2000. Influence of cytochrome P-450 CYP2C9 polymorphisms on warfarin sensitivity and risk of over-anticoagulation in patients on long-term treatment. *Blood* 96:1816–19

154. Tyndale RF, Droll KP, Sellers EM. 1997. Genetically deficient CYP2D6 metabolism provides protection against oral opiate dependence. *Pharmacogenetics* 7: 375–79

155. Vandenbergh DJ, Rodriguez LA, Miller IT, Uhl GR, Lachman HM. 1997. High-activity catechol-O-methyltransferase allele is more prevalent in polysubstance abusers. *Am. J. Med. Genet.* 74:439–42

156. Weinshilboum RM, Otterness DM, Aksoy IA, Wood TC, Her C, Raftogianis RB. 1997. Sulfation and sulfotransferases 1: sulfotransferase molecular biology: cDNAs and genes. *FASEB J.* 11:3–14

157. Whale R, Quested DJ, Laver D, Harrison PJ, Cowen PJ. 2000. Serotonin transporter (5-HTT) promoter genotype may influence the prolactin response to clomipramine. *Psychopharmacology* 150:120–22

158. Whitfield JB, Nightingale BN, Bucholz KK, Madden PA, Heath AC, et al. 1998. ADH genotypes and alcohol use and dependence in Europeans. *Alcohol. Clin. Exp. Res.* 22:1463–69

159. Wong RH, Wang JD, Hsieh LL, Du CL, Cheng TJ. 1998. Effects on sister chromatid exchange frequency of aldehyde dehydrogenase 2 genotype and smoking in vinyl chloride workers. *Mutat. Res.* 420:99–107

160. Wormhoudt LW, Commandeur JN, Vermeulen NP. 1999. Genetic polymorphisms of human N-acetyltransferase, cytochrome P450, glutathione-S-transferase, and epoxide hydrolase enzymes: relevance to xenobiotic metabolism and toxicity. *Crit. Rev. Toxicol.* 29:59–124

161. Yates CR, Krynetski EY, Loennechen T, Fessing MY, Tai HL, et al. 1997. Molecular diagnosis of thiopurine S-methyltransferase deficiency: genetic basis for azathioprine and mercaptopurine intolerance. *Ann. Intern. Med.* 126:608–14

162. Yu YW, Tsai SJ, Lin CH, Hsu CP, Yang KH, Hong CJ. 1999. Serotonin-6 receptor variant (C267T) and clinical response to clozapine. *NeuroReport* 10:1231–33

163. Zhou HH, Koshakji RP, Silberstein DJ, Wilkinson GR, Wood AJ. 1989. Altered sensitivity to and clearance of propranolol in men of Chinese descent as compared with American whites. *N. Engl. J. Med.* 320:565–70

164. Zill P, Baghai TC, Zwanzger P, Schule C, Minov C, et al. 2000. Evidence for an association between a G-protein beta 3-gene variant with depression and response to antidepressant treatment. *NeuroReport* 11:1893–97

Annu. Rev. Genomics Hum. Genet. 2001. 2:41–68

DNA DAMAGE PROCESSING DEFECTS AND DISEASE

Robb E. Moses

Department of Molecular and Medical Genetics, Oregon Health Sciences University,
Portland, Oregon 97201; e-mail: mosesr@ohsu.edu

Key Words   DNA repair, recombination, cancer, mutagenesis, genomic stability

■ **Abstract**   Inherited defects in DNA repair or the processing of DNA damage can lead to disease. Both autosomal recessive and autosomal dominant modes of inheritance are represented. The diseases as a group are characterized by genomic instability, with eventual appearance of cancer. The inherited defects frequently have a specific DNA damage sensitivity, with cells from affected individuals showing normal resistance to other genotoxic agents. The known defects are subtle alterations in transcription, replication, or recombination, with alternate pathways of processing permitting cellular viability. Distinct diseases may arise from different mutations in one gene; thus, clinical phenotypes may reflect the loss of different partial functions of a gene. The findings indicate that partial defects in transcription or recombination lead to genomic instability, cancer, and characteristic disease phenotypes.

INTRODUCTION

Genomes are stable but plastic. The advantages of multicellularity and organ specialization require multiple messages to be expressed correctly for place and time. Genetic information must be protected for expression within the cell and transfer between generations. Natural events lead to changes in information content during the duplication of genes and during the life of the cell, with a balance existing between the generation of diversity and genomic integrity. To achieve genomic stability, cells have developed DNA repair pathways.

The focus of this review is to summarize progress in understanding the relationship of DNA repair defects and human diseases. Some diseases are clearly the result of primary defects in the DNA repair process. Other diseases seem to be related to DNA repair or are related to genomic stability and proper distribution of the genomic elements. In order to recognize that some defects may lie directly in DNA repair functions, but others may relate to manipulation of intermediates of DNA processing, the term "DNA damage processing defects" will be used (65, 156). This review will summarize processing defects recognized to cause disease and present known deficiencies.

CATEGORIES OF DNA REPAIR

DNA repair responses react to a variety of types of damage produced by chemical and physical factors. Alterations resulting from replication also must be addressed. DNA damage may occur from interactions with radiation, chemicals that form adducts with the bases of DNA, structural impediments to transcription and replication, and spontaneous loss of bases. Reactive oxygen species are endogenous damaging agents, causing modification of bases and strand breakage. Cells are programmed to respond to those insults with reactions capable of recognizing the damage and initiating repair (reviewed in 180). Nomenclature in DNA repair for mammals reflects a diverse origin of gene discovery with some functions named for fungal genes, some for complementation of rodent cell lines, and some for mammalian enzymes. In general, only one term will be used in this review for a gene or protein.

Mechanistically, several categories of repair in mammalian cells have been identified. The excision of modified nucleotides or bases (BER) from DNA begins with recognition of the adduct. DNA glycosylases (eight are recognized in humans) with varying specificity excise the modified base by cleavage of the N-glycosilic bond act, leaving an apurinic or apyrimidinic (AP) site that is targeted by apurinic endonuclease (APE1), DNA polymerase β (39), and DNA ligase I (LIG1) to produce a short patch of repair. The proteins of this assembly appear to function coordinately (reviewed in 110, 180).

Nucleotide excision repair (NER) removes more than one base in response to helix-distorting adducts. The prototypic adduct for NER is the pyrimidine dimer resulting from UV radiation. The process requires recognition of the damage, incision of the DNA strand, removal and resynthesis followed by ligation. The reaction is performed by a multimeric protein complex (reviewed in 179). The incision step results from action by the XPG and XPF (complexed with ERCC1) proteins, with localization and assembly requiring XPA, XPC (in concert with HHR23B), RPA protein, and the TFIIH complex of transcription machinery. Two subunits of the TFIIH complex, XPB and XPD, are helicases that appear to "open" the site. The adduct is removed in an oligomer with an average length of 25 nucleotides. The resulting single strand gap is filled by the DNA polymerases, δ or ε (181), with interaction with PCNA (178), and the strand is covalently sealed by LIGI. NER differs in regions of the genome being transcribed from those not being transcribed (64). Transcription-coupled repair (TCR) may be viewed as a subset of NER or a distinct pathway of repair, a position supported by findings in Cockayne syndrome. Although XPC appears to be needed for damage recognition in NER, damage recognition mechanisms remain unclear (189). There are several options, which include halting of transcription that serves as a signal, recognition by a discrete factor (perhaps XPC), or interaction of UV-DNA damage binding factors (UV-DDB) with XPE.

Alternatively, repair may follow a different general scheme and respond to damage or strand breaks by use of components of DNA recombination pathways (reviewed in 62). This general path may act on single- or doublestrand breaks and may join homologous or nonhomologous strands. More than one pathway exists for homologous recombination, but each depends on helicases and exonucleases, with endonucleases sometimes required, and products of the *RAD* genes, for example, *RAD51*, followed by ligase action after gap filling (181). Nonhomologous end-joining (NHEJ) requires the end-binding proteins, Ku70p and Ku80p, ligase IV, XRCC4 protein, a DNA-dependent kinase, DNA-PKcs, the products of *RAD* genes, such as *RAD52* (95, 159), and DNA polymerase activity (71). The several pathways of DNA recombination may compete for the same intermediates. Additionally, recombination is opportunistic and capable of acting on intermediates generated by NER, thus representing a branch point in the pathway of repair.

Synthesis past unrepaired adducts appears to utilize recently identified DNA polymerases that do not display "classical" motifs (49). These polymerases may have a higher error rate than replicative polymerases, so DNA repair must occur prior to DNA replication to prevent mutagenesis. Mutagenesis therefore may be viewed as the outcome of a race between repair and DNA synthesis. If synthesis reaches an adduct or a partially repaired site, a bypass polymerase function would be used. The bypass polymerases are needed for optimal survival after UV damage, at apurinic sites, and with interstrand crosslinks (ICL), thus their decreased stringency of base pairing is advantageous.

Errors occuring during DNA replication are also monitored. Misincorporation of bases occurs at around 1 per 10^{10} events. This can be accounted for by the actual misincorporation of nucleotides by polymerase holoenzymes (about 1 in 100,000), the efficiency of editorial functions in replicative mechanisms (about 99.9%), and the efficiency of mismatch repair (about 99%). The mismatch repair system (16) in mammals is complex, consisting of several homologues to the MutS and MutL proteins of bacteria. There are six known MutS homologues and three MutL homologues. These proteins interact in oligomeric complexes during MMR. Recognition of the parental or "correct" strand of DNA is critical for preservation of information in mismatch repair. The basis for strand recognition and mismatch recognition is not known in mammals, whereas in *E. coli*, recognition is based on strand methylation. Recognition in mammals may be coupled to polarity of replication structures.

Other mechanistic categories of DNA repair exist. Alkylation repair responds to bases that have been exposed to alkyl reagents and modified by methylation (reviewed in 178). The response is specific and the bulky methyl group is removed *in situ*. Photoreactivation of DNA is apparently an ancient process, specific for thymine dimers (43, 127). The repair enzyme is activated by photons and is inactive after reversing the dimer in situ, using FAD as a factor for electron transfer. Some bases in DNA can be oxygenated, and repair systems specific to repair of oxygenated bases respond. Reactive oxygen species can produce AP sites and strand breaks as well as oxidized bases (10, 17). The DNA glycosylase pathway acts on methylated or oxygenated bases, but 8-oxoguanine is also acted on by the OGG1 protein, which is a glycosylase and AP lyase initiating BER (115, 158).

The responses to DNA damage mentioned in this paragraph are not known to be the basis of diseases and will not be considered further.

GENETIC CATEGORIES OF REPAIR

Although genetic definition of repair pathways (epistasis groups) is well established in bacteria and fungi, relatively less is known regarding genetic grouping of repair activities in mammals. When two DNA repair defects are combined, if the sensitivity to a damaging agent is not increased, then the two genes are epistatic, or in the same pathway. If the sensitivity is increased, then they are non-epistatic and in different pathways of repair. This test is a benchmark underlying the biochemical analysis of repair mechanisms, as epistasis groups define control of repair pathways and gene interactions. In *S. cerevisiae*, three groupings of the genes are required for normal survival after DNA damage: excision repair (*RAD3* group), postreplication (*RAD6* group), and recombination (*RAD52* group). The epistasis groups establish separate pathways of response, based on repair defective mutants but also based on the DNA damaging agent. Most of the analysis of the genetic groups has been done with ionizing or ultraviolet radiation (IR and UV, respectively). Although relatively less complete analysis has been made with other agents, DNA crosslinking agents also demonstrate three genetic groupings for repair (K. Grossmann & R. Moses, unpublished results).

Epistasis analysis of repair genes for mammals is not definitive. The same functions may be present for recombination, excision, and postreplication repair, as in fungi. However, additional functions are likely, as shown by Fanconi anemia for interstrand crosslink repair (see below). Epistasis analysis can be derived from matings of mice with repair defects. However, some of the genes, for example, *RAD51*, are early embryonic lethal, making the analysis more difficult.

CELL CYCLE CHECKPOINTS

Cell cycle checkpoints are a consideration in DNA processing defects. The concept of checkpoints arose from studies that used cell cycle and DNA damage-sensitive mutants in *S. cerevisiae*. Genes required for regulation of progression of the cell cycle were identified, and it became apparent that some of the genes were required for normal survival after DNA damage (44, 171, 189). Additionally, some of the gene products are required for normal mitosis or meiosis. The model that evolved is one in which the cell cycle is genetically regulated and responds to functions that sense the completeness of DNA repair or processing of DNA intermediates. Thus cell cycle checkpoints are pauses, permitting assessment and completion of DNA processing. Defects in this quality control allow the products of unrepaired DNA or incomplete processing to pass into the next stage of the cycle. This would be detrimental to genomic integrity.

A checkpoint must require sensing mechanisms (presumably ones that differ for various DNA damage), transducers to signal to the cell cycle engine, effectors to arrest the cycle and process the damage, and finally, signalling to allow resumption of cell cycle progression. In bacteria, damage exceeding the cell's ability to repair is met with a tolerance, or error-acceptance response, the SOS pathway. This allows continued replication, accepting mutagenesis as a price for survival. In contrast, within eukaryotes the cell pauses for damage repair, and if it is not achieved within an appropriate time, the cell enters apoptosis. Thus in a multicellular organism, cell death is favored over genomic instability.

AUTOSOMAL RECESSIVE DNA DAMAGE PROCESSING DISEASES

Xeroderma Pigmentosum

Cleaver's studies demonstrated that a DNA repair deficiency could be the cause of disease, with the finding that xeroderma pigmentosum (XP) patients lacked normal excision repair (23). XP patients show increased photosensitivity and formation of a variety of skin cancers, including squamous cell carcinomas, basal cell carcinomas, and melanomas, primarily in exposed areas of the skin. XP is frequently noted in the first two years of life. The clinical picture is one of accelerated aging of the skin: freckling, followed by leukoplakia, followed by "age spots," or actinic keratoses, followed by carcinoma (85). Ocular findings include atrophy of the lids, conjunctivitis, and corneal opacity (86, 87). Some XP patients have neurological findings, including mental retardation and movement disorders (22, 85). XP patients may have telangiectasia—dilated and tortuous small blood vessels—on the eyelids or conjunctiva. XP patients show an increase in cancers other than skin cancer (86).

At the cellular level, XP patients demonstrate apparently normal chromosome stability. However, cell survival after UV radiation is decreased remarkably (5, 22, 87), and abnormalities in chromosomes increase (34, 176). Reactivation of UV-treated viruses is decreased in XP cells, and UV mutagenesis is increased (67, 68, 120). Cell-cell fusions showed that more than one complementation group, and therefore more than one enzymatic defect, exist for XP (22, 33). Eventually seven complementation groups were defined (22, 23, 177), reflecting a failure in NER.

The gene products isolated for XP (Table 1) fall into several categories (reviewed in 128, 178–180). Homologues to the human XP genes are found in S. cerevisiae among the RAD genes, and conserved orthologs are found in other mammals. The ERCC, excision repair cross-complementing, genes were defined by isolation of human genes that complement repair-defective rodent cells. Subsequently, some were identified to be defective in XP. One category of proteins acts to recognize and bind the UV damage, possibly by duplex bending recognition. A second

TABLE 1

Protein	Function	Pathway
XPA	Damage recognition/binding at adduct	NER
XPB	Helicase in transcription complex	NER
XPC	Damage recognition/binding at adduct	NER
XPD	Helicase in transcription complex	NER
XPE	DNA binding at adduct	NER
XPF	Endonuclease (SS), cleaving 5′ to adduct	NER
XPG	Endonuclease (SS), cleaving 3′ to adduct	NER
POLη	Translesion synthesis at adduct	NER
CSA	Transcriptional activator/transcription coupling	TCR
CSB	Transcriptional activator/transcription coupling	TCR
BLM	3′-5′ helicase/replication	(Recombination)
WRN	3′-5′ helicase/3′-5′ exonuclease/ replication	(Recombination)
RTS1	3′-5′ helicase/replication	(Recombination)
ATM	Kinase acting on NBS1, BRCA1, p53, CHK2	DSB repair
MRE11	Unknown, DSB repair, translesion synthesis	DSB repair
NBS1	Complexes with RAD50, MRE11	DSB repair
FANCA-FANCG	Unknown (structural/replication/ recombination)	DSB repair/ recombination
MSH2, MSH3, MSH4, MSH 5, MSH6, PMS1, PMS2, MLH1	DNA binding/ATPase	MMR
BRCA1 (BRCA2)	Unknown (ubiquitin ligase/structural/ recombination)	DSB repair/ recombination

The protein names are according to common usage, and "h" is not used to denote human, nor are proteins identified as homologs. SS, single-stranded; DSB, doublestrand break; ( ), unproven.

category represents specific endonucleases acting at the adduct. Other XP genes represent transcription helicases, presumably serving to create an "open bubble" in the duplex. Helicases are DNA or RNA-dependent ATPases that unwind duplex helices.

Involvement of transcription proteins in repair suggests a function common for two cellular processes. An alternate possibility is that the transcription component is a multifunctional protein and only the functions relating to repair are defective in XP, whereas other transcription functions are intact. Additional proteins that act in the NER process are reflected in XP: Replicative protein A (RPA) is essential for damage binding by XPA (113); HHR23B complexes with XPC for damage binding; transcription factor TFIIH, containing XPB and XPD along with 4–7 additional subunits, is also required; XPF complexes with the ERCC1 homolog, and an XPE complex with DNA damage binding protein (DDB) is also necessary.

A variant group of XP, manifesting apparently normal NER, is caused by a deficiency in a DNA polymerase from a recently recognized family of such enzymes (reviewed in 49). This family does not display "classical" polymerase motifs (77, 109). The DNA polymerase, product of the *POLH* gene (*hRad30*), is termed η. Ordinarily, UV produces arrest of S-phase, dependent on p53, acting via p21. A rapid response operates directly to destroy cyclins and release p21 prior to additional transcription (2), effecting a cell cycle checkpoint. A complex with the MRE11 protein may contain DNA polymerase η at the site of replication arrest after DNA damage (104). This complex may be involved in bypass DNA synthesis during repair, playing a role in a specialized function for inserting nucleotides when the replicative fork reaches the adduct (translesion synthesis). Polymerase η appears inaccurate and acts in a nonprocessive manner on damaged templates, utilizing undamaged templates poorly, if at all. A variety of damages may be bypassed by polymerase η (66). A conserved ortholog in yeast, the product of the *RAD30* gene, serves a similar purpose in repair of UV products, but it does not appear to act in crosslink repair (K. Grossmann & R. Moses, unpublished).

XP serves as a prototype DNA damage processing disease in which partial defects in a required cellular process that are used for specialized repair functions result in disease but are compatible with viability. If the proteins of the pathway have common roles in transcription and repair is still unknown. Alternatively, discrete functions of the proteins could serve either the repair process or transcription. Evidence is accumulating that favors the latter case (148, 168, 175). XP also serves as a model for the development of cancer, secondary to a deficiency to respond normally to a known mutagen.

Cockayne Syndrome

Cockayne syndrome (CS) is intriguing because of the similarity, though generally without defects in the XP proteins, in UV sensitivity of cells isolated from CS patients to those of XP (165). CS patients manifest extreme dwarfism, severe mental retardation, characteristic retinal pigmentation, photosensitivity, and deafness (22, 63). Patients have a recognizable bird-like facial appearance, due in part to lack of subcutaneous fat. They apparently are not predisposed to cancer, but most do not survive beyond the second decade. Cells from CS patients manifest UV sensitivity similar to XP cells (6, 131, 164), without the same NER deficiency, but the recovery of RNA and DNA synthesis after UV is slower than normal (98). Cellular complementation studies show two CS groups, with rare patients manifesting both CS and XP findings (98). The combination patients have the cutaneous findings of XP, including cancers, but with the mental retardation and retinal degeneration of CS (63). The XP/CS patients fall into the XPB, XPD, and XPG groups (reviewed in 15, 63, 157, 163). Certain *XPG* mutations (truncations) can be associated with the CS phenotype, whereas most others result in the XP phenotype (25).

CS patients fall into two groups, *CSA* or *CSB*. *CSB* (157) was identified originally as *ERCC6*. The CSB protein may serve as a transcription coupling factor and contains motifs that make it a member of the DNA-dependent ATPase, SNF2 family (137). CSB stimulates transcription elongation by RNA polymerase II (137). CSA, a WD repeat family member, interacts with CSB as well as the TFIIH factor (69). Thus, both CS factors appear to have a role in transcription.

The delayed recovery of transcription after UV radiation in CS is reflective of the underlying defect. Transcribed genes are repaired more rapidly (5–10 times) than nontranscribed genes (63, 64, 112). TCR is defective in both CSA and CSB, whereas global NER remains normal (97), suggesting the two proteins serve to couple DNA repair to transcription. Hence, CS is a failure of TCR because of a deficiency in CSA or CSB (excepting the particular XPB, XPD, or XPG patients). Thus, TCR differs from either BER or NER in some respects. Evidence points to the conclusion that CS results from defective TCR of oxidative products in DNA, with TCR occurring as a result of TFIIH, CSA, and CSB function (102). The XPC damage recognition factor is also not essential for TCR, perhaps indicating that stalled RNA polymerase II signals DNA damage.

Trichothiodystrophy

Trichothiodystrophy (TTD) adds yet another twist to excision repair defects. TTD is a rare disease, with symptoms that include brittle hair, dry skin, dysmorphic face, and mental retardation. Some patients manifest photosensitivity, which led to an investigation of the repair capability of TTD cells. NER does appear compromised (reviewed in 24). The defect lies mostly in the *XPD* gene and less often in the *XPB* gene. In some TTD patients, other transcription factors may be involved. TTD illustrates the potential for variable outcome with different mutations in a single component of the transcription machinery, showing that different clinical phenotypes (TTD or XP) can arise from defects in the same gene. In TTD there is correlation of the specific mutation and the clinical phenotype (153). Gene disruptions of *XPD* in mice are lethal early at implantation (32). However, a model of TTD can be built in mice by duplicating a disease-associated mutation in *XPD* (31).

Disruption of the helicase domains of the XPD protein allows transcription initiation, but it produces defective repair (175). Taken together, the observations suggest that discrete functional domains within XPD act in distinct roles. Three different diseases, XP, CS, and TTD, are associated with mutations in XPD (or XPB) helicase activity. Could it be that relative abundance of disease associated with XPD may suggest that the helicase has a minor role in transcription? The variable presentation of disease also raises the question of whether specific targets of transcription are affected in each case. That is, do only some genes require the action of different portions of the XPD protein for expression, and therefore, various mutations affect expression of these genes independently?

Bloom Syndrome

Bloom Syndrome (BS) is a disease of small stature, characteristic facial erythema, immunologic defects, decreased fertility, increased cancer, and normal intelligence (56). Cells from BS patients show chromosome instability with chromosome breaks and sister chromatid exchanges (SCEs) (57). BS cells are also hypermutable (61, 164, 169). Reports on BS cell sensitivity to DNA damage are variable and may reflect cell cycle effects (60, 72). A delay in DNA chain maturation has been noted (58).

The gene encoding the Bloom protein (*BLM*) has homology to the product of the *recQ* helicase gene of *E. coli* (45). There is a distinct polarity to a helicase, defined by whether the unwinding action proceeds $5' \rightarrow 3'$ or $3' \rightarrow 5'$. The RecQ protein is a $3' \rightarrow 5'$ helicase and participates in homologous recombination and recombination required during replication, dependent on *recF*, in *E. coli* (reviewed in 19). One possible function for RecQ protein is as part of a complex that promotes bypass of the replisome beyond an adduct encountered in replication. RecQ also may play a role in postreplication gap filling (27, 83) and in stable DNA replication (108). Thus, RecQ protein may act on intermediates of recombination and replication in prokaryotes. The BLM protein is a $3' \rightarrow 5'$ helicase (79) that promotes Holliday junction branch migration (80). Fission yeast lacking a RecQ homologue, the product of *RQH1*, show delayed replication recovery after fork stalling (143). Complicating interpretation of the defect in BS is the fact that there are multiple (five) *RECQ* homologues in humans, allowing the possibility of functional redundancy, obscuring the "pure" phenotype of a defect.

The basis for the chromosomal instability in BS is not understood. Perhaps the helicase defect prevents correct interaction of other proteins required for chromosome modeling, such as topoisomerases, given the genetic evidence of interaction of helicases and topoisomerases (51). It may be that these two activities act in oppositional concert in maintaining genome stability. Alternatively, the BLM helicase may act directly in recombination during replication, as inactivation of a BLM ortholog leads to inhibition of DNA replication in frog egg lysates (103).

Werner Syndrome

The *WRN* gene, defective in Werner Syndrome (WS), has been identified and is a member of the *recQ* family (187). WS patients show accelerated aging, including hair graying, osteoporosis, cataracts, atherosclerosis, and loss of subcutaneous fat. They also show an increased incidence of diabetes, a characteristic high-pitched voice, and delayed sexual maturity. There is an increased incidence of cancer in WS (46). WS cells have a diminished replicative lifespan in culture and exhibit chromosomal translocations and increased mutation rates (50, 126). WS cells are hypersensitive to 4-nitroquinoline-1-oxide (4-NQO) and manifest increased chromosome breakage after 4-NQO treatment (55; H. Saito & R. Moses, unpublished results). The WRN helicase shows *recQ* family motifs, including an ATP-binding box, a DExH-box, and seven helicase repeats (19); and it has $3' \rightarrow 5'$ helicase activity and a $3' \rightarrow 5'$ exonuclease function (139, 140). As with the BLM protein, WRN has a nuclear localization signal (NLS) in the carboxy-terminal region. Most *WRN* mutations appear to cause truncations with loss of NLS, whereas BLM mutations appear to cause loss of helicase activity. WRN may act to recruit DNA polymerase activity (152) for specific replication function. Thus, WRN may play a role in recombinational repair or bypass at the replicative fork, and it is important for genetic stability.

Rothmund-Thomson Syndrome

Rothmund-Thomson syndrome (RTS) is characterized by poikiloderma (a patchy coloration of the skin, mimicking the pattern seen in an infant who is cold), skeletal abnormalities (particularly the thumb), cataracts at an early age, signs of accelerated aging, and increased incidence of cancer (161, 162). The gene (*RECQ4*) responsible for some of the cases of RTS is a helicase belonging to the *recQ* family (81, 82), the same family defective in Werner syndrome and Bloom syndrome. Thus, these three syndromes all share a defect in a single family of proteins with the activity of unwinding DNA, yet the phenotypes are distinct. The syndromes therefore present evidence for specificity of function rather than redundancy.

A defect in DNA repair or processing has not been noted yet as a hallmark of RTS. However, mutations in orthologs of the *recQ* helicase family in budding yeast (*SGS1*) lead to defects in homologous recombination and have increased sensitivity to DNA-damaging agents (19, 52).

The comparison of XP (notably XPD and XPB), CS, BS, RTS, and WS presents a bewildering combination of findings, tied by the common theme of helicase and transcription defects (106). The findings of the various syndromes are distinct, yet with some common ground, especially sensitivity to DNA damage. The activities of the affected proteins are present in transcription and recombination or replication. Whether each of the helicases has more than one role is still unknown. It is clear that an assay assessing whether a protein is a helicase and whether it has a given polarity gives insufficient information regarding its in vivo activity.

Ataxia Telangiectasia

Ataxia telangiectasia (AT) is a disorder characterized by immune deficiencies, lack of coordination, increased incidence of tumors, chiefly lymphomas, and telangiectasias (reviewed in 141). Patients have sometimes been recognized by their development of secondary cancers at the sites of radiotherapy for a primary tumor, but occasionally they are brought to attention because of dysarthria. There is cell death in the cerebellum, apparently explaining the ataxia and dysarthria. Hypoplastic ovaries are frequent in female AT patients. The telangiectasias are present on ears, malar regions, the dorsum of the hands, and the sclerae. There is overlap with XP of cutaneous manifestations (85): XP may manifest telangiectasia, AT and XP each have hypo-and hyperpigmentation, both manifest dry skin, and both show cutaneous atrophy. The ataxia is progressive in AT, and patients are frequently wheelchair-bound by the end of the first decade (85, 141). The reasons for concentration of neurological findings in the cerebellum are not known.

The *ATM* (ataxia telangiectasia mutated) gene may be responsible for a large amount of background genetic influence on the incidence of cancer in the population (7, 149–159). This remains a controversial conclusion.

Cells from AT patients have two characteristic findings that indicate a defect in DNA damage processing: sensitivity to ionizing radiation (IR) and chromosome instability (84, 85, 111, 119, 141). The instability manifests as breaks and translocations, frequently involving chromosome 14q (84, 85). Cells from AT patients are radiosensitive (84, 85, 170), and radiation increases the chromosome aberrations (reviewed in 111, 141). AT cells are also sensitive to bleomycin (100), which is an IR-mimetic. DNA synthesis normally is inhibited after IR, but AT cells manifest a "radio-resistant" DNA synthesis (RDS) (99, 116) after IR, showing less of a decrease or pause in DNA synthesis than normal cells. Although AT patients' cells are sensitive to IR compared to normal cells, they do not show increased sensitivity to UV, alkylation, or crosslinking compounds. Thus the repair defect is specific, although it is not clear which product(s) of IR AT cells fail to process correctly.

It was thought that more than one complementation group existed for AT (141), but localization (54) and identification of the *ATM* gene showed there was only one locus (129). The *ATM* gene is a cell cycle regulatory gene; its product controls entry into S-phase. Loss of function allows the cell cycle to progress without proper quality control, leading to genomic instability. ATM protein has kinase activity (reviewed in 114, 141), which apparently acts in response to IR damage to phosphorylate targets (CHK2, p53, NBS1, and BRCA1, among others), some of which are needed for checkpoint function, particularly at the G1/S boundary. The phosphorylation leads to accumulation of p53, leading to transcriptional activation of genes that produce p21, GADD45, and p53R2, inhibiting cell cycle progression (44, 154). The kinase activity of ATM also promotes activation of DNA repair, because phosphorylation is linked to assembly of the NBS1/MRE11/RAD50 complex (114). Loss of cell cycle checkpoint control fits with a lack of inhibition of DNA synthesis after IR (mentioned above). ATM does not phosphorylate its targets after other types of DNA damage (e.g., UV or HU); but the targets are still phosphorylated, so an independent mechanism exists. Yeast homologues to *ATM* are known, *MEC1* and *TEL1*, and partial inactivation of *MEC1* leads to genomic instability.

Nijmegen Breakage Syndrome

Nijmegen breakage syndrome (NBS) shares features with other chromosome instability syndromes, including predisposition to cancer, immune deficiency, developmental defects, and chromosome breaks (29, 141, 160). Chromosome translocations frequently occur and involve chromosome 14. However, telangiectasias are not a prominent finding, nor is ataxia. Cells from patients with NBS show increased sensitivity to IR and RDS, leading to consideration of the disease as a variant of AT (29). However, mapping placed the *NBS1* gene on chromosome 8, whereas *ATM* is on 11. The *NBS1* product, Nibrin protein or NBS1, is 95 kDa and is needed for doublestrand break repair (160). The NBS1 protein is specifically phosphorylated at serine residues by ATM in response to IR (182, 188). NBS1 colocalizes with MRE11 and RAD50 in complexes formed in response to DNA damage. Thus the overlap in cellular characteristics between AT and NBS may reflect the function of both gene products in a common pathway of recombinational repair. This would help explain the specificity of sensitivity to DNA damage in AT. A homolog to ATM, ATR kinase, is not yet linked to disease.

Underscoring the funtional linkage of the components of the MRE11/RAD50/NBS1 complex that appears after DNA damage is the observation of an AT-like disorder (AT-LD) that has a mutation in MRE11 (144). ATM also phosphorylates *BRCA1*, the product of a familial breast cancer gene (see below). Thus ATM is an actor in more than one pathway of repair, making it an "upstream" function (144). What might be the common thread? The repair functions descending from ATM appear to share repair of DSB as a common theme. This would be reasonable in view of AT sensitivity to IR. Moreover, DSB repair appears to involve recombination with DNA replication, as noted above, a combination of activities that have direct coupling (13).

Fanconi Anemia

Fanconi Anemia (FA) is a rare DNA damage processing defect that produces malformations, including defects in the radius and its associated structures, somewhat darkened skin pigmentation, pancytopenia of the marrow elements, and an increased incidence of leukemia (reviewed in 4, 9, 30). FA is a phenotype that results from a defect in any of at least eight genes, indicating a complex process. There is evidence of somatic mosaicism in FA patients (9, 107).

At the cellular level, the patients are characterized by hypersensitivity to DNA crosslinking agents, with increased cell death and increased abnormal chromosome forms, including breaks and quadraradials that occur spontaneously but are also increased following treatment with DNA crosslinkers (8, 73, 132, 133). The G2-phase of the cell cycle appears prolonged in FA cells after treatment with crosslinkers, but it also appears to be intrinsically prolonged (42, 91, 138). However, FA cells respond normally to inhibitors of the cell cycle (78). FA cells show oxygen sensitivity (90, 123, 125, 130, 186), but this may be a secondary defect (124, 174). FA cells have been reported to be either hypo- or hypermutable in response to crosslinkers (9, 117). FA cells do incise DNA after crosslinking treatment, and repair crosslinks, but apparently more slowly than normal cells (3, 92, 147). After ICL, the G2 prolongation seems to reflect an intact checkpoint responding to slower DNA damage processing. Recombination may also be defective in FA (94). Given the likelihood that recombination is required for crosslink repair (72) and may have a function in rescuing stalled replication forks (28), it is quite possible that the *FANC* gene products serve in recombination during replication.

An open question is whether the *FANC* genes are primary players in crosslink repair, and if that is the case, if all *FANC* genes act in that process. FA cells show apparently normal sensitivity to a wide variety of DNA damaging agents other than crosslinkers. Thus the *FANC* gene products define a response to interstrand crosslink damage but not to UV or alkylation. Presumably the heightened sensitivity of FA cells to DNA interstrand crosslinkers reflects a deficiency in processing a normal intermediate for DNA, which over time leads to genomic instability, chromosomal aberrations, and cancer. There are at least eight complementation groups (76), with the group H cells actually being group A (75) and the prototypic D cell line (HSC62) not complementing cell line PD20, which was used to isolate the *FANCD2* gene (156a). The six identified *FANC* genes have been isolated by positional as well as complementation cloning (35–37, 48, 146, 155, 156a). The gene products do not show meaningful homology with genes from other systems, although there may be a homolog to *FANCD2* in plants (Timmers et al. 2001). Partial conservation with the bacterial RNA-binding protein ROM was shown for *FANCF* (36). Whether the *FANC* genes act in concert or sequentially is unknown, although there is evidence for interaction of several of the gene products (e.g., FANCG with FANCA, FANCC, and FANCF) (53, 93, 166; M. Matkovic & R. Moses, unpublished data). FANCD2 complexes with RAD51 and BRCA1 (53a). Some of the FANC proteins appear localized to the cytoplasm, but repair functions occur in the nucleus. Localization of FANC proteins to the nucleus, which may be cell cycle–dependent, is also seen (70, 88, 89, 93, 183–185). Posttranslational modification of the proteins may be needed for this (184). FANCD2 is monoubiquitinated after DNA damage, and this is required for interaction with BRCA1

(53a). This raises the possibility of concerted action in recombinational repair with BRCA1.

Mice deficient in FANCC have been constructed (21, 173). The mice are viable and show no obvious birth defects. Unfortunately, they do not mimic the appearance of anemia or cancer seen in humans. They do show some common findings with people, however, including a decrease in fertility in males, with decreased cellularity in the testes, and a cellular sensitivity to crosslinkers, including chromosome instability. Localization of FANC proteins in telomerases and synaptosomes and their function in ICL repair, as well as colocalization with RAD51, argue for a function in recombination. Synaptonemal complex formation does not appear normal in *FANCC* disruption mice, with aberrant pairing of chromatids (S. Meyn, personal communication).

Although current results do not exclude *FANC* genes from being limited to mammals (except for *FANCD2*), mutants specifically sensitive to DNA crosslink damage do exist in yeast, despite yeast not having homologues for any of the identified *FANC* genes. A mammalian ortholog exists for one yeast gene that appears specifically dedicated to crosslink repair, the *SNM1* gene. Disruption of the *SNM1* murine ortholog produces mice with normal viability and reproduction, but decreased survival after mitomycin C (MMC) and increased cellular sensitivity to MMC (40). Combined findings in yeast and FA lead to the conclusion that multiple pathways for repair of ICL exist in mammals. At least three separate pathways exist for ICL repair in yeast (K. Grossmann & R. Moses, unpublished). The existence of multiple epistasis groups for crosslink repair means that multiple mechanisms for dealing with the interstrand crosslink exist. Moreover, because sensitivities are additive, the pathways cannot share components. Whether this observation will hold for mammals is unknown. Because FA cells have normal NER but incisions are required for ICL repair, other incision activities must exist.

The genotype/phenotype correlation for FA can be rationalized as reflecting loss of processing of a normal recombination substrate, which is duplicated in the repair of ICL. A defect in rate of processing, due to collapse of one of several parallel pathways, might be such a subtle defect. Delayed processing of a product that arises during DNA replication, requiring recombination or sharing a function with ICL repair, might cause arrest at the G2/M checkpoint, with slow processing occasionally forcing the cell to proceed either via apoptosis or entry into M-phase without finished processing. The latter would lead to genomic instability.

Finally, the existence of eight gene products, at least some of which interact, with unknown function leaves a riddle. Conceivably, FANC proteins could play a structural role, serving as a scaffold for maintenance of configuration during ICL repair or processing of DSBs. Alternatively, some FANC proteins might function in a complex with BRCA1 and RAD51 (see below) directly in the processing of the ICL, or they may serve in signaling for recruitment of repair enzymes and cell cycle arrest.

DOMINANT OR SPORADIC DNA DAMAGE PROCESSING DISEASES

Hereditary Nonpolyposis Colorectal Cancer

Hereditary nonpolyposis colorectal cancer (HNPCC) arises from defects in mismatch repair (MMR). MMR represents a different path of repair from BER, NER, or recombination. HNPCC, or Lynch syndrome, accounts for 5%–7% of colon cancers. Heritable cases of HNPCC have early onset, and the patients have increased risk of cancers elsewhere, including in the reproductive organs (121). There is clinical overlap with Muir-Torre and Turcot syndromes, and defects in MMR also exist for those syndromes. The identification of the tumor suppressor function of MMR that is linked to the occurrence of colorectal cancer depended on an assay for instability in DNA sequences with oligonucleotide repeats, microsatellite instability (MSI or MIN). In yeast, MMR mutations led to MSI (145). MSI had been found in some human colon cancers (1, 122). Therefore, deficiencies in MMR were sought and found in HNPCC (14, 96). The MIN was limited to the tumor and was not found elsewhere in the patient's tissues.

MMR depends on the action of complexes of multiple homologs of the bacterial *mutS* and *mutL* genes. There are at least six homologs of *mutS* and three of *mutL*, which appear to function in mismatch recognition (16). Both the *mutS* and the *mutL* homologs have ATPase activity, but the requirement for such activity in MMR is not clear (16, 142, 143). The MSH and MLH1/PMS complexes seem to interact with proliferating cell nuclear antigen (PCNA) for recognition of mismatches (16, 47) and lead to phosphorylation of p53 (41). The basis for the formation of the individual heterocomplexes and the specific function of each are not clear. Mutations have been found in most of the subunits in association with HNPCC. HNPCC patients are heterozygous for a MMR gene defect, but tumor tissues have loss of the active allele by mutation or loss of heterozygosity (LOH), producing a complete defect in MMR. This accounts for the apparent autosomal dominant inheritance.

MSI (or MIN) is one form of genomic instability associated with cancer. More commonly, tumors show aneuploidy with varying numbers of chromosomes per cell. This type of genomic instability leads to loss of chromosomes with loss of DNA processing genes (18, 101), which produces an acceleration of genomic instability. Therefore, MIN appears to be the minority finding compared to aneuploidy or chromosome instability (CIN).

HNPCC establishes yet another model for failure of DNA repair being causative in cancer. The observations support the hypothesis that a mutator phenotype is a facilitating factor in the development of cancer (105). Following failure of MMR, the increase in mutation incidence would cause the appearance of "second hits" in a carcinogenesis path. If carcinogenesis is a progressive, multi-step process, then the accumulation of sufficient mutations for loss of control of cellular proliferation and inhibition of cell contact would occur at an increased pace.

Breast Cancer

Breast cancer occurs as a nonfamilial disease in most instances, but a subset is seen in a familial pattern. Two genes have been identified that are associated with familial breast cancer: *BRCA1* and *BRCA2* (reviewed in 38, 172), accounting for somewhat less than 10% of breast cancers. Mutations in *BRCA1* or *BRCA2* are found in familial breast or ovarian cancers, so the genes qualify as tumor suppressor genes, with LOH for the genes frequently found in tumors. In the remaining majority of breast cancers, mutations in *BRCA1*, *BRCA2*, or interacting proteins are uncommon, aside from p53.

BRCA1 has been implicated in DNA repair. ES cells deficient in BRCA1 protein are defective in TCR of oxidative damage (59) and are sensitive to interstrand crosslinks (12). Additionally, targeted disruption of *BRCA1* in ES cells or MEF leads to increased sensitivity to hydrogen peroxide or IR. The BRCA1 protein has domains that interact with the RAD51 protein, RAD50 protein, Rb protein, and p53 protein, among others (20, 135). Thus, the BRCA1 protein interacts with proteins that function in DNA repair and recombination, or cell cycle checkpoints. The results suggest that BRCA1 (and perhaps BRCA2) functions in the repair of doublestrand breaks and may act in a pathway that involves homologous recombination. BRCA1 colocalizes with RAD51 and BRCA2 in the nucleus after DNA damage (134) in a complex that may be located at the site of DNA replicative repair synthesis. BRCA1 also colocalizes with RAD50, MRE11, and NBS1 in complexes after DNA damage.

BRCA1 is phosphorylated by ATM after IR DNA damage (26). Targeted deletions and disruptions in *BRCA1* suggest BRCA1 also may play a role in the G2/M checkpoint after DNA damage. The transcriptional activator function of BRCA1 may play a role in DNA repair, as BRCA1 activates *GADD45* expression. Taken together, the findings implicate BRCA1 (and BRCA2) as directly involved in proximate DNA repair steps. The results also suggest that BRCA1 may serve as a factor in several distinct repair pathways involved in repair of doublestrand breaks and requiring replication, that is, the NBS1 pathway and the RAD51 pathway. Perhaps BRCA1 serves as an interface or common interactor between the RAD50 and RAD51 functions. As may be the case for the FANC proteins, BRCA1 might also act as a structural component in DSB repair (167).

BRCA1 is also notable for an N-terminal ring finger domain, a possible site for dimerization with other proteins, and recently shown to function as an E3 ubiquitin ligase, giving specificity to ubiquitin conjugating E2 activities (reviewed in 74). This raises the possibility that mono- or oligo-ubiquitination of BRCA1 or its interacting proteins may serve an important regulatory function in DNA repair. The interaction with FANCD2 and the mono-ubiquitination of that protein following DNA damage may reflect the central role of BRCA1 in DNA repair.

Related Diseases

Related diseases merit analysis, although not included in this review. Tumor suppressor gene inactivations in association with human disease include von Hippel Lindau disease, neurofibromatosis, and the Cowden/Bannayan-Zonana syndrome, as well as familial adenomatous polyposis (FAP) syndrome. Although leading to tumors, these inactivations are not yet clearly linked to DNA processing defects.

Another category of diseases is represented by retinoblastoma, Li-Fraumeni syndrome, and Nevoid basal cell carcinoma syndrome. The patched gene, PCTH, is mutated in nevoid basal cell carcinoma syndrome. The retinoblastoma protein, p105-Rb, functions in cell cycle progression, at entry into S-phase and M-phase. The p53 protein, defective in Li-Fraumeni syndrome, is a key regulatory element in entry into S-phase, and it appears to be a control point in assessment of DNA damage prior to replication. Cells from patients with these diseases show increased sensitivity to DNA damage agents (ionizing radiation for nevoid basal cell carcinoma), increased mutation rates, or chromosome rearrangements. However, a distinct failure of DNA processing has not been well established. Rather, they appear to represent syndromes of genomic instability. The p53 protein is apparently abnormal in half of all tumors, suggesting degradation of its function is important to tumor promotion. The CHK2 protein also is defective in some Li-Fraumeni cases (11). For Li-Fraumeni and retinoblasoma, the sensitivity to DNA damage and increased risk of cancer can be modeled on the basis of premature cell cycle progression, without normal checkpoint function. Thus, replication of DNA or mitosis could proceed without completion of the repair and safety checks that ensure genomic integrity.

CONCLUSIONS AND OUTLOOK REGARDING DNA PROCESSING AND DISEASE

■ Genomic instability is the hallmark of processing defects that lead to tumors. Presumably, the genomic instability results from a partial defect in processing a normal intermediate that is duplicated in the processing of damages.

■ DNA damage processing diseases represent partial defects in replication, recombination, or transcription, with secondary pathways supplying vital functions that permit cellular survival.

■ DNA damage processing defects show that complex pathways of repair with multiple genes can lead to one clinical syndrome. Yet one gene (e.g., *XPB*) can be defective in three different clinical disorders.

■ The individual molecular defects in DNA repair associated with disease produce specific damage sensitivities at the cellular level and specific disease phenotypes.

- Mutagenesis can result from error-prone DNA synthesis occurring prior to repair, thus a slowed rate of damage processing favors mutation.

- Because some recombination defects appear to involve proteins that localize to sites of replication, it is likely that some DNA damage processing diseases represent failure of recombination or DSB repair, which occurs as a normal part of replication. Possibilities include AT, NBS, FA, WS, BC, and RTS.

- A structural complex, possibly utilizing BRCA1 and/or some FANC proteins, may serve in recombination.

- Regulation of proteins acting in DNA damage and cell cycle progression includes phosphorylation. Ubiquitination at low levels also may play a role in regulation of processing activities.

ACKNOWLEDGMENTS

Results from the author's laboratory were supported by 5P01HL48546 or 5R01-GM58186. Prepublication data were shared by M. Grompe, A. D'Andrea, S. Meyn, and Y. Akkari. The author thanks J. Blouse for assistance.

Visit the Annual Reviews home page at www.AnnualReviews.org

LITERATURE CITED

1. Aaltonen LA, Peltomaki P, Leach F, Sistonen P, Pylkkanen SM, et al. 1993. Clues to the pathogenesis of familial colorectal cancer. *Science* 260:812–16

2. Agami R, Bernards R. 2000. Distinct initiation and maintenance mechanisms cooperate to induce G1 cell cycle arrest in response to DNA damage. *Cell* 102:55–66

3. Akkari YM, Bateman RL, Reifsteck CA, Olson SB, Grompe M. 2000. DNA replication is required to elicit cellular responses to psoralen-induced DNA interstrand cross-links. *Mol. Cell. Biol.* 20:8283–89

4. Alter BP, Caruso JP, Drachtman RA, Uchida T, Vlagaleti GV, Elghetany MT. 2000. Fanconi anemia: myelodysplasia as a predictor of outcome. *Cancer Genet. Cytogenet.* 117:125–31

5. Andrews AD, Barrett SF, Robbins JH. 1978. Xeroderma pigmentosum neu-
rological abnormalities correlate with colony-forming ability after ultraviolet radiation. *Proc. Natl. Acad. Sci. USA* 75:1984–88

6. Andrews AD, Barrett SF, Yoder FW, Robbins JH. 1978. Cockayne's syndrome fibroblasts have increased sensitivity to ultraviolet light but normal rates of unscheduled DNA synthesis. *J. Invest. Dermatol.* 70:237–39

7. Athma P, Rappaport R, Swift M. 1996. Molecular genotyping shows that ataxia-telangiectasia heterozygotes are predisposed to breast cancer. *Cancer Genet. Cytogenet.* 92:130–34

8. Auerbach AD, Adler B, Chaganti RS. 1981. Prenatal and postnatal diagnosis and carrier detection of Fanconi anemia by a cytogenetic method. *Pediatrics* 67:128–35

9. Auerbach AD, Buchwald M, Joenje H. 1998. Fanconi anemia. In *The Genetic*

Basis of Human Cancer, ed. B Vogelstein, KW Kinzler, pp. 317–32. New York: McGraw-Hill

10. Beckman KB, Ames BN. 1997. Oxidative decay of DNA. *J. Biol. Chem.* 272:19633–36

11. Bell DW, Varley JM, Szydlo TE, Kang DH, Wahrer DC, et al. 1999. Heterozygous germ line hCHK2 mutations in Li-Fraumeni syndrome. *Science* 286:2528–31

12. Bhattacharyya A, Ear US, Koller BH, Weichselbaum RR, Bishop DK. 2000. The breast cancer susceptibility gene *BRCA1* is required for subnuclear assembly of Rad51 and survival following treatment with the DNA cross-linking agent cisplatin. *J. Biol. Chem.* 275:23899–903

13. Borde V, Goldman ASH, Lichten M. 2000. Direct coupling between meiotic DNA replication and recombination initiation. *Science* 290:806–9

14. Bronner CE, Baker SM, Morrison PT, Warren G, Smith LG, et al. 1994. Mutation in the DNA mismatch repair gene homologue *hMLH1* is associated with hereditary non-polyposis colon cancer. *Nature* 368:258–61

15. Broughton BC, Thompson AF, Harcourt SA, Vermeulen W, Hoeijmakers JH, et al. 1995. Molecular and cellular analysis of the DNA repair defect in a patient in xeroderma pigmentosum complementation group D who has the clinical features of xeroderma pigmentosum and Cockayne syndrome. *Am. J. Hum. Genet.* 56:167–74

16. Buermeyer AB, Deschenes SM, Baker SM, Liskay RM. 1999. Mammalian DNA mismatch repair. *Annu. Rev. Genet.* 33:533–64

17. Cadet J, Berger M, Douki T, Ravanat JL. 1997. Oxidative damage to DNA: formation, measurement, and biological significance. *Rev. Physiol. Biochem. Pharmacol.* 131:1–87

18. Cahill DP, Kinzler KW, Vogelstein B, Lengauer C. 1999. Genetic instability and darwinian selection in tumours. *Trends Cell. Biol.* 9:M57–60

19. Chakraverty RK, Hickson ID. 1999. Defending genome integrity during DNA replication: a proposed bioessays role for RecQ family helicases. *BioEssays* 21:286–94

20. Chen J, Silver DP, Walpita D, Cantor SB, Gazdar AF, et al. 1998. Stable interaction between the products of the *BRCA1* and *BRCA2* tumor suppressor genes in mitotic and meiotic cells. *Mol. Cells* 2:317–28

21. Chen M, Tomkins DJ, Auerbach W, McKerlie C, Youssoufian H, et al. 1996. Inactivation of Fac in mice produces inducible chromosomal instability and reduced fertility reminiscent of Fanconi anaemia. *Nat. Genet.* 12:448–51

22. Cleaver JE, Kraemer KH. 1995. Xeroderma pigmentosum and Cockayne syndrome. In *The Metabolic and Molecular Bases of Inherited Disease*, ed. CR Scriver, AL Beaudet, WS Sly, D Valle, JB Stanbury, et al. 148:4393–19. New York: McGraw-Hill

23. Cleaver JE. 1968. Defective repair replication of DNA in xeroderma pigmentosum. *Nature* 218:652–56

24. Cleaver JE. 1998. Hair today, gone tomorrow: transgenic mice with human repair deficient hair disease. *Cell* 93:1099–102

25. Cooper PK, Nouspikel T, Clarkson SG, Leadon SA. 1997. Defective transcription-coupled repair of oxidative base damage in Cockayne syndrome patients from XP group G. *Science* 275:990–93

26. Cortez D, Wang Y, Qin J, Elledge SJ. 1999. Requirement of ATM-dependent phosphorylation of BRCA1 in the DNA damage response to double-strand breaks. *Science* 286:1162–66

27. Courcelle J, Carswell-Crumpton C, Hanawalt PC. 1997. *recF* and *recR* are required for the resumption of replication at DNA replication forks in *Escherichia coli*. *Proc. Natl. Acad. Sci. USA* 94:3714–19

28. Cox MM, Goodman MF, Kreuzer KN, Sherratt DJ, Sandler SJ, Marians KJ.

2000. The importance of repairing stalled replication forks. *Nature* 404:37–41

29. Curry CJ, O'Lague P, Tsai J, Hutchison HT, Jaspers NG, et al. 1989. ATFresno: a phenotype linking ataxia-telangiectasia with the Nijmegen breakage syndrome. *Am. J. Hum. Genet.* 45:270–75

30. D'Andrea AD, Grompe M. 1997. Molecular biology of Fanconi anemia: implications for diagnosis and therapy. *Blood* 90:1725–36

31. de Boer J, de Wit J, van Steeg H, Berg RJ, Morreau H, et al. 1998. A mouse model for the basal transcription/DNA repair syndrome trichothiodystrophy. *Mol. Cells* 1:981–90

32. de Boer J, Donker I, de Wit J, Hoeijmakers JH, Weeda G. 1998. Disruption of the mouse xeroderma pigmentosum group D DNA repair/basal transcription gene results in preimplantation lethality. *Cancer Res.* 58:89–94

33. De Weerd-Kastelein EA, Keijzer W, Bootsma D. 1972. Genetic heterogeneity of xeroderma pigmentosum demonstrated by somatic cell hybridization. *Nat. New Biol.* 238:80–83

34. De Weerd-Kastelein EA, Keijzer W, Rainaldi G, Bootsma D. 1977. Induction of sister chromatid exchanges in xeroderma pigmentosum cells after exposure to ultraviolet light. *Mutat. Res.* 45:253–61

35. de Winter JP, Leveille F, van Berkel CG, Rooimans MA, van Der Weel L, et al. 2000. Isolation of a cDNA representing the Fanconi anemia complementation group E gene. *Am. J. Hum. Genet.* 67:1306–8

36. de Winter JP, Rooimans MA, van Der Weel L, van Berkel CG, Alon N, et al. 2000. The Fanconi anaemia gene *FANCF* encodes a novel protein with homology to ROM. *Nat. Genet.* 24:15–16

37. de Winter JP, Waisfisz Q, Rooimans MA, van Berkel CG, Bosnoyan-Collins L, et al. 1998. The Fanconi anaemia group G gene *FANCG* is identical with XRCC9. *Nat. Genet.* 20:281–83

38. Deng C-X, Brodie SG. 2000. Roles of BRAC1 and its interacting proteins. *BioEssays* 22:728–37

39. Dianov GL, Prasad R, Wilson SH, Bohr VA. 1999. Role of DNA polymerase beta in the excision step of long patch mammalian base excision repair. *J. Biol. Chem.* 274:13741–43

40. Dronkert ML, de Wit J, Boeve M, Vasconcelos ML, van Steeg H, et al. 2000. Disruption of mouse SNM1 causes increased sensitivity to the DNA interstrand cross-linking agent mitomycin C. *Mol. Cell. Biol.* 20:4553–61

41. Duckett DR, Bronstein SM, Taya Y, Modrich P. 1999. hMutS alpha- and HmutL alpha-dependent phosphorylation of p53 in response to DNA methylator damage. *Proc. Natl. Acad. Sci. USA* 96:12384–88

42. Dutrillaux B, Aurias A, Dutrillaux AM, Buriot D, Prieur M. 1982. The cell cycle of lymphocytes in Fanconi anemia. *Hum. Genet.* 62:327–32

43. Eker AP, Yajima H, Yasui A. 1994. DNA photolyase from the fungus *Neurospora crassa*. Purification, characterization and comparison with other photolyases. *Photochem. Photobiol.* 60:125–33

44. Elledge SJ. 1996. Cell cycle checkpoints: preventing an identity crisis. *Science* 274:1664–72

45. Ellis NA, Groden J, Ye TZ, Straughen J, Lennon DJ, et al. 1995. The Bloom's syndrome gene product is homologous to RecQ helicases. *Cell* 83:655–66

46. Epstein CJ, Martin GM, Schultz AL, Motulsky AG. 1966. Werner's syndrome: a review of its symptomatology, natural history, pathologic features, genetics and relationship to the natural aging process. *Medicine* 45:177–221

47. Flores-Rozas H, Clark D, Kolodner RD. 2000. Proliferating cell nuclear antigen and Msh2p-Msh6p interact to form an active mispair recognition complex. *Nat. Genet.* 26:375–78

48. Foe JR, Rooimans MA, Bosnoyan-Collins L, Alon N, Wijker M, et al. 1996. Expression cloning of a cDNA for the major Fanconi anaemia gene, FAA. *Nat. Genet.* 14:320–23

49. Friedberg EC, Feaver WJ, Gerlach VL. 2000. The many faces of DNA polymerases: strategies for mutagenesis and for mutational avoidance. *Proc. Natl. Acad. Sci. USA* 97:5681–83

50. Fukuchi K, Martin GM, Monnat RJ Jr. 1989. Mutator phenotype of Werner syndrome is characterized by extensive deletions. *Proc. Natl. Acad. Sci. USA* 86:5893–97

51. Gangloff S, McDonald JP, Bendixen C, Arthur L, Rothstein R. 1994. The yeast type I topoisomerase Top3 interacts with Sgs1, a DNA helicase homolog: a potential eukaryotic reverse gyrase. *Mol. Cell. Biol.* 14:8391–98

52. Gangloff S, Sousteele C, Fabre F. 2000. Homologous recombination is responsible for cell death in the absence of the Sgs1 and Srs2 helicases. *Nat. Genet.* 25:192–94

53. Garcia-Higuera I, Kuang Y, Naf D, Wasik J, D'Andrea AD. 1999. Fanconi anemia proteins FANCA, FANCC, and FANCG/XRCC9 interact in a functional nuclear complex. *Mol. Cell. Biol.* 19:4866–73

53a. Garcia-Higuera I, Taniguchi T, Ganesan S, Meyn MS, Timmers C, et al. 2001. Interaction of the Fanconi anemia proteins and *BRCA1* in a common pathway. *Mol. Cell* 7:249–62

54. Gatti RA, Berkel I, Boder E, Braedt G, Charmley P, et al. 1988. Localization of an ataxia-telangiectasia gene to chromosome 11q22–23. *Nature* 336:577–80

55. Gebhart E, Bauer R, Raub U, Schinzel M, Ruprecht KW, Jonas JB. 1988. Spontaneous and induced chromosomal instability in Werner syndrome. *Hum. Genet.* 80:135–39

56. German J. 1993. Bloom syndrome: a mendelian prototype of somatic mutational disease. *Medicine* 72:393–406

57. German J, Crippa LP, Bloom D. 1974. Bloom's syndrome. III. Analysis of the chromosome aberration characteristic of this disorder. *Chromosoma* 48:361–66

58. Gianneli F, Benson PF, Pawsey SA, Polani PE. 1977. Ultraviolet light sensitivity and delayed DNA-chain maturation in Bloom's syndrome fibroblasts. *Nature* 265:466–69

59. Gowen LC, Avrutskaya AV, Latour AM, Koller BH, Leadon SA. 1998. BRCA1 required for transcription-coupled repair of oxidative DNA damage. *Science* 281:1009–12

60. Gupta PK, Sirover MA. 1984. Altered temporal expression of DNA repair in hypermutable Bloom's syndrome cells. *Proc. Natl. Acad. Sci. USA* 81:757–61

61. Gupta RS, Goldstein S. 1980. Diphtheria toxin resistance in human fibroblast cell strains from normal and cancer-prone individuals. *Mutat. Res.* 73:331–38

62. Haber JA. 2000. Partners and pathways repairing a double strand break. *Trends Genet.* 16:259–64

63. Hanawalt P. 2000. The bases for Cockayne syndrome. *Nature* 405:415–16

64. Hanawalt PC. 1994. Transcription-coupled repair and human disease. *Science* 266:1957–58

65. Hanawalt PC, Sarasin A. 1986. Cancer-prone hereditary diseases with DNA processing abnormalities. *Trends Genet.* 2:124–29

66. Haracska L, Yu S-L, Johnson RE, Prakash L, Prakash S. 2000. Efficient and accurate replication in the presence of 7,8-dihydro-8-oxoguanine by DNA polymerase eta. *Nat. Genet.* 25:458–61

67. Henderson EE. 1978. Host cell reactivation of Epstein-Barr virus in normal and repair-defective leukocytes. *Cancer Res.* 38:3256–63

68. Henderson EE, Long WK. 1981. Host cell reactivation of UV- and X-ray-damaged herpes simplex virus by Epstein-Barr

virus (EBV)-transformed lymphoblastoid cell lines. *Virology* 115:237–48

69. Henning KA, Li L, Iyer N, McDaniel LD, Reagan MS, et al. 1995. The Cockayne syndrome group A gene encodes a WD repeat protein that interacts with CSB protein and a subunit of RNA polymerase II TFIIH. *Cell* 82:555–64

70. Hoatlin ME, Christianson TA, Keeble WW, Hammond AT, Zhi Y, et al. 1998. The Fanconi anemia group C gene product is located in both the nucleus and cytoplasm of human cells. *Blood* 91:1418–25

71. Holmes AM, Haber JE. 1999. Double-strand break repair in yeast requires both leading and lagging strand DNA polymerases. *Cell* 96:415–24

72. Inoue T, Hirano K, Yokoiyama A, Kada T, Kato H. 1977. DNA repair enzymes in ataxia telangiectasia and Bloom's syndrome fibroblasts. *Biochim. Biophys. Acta* 479:497–500

73. Ishida R, Buchwald M. 1982. Susceptibility of Fanconi's anemia lymphoblasts to DNA-cross-linking and alkylating agents. *Cancer Res.* 42:4000–6

74. Joazeiro CAP, Weissman AM. 2000. RING finger proteins: mediators of ubiquitin ligase activity. *Cell* 102:549–52

75. Joenje H, Levitus M, Waisfisz Q, D'Andrea A, Garcia-Higuera I, et al. 2000. Complementation analysis in Fanconi anemia: assignment of the reference FA-H patient to group A. *Am. J. Hum. Genet.* 67:759–62

76. Joenje H, Oostra AB, Wijker M, di Summa FM, van Berkel CG, et al. 1997. Evidence for at least eight Fanconi anemia genes. *Am. J. Hum. Genet.* 61:940–44

77. Johnson RE, Kondratick CM, Prakash S, Prakash L. 1999. hRAD30 mutations in the variant form of xeroderma pigmentosum. *Science* 285:263–65

78. Johnstone P, Reifsteck C, Kohler S, Worland P, Olson S, Moses RE. 1997. Fanconi anemia group A and D cell lines respond normally to inhibitors of cell cycle regulation. *Somat. Cell. Mol. Genet.* 23:371–77

79. Karow JK, Chakraverty RK, Hickson ID. 1997. The Bloom's syndrome gene product is a 3′-5′ DNA helicase. *J. Biol. Chem.* 272:30611–14

80. Karow JK, Constantinou A, Li JL, West SC, Hickson ID. 2000. The Bloom's syndrome gene product promotes branch migration of holliday junctions. *Proc. Natl. Acad. Sci. USA* 97:6504–8

81. Kitao S, Ohsugi I, Ichikawa K, Goto M, Furuichi Y, Shimamoto A. 1998. Cloning of two new human helicase genes of the RecQ family: biological significance of multiple species in higher eukaryotes. *Genomics* 54:443–52

82. Kitao S, Shimamoto A, Goto M, Miller RW, Smithson WA, et al. 1999. Mutations in *RECQL4* cause a subset of cases of Rothmund-Thomson syndrome. *Nat. Genet.* 22:82–84

83. Kogoma T. 1997. Is RecF a DNA replication protein? *Proc. Natl. Acad. Sci. USA* 94:3483–84

84. Kojis TL, Gatti RA, Sparkes RS. 1991. The cytogenetics of ataxia telangiectasia. *Cancer Genet. Cytogenet.* 56:143–56

85. Kraemer KH. 1977. Progressive degenerative diseases associated with defective DNA repair: Xeroderma pigmentosum and ataxia telangiectasia. In *DNA Repair Processes*, ed. WW Nichols, DG Murphy, pp. 37–73. Miami, FL: Symp. Specialists, Inc.

86. Kraemer KH, Lee MM, Scotto J. 1984. DNA repair protects against cutaneous and internal neoplasia: evidence from xeroderma pigmentosum. *Carcinogenesis* 5:511–14

87. Kraemer KH, Lee MM, Scotto J. 1987. Xeroderma pigmentosum. Cutaneous, ocular, and neurologic abnormalities in 830 published cases. *Arch. Dermatol.* 123:241–50

88. Kruyt FA, Waisfisz Q, Dijkmans LM, Hermsen MA, Youssoufian H, et al. 1997. Cytoplasmic localization of a functionally active Fanconi anemia group A-green

fluorescent protein chimera in human 293 cells. *Blood* 90:3288–95

89. Kruyt FA, Youssoufian H. 1998. The Fanconi anemia proteins FAA and FAC function in different cellular compartments to protect against cross-linking agent cytotoxicity. *Blood* 92:2229–36

90. Kruyt FAE, Hoshino T, Liu JM, Joseph P, Jaiswal AK, et al. 2000. Abnormal microsomal detoxification implicated in Fanconi anemia group C by interaction of the FAC protein with NADPH cytochrome P450 reductase. *Blood.* In press

91. Kubbies M, Schindler D, Hoehn H, Schinzel A, Rabinovitch PS. 1985. Endogenous blockage and delay of the chromosome cycle despite normal recruitment and growth phase explain poor proliferation and frequent edomitosis in Fanconi anemia cells. *Am. J. Hum. Genet.* 37:1022–30

92. Kumaresan KR, Lambert MW. 2000. Fanconi anemia, complementation group A, cells are defective in ability to produce incisions at sites of psoralen interstrand cross-links. *Carcinogenesis* 21:741–51

93. Kupfer GM, Naf D, Suliman A, Pulsipher M, D'Andrea AD. 1997. The Fanconi anaemia proteins, FAA and FAC, interact to form a nuclear complex. *Nat. Genet.* 17:487–90

94. Laquerbe A, Moustacchi E, Fuscoe JC, Papadopoulo D. 1995. The molecular mechanism underlying formation of deletions in Fanconi anemia cells may involve a site-specific recombination. *Proc. Natl. Acad. Sci. USA* 92:831–35

95. Le S, Moore JK, Haber JE, Greider CW. 1999. *RAD50* and *RAD51* define two pathways that collaborate to maintain telomeres in the absence of telomerase. *Genetics* 152:143–52

96. Leach FS, Nicolaides NC, Papadpoulos N, Liu B, Jen J, et al. 1993. Mutations of a mutS homolog in hereditary nonpolyposis colerectal cancer. *Cell* 75:1215–25

97. Leadon SA, Cooper PK. 1993. Preferential repair of ionizing radiation-induced damage in the transcribed strand of an active human gene is defective in Cockayne syndrome. *Proc. Natl. Acad. Sci. USA* 90:10499–503

98. Lehmann AR. 1982. Three complementation groups in Cockayne syndrome. *Mutat. Res.* 106:347–56

99. Lehmann AR, Arlett CF, Burke JF, Green MH, James MR, Lowe JE. 1986. A derivative of an ataxia-telangiectasia (A-T) cell line with normal radiosensitivity but A-T-like inhibition of DNA synthesis. *Int. J. Radiat. Biol.* 49:639–43

100. Lehmann AR, Stevens S. 1979. The response of ataxia telangiectasia cells to bleomycin. *Nucleic Acids Res.* 6:1953–60

101. Lengauer C, Kinzler KW, Vogelstein B. 1998. Genetic instabilities in human cancers. *Nature* 396:643–49

102. Le Page F, Klungland A, Barnes DE, Sarasin A, Boiteux S. 2000. Transcription coupled repair of 8-oxoguanine in murine cells: the Ogg1 protein is required for repair in nontranscribed sequences but not in transcribed sequences. *Proc. Natl. Acad. Sci. USA* 97:8397–402

103. Liao S, Graham J, Yan H. 2000. The function of *Xenopus* Bloom's syndrome protein homolog (xBLM) in DNA replication. *Genes Dev.* 14:2570–75

104. Limoli CL, Giedzinski E, Morgan WF, Cleaver JE. 2000. Polymerase H deficiency in the xeroderma pigmentosum variant uncovers an overlap between the S phase checkpoint and double-strand break repair. *Proc. Natl. Acad. Sci. USA* 97:7939–46

105. Loeb LA. 1991. Mutator phenotype may be required for multistage carcinogenesis. *Cancer Res.* 2:169–74

106. Lombard DB, Guarente L. 1996. Cloning the gene for Werner syndrome: a disease with many symptoms of premature ageing. *Trends Genet.* 12:283–86

107. Lo Ten Foe JR, Kwee ML, Rooimans MA, Oostra AB, Veerman AJ, et al. 1997. Somatic mosaicism in Fanconi anemia:

molecular basis and clinical significance. *Eur. J. Hum. Genet.* 5:137–48

108. Magee TR, Kogoma T. 1991. Rifampin-resistant replication of pBR322 derivatives in *Escherichia coli* cells induced for the SOS response. *J. Bacteriol.* 173:4736–41

109. Masutani C, Kusumoto R, Yamada A, Dohmae N, Yokoi M, et al. 1999. The XPV (xeroderma pigmentosum variant) gene encodes human DNA polymerase eta. *Nature* 399:700–4

110. McCullough AK, Dodson ML, Lloyd RS. 1999. Initiation of base excision repair: glycosylase mechanisms and structures. *Annu. Rev. Biochem.* 68:255–85

111. McKinnon PJ. 1987. Ataxia-telangiectasia: an inherited disorder of ionizing-radiation sensitivity in man. Progress in the elucidation of the underlying biochemical defect. *Hum. Genet.* 75:197–208

112. Mellon I, Spivak G, Hanawalt PC. 1987. Selective removal of transcription-blocking DNA damage from the transcribed strand of the mammalian DHFR gene. *Cell* 51:241–49

113. Mer G, Bochkarev A, Gupta R, Bochkareva E, Frappier L, et al. 2000. Structural basis for the recognition of DNA repair proteins UNG2, XPA, and RAD52 by replication factor RPA. *Cell* 103:449–56

114. Michelson RJ, Weinert T. 2000. Closing the gaps among a web of DNA repair disorders. *BioEssays* 22:966–69

115. Nash HM, Bruner SD, Scharer OD, Kawate T, Addona TA, et al. 1996. Cloning of a yeast 8-oxoguanine DNA glycosylase reveals the existence of a base-excision DNA-repair protein superfamily. *Curr. Biol.* 6:968–80

116. Painter RB, Young BR. 1980. Radio sensitivity in ataxia-telangiectasia: a new explanation. *Proc. Natl. Acad. Sci. USA* 77:7315–17

117. Papadopoulo D, Guillouf C, Mohren-weiser H, Moustacchi E. 1990. Hypo-mutability in Fanconi anemia cells is associated with increased deletion frequency at the HPRT locus. *Proc. Natl. Acad. Sci. USA* 87:8383–87

118. Paques F, Haber JE. 1999. Multiple pathways of recombination induced by double-strand breaks in *Saccharomyces cerevisiae. Microbiol. Mol. Biol. Rev.* 63:349–404

119. Paterson MC, Smith BP, Lohman PH, Anderson AK, Fishman L. 1976. Defective excision repair of gamma-ray-damaged DNA in human (ataxia telangiectasia) fibroblasts. *Nature* 260:444–47

120. Patton JD, Rowan LA, Mendrala AL, Howell JN, Maher VM, McCormick JJ. 1984. Xeroderma pigmentosum fibroblasts including cells from XP variants are abnormally sensitive to the mutagenic and cytotoxic action of broad spectrum simulated sunlight. *Photochem. Photobiol.* 39:37–42

121. Peltomaki P, de la Chapelle A. 1997. Mutations predisposing to hereditary nonpolyposis colorectal cancer. *Adv. Cancer Res.* 71:93–119

122. Peltomaki P, Lothe RA, Aaltonen LA, Pylkkanen L, Nystrom-Lahti M, et al. 1993. Microsatellite instability is associated with tumors that characterize the hereditary non-polyposis colorectal carcinoma syndrome. *Cancer Res.* 53:5853–55

123. Poot M, Grob O, Epe B, Pflaum M, Hoehn H. 1996. Cell cycle defect in connection with oxygen and iron sensitivity in Fanconi anemia lymphoblastoid cells. *Exp. Cell. Res.* 222:262–68

124. Saito H, Hammond AT, Moses RE. 1993. Hypersensitivity to oxygen is a uniform and secondary defect in Fanconi anemia cells. *Mutat. Res.* 294:255–62

125. Saito H, Hammond AT, Moses RE. 1995. The effect of low oxygen tension on the in vitro-replicative life span of human diploid fibroblast cells and their transformed derivatives. *Exp. Cell. Res.* 217:272–79

126. Salk D, Au K, Hoehn H, Martin GM. 1981. Cytogenetics of Werner's syndrome cultured skin fibroblasts: variegated

translocation mosaicism. *Cytogenet. Cell. Genet.* 30:92–107

127. Sancar A. 1994. Structure and function of DNA photolyase. *Biochemistry* 33:2–9

128. Sancar A. 1995. Excision repair in mammalian cells. *J. Biol. Chem.* 270:15915–18

129. Savitsky K, Bar-Shira A, Gilad S, Rotman G, Ziv Y, et al. 1995. A single ataxia telangiectasia gene with a product similar to PI-3 kinase. *Science* 268:1749–53

130. Schindler D, Hoehn H. 1988. Fanconi anemia mutation causes cellular susceptibility to ambient oxygen. *Am. J. Hum. Genet.* 43:429–35

131. Schmickel RD, Chu EH, Trosko JE, Chang CC. 1977. Cockayne syndrome: a cellular sensitivity to ultraviolet light. *Pediatrics* 60:135–39

132. Schroeder TM, Anschutz F, Knopp A. 1964. Spontaneous chromosome aberrations in familial panmyelopathy. *Humangenetik* 1:194–96

133. Schroeder-Kurth TM, Zhu TH, Hong Y, Westphal I. 1989. Variation in cellular sensitivities among Fanconi anemia patients, non-Fanconi anemia patients, their parents and siblings, and control probands. In *Fanconi Anemia. Clinical, Cytogenetic and Experimental Aspects*, ed. TM Schroeder-Kurth, AD Auerbach, G Obe, pp. 105–36. Berlin: Springer-Verlag

134. Scully R, Chen JJ, Ochs RL, Keegan K, Hoekstra M, et al. 1997. Dynamic changes of BRCA1 subnuclear location and phosphorylation state are initiated by DNA damage. *Cell* 90:425–35

135. Scully R, Chen JJ, Plug A, Xiao YH, Weaver D, et al. 1997. Association of BRCA1 with Rad51 in mitotic and meiotic cells. *Cell* 88:265–75

136. Scully RE, Mark EJ, McNeely WF, McNeely BU. 1987. Case records of the Massachusetts General Hospital, Case 2. *N. Engl. J. Med.* 316:91–100

137. Selby CP, Sancar A. 1997. Cockayne syndrome group B protein enhances elonga-

tion by RNA polymerase II. *Proc. Natl. Acad. Sci. USA* 94:11205–9

138. Seyschab H, Bretzel G, Friedl R, Schindler D, Sun Y, Hoehn H. 1994. Modulation of the spontaneous G2 phase blockage in Fanconi anemia cells by caffeine: differences from cells arrested by X-irradiation. *Mutat. Res.* 308:149–57

139. Shen J-C, Gray MD, Oshima J, Kamath-Loeb AS, Fry M, et al. 1998. Werner syndrome protein. I. DNA helicase and DNA exonuclease reside on the same polypeptide. *J. Biol. Chem.* 273:34139–44

140. Shen J-C, Gray MD, Oshima J, Loeb LA. 1998. Characterization of Werner syndrome protein DNA helicase activity: directionality, substrate dependence and stimulation by replication protein A. *Nucleic Acids Res.* 26:2879–85

141. Shiloh Y. 1997. Ataxia-telangiectasia and the Nijmegen breakage syndrome: related disorders but genes apart. *Annu. Rev. Genet.* 31:635–62

142. Spampinato C, Modrich P. 2000. The MutL ATPase is required for mismatch repair. *J. Biol. Chem.* 275:9863–69

143. Stewart E, Chapman CR, Al-Khodairy F, Carr AM, Enoch T. 1997. rqh1+, a fission yeast gene related to the Bloom's and Werner's syndrome genes, is required for reversible S phase arrest. *EMBO J.* 16:2682–92

144. Stewart GS, Maser RS, Stankovic T, Bressan DA, Kaplan MI, et al. 1999. The DNA double-strand break repair gene hMRE11 is mutated in individuals with an ataxia-telangiectasia-like disorder. *Cell* 99(6):577–87

145. Strand M, Prolla TA, Liskay RM, Petes TD. 1993. Destabilization of tracts of simple repetitive DNA in yeast by mutations affecting DNA mismatch repair. *Nature* 365:274–76

146. Strathdee CA, Gavish H, Shannon WR, Buchwald M. 1992. Cloning of cDNAs for Fanconi's anaemia by functional complementation. *Nature* 356:763–67

147. Sun Y, Moses RE. 1991. Reactivation

of psoralen-reacted plasmid DNA in Fanconi anemia, xeroderma pigmentosum, and normal human fibroblast cells. *Somat. Cell. Mol. Genet.* 17:229–38

148. Svejstrup JQ, Wang Z, Feaver WJ, Wu X, Bushnell DA, et al. 1995. Different forms of TFIIH for transcription and DNA repair: holo-TFIIH and a nucleotide excision repairosome. *Cell* 80:21–28

149. Swift M, Morrell D, Cromartie E, Chamberlin AR, Skolnick MH, Bishop DT. 1986. The incidence and gene frequency of ataxia-telangiectasia in the United States. *Am. J. Hum. Genet.* 39:573–83

150. Swift M, Morrell D, Massey RB, Chase CL. 1991. Incidence of cancer in 161 families affected by ataxia-telangiectasia. *N. Engl. J. Med.* 325:1831–36

151. Swift M, Reitnauer PJ, Morrell D, Chase CL. 1987. Breast and other cancers in families with ataxia-telangiectasia. *N. Engl. J. Med.* 316:1289–94

152. Szekely AM, Chen Y-H, Zhang C, Oshimo J, Weissman SM. 2000. Werner protein recruits DNA polymerase δ to the nucleolus. *Proc. Natl. Acad. Sci. USA* 97:11365–70

153. Takayama K, Salazar EP, Broughton BC, Lehmann AR, Sarasin A, et al. 1996. Defects in the DNA repair and transcription gene ERCC21(XPD) in trichothiodystrophy. *Am. J. Hum. Genet.* 58:263–70

154. Tanaka H, Arakawa H, Yamaguchi T, Shiraishi K, Fukuda S, et al. 2000. A ribonucleotide reductase gene involved in a p53-dependent cell-cycle checkpoint for DNA damage. *Nature* 404:42–49

155. The Fanconi Anaemia/Breast Cancer Consortium. 1996. Positional cloning of the Fanconi anaemia group A gene. *Nat. Genet.* 14:324–28

156. Timme TL, Moses RE. 1988. Review: diseases with DNA damage-processing defects. *Am. J. Med. Sci.* 295:40–48

156a. Timmers C, Taniguchi T, Hejna J, Reifsteck C, Lucas L, et al. 2001. Positional cloning of a novel Fanconi anemia gene, *FANCD2*. *Mol. Cell* 7:241–48

157. Troelstra C, van Gool A, de Wit J, Vermeulen W, Bootsma D, Hoeijmakers JH. 1992. ERCC6, a member of a subfamily of putative helicases, is involved in Cockayne's syndrome and preferential repair of active genes. *Cell* 71:939–53

158. van der Kemp PA, Thomas D, Barbey R, de Oliveira R, Boiteux S. 1996. Cloning and expression in *Escherichia coli* of the OGG1 gene of *Saccharomyces cerevisiae*, which codes for a DNA glycosylase that excises 7,8-dihydro-8-oxoguanine and 2,6-diamino-4-hydroxy-5-N-methylformamidopyrimidine. *Proc. Natl. Acad. Sci. USA* 93:5197–202

159. Van Dyck E, Stasiak AZ, Stasiak A, West SC. 1999. Binding of double-strand breaks in DNA by human Rad52 protein. *Nature* 398:728–31

160. Varon R, Vissinga C, Platzer M, Cerosaletti KM, Chrzanowska KH, et al. 1998. Nibrin, a novel DNA double-strand break repair protein, is mutated in Nijmegen breakage syndrome. *Cell* 93:467–76

161. Vennos EM, Collins M, James WD. 1992. Rothmund-Thomson syndrome: review of the world literature. *J. Am. Acad. Dermatol.* 27:750–62

162. Vennos EM, James WD. 1995. Rothmund-Thomson syndrome. *Dermatol. Clin.* 13:143–50

163. Vermeulen W, Jaeken J, Jaspers NG, Bootsma D, Hoeijmakers JH. 1993. Xeroderma pigmentosum complementation group G associated with Cockayne syndrome. *Am. J. Hum. Genet.* 53:185–92

164. Vijayalaxmi KK, Evans HJ, Ray JH, German J. 1983. Bloom's syndrome: evidence for an increased mutation frequency in vivo. *Science* 221:851–53

165. Wade MH, Chu EH. 1979. Effects of DNA damaging agents on cultured fibroblasts derived from patients with Cockayne syndrome. *Mutat. Res.* 59:49–60

166. Waisfisz Q, de Winter JP, Kruyt FA, de Groot J, van der Weel L, et al. 1999. A

physical complex of the Fanconi anemia proteins FANCG/XRCC9 and FANCA. *Proc. Natl. Acad. Sci. USA* 96:10320–25

167. Wang Y, Cortez D, Yazdi P, Neff N, Elledge SJ, et al. 2000. BASC, a super complex of BRCA1-associated proteins involved in the recognition and repair of aberrant DNA structures. *Genes Dev.* 14:927–39

168. Wang Z, Buratowski S, Svejstrup JQ, Feaver WJ, Wu X, et al. 1995. The yeast *TFB1* and *SSL1* genes, which encode subunits of transcription factor IIH, are required for nucleotide excision repair and RNA polymerase II transcription. *Mol. Cell. Biol.* 15(4):2288–93

169. Warren ST, Schultz RA, Chang CC, Wade MH, Trosko JE. 1981. Elevated spontaneous mutation rate in Bloom syndrome fibroblasts. *Proc. Natl. Acad. Sci. USA* 78:3133–37

170. Weichselbaum RR, Nove J, Little JB. 1978. Deficient recovery from potentially lethal radiation damage in ataxia telengiectasia and xeroderma pigmentosum. *Nature* 271:261–62

171. Weinert T. 1998. DNA damage checkpoints update: getting molecular. *Curr. Opin. Genet. Dev.* 8:185–93

172. Welcsh P, Owens KN, King M-C. 2000. Insights into the functions of BRAC1 and BRAC2. *Trends Genet.* 16:69–74

173. Whitney MA, Royle G, Low MJ, Kelly MA, Axthelm MK, et al. 1996. Germ cell defects and hematopoietic hypersensitivity to gamma-interferon in mice with a targeted disruption of the Fanconi anemia C gene. *Blood* 88:49–58

174. Will O, Schindler D, Boiteux S, Epe B. 1998. Fanconi's anaemia cells have normal steady-state levels and repair of oxidative DNA base modifications sensitive to Fpg protein. *Mutat. Res.* 409:65–72

175. Winkler GS, Araujo SJ, Fiedler U, Vermeulen W, Coin F, et al. 2000. TFIIH with inactive XPD helicase functions in transcription initiation but is defective in DNA repair. *J. Biol. Chem.* 275:4258–66

176. Wolff S, Rodin B, Cleaver JE. 1997. Sister chromatid exchanges induced by mutagenic carcinogens in normal and xeroderma pigmentosum cells. *Nature* 265:347–49

177. Wood RD. 1991. DNA repair. Seven genes for three diseases. *Nature* 350:190

178. Wood RD. 1996. DNA repair in eukaryotes. *Annu. Rev. Biochem.* 65:135–67

179. Wood RD. 1997. Nucleotide excision repair in mammalian cells. *J. Biol. Chem.* 272:23465–68

180. Wood RD, Lindahl T. 1999. Quality control by DNA repair. *Science* 286:1897–905

181. Wood RD, Shivji MK. 1997. Which DNA polymerases are used for DNA-repair in eukaryotes? *Carcinogenesis* 18:605–10

182. Wu X, Ranganathan V, Weisman DS, Heine WF, Ciccone DN, et al. 2000. ATM phosphorylation of Nijmegen breakage syndrome protein is required in a DNA damage response. *Nature* 405:477–82

183. Yamashita T, Barber DL, Zhu Y, Wu N, D'Andrea AD. 1994. The Fanconi anemia polypeptide FACC is localized to the cytoplasm. *Proc. Natl. Acad. Sci. USA* 91:6712–16

184. Yamashita T, Kupfer GM, Naf D, Suliman A, Joenje H, et al. 1998. The Fanconi anemia pathway requires FAA phosphorylation and FAA/FAC nuclear accumulation. *Proc. Natl. Acad. Sci. USA* 95:13085–90

185. Youssoufian H. 1994. Localization of Fanconi anemia C protein to the cytoplasm of mammalian cells. *Proc. Natl. Acad. Sci. USA* 91:7975–79

186. Youssoufian H, Li Y, Martin ME, Buchwald M. 1996. Induction of Fanconi anemia cellular phenotype in human 293 cells by overexpression of a mutant FAC allele. *J. Clin. Invest.* 97:957–62

187. Yu CE, Oshima J, Fu YH, Wijsman EM, Hisama F, et al. 1996. Positional cloning

of the Werner's syndrome gene. *Science* 272:258–62

188. Zhao S, Weng YC, Yuan SS, Lin YT, Hsu HC, et al. 2000. Functional link between ataxia-telangiectasia and Nijmegen breakage syndrome gene products. *Nature* 405:473–77

189. Zhou B-BS, Elledge SJ. 2000. The DNA damage response: putting checkpoints in perspective. *Nature* 408:433–39

Annu. Rev. Genomics Hum. Genet. 2001. 2:69–101

HUMAN GENETICS: Lessons from Quebec Populations[1]

Charles R. Scriver

Departments of Human Genetics, Pediatrics, and Biology, McGill University, Montreal, Quebec, Canada H3G 1Y6; e-mail: cscriv@po-box.mcgill.ca

Key Words population structure, demography, rare alleles, founder effect, genetic drift, linkage disequilibrium, Mendelian disease

■ **Abstract** The population of Quebec, Canada (7.3 million) contains ∼6 million French Canadians; they are the descendants of ∼8500 permanent French settlers who colonized Nouvelle France between 1608 and 1759. Their well-documented settlements, internal migrations, and natural increase over four centuries in relative isolation (geographic, linguistic, etc.) contain important evidence of social transmission of demographic behavior that contributed to effective family size and population structure. This history is reflected in at least 22 Mendelian diseases, occurring at unusually high prevalence in its subpopulations. Immigration of non-French persons during the past 250 years has given the Quebec population further inhomogeneity, which is apparent in allelic diversity at various loci. The histories of Quebec's subpopulations are, to a great extent, the histories of their alleles. Rare pathogenic alleles with high penetrance and associated haplotypes at 10 loci (*CFTR, FAH, HBB, HEXA, LDLR, LPL, PAH, PABP2, PDDR,* and *SACS*) are expressed in probands with cystic fibrosis, tyrosinemia, β-thalassemia, Tay-Sachs, familial hypercholesterolemia, hyperchylomicronemia, PKU, oculopharyngeal muscular dystrophy, pseudo vitamin D deficiency rickets, and spastic ataxia of Charlevoix-Saguenay, respectively) reveal the interpopulation and intrapopulation genetic diversity of Quebec. Inbreeding does not

---

[1]Abbreviations and gene symbols used in text: *ARSACS*, autosomal recessive spinal atrophy of Charlevoix-Saguenay; *CFTR*, symbol for cystic fibrosis gene; Ch-SLSJ, Charlevoix-Saguenay Lac St-Jean region; *FAH*, gene symbol for fumarylacetoacetate hydrolase, the defective enzyme in hereditary tyrosinema Type 1; *HBB*, β-globin locus that is involved in β-thalassemia; *HEXA*, gene symbol for hexosaminidase A, the defective enzyme in Tay-Sachs disease; *LDLR*, gene symbol for low density lipoprotein receptor, defective in familial hypercholesterolemia; *LPL*, gene symbol for lipoprotein lipase, the defective enzyme in familial hyperchylomicronemia; OPMD, oculopharyngeal muscular dystrophy; *PAH*, gene symbol for phenylalanine monooxygenase (hydrolase), the enzyme defective in phenylketonuria (PKU) and related forms of hyperphenylalaninemia (HPA); *PABP2*, gene for polyadenylation binding protein 2, involved in OPMD; *PDDR*, gene locus involved in pseudodeficiency vitamin D rickets; *SACS*, gene for the sacsin protein affected in ARSACS; SNP, single nucleotide polymorphism.

1527-8204/01/0728-0069$14.00 **69**

explain the clustering and prevalence of these genetic diseases; genealogical reconstructions buttressed by molecular evidence point to founder effects and genetic drift in multiple instances. Genealogical estimates of historical meioses and analysis of linkage disequilibrium show that sectors of this young population are suitable for linkage disequilibrium mapping of rare alleles. How the population benefits from what is being learned about its structure and how its uniqueness could facilitate construction of a genomic map of linkage disequilibrium are discussed.

HISTORY: SETTLEMENT, MIGRATION, DEMOGRAPHY

Introduction

Patterns of migration and admixture over the last 100,000 years are determinants of the recent microevolution of human populations (13); 500 years ago, range expansion and colonization began to be important processes by which Europeans established new populations overseas (35). The large-scale analyses by Cavalli-Sforza et al. (28) and the further reflections by Fix (65) provide depth and breadth to our understanding of human migrations over the past many millennia; their work serves as a palimpsest for this selective analysis of a New World population. However, any attempt to study the Quebec example of human population genetics without taking into consideration associated cultural features is likely to lead to confusion (57); accordingly, how colonization, immigration, migration, and expansion by natural increase occurred in Quebec is described in some detail here because it will explain why there are subpopulations.

Modern continental Europeans have population structures that resemble networks more than trees (29), and the population of France has a particularly dense network configuration. In the seventeenth century, emigration to overseas colonies occurred from particular regions of France, with selective sampling of the French population structure. From these initial expansions were created two separate populations of French settlers in the New World (Figure 1, see color insert). The founders of Nouvelle France (which was later incorporated into Lower Canada and then became Quebec) established French enclaves along the banks of the St. Lawrence River [Charbonneau & Robert, Plate 45 in (82)]. Another population of settlers from France established an independent North American colony in the seventeenth century; it was situated on the southern shore of the Bay of Fundy [Daigle, Plate 29 in (82)]. The latter were the Acadians; they flourished until their colony was dispersed by the British in the 1750s [Daigle & LeBlanc, Plate 30 in (82)]. The Acadians retained a lively culture and identity, and their forced diaspora had genetic consequences in the twentieth century, notably, clustering of Mendelian disorders among the Acadians, among those who settled anew in Louisiana (the Cajuns), and those who resettled in Canada, some 4000 of them in Quebec (31).

Migration is a recurrent theme of this review. The "unit of migration," as it is perceived by Alan Fix (65) and discussed elsewhere (110), was probably small, particularly in the geographic loci of settlement in the New World by the original French colonizers of the seventeenth and eighteenth centuries. Again it was small when some of the French Canadians began a second internal migration in the nineteenth century from Charlevoix County (on the northern shore of the St. Lawrence River downstream from Quebec City) into the region of Saguenay Lac St. Jean in northeastern Quebec. A kin-structured migration (65, 110), involving shared genetic ancestry, was highly probable in the event and would have increased the chance of founder effect and genetic drift. It follows that awareness of the cultural and social behavior of the settlers and their descendants in the New World is necessary if we are to explain the notable clusterings of Mendelian diseases in contemporary Quebec.

The Acadian theme receives little further attention here, but on behalf of modern-day Acadians in Canada and in Louisiana, it deserves extensive formal treatment at some future time. The genetic contribution by the Native Americans is also a regretted omission; they were significant in the success of the early colonists of Nouvelle France.

Themes and Approach

CONTEXTS   The histories of populations and the histories of their alleles can reflect each other (133). In the particular case of French Canadians, demographic and genetic histories may begin to provide the history of a genome (16).

MIGRATION AND DEMOGRAPHIC HISTORY   The colonization of Nouvelle France by (largely) French settlers[2] in the seventeenth and eighteenth centuries, their subsequent natural increase from founders in relative genetic isolation, and their own interregional migrations underlie the present-day regional clustering of at least 22 hereditary diseases in Quebec (38, 39, 41) (Table 1). Migration to Quebec continued after the French colony became British in 1759; it mainly comprised

---

[2]Definitions and descriptions for this article: (*a*) Settler: an individual who immigrated to North America and participated in establishing the colony of Nouvelle France. Of the estimated 8500 settlers, 1600 were women; 90% of the settlers produced an offspring in the colony (41, 42). (*b*) Founder: an individual who settled in Nouvelle France, who had at least one grandchild capable of contributing to the gene pool of the French Canadian population (41). This particular definition accommodates a more comprehensive description in Falconer (62a, p. 83): "The founding members of the new subpopulation may be very few in numbers causing a substantial amount of random drift in the first generation. This is called the founder effect. If the subpopulation then expands, its difference from the main population may seem much too great to be consistent with its present numbers. To attribute the difference to a founder effect may often be plausible but, in the absence of pedigree records, can seldom be other than a guess." Access to linked genealogical records for the Ch-SLSJ populations (18) allows founder effects to be identified. (*c*) Isolate: a closed population within which mating occurs. Expansion of the French Canadian population was mainly by natural increase, first from founders of Nouvelle France then from founders of the Ch-SLSJ population, and took place in geographic isolation.

TABLE 1   Mendelian phenotypes and geographic distribution in Quebec (39, 42)[h]

| Disease | OMIM | Regional distributions | |
		Ch-SLSJ and probands/cases (39)	Other
Congenital liver fibrosis		+4/4	
*Cystic fibrosis	219700	(+)97/114	+[a]
Cystinosis	219800	+7/8	
Cytochrome oxidase defic.	220111	+27/33	
*Familial hypercholesterolemia	143890	+	+[b]
Friedreich ataxia	239300		+[c]
Hemochromatosis	235200	+28/54	
Histidinemia	235800	+28/54	
*Hyperphenylalaninemia (PKU and non-PKU HPA)	261600	(+)	+[a]
Intestinal atresia (multiple)	243150	+2/2	
*Lipoprotein lipase defic.	238600	+22/33	+[d]
Mucolipidosis II	252500	+9/9	
Myotonic dystrophy	160900		
*Oculopharyngeal muscular dystrophy	164300		+[c, e]
Polyneuropathy sensorimotor, +/− agenesis of corpus callosum	218000	+76/94	
*Pseudo-vitamin D deficiency rickets	264700	+36/52	
Sarcosinemia	268900	+21/22	
*Spastic ataxia of Charlevoix-Saguenay (ARSACS)	270550	+116/201	
*Tay-Sachs disease	272800	(+)	+[f]
*β-thalassemia	141900		+[g]
*Tyrosinemia type 1	276700	+88/114	
Zellweger syndrome	214100	+5/5	

*Discussed in this article.

+Specific alleles in the region.

[a]Allelic heterogeneity stratifies in Quebec.

[b]Multiple origins.

[c]Southeast Quebec.

[d]Codon 188 and 207 alleles segregate toward western and eastern Quebec regions, respectively.

[e]Southwest Quebec.

[f]Unusual allele in French Canadians (Rimouski region).

[g]β-thalassemia minor allelès in French Canadians (Portneuf County near Quebec City).

[h]See Addendum, p. 101.

non-French persons. These latter day immigrants brought other rare alleles that now account for some clusterings of genetic diseases in Quebec (Table 1). These unique demographic and social histories and the relative youth of this New World population account for differences in its genetic structure compared with that of Finland, for example (see J. Kere, Human Population Genetics: Lessons from Finland, this Volume).

REGIONAL DISTRIBUTIONS   Alleles are not randomly distributed in the geographic regions of Quebec and they reveal both intrapopulation and interpopulation distributions that reflect the patterns of migration and population growth.

FOUNDER EFFECT   The high prevalence of several autosomal recessive diseases in Quebec (Table 1) might be a result of inbreeding, although the evidence indicates otherwise. The degree of inbreeding is not exceptional, as shown by Catholic records of dispensation (51), measures of homozygosity at the *PAH* locus (26), and by analysis of consanguinity and kinship in a non-Mendelian genetic disease (Down syndrome) (48). The more likely explanation for certain disease prevalences and clustering in Quebec is founder effect and/or genetic drift.

ALLELES   Highly penetrant pathogenic alleles, usually rare in frequency elsewhere, appear at elevated frequencies in Quebec and manifest themselves as autosomal recessive or dominant diseases. Although the origins of these alleles apparently lie in founder effects and/or genetic drift (during rapid expansion in relative isolation), allelic heterogeneity is observed repeatedly in the disease clusters. Phenotypic heterogeneity also occurs in several of the Quebec diseases, which could be the result of the allelic heterogeneity but also the result of modifier loci (in which case, studies of genotype/phenotype correlations in Quebec families could be informative). Whether the Quebec model will be equally informative for genomic searches of prevalent alleles with low penetrance and associated with common multifactorial disease is a question yet unanswered and certainly of timely interest.

SOME LESSONS FROM QUEBEC   To know how Quebec was peopled and how its population increased over 400 years is to appreciate events reflected in the distribution of some rare penetrant alleles, as it were, into subpopulations. Ten loci are used to develop the themes as follows: The *PAH* locus illustrates the effects of demography on interpopulation and regional allelic diversity; the *CFTR* locus also has those features but also shows a founder effect in Ch-SLSJ. Two dyslipidemias illustrate (through the *LDLR* and *LPL* loci) intrapopulation allelic heterogeneity and regional diversity among French Canadians. Two loci (*HEXA, HBB*) remind us that the appearance of a typical ethnic disease (e.g. Tay-Sachs, β-thalassemia) in a nontypical population, as a transpopulation event, need not result from its introduction by someone from the typical community. Four loci (*FAH, PDRR, SACS, and PAPB2*) reflect founder effects and/or genetic drift and show how populations in eastern Quebec, in which the number of historical meioses is very large, can serve the mapping of genes by linkage disequilibrium, even in a young population.

The Peopling of Quebec

OVERVIEW   The province of Quebec is a vast modern geopolitical entity. It occupies a region of the North American continent easily visible from space (Figure 1, see color insert). In 1998, there were 7.3 million persons in the Province, 80% distributed in the St. Lawrence Valley, with 3.4 million living in the Montreal region and 1.62 million in the northeast (administrative regions 01, 02, 03, and 09). The 6 million French Canadians of Quebec consider themselves to be descendants of the original colonists. The anglophone population, comprising descendants from earlier settlers and more recent immigrants, is only 0.6 million; the Native American population is smaller still, ~72,000. The balance of the population is allophone, reflecting non-French, non-British immigration, comprising 225,000 of Italian origin, 100,000 Jewish, 90,000 German, 60,000 Greek, 40,000 Portuguese, and similar numbers of Polish and Arab-speaking nationals.

With the exception of its Native Americans, all citizens of Quebec today are either immigrants themselves or descendants of recent (within the past 400 years) immigrants. With only modern exceptions, those settlers and immigrants are of European origin, and they account for population growth and changes in its components during the past four centuries (Figure 2).

Modern Quebec, known as Nouvelle France before 1759 and as Lower Canada after 1841 (165), became a province of Canada at Confederation in 1867. By the mid-nineteenth century, the formerly French (and francophone) colony had significantly changed by virtue of non-French immigration. The anglophone population (221,000 persons) became 25% of the total, itself comprising people of Irish (60%), English (29%), and Scottish (11%) origins (144).

The large non-French immigration to Quebec mainly took place after 1820, the majority settling in the Montreal region (170). However, the ethnic origin could be indistinct; for example, among the Catholic Irish, intermarriage with French Canadians was frequent, and when the mother was Irish, the offspring would be counted as French in the Quebec census. By 1931, one in seven Quebecois of Irish origin spoke French as the mother tongue (144). Such details are important when inferring, for example, the origins of particular *PAH* alleles in Quebec families today (162).

In the 1890s, Canada developed an immigration policy that has remained relatively constant (81). In the late twentieth century, Quebec was accepting 25,000 immigrants annually. A total of 600,000 new citizens arrived from Europe, Africa, Latin America, and Asia during that century.

NOUVELLE FRANCE 1608–1759   In 1535, Jacques Cartier and his crew wintered over at Stadacona near what would later become Quebec City [Trigger, Plate 33 in (82)]. Native tales of gold in the Kingdom of Saguenay persuaded Francis I, King

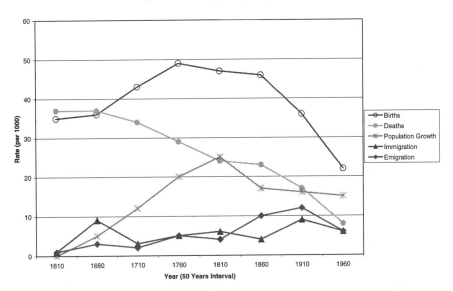

Figure 2   Mean annual rates (per 1000, by 50 year intervals) for components of total population growth in Quebec [redrawn from Charbonneau (31)].

of France, to send Cartier back in 1541 to form a colony. The venture failed. In the 1580s, fur trade became a new focus of interest and traders reached the Island of Montreal in 1600 where they built a post and wintered over for the year. At the time, excepting the few possible survivors of Raleigh's Virginia Colony, these were the only Europeans to live through a winter on the American continent north of Florida (127).

In 1603, Samuel de Champlain sailed the St. Lawrence River and then returned to France (127). The following year he returned, settled temporarily on the Island of St. Croix off the coast of New Brunswick, and in 1605 explored the coast that would soon host the Puritan colony at Plymouth. (Champlain did not claim this land for France; the "what-if" school of history pauses to reflect.) Champlain returned to the St. Lawrence River in 1608 and established a permanent post on the future site of Quebec City. The French colony of Nouvelle France had begun.

A post at Trois-Rivières was established in 1634 where the St. Francis river from the south and the St. Maurice River from the north meet the St. Lawrence [Harris, pp. 113–17 in (82)]; the site would divide Quebec into Western and Eastern halves of settlement. Champlain died in his colony in 1635; at the time, some 200 settlers were established in Nouvelle France. In 1642, a religious impulse, led by the soldier Sieur de Maisonneuve and the dynamic Ursuline nun Jeanne Mance, created a permanent settlement at Montreal (125).

The settlers of Nouvelle France and their descendants are among the best documented populations in the world (121). The archives, which cover births, marriages, and deaths in the Catholic parishes and reflect civic censuses, reveal who came as a colonist, from where in France, who returned permanently from the New World to France, who settled where in the St. Lawrence Valley, and how the population grew by immigration and natural increase. There were fewer than 10,000 permanent settlers [some say only 8483 settlers actually contributed to population expansion (41)] in the 150 years of this European colonization (31) [Charbonneau & Robert, Plate 45 in (82)], but by the mid-eighteenth century, natural increase had brought the French Canadian population to 60,000 inhabitants [Harris, pp. 113–17 in (82)]. The records also describe settlement, first along the northern shore of the river east of Quebec City in the region of Charlevoix from where, because of population expansion, there was migration up the Saguenay fiord followed by settlement after 1832 in the Saguenay Lac St. Jean region (SLSJ). The latter, a form of kin-structured migration (65, 110), introduced a second bottleneck and potential for founder effect. The pertinent archives for the Charlevoix and SLSJ regions now exist as a computerized genealogical database (18, 27).

The Seven Years War (1756–1763) engulfed the European world and its colonies. After a long siege, on September 12, 1759, British troops led by Wolfe climbed the cliffs of Quebec at night; in the morning, they faced Montcalm on the Plains of Abraham. By evening, the colony of Nouvelle France had become a British possession (123); the Treaty of Paris in 1763 confirmed this. Yet, by 1776, the British Crown held only a foothold in North America, and this only by virtue of its acquired Canadian colony. It had lost what would become the United States of America, and it left the continent to its Native Americans, Spanish colonists, English religious dissidents, and other European emigrants.

LOWER CANADA/QUEBEC 1763–1900   The demography of Quebec changed after the Treaty of Paris in 1763. French immigration essentially ceased and was replaced in the eighteenth century by settlers from Great Britain [Harris, pp. 113–17 in (82)] and by Loyalist emigrants leaving the newly created United States of America [Gentilcore et al., Plate 7 in (71)].

By the time of the Canadian Confederation in 1867, 93% of the inhabitants of Lower Canada were Canadian-born [Robert, pp. 77–79 (71)] and comprised three groups: Native Americans, French Canadians, and descendants either of Loyalists from the United States [Gentilcore et al., Plate 7 in (71)] or British settlers [Robert, pp. 21–23 (71)]. While the Native American population was in decline, the French Canadians were still expanding. In spite of almost no new francophone immigration to Quebec (the 4000 Acadians were a small exception), the French Canadian population doubled every 30 years between 1750 and 1875 [Robert, pp. 21–23 (71)]. The increase (Figure 2) was attributed to the high birth and fertility rates of women who married early [McInnis, Plate 30 (71)] (31). Expansion of the French Canadian population took place largely in rural areas, yet it also required spatial expansion to accommodate it in the Charlevoix region, which ultimately led to internal migration and the founding of the SLSJ enclave centered on Chicoutimi (Figure 1) (16). There was also migration to the cities and, during the economic depression of 1873, movement to the United States [Thornton, Plate 31 (71)], with implications for a diaspora of French Canadian alleles inside and beyond Quebec.

QUEBEC 1900–PRESENT   The twentieth century saw the trials and triumphs of materialism in Canada (32). The prospects were sufficient in the first decade for its Prime Minister to propose that this was to be "Canada's Century." Immigrants came in large waves (81) in the first three decades and again after World War II [Cartwright & MacPherson, Plate 9; McInnes, Plates 27, 28 in (102)]. In the final four decades of the twentieth century, the Canadian population more than doubled, with one fifth of the population (30.7 million in the year 2000) as new immigrants. Between 1946 and 1971, of the 3.5 million people who entered Canada, 15% settled in Quebec; only 5% came from francophone countries. The immigration policy fractured the traditional identity of Canada (and Quebec), and the country became "a polity in search of a nation" (81). The lingering effect of that policy may now be contributing to francophone Quebec's renewed search to retain its identity through political independence. Immigration brought to Quebec a collective genetic heritage quite different from the one initiated four centuries earlier. Each part of that heritage now echoes in the spectrum of alleles segregating in Quebec's population.

Origins of Hereditary Disease Clusters in Quebec

Demographic histories will explain not only the clustering of over 20 Mendelian recessive diseases in Ch-SLSJ, but also distributions of other pathogenic alleles in other regions (Table 1).

THE CH-SLSJ MODEL   The Ch-SLSJ population has expanded to $\sim$300,000 persons (39), largely by natural increase in isolation over 150 years. The founders were 30,000 migrants, half of whom came from the Charlevoix county. Genealogies for 180,000 people married in the Ch-SLSJ region have been reconstructed (15).

The unusually high prevalence of certain autosomal recessive diseases in the Ch-SLSJ region (39) (Table 1) might be explained by one of the two following mechanisms:

INBREEDING   The coefficients for inbreeding (44) and kinship relationships (45) among 567 probands in the Ch-SLSJ region and affected by 17 different Mendelian disorders (range of proband size, 2–116) were calculated and compared with those for 1701 matched controls. The mean inbreeding coefficient for the probands was 17.7×10^{-4} (range 0–91 $\times 10^{-4}$) and for controls 6.5×10^{-4} (range 0–32 $\times 10^{-4}$). The average depth in generations was 4.3 and 4.1 in disease and control groups, respectively. However, only 13% of the disease probands were offspring of consanguinous matings (6.4% in controls) (44). Inbreeding is not a comprehensive mechanism to explain the elevated frequencies of Mendelian diseases in the Ch-SLSJ population. The corresponding estimates of kinship coefficients (9.8×10^{-4} and 1.8×10^{-4} in disease and control groups, respectively) showed that almost half the probands were only third degree cousins (45).

FOUNDER EFFECT   The search for founder effect in the Quebec population was addressed through some of the hereditary disorders (41). Twenty diseases occur mainly in northeastern Quebec, and 24 are aligned east of Quebec city, either north or south of the St. Lawrence river. Genealogical reconstructions for 21 diseases (41) indicated that fewer than 250 settlers in Nouvelle France served as common founders. Moreover, tracing these founders back in time and place to seventeenth century France identified centers of diffusion in its northwestern region. These ancestors contributed disproportionately to population growth in Nouvelle France, and 34 founders apparently contributed to more than one Mendelian disease (41, 42). This phenomenon is explained by (*a*) the early dominance in the seventeenth century of the founders who settled the Charlevoix region, (*b*) their particular contribution to natural increase in the eighteenth century, which created a large effective population (41), and (*c*) internal migration and population expansion, notably in the Ch-SLSJ region in the nineteenth and twentieth centuries. Yet how could so few settlers from one region of Old World France explain the regional clustering, particularly in Ch-SLSJ, of so many genetic diseases in their descendants in the New World? Heyer and colleagues (8, 87–90) addressed this important question.

Heyer (87) used the BALSAC genealogical database (18) to show that founders who entered the SLSJ region from Charlevoix before 1870 contributed 45% of the average expected contribution to the gene pool in the northeastern region, although they were only 15% of the 20,000 migrants whose descendants were born in the region between 1950 and 1971. Half the genes introduced by founders were lost, but even if only 2% (i.e. 68 persons) of the first founders were carriers of a pathogenic allele, that was sufficient to account for the observed frequency of the corresponding Mendelian disease in the Ch-SLSH region (87).

Heyer & Tremblay (89) next reconstructed genealogies for control families and for those harboring one of five Mendelian diseases (hereditary tyrosinemia, spastic ataxia, sensorimotor polyneuropathy, cystic fibrosis, and hemochromatosis). From 545 genealogies, they showed that 80% of the gene pool could have descended from the seventeenth century founders of Nouvelle France. Moreover, as few as 15% of the founders could account for 90% of the total genetic contribution from the pool of founders (89). However, the latter interpretation still begged an explanation. Austerlitz & Heyer (8) found it in the variance and correlation of effective family size (EFS) from one generation to the next. They used real demographic data and a branching process method. When the variance of EFS alone was taken into account, the estimated allele frequencies did not fit observed frequencies. But when the EFS correlation between generations was introduced into the model, estimates fit observed frequencies. Reproductive behavior and EFS in one generation would be copied in the next, as evidence for social transmission of a demographic behavior in a population (8). Thus, the transmitting effective population size does not need to be large when the intergenerational EFS correlation is high.

Heyer (88) then used four of the SLSJ recessive diseases that have high known carrier frequencies (3.5%–5%) to test and support another hypothesis, that under the conditions described above, one could assume that one founder would introduce one disorder into the population. Finally, Heyer et al. (90) used the most up-to-date genealogical records for 673 probands with six different hereditary diseases, and by taking several factors into consideration, they found many possible candidate centers of diffusion for founders both in France (at least six regions in northwest France) and outside France (Switzerland and Britain among them). Moreover, there were notable differences in the origins of male and female founders. Most founders common to probands of one disease were also common to at least one other disease, with 29 founders common to over 95% of contemporary probands in both the disease and control groups.

Austerlitz & Heyer (9) also used multiallelic markers at the PDDR locus (see below) and Ch-SLSJ demographic data to show that population growth rate has a higher impact on estimates of recombination rate than the shape of the demographic distribution. Adaptation of the branching process to growing populations would allow allele frequencies to be integrated in simulation when dealing with the biallelic polymorphic alleles (9).

MENDELIAN MODELS OF QUEBEC POPULATION STRUCTURE

Several approaches reveal interpopulation and intrapopulation genetic diversity in Quebec. Pathogenic alleles at the *PAH* locus are distributed in ways that reflect the complex ancestry of contemporary populations in Quebec. Cystic fibrosis alleles, another ubiquitous disease of Caucasians, reveal both interpopulation variation and a founder effect among French Canadians in the northeastern region of the Province. Genetic inhomogeneity of the French Canadian population is also revealed by intrapopulation distributions of alleles at the *LDLR* and *LPL* loci. Tay-Sachs disease and β-thalassemia, usually not considered to be diseases of French Canadians, both appear there as independent transpopulation events not introduced by ethnic immigrants in the New World.

Then there are the models of founder effect, explaining the large clusters of hereditary tyrosinemia, pseudodeficiency vitamin D rickets, and spastic ARSACS; these three diseases illustrate a process of founder effect and genetic drift in northeastern Quebec. OPMD serves to illustrate the process behind a regional cluster in southeastern Quebec. The latter complements the evidence from polymorphic haplotypes at the *PAH* locus that the north and south shore communities in eastern

Quebec were settled independently by different sets of ancestors whose descendants increased independently in relative isolation.

The *PAH* Locus Reflects Interpopulation Diversity

The *PAH* locus is rich in alleles (155), and neutral and polymorphic alleles form haplotypes in various degrees of linkage disequilibrium that in turn show associations with particular pathogenic alleles (52, 63, 104, 115, 154). Over 400 annotated *PAH* alleles are recorded in a locus-specific mutation database (http://www.mcgill. ca/pahdb). The pathogenic alleles are a cause of hyperphenylalaninemia, notably phenylketonuria (OMIM 261600).

PAH alleles, both pathogenic and polymorphic, are being intensively sampled worldwide (152, 155). Probands have been found in all Quebec populations, and the corresponding *PAH* alleles have been uniformly sampled across the whole population because an efficient newborn screening program for hyperphenylalaninemia (111) has been in place in Quebec since 1970 as part of its universal health care system (156). *PAH* mutations in a particular cohort (1973–1990), containing on the order of 1.5 million newborns, have been analyzed with ∼95% efficiency (26).

The mean incidence (cases per million live births) of PKU in Europe varies from 60 (Italy) to 190 (Scotland) and higher still in Turkey (155). The incidence for nonPKU hyperphenylalaninemia is usually lower (15–75 cases per million). Incidences for PKU in Quebec are 40, and 41 for the non-PKU variant (C. Laberge, personal communication).

The *PAH* locus serves as a prototype to analyze populations, and it is presented here for several reasons: (*a*) When an international mutation analysis consortium exists (152), it provides a worldwide perspective against which the particular regional (e.g. Quebec) alleles can be compared. (*b*) Each allele can be named under a standardized taxonomy (157) according to established conventions (5, 58) and can be integrated into a genomic database of alleles (157). (*c*) Guidelines exist for classifying alleles as pathogenic or otherwise (33). (*d*) Genotype-phenotype correlations (80, 100) are always of interest, and the exceptions to predictions may reveal modifiers or other features (159).

In Quebec populations, the *PAH* locus harbors at least 45 different alleles (other than polymorphic haplotype markers), 7 of which are not likely to be pathogenic (26). The maximum relative frequency of a pathogenic *PAH* allele in this population is 0.17, corresponding to a population frequency of ∼0.0002. Ten alleles were identified first in Quebec; of these, 5 have yet to be discovered elsewhere (26). The subset of *PAH* alleles transmitted from the original French settlers is distinctive. Whereas 54 different pathogenic alleles occur in France today, or in other French populations (1), 33 have not been found in the French Canadian population.

Only 7 of 38 pathogenic *PAH* alleles account for half the relative frequency in Quebec, and 12 alleles account for two thirds of the total. This distribution between frequent and rare (or private) categories fits the pattern for highly penetrant pathogenic alleles noticed by Weiss (167) at other loci, in other populations.

There is regional differentiation of the 12 most prevalent alleles in Quebec, with different subsets in the eastern, western, and Montreal regions of the province (26). *M1V* and *S249P* predominate in the eastern region, *F39L* and *F299C* in the western region, and *R261Q* in the Montreal region. *IVS12nt1*, which is broadly distributed in contemporary northern European populations (155), appears in all Quebec regions, as does *R408W*, the most prevalent *PAH* allele in Europe. This stratification in the regional distributions of *PAH* alleles in Quebec reflects the history of settlement and migration.

One *PAH* allele (*M1V*) segregates only in French Canadians, where it behaves as a recessive, probably introduced by one founder, and has survived in the population. Discovered first in Quebec (92), it is rare in France (1) and is unknown in Europe outside of France. Genealogical reconstructions reveal time and space clusters for ancestors in the seventeenth century living in the regions of northwestern France from which the colonists of Nouvelle France came (118).

The frequency of homoallelic *PAH* genotypes in French Canadians is no higher than elsewhere in Europe or North America (79, 80, 100). The expected homozygosity value (j) at the *PAH* locus calculated for Quebec as a whole is 0.06 (26). The values for eastern (French-Canadian), western (French-Canadian and mixed), and Montreal (multiethnic) regions of Quebec are 0.08, 0.05, and 0.08, respectively (26); values for most European populations are higher (79).

In Quebec, the *R408W* allele occurs more often on *PAH* haplotype 1 than on haplotype 2 (26, 93). The association between mutation and haplotype 1 segregates in pedigrees with Celtic ancestry (162). The *R408W* allele, when it occurs on haplotype 1 in Europe, is found predominantly in populations inhabiting the northwestern fringe of the European continent. The *R408W* allele on haplotype 2 has its center of diffusion in central and northeastern Europe (59). This previously unrecognized regional differentiation in *PAH* haplotype and *R408W* associations in Europe is compatible with what has been surmised about their population structure (28, 29).

A CpG dinucleotide sequence is involved in the R408 codon. Its association with two different *PAH* core haplotypes indicates a probable recurrent event at this hypermutable dinucleotide sequence (25). Accordingly, *R408W* is identical by state across the different haplotypes but not identical by descent. [Note: The worldwide *R408W* allele occurs on seven different haplotypes (155), and recurrent mutation is likely to have occurred on four different haplotypes.]

Three other PKU-causing *PAH* mutations were found on unconventional haplotypes in Quebec (26). In each case, the finding could be attributed to a single intragenic recombination and taken as possible evidence for intragenic recombination within the *PAH* gene, which spans only 100 kb. [Note: Because reports of *PAH* mutations often do not describe the polymorphic haplotypes, the frequency of putative intragenic recombination at this locus has not been adequately interrogated.]

POLYMORPHIC PAH HAPLOTYPES AND POPULATION STRUCTURES IN QUEBEC   Restriction enzyme–defined biallelic markers (55, 116), two multiallelic sites [a tetra-nucleotide STR (76) and a VNTR series (75)], and numerous single nucleotide polymorphisms (SNPs) exist in the *PAH* locus (155). Sets of these polymorphic markers in linkage disequilibrium have been used to study human microevolution (52, 104), and they provide evidence to support an "out-of-Africa" hypothesis for the diaspora of *H. sapiens*. Whereas the latter studies were designed to address population structure across many generations and large geographic distances, the *PAH* locus appears to have sufficient power in Quebec to interrogate young French Canadian subpopulations separated by a river. The *PAH* locus was analyzed in persons whose ancestries placed their origins 10–12 generations earlier in different settlements either northeast or southeast of the St. Lawrence River (24). The frequency distributions and content of *PAH* haplotypes in the two subpopulations differ sufficiently to imply that their structures are different. The haplotype data reflect what we know about early settlement and subsequent population expansion in French Canada.

The CFTR Locus Reflects Both Inter- and Intrapopulation Diversity

Cystic fibrosis (OMIM 219700) is the most prevalent, fatal autosomal recessive disease affecting Caucasian populations (168). *CFTR* has been mapped to chromosome 7q31.2, cloned, and characterized as a cAMP-dependent activator of an outward oriented transmembrane apical chloride channel protein (1480 amino acids) in various epithelial cell types (137, 139). The large gene (230 kb, 27 coding exons) harbors over 900 population-related pathogenic mutations and polymorphic markers (36, 101, 153) (http://www.genet.sickkids.on.ca/cftr/). The major disease-causing mutation ($\Delta F508$) is a three–base pair deletion that removes phenylalanine residue 508 (36), which accounts for about 70% of all pathogenic CFTR mutations in Northern Europeans. A case has been made (128) and disputed (98) for the age of this mutation (52,000 years bp) and its origins (IndoEuropean) as well as for selective advantage in the heterozygote (133a).

Incidence of CF probands identified by clinical diagnosis in Caucasian populations (131) is ~1 in 2500 (slightly lower, 1 in 3200, when ascertained by newborn screening), with ranges from 1 in 1700 to 1 in 6500 in different European populations. The disease is much rarer in non-Caucasian populations and is rare in Finland. An unusually high incidence of CF is found in populations where founder effect and genetic drift could have occurred, for example, in a subset of Afrikaners, in a region of Brittany, among the Old Order Amish in Ohio, and in the Hutterite Brethren of Alberta (168).

The province-wide incidence of CF in Quebec is ~1 in 2500 (R. Rozen, personal communication), but regional disparity exists. Only certain mutations are apparent in French Canadians (36) and the SLSJ population is one of those in which incidence is high (37). A detailed study of 163 patients in 143 families born between 1975 and 1995 in the region showed the average incidence to be 1 in 936 births (138). Assuming Hardy-Weinberg equilibrium, the carrier rate in the region is 1 in 15.

Mutation analysis for *CFTR* alleles has been carried out in French Canadians in Quebec City, SLSJ, and in major urban centers. Although 18 different alleles account for over 90% of the mutant CF chromosomes in the Quebec patients, the *ΔF508* mutation is predominant at ~70% relative frequency (R. Rozen, personal communication). The *L206W* allele (with a mild phenotypic effect) reflects a particular French Canadian heritage (142), whereas *W1282X* and *G542X* are prominent in Ashkenazi Jews (2, 145), which reflects corresponding twentieth century immigrations into Quebec.

The *CFTR* locus further reveals particular distributions of its alleles in French Canadians (141, 143). The *ΔF508* allele is present on only 55% of the CF chromosomes in the SLSJ population but on 71% in the Quebec City region (weighted estimates). SLSJ and Quebec City have different histories of settlement and demography and this is reflected in different distributions of D7S23 haplotype–CF mutation combinations in their CF families. The B version of the haplotype (101) occurs on 86% of CF chromosomes without the *ΔF508* allele in the SLSJ population, compared with 31% in Quebec City families. When a profile of 10 different CFTR mutations is taken into account, SLSJ probands are shown to differ from other French Canadian probands (50, 141). Three mutations (*ΔF508*, 60%; 621 + 1, 25.5%; A455E, 8.5%) account for 94% of CF chromosomes in the SLSJ population, with 42% of probands being homoallelic (50). Another set of 3 alleles (*ΔF508*, 71%; *711 + 1*, 9%; *621 + 1*, 5%) occurs on 85% of the French Canadian chromosomes elsewhere (141). A set of 11 CFTR mutations accounts for 100% of the CF-causing alleles in the SLSJ population (50). The *A455E* allele, prominent in SLSJ, has a mild effect and confers pancreatic sufficiency (138). Estimates of inbreeding and kinship coefficients in the SLSJ families harboring *ΔF508, 621 + 1*, and *A455E* alleles are somewhat higher than in the general population, and a putative common ancestor for carriers of the *A455E* allele is claimed (40).

Two Dyslipidemias (FH and LDL Deficiency): Models of Intrapopulation Diversity in French Canadians

Autosomal codominant familial hypercholesterolemia (FH) (OMIM 143890) is the most prevalent inborn error of metabolism in the human species (74). It is caused by mutations in the *LDLR* gene on chromosome 19p13.1-13.3 (34). Familial hyperchylomicronemia (FHC) (OMIM 238600) is a very rare autosomal recessive inborn error of metabolism (23), caused by mutation in the *LPL* gene on chromosome 8p22. Both diseases show elevated prevalence in the French Canadian population, and each has a particular explanation for this phenomenon.

Studying a population in Seattle, Washington, Goldstein et al (73) found that hyperlipidemias are important factors in early-onset coronary heart disease. Twenty years later, a similar epidemiological study in a Boston population found that at least half the patients with premature coronary artery disease, confirmed by angiography, had a familial dyslipidemia (70). Some of the Boston probands might have been descendants of or among those French Canadians who had emigrated from Quebec to New England. A study focused directly on French Canadians in Quebec, investigating cardiovascular disease as the primary form of premature mortality, described very high relative risks for FH, 6.6, and for FHC, 130, in the SLSJ population (69). Incidences of these two hyperlipidemias are higher in this region of Quebec than almost anywhere else in the world (84).

FAMILIAL HYPERCHOLESTEROLEMIA   An epidemiological study of French Canadian FH patients gave estimated carrier rates of 1 in 167 in the lower St. Lawrence, 1 in 122 in the SLSJ region, and 1 in 81 in the small communities of the north shore east of the Saguenay. These prevalences compare with 1 in 270 for Quebec province as a whole and 1 in 900 in the Montreal region (126). Mutation analysis at the LDLR locus on 1460 chromosomes, almost all from French Canadian FH patients, led to the discovery of only 6 different mutations (91, 113, 119, 164), accounting for 85% of the FH-causing alleles in Quebec. When ranked by relative contribution to the pool of LDLR alleles in French Canada, along with the region in which they cluster (164), the alleles can be ordered as follows: (*a*) A large (>15 kb) 5′ deletion (91) (~56% of the total alleles), with a center of diffusion in Kamouraska County near the Gaspé (94). Because of internal migration, this allele is found throughout the province. (*b*) A missense allele (*W66G*) (113) in exon 3 (~32% of total) clustering in the SLSJ population but also found across the province. (*c*) A missense allele (*C646Y*) (113) in exon 14 (~4% of total) mainly found in central Quebec. (*d*) A missense allele (*E207K*) in exon 4 (113) (~4% of total) that is more frequent in the SLSJ and Gaspé region. (*e*) A stop codon in exon 10 (~2% of total) found mainly in the Chaudière region south of Quebec City. (*f*) A 5-kb deletion (119), affecting exons 2 and 3, (~2% of total) found in the Montreal region. The first two alleles alone, each with a recognized geographic center of diffusion in Quebec, account for the great majority of all FH alleles in French Canadians. A common haplotype at the LDLR locus, flanked by the markers D19S865 and D19S221, is associated with the *W66G* allele (34).

The 15-kb deletion allele (91) has been identified in one French FH patient residing in the Saumur region of France (67, 84). Genealogical reconstructions done earlier on French Canadian patients identified 14 French founders in the seventeenth century, 7 of whom were resident in the Perche and Poitu regions, which embrace Saumur (94). The large deletion allele, which resides on only one polymorphic haplotype in French Canadians (11), provides molecular evidence to support a founder hypothesis.

Four homozygous FH patients with the 15-kb deletion exhibited striking phenotypic diversity despite allelic homozygosity (91). One patient died at 3 years of age, while another was still alive at 33 years of age and had not become symptomatic until age 19; the effect of a genetic modifier seems likely (74). The FH

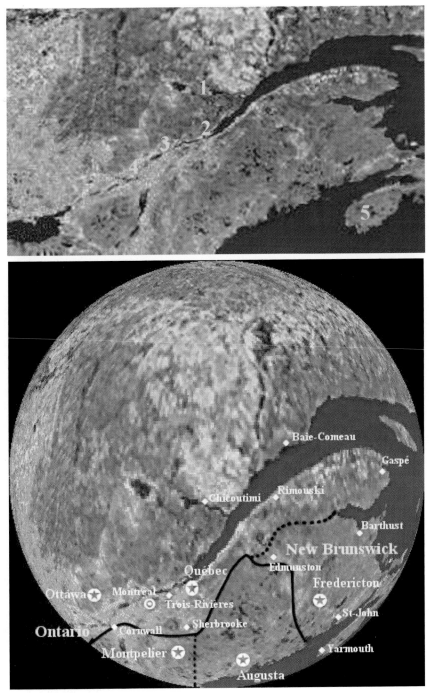

See legend page C-2

Figure 1 (see figure page C-1) (*Top*) Satellite image centered on the St. Lawrence Valley region of Quebec, the Maritime region of Canada is also shown. Numbered regions: (*1*) Saguenay-Lac St. Jean (the lake is clearly visible), (*2*) Quebec City, (*3*) Trois-Rivières, (*4*) Montréal, (*5*) site of the original seventeenth century Acadian settlement in Nova Scotia. Regions 2, 3, and 4 comprised Nouvelle France in the seventeenth century. (Taken from http://www.fourmilab.ch/earthview/satellite.html; earth view by J. Walker.) (*Bottom*) Conventional geopolitical map of the populated regions of Quebec, showing international and interprovincial boundaries. Cities in Quebec mentioned in text are shown. Starred cities are the capitals of Vermont (Montpelier), Maine (Augusta), New Brunswick (Fredrickton), and Canada (Ottawa). Loyalist immigrants entered from the south to cross the frontier into Lower Canada; Acadians usually came from New Brunswick.

phenotype in French Canadians responds to therapy with the statin class of HMG
Co A reductase inhibitors (99).

LIPOPROTEIN LIPASE DEFICIENCY   Incidence of the disease reaches 1 in 6000 in the
Ch-SLSJ region, with an estimated carrier rate of 1 in 40 (56, 68), the highest known
frequencies. Clustering is attributed to a founder effect in the region (43); but in
the course of this particular study, it was noticed that the birthplace of obligate
carriers actually occurred in three different regions of Quebec, the Ch-SLSJ region,
central Quebec, and the Trois-Rivières region. Genealogical studies identified three
sets of founders, one for each region harboring FHC probands. However, the
pool of founders in the southeast and northeast regions overlapped, whereas those
in the Trois-Rivières region emerged as an independent set. Thus, investigators
noted the possibility of more than one major pathogenic LPL allele in French
Canadians.

Mutation analysis demonstrates that there are two predominant missense *LPL*
alleles in French Canadians. A *G188E* allele accounts for ~30% of the total;
P207L accounts for most of the remainder. Allele *G188E* (124) is prevalent elsewhere in the world and is probably an ancient mutation (84). In Quebec, this allele
is distributed westward of Quebec City (10). The *P207L* allele is distributed in the
eastern region of the province (130), and the estimated pool of its heterozygotes
is 31,000 persons in the Ch-SLSJ population. Analysis of specific microsatellite haplotypes at the *LPL* locus subsequently identified two different haplotypes
in tight linkage disequilibrium with the two different pathogenic alleles (169).
Molecular evidence indicates two founder effects that are compatible with the historical record of independent settlement of regions along the St. Lawrence Valley
by the original French colonists.

*Trans*population Models of Diversity: *HEXA* and *HBB* Alleles

Pathogenic alleles at the *HEXA* locus cause Tay-Sachs disease (TSD) (OMIM
272800) (77); alleles at the *HBB* locus cause β-thalassemia (166) (OMIM 141900).
The infantile form of TSD is a neurodegenerative disease fatal in early childhood
that segregates predominantly in the Ashkenazic Jewish population. β-thalassemia
is a heterogeneous hereditary disorder of globin production and is fatal without
intensive transfusion and iron chelation therapy; it segregates mainly in populations
from regions where malaria is (or was) prevalent. The appearance of large numbers
of patients with TSD and β-thalassemia in Quebec is largely, but not solely, a
reflection of twentieth century migrations of Jewish, Mediterranean, and Asian
persons.

TSD also occurs in French Canadians, with southeastern Quebec being a center
of diffusion for affected families. β-thalassemia minor (the codominant carrier
state) also occurs in French Canadians, with its center of diffusion in Portneuf
County on the north shore of the St. Lawrence, near Quebec City.

HEXA ALLELES   The *HEXA* alleles in Quebec's Jewish community are characteristic of those found in European Jews (64). The predominant French Canadian *HEXA* allele, a 7.5-kb deletion (129), has not yet been found in any Jewish population, and presumably it was not introduced into the French Canadian deme by a Jewish immigrant. It occurs most often in the homozygous state, accounts for 19 of 22 *HEXA* mutations in French Canadian probands, and is the source of TSD heterozygosity at 5%–7% frequency in the southeastern region (4). Genealogical reconstructions suggest that the 7.5-kb deletion allele could have originated in a person born in Nouvelle France (47). This *HEXA* allele also occurs elsewhere in the Province, including the SLSJ region (Table 1), presumably as a result of interregional migration by French Canadians (86). A second unusual *HEXA* allele causing TSD (*IVS7 + 1*) also occurs in the French Canadians. A "Jewish" *HEXA* allele (exon 11, 4 bp insertion) was found in a single French Canadian family (85).

HBB ALLELES   Extensive heterogeneity of thalassemia-causing *HBB* alleles exists in the non-French Canadian populations of Quebec (96), with allelic profiles in these communities reflecting the regions from which members emigrated. However, two typical European *HBB* alleles also occur in French Canadians (97) and account for the occurrence of β-thalassemia minor (54) at a population frequency of 1% in the Portneuf region (135). Genealogical reconstructions point to ancestors from Southern France (Languedoc) as one source of the alleles and also from northern France where β-thalassemia alleles may be more frequent than realized.

AN UNUSUAL *HBB* ALLELE   A rare β-thalassemia allele (131) coexists in a Jewish family in Montreal and in a Jewish family that recently emigrated from Russia to Israel (120). The allele in association with an extended haplotype is identical both by state and by descent in the two families, and it has led to the discovery of unrecognized survivors of twentieth century horrors and of mutual ancestors in early nineteenth century Europe (120).

MODELS OF FOUNDER EFFECTS

Hereditary Tyrosinemia (*FAH* gene)

Hereditary tyrosinemia (OMIM 276700), a lethal autosomal recessive disease (122), is prevalent in French Canadians living in the Ch-SLSJ region of northeast Quebec.

The regional cluster of patients exceeds 150, where the majority have been identified by newborn screening since 1970 and the remainder by astute clinical observation (112). The estimated live birth incidence (per million births) is 8–10 cases worldwide, 60 cases in the Quebec province as a whole ($\sim$85,000 births/yr), and $\sim$500 cases in the Ch-SLSJ region. The estimated carrier frequency in the latter region is $\sim$1 in 22 (122).

Symptoms of the disease are quite variable, but they are usually first manifest in infancy; they include acute liver failure, hepatic cirrhosis, a renal Fanconi syndrome, and porphyria-like crises. Infants who survive go on to develop complications of the Fanconi syndrome and neurological crises. Hepatocellular carcinoma is a late final stage [and before 2-(2-nitro-4-trifluoromethylbenzoyl)-1, 3-cyclohexanedione (NTBC) therapy, the patients placed a large demand on liver transplantation]. The metabolic phenotype includes hypertyrosinemia due to (secondary) inhibition (by succinylacetone) of an enzyme, which at one time was thought to be the primary enzyme defect in the disease (146). When succinylacetone was found in affected patients, it revealed the primary deficiency of FAH enzyme function (117). The dismal course of the disease has changed for the better by treatment with NTBC, a triketone that inhibits an upstream enzyme in the tyrosine catabolic pathway and reduces production of the downstream toxic metabolites (122). The human *FAH* gene has been obtained by functional cloning and mapped to chromosome 15q23-q25 (3, 132).

In an early landmark study, Laberge (107) traced affected probands to a common founder couple. Aided by the large computerized genealogical register for the Ch-SLSJ population, a larger number of putative founders (including the couple identified by Laberge) was later revealed (17). Further pursuit of the genealogical traces identified a time and space cluster of ancestors in northwest France from which migration to Nouvelle France had occurred in the seventeenth century (16).

The *FAH* gene contains polymorphic restriction sites from which core haplotypes can be identified (53). Haplotype 6, derived from the five most informative restriction sites, segregates on 96% of the mutant HT1 chromosomes in the Ch-SLSJ population (53). The same haplotype is found in only 18% of normal alleles in a control group of French Canadians and occurs at much lower frequency in European patients (140).

Mutation analysis offers the strongest evidence for a founder effect for hereditary tyrosinemia in the Ch-SLSJ region (78). There is strong allelic homogeneity in the Ch-SLSJ patients (78, 134). One mutation (*IVS12 + 5 g->a*) is present in 100% of patients in the region; it accounts for over 95% of the *FAH* mutations (in patients and carriers), and 80% of the patients are homozygous for it in the region (78). The same *FAH* allele accounts for only 60% in European probands (163). *N16I* and *E381G* alleles were first identified, each only once, in the Quebec patients, and they have not been reported elsewhere. Of 34 known *FAH* alleles in the world population of HT1 patients (122), only five (*N16I, IVS12 + 5, E357X, E364X, E381G*) occur in the French-Canadian population.

Pseudo Vitamin D Deficiency Rickets (PDDR)

PDDR (OMIM 264700) illustrates an important general theme: Heritability of a multifactorial phenotype will increase when its environmental causes are removed. Nutritional rickets due to vitamin D deficiency was prevalent in Quebec until the 1970s (148). Following elimination of the environmental cause, the heritability of rickets increased in the Quebec population (158), and a hereditary form of rickets (PDDR) became apparent (7).

Autosomal recessive selective simple deficiency of the hormonal product of vitamin D [1α, 25-dihydroxy vitamin D; $1\alpha,25(OH)_2D$] (114) occurs at a high prevalance in the French Canadian population of Ch-SLSJ. The condition is caused by mutations in the *PDDR* gene encoding the renal mitochondrial cytochrome P_{450} enzyme, and it confers a dependency on pharmacological doses of vitamin D (147) or on replacement of $1\alpha,25(OH)_2D$ in physiological doses (66).

Deficient activity of the hydroxylase enzyme has been demonstrated (72), and pathogenic mutations map to the *PDDR* disease locus on chromosome 12q13-q14. The candidate gene has been cloned, and it maps to chromosome 12q13.1-q13 (160) in French Canadian and other families.

Observed incidence of the disease in Ch-SLSJ is ~1 in 2900 births, and assuming Hardy-Weinberg equilibrium, the carrier frequency is 1 in 27 (49). Genealogical reconstructions from the BALSAC database have identified founders who settled in Nouvelle France in the seventeenth century (17).

A complex polymorphic haplotype in linkage disequilibrium with the PDDR phenotype was used to estimate the depth of the founder effect, which was anticipated at ~12 generations (109). These haplotype markers segregating in the Ch-SLSJ population enabled the mapping of the PDDR disease locus, thus showing that mapping by linkage disequilibrium was feasible in the young Ch-SLSJ population.

Spastic Ataxia of Charlevoix-Saguenay (ARSACS)

ARSACS (OMIM 270550) is a clinically homogeneous early-onset neurodegenerative disease with high prevalence (over 300 patients) in the Ch-SLSJ region where the carrier frequency is estimated to be 1 in 22 (46). The disease locus was mapped to chromosome 13q11, when increased shared homozygosity at locus D13S787 was observed in patients (136). The majority of ARSACS patients (96%) carry a single haplotype, defined by the D13S232 allele, and two SNPs across 11.1 cMs in the region flanked by D13S1236 and D13S1285 (136); the remainder carry the second haplotype, both of which are considered to be ancestral. After obtaining a high resolution physical and transcript map of the candidate region (61), investigators cloned the *SACS* gene. Its open reading frame encodes a single exon, the largest yet identified in vertebrate organisms, in a protein called sacsin (60). The *SACS* gene is expressed in the central nervous system and other tissues. Two protein-truncating mutations have been identified in ARSACS patients, each on a specific haplotype. Again, the Ch-SLSJ population yielded one of its pathogenic genes to mapping by linkage disequilibrium (60).

Oculopharyngeal Muscular Dystrophy (OPMD)

This autosomal dominant adult-onset disease (OMIM 164300) has a worldwide distribution (21); the highest concentration of patients in the world is in southeastern Quebec (19). The clinical phenotype includes progressive dysphagia, eyelid ptosis, and proximal limb weakness caused by a triplet repeat (GCG) expansion in the *PABP2* gene (encoding polyadenylation binding protein 2). The gene maps to chromosome 14q11 and has been cloned and characterized (20). When the normal $(GCG)_6$ repeat encoding an N-terminus polyalanine tract in the protein expands (7–13 copies), it becomes a cause of OPMD. Phenotypic variation correlates with the number and combination of repeats; $(GCG)_7$ homozygosity produces an autosomal recessive form of OPMD, whereas the $(GCG)_7$ allele in a compound heterozygote genotype acts as a modifier to aggravate the otherwise dominant phenotype (12, 20). The predicted expansion of the polyalanine tract is associated with intranuclear aggregates (12).

The *PAPB2* gene was identified in a genome-wide search, using a large suite of polymorphic markers in French Canadian families. A candidate region was identified by linkage analysis in a 5 cM region of chromosome 14q11.2-q13. The candidate interval was then narrowed using 11 large families. Thereafter, the gene was isolated from a 217 kb candidate interval containing the $(GCG)_6$ repeat, which was expanded to 7–13 copies in affected family members.

Analysis of haplotypes in 42 OPMD families of French Canadian descent pointed to a single OPMD-associated haplotype in the patients (19). The core French Canadian haplotype was found in a French family but not in other OPMD families in Europe. Genealogical reconstructions suggest that three sisters introduced the OPMD allele to Nouvelle France in the mid-seventeenth century. Again, it was this particular population structure in Quebec and the archival records that made possible the genealogy-based estimate of historical meiosis from which the distance between markers in linkage disequilibrium and the disease locus were effectively calculated (19).

GIVING BACK AND TAKING FROM: SHOWING RESPECT FOR POPULATIONS

Tension can exist between geneticists and persons in the populations they study. The former take samples and analyze them in ways that can be interpreted as appropriating identities. What does the population receive in return and what particular assets does it have to interest the geneticists?

Giving Back: Genetic Health Care for a Population

Although Canada has a Federal system of universal health insurance (62), the processes of health care and distribution of services are under provincial authority. The Quebec Network of Genetic Medicine (156) was created in 1971 by the Quebec government as a program for "public health genetics." The Network's newborn screening programs bring patients with phenylketonuria or hereditary tyrosinemia to early diagnosis and effective treatment. They also identify patients with histidinemia and sarcosinemia (Table 1), which are essentially harmless traits. The screening programs also provide data on incidence and regional distribution of patients so that services can be targeted efficiently.

The Network has voluntary programs for carrier screening and counselling on reproductive options. These programs were developed in response to requests from the communities where Tay-Sachs disease and β-thalassemia major cluster, the result being that the incidence of both diseases in Quebec has declined by over 90% in one generation (123). The same programs facilitated the studies in Quebec of allelic diversity at the *HEXA* and *HBB* loci.

Cascade screening, or genetic testing in the extended family, is now feasible by molecular (mutation) analysis for alleles segregating, for example, in the French Canadian families at elevated risk for lipoprotein lipase deficiency or familial hypercholesterolemia. As there are geographic centers of diffusion for the alleles causing these diseases in Quebec, a combination of molecular and family medicine becomes the application of genetic epidemiology to individuals, families, and communities (149, 150). These initiatives in public health genetics (103) (or community genetics) in Quebec have so far mainly addressed patients and diseases associated with rare penetrant alleles. However, genetic medicine will increasingly need to address adult-onset common diseases associated with prevalent alleles of low penetrance. Does Quebec offer opportunities to learn how that can be done?

Taking Lessons From the Quebec Population Experience

The high prevalence and clustering of many Mendelian diseases in Quebec is attributed to founder effects and/or genetic drift, to migrations in a population with a unique modern geopolitical identity (Quebec in Canada in North America), and to a unique set of complex demographic histories compressed into four centuries. Hypermutability at the relevant loci, inbreeding in the population, or positive selection are, in no case, comprehensive explanations for prevalence and clustering in Quebec. The particular stories of phenylketonuria and cystic fibrosis (for which selection scenarios acting during human history have been offered) fit nicely into their corresponding demographic models in Quebec, and as said elsewhere, "there is no need to postulate positive selection with respect to the common disease-associated alleles for such entities as phenylketonuria or cystic fibrosis" (161).

Contrary to stated reservations about its power for the relevant studies (83, 95), the Quebec population, and its French Canadian component in eastern Quebec in particular, may be well suited to fine mapping of genes by linkage disequilibrium (19), where linkage disequilibrium means the nonrandom association of alleles at two or more loci. Three recurrent demographic features facilitate population genetic studies in Quebec: (*a*) a relatively small number of founders, (*b*) population expansion in relative isolation, and (*c*) sustained and great expansion by natural increase. Access to the extraordinary archives documenting the demography is an additional important feature.

The opportunity exists to combine social (historical and cultural) and genetic (molecular) approaches to mutual advantage in Quebec. Genealogical reconstructions from (computerized) archives can reveal the extent to which an ancestral couple is shared by contemporary patients and controls, in and between regions and communities. Molecular analysis of alleles and haplotypes can reveal how often they are shared among probands and with controls. The genealogical reconstructions (in Quebec and elsewhere) may indicate who could have introduced the allele and when, but only haplotype analysis can establish whether a single copy of the allele was introduced. As shown by the mapping of *PABP2*, *PDDR*, and *SACS* genes, each existing within a subset of the Quebec population (which has also experienced extraordinary natural growth in isolation and accumulated the corresponding number of historical meioses), these young populations are indeed suitable for mapping genes by linkage disequilibrium.

The physical extent of linkage disequilibrum will determine the efficiency with which prevalent genes of low penetrance that predispose to disease can be mapped (14). This challenge is rather different from that met when mapping highly penetrant disease-causing alleles. If Quebec populations were to serve complex trait mapping, two recently recognized facets would need to be taken into consideration (14). (*a*) Not all populations are equally informative about the extent of disequilibrium. Quebec is made up of subpopulations, even among French Canadians. (*b*) Not all regions of the genome are equally informative. For example, to what extent the tentative evidence for intragenic combination at the *PAH* locus may be a special case is unknown. If all genetic isolates are not equal (105), it would be useful to create a disequilibrium map of the human genome covering both populations and isolates as well as regions of the genome to document variation in the extent of disequilibrium (106). Whereas a useful level of linkage disequilibrum may not extend beyond an average distance of ~3 kb in the general population (106), the distance is apparently greater at several loci in subpopulations of Quebec.

Lessons from Iceland are echoing in Quebec. The Icelandic population is not as genetically homogeneous as initially predicted. Analyses of blood groups, microsatellite markers, and D-loop mitochondrial DNA variation reveal its heterogeneity (6). As described in the present article, the Quebec population is also manifestly heterogeneous in structure. On the other hand, some of its subsets are less internally heterogeneous than others; the Ch-SLSJ subpopulation has already been recognized in this regard as a small isolate with some potential disadvantages (immigration and admixture) but also with many advantages for linkage disequilibrium mapping (30). Each human population isolate has its own unique evolutionary history (30), and such a history is known in great depth and detail by many of the Quebec subpopulations. Accordingly, they could be participants in creating a particular linkage disequilibrium map of the human genome (108).

ACKNOWLEDGMENTS

This work has been supported by operating funds and a career award from the Medical Research Council of Canada (now the Canadian Institutes for Health Research) and by operating funds from the Quebec Network for Applied Genetics and the Canadian Genetic Diseases Network (Networks of Centers of Excellence). I thank Gerard Bouchard, Bernard Brais, Mary Fujiwara, Claude Laberge, and Ken Morgan for their collegial guidance. Any skewed perspectives and errors in this article are the fault of the author alone. I thank Lynne Prevost for her computer literacy, and David Côté for finding the maps on the web.

Visit the Annual Reviews home page at www.AnnualReviews.org

LITERATURE CITED

1. Abadie V, Lyonnet S, Melle D, Berthelon M, Caillaud C, et al. 1993. Molecular basis of phenylketonuria in France *Dev. Brain Dysfunct.* 6:120–26
2. Abeliovich D, Lavon IP, Lerer I, Cohen T, Springer C, et al. 1992. Screening for five mutations detects 97% of CF chromosomes and predicts a carrier frequency of 1:29 in the Jewish Ashkenazi population. *Am. J. Hum. Genet.* 51:951–56
3. Agsteribbe E, van Faassen H, Hartog MV, Reversma T, Taanman J-W, et al. 1990. Nucleotide sequence of cDNA encoding human fumarylacetoacetase. *Nucleic Acids Res.* 18:1887
4. Andermann E, Scriver CR, Wolfe LS, Dansky L, Andermann F. 1977. Genetic variants of Tay-Sachs disease: Tay-Sachs disease and Sandhoff's disease in French Canadians, juvenile Tay-Sachs disease in Lebanese Canadians, and a Tay-Sachs screening program in the French-Canadian population. *Prog. Clin. Biol. Res.* 18:161–88
5. Antonarakis SE, Nomencl. Work. Group. 1998. Recommendations for a nomenclature system for human gene mutations. *Hum. Mutat.* 11:1–3
6. Arnason E, Sigurgislason H, Benedikz E. 2000. Genetic homogeneity of Icelanders: fact or fiction? *Nat. Genet.* 25:373–74

7. Arnaud C, Maijer R, Reade T, Scriver CR, Whelan DT. 1970. Vitamin D dependency: an inherited postnatal syndrome with secondary hyperparathyroidism. *Pediatrics* 46:871–80
8. Austerlitz F, Heyer E. 1998. Social transmission of reproductive behaviour increases frequency of inherited disorders in a young expanding population. *Proc. Natl. Acad. Sci. USA* 95:15140–44
9. Austerlitz F, Heyer E. 1999. Impact of demographic distribution and population growth rate on haplotypic diversity linked to a disease gene and their consequences for the estimation of recombination rate: example of a French Canadian population. *Genet. Epidemiol.* 16:2–14
10. Bergeron J, Normand T, Bharucha A, Ven Murthy MR, Julien P, et al. 1992. Prevalence, geographical distribution and genealogical investigations of mutations 188 of lipoprotein lipase gene in the French Canadian population of Quebec. *Clin. Genet.* 41:206–10
11. Betard C, Kessling AM, Roy M, Chamberland A, Lussier-Cancan S, Davignon J. 1992. Molecular genetic evidence for a founder effect in familial hypercholesterolemia among French Canadians. *Hum. Genet.* 88:529–36
12. Blumen SC, Brais B, Korczyn AD, Medinsky S, Chapman J, et al. 1999.

Homozygotes for oculopharnygeal muscular dystrophy have a severe form of the disease. *Ann. Neurol.* 46:115–18

13. Bodmer WF. 2001. Population genetics. See Ref. 151, pp. 299–310

14. Boehnke M. 2000. A look at linkage disequilibrium. *Nat. Genet.* 25:246–47

15. Bouchard G. 1988. Sur la distribution spatiale des gènes délétères dans la région du Saguenay (xixᵉ – xxᵉ siècles). *Cah. Géogr. Québec* 32:27–47

16. Bouchard G, De Braekeleer M, eds. 1991. *Histoire d'un Gènome. Population et Génétique dans L'est du Québec.* Sillery, Québec: Presses Univ. Qué. 607 pp.

17. Bouchard G, Laberge C, Scriver CR. 1992. Comportements démographiques et effets fondateurs dans la population du Québec (XVIIe-XXe siècles). In *Anonymous Societé Belge de Demographie Historiens et Populations: Liber Amicorum Etienne Hélin.* Louvain-la-Neuve: Academia

18. Bouchard G, Roy R, Casgrain B, Hubert M. 1995. Computer in human sciences: from family reconstitution to population reconstruction. In *From Information to Knowledge. Conceptual and Content Analysis by Computer,* ed. E Nissan, KM Schmidt. Oxford: Intellect. 201 pp.

19. Brais B. 1998. *Oculopharyngeal muscular dystrophy: from phenotype to genotype.* PhD thesis. McGill Univ., Montréal, Qué.

20. Brais B, Bouchard J-P, Xie YG, Rochefort DL, Tomé FM, et al. 1998. Short GCG expansions in the *PABP2* gene cause oculopharyngeal muscular dystrophy. *Nat. Genet.* 18:164–67

21. Brais B, Rouleau GA, Bouchard J-P, Fardeau M, Tomé FM. 1999. Oculopharyngeal muscular dystrophy. *Semin. Neurol.* 19:59–66

22. Brown C, ed. 1987. *The Illustrated History of Canada.* Toronto: Lester Orpen-Dennys

23. Brunzell JD, Deeb SS. 2001. Familial lipoprotein lipase deficiency, Apo C-II deficiency and hepatic lipase deficiency. See Ref. 151, pp. 2789–816

24. Byck S, Morgan K, Blanc L, Scriver CR. 1996. The *PAH* locus and population genetic variation: the Quebec example. *Am. J. Hum. Genet.* 59:A33(Abstr.)

25. Byck S, Morgan K, Tyfield L, Dworniczak B, Scriver CR. 1994. Evidence for origin, by recurrent mutation, of the phenylalanine hydroxylase *R408W* mutation on two haplotypes in European and Quebec populations. *Hum. Mol. Genet.* 3:1675–77

26. Carter KC, Byck S, Waters PJ, Richards B, Nowacki PM, et al. 1998. Mutation at the phenylalanine hydroxylase gene (*PAH*) and its use to document population genetic variation: the Quebec experience. *Eur. J. Hum. Genet.* 6:61–70

27. Casgrain B, Hubert M, Bouchard G, Roy R. 1991. Structure de gestion et d'exploitation du fichier-reseau BALSAC. See Ref. 16, pp. 47–71

28. Cavalli-Sforza L, Menozzi P, Piazza A. 1994. *The History and Geography of Human Genes.* Princeton, NJ: Princeton Univ. Press

29. Cavalli-Sforza LL, Piazza A. 1993. Human genomic diversity in Europe: a summary of recent research and prospects for the future. *Eur. J. Hum. Genet.* 1:3–18

30. Chapman NH, Thompson EA. 2001. Linkage disequilibrium mapping: the role of population history, size and structure. *Adv. Genet.* 42:413–37

31. Charbonneau H. 1984. Essai sur l'évolution démographique du Québec de 1534 à 2034. *Cah. Québ. Démogr.* 13:5–21

32. Cook R. 1987. The triumph and trials of materialism, 1900–1945. See Ref. 22 pp. 375–466

33. Cotton RGH, Scriver CR. 1998. Proof of "disease-causing" mutation. *Hum. Mutat.* 12:1–3

34. Couture P, Morrissette J, Gaudet D, Vohl M-C, Gagné C, et al. 1999. Fine mapping of low-density lipoprotein receptor gene by genetic linkage on chromosome

19p13.1–p13.3 and study of the founder effect of four French Canadian low-density lipoprotein receptor gene mutations. *Atherosclerosis* 143:145–51

35. Crosby AW. 1986. *Ecological Imperialism. The Biological Expansion of Europe 900-1900*, pp. 1–368. Cambridge, UK: Cambridge Univ. Press

36. Cystic Fibrosis Genet. Anal. Consort. 1994. Population variation of common cystic fibrosis mutations. *Hum. Mutat.* 4:167–77

37. Daigneault J, Aubin G, Simard F, De Braekeeler M. 1991. Genetic epidemiology of Cystic Fibrosis in Saguenay-Lac-St-Jean (Quebec, Canada). *Clin. Genet.* 40:298–303

38. De Braekeleer M. 1991. Hereditary disorders in Saguenay-Lac-St-Jean (Quebec, Canada). *Hum. Hered.* 41:141–46

39. De Braekeeler M. 1995. Geographic distribution of 18 autosomal recessive disorders in the French Candian population of Saguenay-Lac-Saint-Jean, Quebec. *Ann. Hum. Biol.* 22:111–22

40. De Braekeeler M, Daigneault J, Allard C, Simard F, Aubin G. 1996. Genealogy and geographic distribution of CFTR mutations in Saguenay-Lac-Saint-Jean (Quebec, Canada). *Ann. Hum. Biol.* 23:345–52

41. De Braekeeler M, Dao TN. 1994. Hereditary disorders in the French Canadian population of Quebec. I. In search of founders. *Hum. Biol.* 66:205–23

42. De Braekeeler M, Dao TN. 1994. Hereditary disorders in the French Canadian population of Quebec. II. Contribution of Perche. *Hum. Biol.* 66:225–49

43. De Braekeeler M, Dionne C, Gagné C, Julien P, Brun D, et al. 1991. Founder effect in familial hyperchylomicronemia among French Canadians of Quebec. *Hum. Hered.* 41:168–73

44. De Braekeeler M, Gauthier S. 1996. Autosomal recessive disorders in Saguenay-Lac-Saint-Jean (Quebec, Canada): a study of inbreeding. *Ann. Hum. Genet.* 60:51–56

45. De Braekeeler M, Gauthier S. 1996. Autosomal recessive disorders in Saguenay-Lac-St-Jean, Quebec: study of kinship. *Hum. Biol.* 68:371–81

46. De Braekeeler M, Giasson F, Mathiew J, Roy M, Bouchard J-P, Morgan K. 1993. Genetic epidemiology of autosomal recessive spastic ataxia of Charlevoix-Saguenay in northeastern Quebec. *Genet. Epidemiol.* 10:17–25

47. De Braekeeler M, Hechtman P, Andermann E, Kaplan F. 1992. The French Canadian Tay-Sachs disease deletion mutation: identification of probable founders. *Hum. Genet.* 89:83–87

48. De Braekeeler M, Landry T, Cholette A. 1994. Consanguinity and kinship in Down Syndrome in Saguenay-Lac-Saint-Jean (Quebec). *Ann. Genet.* 37:86–88

49. De Braekeeler M, Larochelle J. 1991. Population genetics of vitamin D–dependent rickets in northeastern Quebec. *Ann. Hum. Genet.* 55:283–90

50. De Braekeeler M, Mari C, Verlingue C, Allard C, Leblanc JP, et al. 1998. Complete identification of cystic fibrosis transmembrane conductance regulator mutations in the CF population of Saguenay-Lac-Saint-Jean (Quebec, Canada). *Clin. Genet.* 53:44–46

51. De Braekeeler M, Ross M. 1991. Inbreeding in Saguenay-Lac-St-Jean (Quebec, Canada): a study of Catholic church dispensations 1842–1971. *Hum. Hered.* 41:379–84

52. Degioanni A, Darlu P. 1994. Analysis of the molecular variance at the phenylalanine hydroxylase (PAH) locus. *Eur. J. Hum. Genet.* 2:166–76

53. Demers SI, Phaneuf D, Tanguay RM. 1914. Hereditary Tyrosinemia Type I: strong association with haplotype 6 in French Canadians permits simple carrier detection and prenatal diagnosis. *Am. J. Hum. Genet.* 55:327–33

54. Desjardins L, Rousseau C, Duplain J-M, Vallet J-P, Auger P. 1978. La Thalassémie

chez les Québécois Francophones. *Can. Med. Assoc. J.* 119:709–13

55. DiLella AG, Kwok SCM, Ledley FD, Marvit J, Woo SLC. 1986. Molecular structure and polymorphic map of human phenylalanine hydroxylase gene. *Biochemistry* 25:743–49

56. Dionne C, Gagné C, Julien P, Murthy MR, Lambert M, et al. 1992. Genetic epidemiology of lipoprotein lipase deficiency in Saguenay-Lac-St-Jean (Quebec, Canada). *Ann. Genet.* 35:89–92

57. Dobzhansky T. 1967. On genetic aspects of human evolution. In *Proc. 3rd Int. Congr. Hum. Genet.*, ed. JF Crow, JV Neel, pp. 361–65. Baltimore, MA: Johns Hopkins Press

58. Dunnen JT, Antonarakis SE. 2000. Mutation nomenclature extensions and suggestions to describe complex mutations: a discussion. *Hum. Mutat.* 15:7–12

59. Eisensmith RC, Goltsov AA, O'Neill C, Tyfield LA, Schwartz EI, et al. 1995. Recurrence of the R408W mutation in the phenylalanine hydroxylase locus in Europeans. *Am. J. Hum. Genet.* 56:278–86

60. Engert JC, Bérubé D, Mercier J, Doré C, Lepage P, et al. 2000. ARSACS, a spastic ataxia common in northeastern Quebec, is caused by mutations in a new gene encoding an 11.5-kb ORF. *Nat. Genet.* 24:120–25

61. Engert JC, Doré C, Mercier J, Ge B, Bétard C, et al. 1999. Autosomal recessive spastic ataxia of Charlevoix-Saguenay (ARSACS): high-resolution physical and transcript map of the candidate region in chromosome region 13q11. *Genomics* 62:156–64

62. Evans RG. 1988. "We'll take care of it for you." Health care in the Canadian community. *Daedalus* 117:155–89

62a. Falconer DS. 1989. *Introduction to Quantitative Genetics*. Menlo Park, CA: Addison-Wesley. 3rd ed.

63. Feingold J, Guilloud-Bataille M, Feingold N, Rey F, Berthelon M, Lyonnet S. 1993. Linkage disequilibrium in the human phenylalanine hydroxylase. *Dev. Brain. Dysfunct.* 6:26–31

64. Fernandes MJG, Kaplan F, Clow CL, Hechtman P, Scriver CR. 1992. Specificity and sensitivity of hexosaminidase assays and DNA analysis for the detection of Tay-Sachs disease gene carriers among Ashkenazi Jews. *Genet. Epidemiol.* 9:169–75

65. Fix A. 1999. *Migration and Colonization in Human Microevolution*. New York: Cambridge Univ. Press

66. Fraser D, Kooh SW, Kind HP, Holick MF, Tanaka Y, Deluca HF. 1973. Pathogenesis of hereditary vitamin D–dependent rickets: an inborn error of vitamin D metabolism involving defective conversion of 25-hydroxyvitamin D to $1\alpha,25$-dihydroxyvitamin D. *N. Engl. J. Med.* 289:817–22

67. Fumeron F, Grandchamp B, Fricker J, Krempf M, Wolf L-M, et al. 1992. Presence of the French Canadian deletion in a French patient with familial hypercholesterolemia. *N. Engl. J. Med.* 326:69

68. Gagné C, Brun L-D, Julien P, Moorjani S, Lupien PJ. 1989. Primary lipoprotein lipase activity deficiency: clinical investigation of a French Canadian population. *Can. Med. Assoc. J.* 140:405–11

69. Gaudet D, Tremblay G, Perron P, Moorjani S, Ouadahi Y, Moorjani S. 1995. L'hypercholestérolémie familiale dans l'est du Québec: une problème de santé publique? L'expérience de la clinique des maladies lipidique du Chicoutimi. *Union Méd. Can.* 124:54–60

70. Genest JJ Jr, Martin-Munley SS, McNamara JR, Ordovas JM, Jenner J, et al. 1992. Familial lipoprotein disorders in patients with premature coronary artery disease. *Circulation* 85:2025–33

71. Gentilcore RL, Measner D, Walder RH, Matthews GJ, Moldofsky B, eds. 1993. *Historical Atlas of Canada. Vol. II: The Land Transformed 1800–1891*. Toronto: Univ. Toronto Press

72. Glorieux FH, Arabian A, Delvin EE. 1995. Pseudo-vitamin D deficiency: absence of 25-hydroxyvitamin D 1α-hydroxylase activity in human placenta decidual cells. *J. Clin. Endocrinol. Metab.* 80:2255–58

73. Goldstein JL, Hazzard WR, Schrott WR, Bierman EL, Motulsky AG, et al. 1973. Hyperlipidemia in coronary heart disease. I. Lipid levels in 500 survivors of myocardial infarction. *J. Clin. Invest.* 52:1533–43

74. Goldstein JL, Hobbs HH, Brown MS. 2001. Familial hypercholesterolemia. See Ref. 151, pp. 2863–914

75. Goltsov AA, Eisensmith RC, Konecki DS, Lichter-Konecki U, Woo SLC. 1992. Associations between mutations and a VNTR in the human phenylalanine hydroxylase gene. *Am. J. Hum. Genet.* 51:627–36

76. Goltsov AA, Eisensmith RC, Naughton ER, Jin L, Chakraborty R, Woo SLC. 1993. A single polymorphic STR system in the human *phenylalanine hydroxylase* gene permits rapid prenatal diagnosis and carrier screening for phenylketonuria. *Hum. Mol. Genet.* 2:577–81

77. Gravel RA, Kaback MM, Proia RL, Sandhoff R, Suzuki K. 2001. The GM₂ gangliosidoses. See Ref. 151, pp. 3827–76

78. Grompe M, St-Louis M, Demers SI, Al-Dhalimy M, Leclerc B, Tanguay RM. 1994. A single mutation of the *fumarylacetoacetate hydrolase* gene in French Canadians with hereditary tyrosinemia type I. *N. Engl. J. Med.* 331:353–57

79. Guldberg P, Levy HL, Hanley WB, Koch R, Matalon R, et al. 1996. *Phenylalanine hydroxylase* gene mutations in the United States: report from the Maternal PKU Collaborative Study. *Am. J. Hum. Genet.* 59:84–94

80. Guldberg P, Rey F, Zschocke J, Romano C, Francois B, et al. 1998. A European multicenter study of phenylalanine hydroxylase deficiency: classification of 105 mutations and a general system for genotype-based prediction of metabolic phenotype. *Am. J. Hum. Genet.* 63:71–79

81. Harney RF. 1988. "So great a heritage as ours." Immigration and the survival of the Canadian polity. *Daedalus* 117:51–97

82. Harris RC, Matthews GJ, eds. 1987. *Historical Atlas of Canada. Vol. I. From the Beginning to 1800.* Toronto: Univ. Toronto Press

83. Hastbacka J, de la Chapelle A, Kaitila I, Sistonen P, Weaver A. 1992. Linkage disequilibrium mapping in isolated founder populations: diastrophic dysplasia in Finland. *Nat. Genet.* 2:204–11

84. Hayden MR, De Braekeeler M, Henderson HE, Castelein J. 1992. Molecular geography of inherited lipoprotein metabolism: lipoprotein lipase deficiency and familial hypercholesterolemia. In *Molecular Genetics of Coronary Artery Disease. Candidate Genes and Processes in Atherosclerosis,* ed. AJ Lusis, JI Rotter, RS Sparkes, 14:350–62. Basel: Karger

85. Hechtman P, Boulay B, De Braekeleer M, Andermann E, Melançon S, et al. 1992. The intron 7 donor splice site transition: a second Tay-Sachs disease mutation in French Canada. *Hum. Genet.* 90:402–6

86. Hechtman P, Kaplan F, Bayleran J, Boulay B, Andermann E, et al. 1990. More than one mutant allele causes infantile Tay-Sachs disease in French-Canadians. *Am. J. Hum. Genet.* 47:815–22

87. Heyer E. 1995. Genetic consequences of differential demographic behaviour in the Saguenay region, Quebec. *Am. J. Phys. Anthropol.* 98:1–11

88. Heyer E. 1999. One founder/one gene hypothesis in a new expanding population: Saguenay (Quebec, Canada). *Hum. Biol.* 71:99–109

89. Heyer E, Tremblay M. 1995. Variability of the genetic contribution of Quebec population founders associated to some deleterious genes. *Am. J. Hum. Genet.* 56:970–78

90. Heyer E, Tremblay M, Desjardins B. 1997. Seventeenth century European origins of hereditary diseases in the Saguenay population (Quebec, Canada). *Hum. Biol.* 69:209–25

91. Hobbs HH, Brown MS, Russell DW, Davignon J, Goldstein JL. 1987. Deletion in the gene for the low-density–lipoprotein receptor in a majority of French Canadians with familial hypercholesterolemia. *N. Engl. J. Med.* 317:734–37

92. John SWM, Rozen R, Laframboise R, Laberge C, Scriver CR. 1989. Novel PKU mutation on haplotype 2 in French-Canadians. *Am. J. Hum. Genet.* 45:905–9

93. John SWM, Rozen R, Scriver CR, Laframboise R, Laberge C. 1990. Recurrent mutation, gene conversion, or recombination at the human phenylalanine hydroxylase locus: evidence in French-Canadians and a catalog of mutations. *Am. J. Hum. Genet.* 46:970–74

94. Jomphe M, Bouchard G, Davignon J, De Braekeeler M, Gradie M, et al. 1988. Familial hypercholesterolemia in French Canadians: geographic distribution and centre of origin of an LDL deletion mutation. *Am. J. Hum. Genet.* 43:A216 (Abstr.)

95. Jorde LB. 1995. Linkage disequilibrium as a gene-mapping tool. *Am. J. Hum. Genet.* 56:11–14

96. Kaplan F, Kokotsis G, Capua A, Scriver CR. 1991. Quantification of β-thalassemia genes in Quebec immigrants of Mediterranean, Southeast Asian, and Asian Indian origin. *Clin. Invest. Med.* 14:325–30

97. Kaplan F, Kokotsis G, De Braekeeler M, Morgan K, Scriver CR. 1990. β-thalassemia genes in French-Canadians: haplotype and mutation analysis of Portneuf chromosomes. *Am. J. Hum. Genet.* 46:126–32

98. Kaplan NL, Lewis PO, Weir BS. 1994. Age of the ΔF508 cystic fibrosis mutation. *Nat. Genet.* 8:216–18

99. Karayan L, Qiu S, Betard C, Dufour R, Roederer G, et al. 1994. Response to HMG CoA reductase inhibitors in heterozygous familial hypercholesterolemia due to the 10-kb deletion ("French Canadian mutation") of the LDL receptor gene. *Arteroscler. Thromb.* 14:1258–63

100. Kayaalp E, Treacy E, Waters PJ, Byck S, Nowacki P, Scriver CR. 1997. Human phenylalanine hydroxylase mutations and hyperphenylalaninemia phenotypes: a metanalysis of genotype-phenotype correlations. *Am. J. Hum. Genet.* 61:1309–17

101. Kerem B-S, Rommens JM, Buchanan JA, Markiewicz D, Cox DK, et al. 1989. Identification of the cystic fibrosis gene: genetic analysis. *Science* 245:1073–80

102. Kerr D, Holdsworth DW, Laskin SL, Mathews GJ, eds. 1990. *Historical Atlas of Canada. Vol. III. Addressing the Twentieth Century. 1891–1961.* Toronto: Univ. Toronto Press

103. Khoury MJ, Burke W, Thomson EJ. 2000. Genetics and public health: a framework for the integration of human genetics into public health practice. In *Genetics and Public Health in the 21st Century. Using Genetic Information to Improve Health and Prevent Disease*, ed. MJ Khoury, W Burke, EJ Thomson, pp. 3–23. Oxford: Oxford Univ. Press

104. Kidd JR, Pakstis AJ, Zhao H, Lu R-B, Okonofua FE, et al. 2000. Haplotypes and linkage disequilibrium at the phenylalanine hydroxylase locus, *PAH*, in a global representation of populations. *Am. J. Hum. Genet.* 66:1882–99

105. Kruglyak L. 1999. Genetic isolates: separate but equal? *Proc. Natl. Acad. Sci. USA* 96:1170–72

106. Kruglyak L. 1999. Prospects for whole-genome linkage disequilibrium mapping of common disease genes. *Nat. Genet.* 22:139–44

107. Laberge C. 1969. Hereditary tyrosinemia in a French Canadian isolate. *Am. J. Hum. Genet.* 21:36–45

108. Laberge C. 2000. La médecine génétique en début de siècle. *Rech. Santé* 24:26–29

109. Labuda M, Labuda D, Korab-Laskowska M, Cole DEC, Zietkiewicz E, et al. 1996. Linkage disequilibrium analysis in young populations: pseudo-vitamin D–deficiency rickets and the founder effect in French Canadians. *Am. J. Hum. Genet.* 59:633–43

110. Lahr MM. 2000. Wandering genes. *Science* 289:2057

111. Lambert DM. 1994. *The genetic epidemiology of hyperphenylalaninemia in Quebec.* PhD thesis. McGill Univ., Montreal

112. Larochelle J, Mortezai A, Belanger M, Tremblay M, Claveau JC, Aubin G. 1967. Experience with 37 infants with tyrosinemia. *Can. Med. Assoc. J.* 97:1051–54

113. Leitersdorf E, Tobin EJ, Davignon J, Hobbs HH. 1990. Common low-density lipoprotein receptor mutations in the French Canadian population. *J. Clin. Invest.* 85:1014–23

114. Liberman UA, Marx SJ. 2001. Vitamin D and other calciferols. See Ref. 151, pp. 4223–40

115. Lichter-Konecki U, Schlotter M, Konecki DS. 1994. DNA sequence polymorphisms in exonic and intronic regions of the human phenylalanine hydroxylase gene aid in the identification of alleles. *Hum. Genet.* 94:307–10

116. Lidsky AS, Ledley FD, DiLella AG, Kwok SCM, Daiger SP, et al. 1985. Extensive restriction site polymorphism at the human phenylalanine hydroxylase locus and application in prenatal diagnosis of phenylketonuria. *Am. J. Hum. Genet.* 37:619–34

117. Lindblad B, Lindstedt S, Steen G. 1967. On the enzymic defects in hereditary tyrosinemia. *Proc. Natl. Acad. Sci. USA* 74:4641–45

118. Lyonnet S, Melle D, DeBrakeleer M, Laframboise R, Rey F, et al. 1992. Time and space clusters of the French-Canadian M1V phenylketonuria mutation in France. *Am. J. Hum. Genet.* 51:191–96

119. Ma Y, Betard C, Roy M, Davignon J, Kessling AM. 1989. Identification of a second French-Canadian LDL receptor gene deletion and development of a rapid method to detect both deletions. *Clin. Genet.* 36:219–28

120. Martino T, Kaplan F, Diamond S, Oppenheim A, Scriver CR. 1997. Probable identity by descent and discovery of familial relationships by means of a rare β-thalassemia haplotype. *Hum. Mutat.* 9:86–87

121. McEvedy C, Jones R. 1978. *Atlas of World Population History*, pp. 283–85. London, UK: Penguin

122. Mitchell GA, Grompe M, Lambert M, Tanguay RM. 2001. Hypertyrosinemia. See Ref. 151, pp. 1777–805

123. Mitchell JJ, Capua A, Clow C, Scriver CR. 1996. Twenty-year outcome analysis of genetic screening programs for Tay-Sachs and β-thalassemia disease carriers in high schools. *Am. J. Hum. Genet.* 59:793–98

124. Monsalve MV, Henderson H, Roederer G, Julien P, Deeb S, et al. 1990. A missense mutation at codon 188 of the human lipoprotein lipase gene is a frequent cause of lipoprotein lipase deficiency in persons of different ancestries. *J. Clin. Invest.* 86:728–34

125. Moore C. 1987. Illustrated history of Canada. See Ref. 22, pp. 105–88

126. Moorjani S, Roy M, Gagné C, Davignon J, Brun D, et al. 1989. Homozygous familial hypercholesterolemia among French Canadians in Québec province. *Arteriosclerosis* 9:211–16

127. Morison SE. 1972. *Samuel de Champlain. Father of New France.* Boston, MA: Little, Brown

128. Morral N, Bertranpetit J, Estivill X, Nunes V, Casals T, et al. 1994. The origin of the major cystic fibrosis mutation (ΔF508) in European populations. *Nat. Genet.* 7:169–75

129. Myerowitz R, Hogikyan ND. 1987. A deletion involving *Alu* sequences in the

β-hexosaminidase α-chain of French Canadians with Tay-Sachs disease. *J. Biol. Chem.* 262:15396–99

130. Normand T, Bergeron J, Fernandes-Margallo T, Bharucha A, Ven Murthy MR, et al. 1992. Geographic distribution and genealogy of mutation 207 of the lipoprotein lipase gene in the French Canadian population of Quebec. *Hum. Genet.* 89:671–75

131. Oppenheim A, Oron V, Filon D, Fearon CC, Rachmilewitz EA, et al. 1993. Sporadic alleles, including a novel mutation, characterize β-thalassemia in Ashkenazi Jews. *Hum. Mutat.* 2:155–57

132. Phaneuf D, Labelle Y, Bérubé D, Arden K, Cavanee W, Gagné R. 1991. Cloning and expression of the cDNA encoding human fumarylacetoacetate hydrolase, the enzyme deficient in hereditary tyrosinemia: assignment of the gene to chromosome 15. *Am. J. Hum. Genet.* 48:525–35

133. Piazza A. 1993. Who are the Europeans? *Science* 260:1767–68

133a. Pier GB. 2000. Role of the cystic fibrosis transmembrane conductance regulator in innate immunity of *Pseudomonas aeruginosa* infections. *Proc. Natl. Acad. Sci. USA* 97:8822–28

134. Pourdier J, St-Louis M, Lettre F, Gibson K, Prévost C, et al. 1996. Frequency of the IVS12_5G->A splice mutation of the *fumarylacetoacetate hydroxylase* gene in carriers of hereditary tyrosinemia in the French Canadian population of Saguenay-Lac-St-Jean. *Prenat. Diagn.* 16:59–64

135. Prévost C, Laframboise R, Bardanis M, Clow C, Lancaster G, et al. 1988. Le gène de la β-thalassémie au Canada français: relance dans le comté de Portneuf. *Union Méd. Can.* 118:242–44

136. Richter A, Rioux JD, Bouchard J-P, Mercier J, Mathieu J, et al. 1999. Location score and haplotype analyses of the locus for autosomal recessive spastic ataxia of Charlevoix-Saguenay, in chromosome region 13q11. *Am. J. Hum. Genet.* 64:768–75

137. Riordan JR, Rommens JM, Kerem B-S, Alon N, Rozmahel R, et al. 1989. Identification of the cystic fibrosis gene: cloning and characterization of complementary DNA. *Science* 245:1066–73

138. Rivard SR, Allard C, Leblanc J-P, Milot M, Aubin G, et al. 2000. Correlation between mutations and age in cystic fibrosis in a French Canadian population. *J. Med. Genet.* 37:225–27

139. Rommens JM, Iannuzzi MC, Kerem B-S, Drumm ML, Melmer G, et al. 1989. Identification of the cystic fibrosis gene: chromosome walking and jumping. *Science* 245:1059–66

140. Rootwelt H, Kvittingen EA, Hoie K, Agsteribbe E, Hartog M, et al. 1992. The human *fumaryllacetoacetase* gene: characterisation of restriction fragment length polymorphisms and identification of haplotypes in tyrosinemia type 1 pseudodeficiency. *Hum. Genet.* 89:229–33

141. Rozen R, De Braekeeler M, Daigneault J, Ferreira-Rajabi L, Gerdes M, et al. 1992. Cystic fibrosis mutations in French Canadians: three CFTR mutations are relatively frequent in a Quebec population with an elevated incidence of cystic fibrosis. *Am. J. Hum. Genet.* 42:360–64

142. Rozen R, Ferreira-Rajabi L, Robb L, Colman N. 1995. L206W mutation of the cystic fibrosis gene, relatively frequent in French Canadians, is associated with atypical presentations of cystic fibrosis. *Am. J. Med. Genet.* 57:437–39

143. Rozen R, Schwartz RH, Hilman BC, Stanislovitis P, Horn GT, et al. 1990. Cystic fibrosis mutations in North American populations of French ancestry: analysis of Quebec French-Canadian and Louisiana Acadian families. *Am. J. Hum. Genet.* 47:606–10

144. Rudin R. 1985. *The Forgotten Quebecers. A History of English-Speaking Quebec. 1759–1980.* Quebec: Inst. Que. Rech. Cult.

145. Schoshani T, Augarten A, Gazit E, Bashan N, Yahav Y, et al. 1992. Association of a nonsense mutation (W1282X), the most common mutation in the Ashkenazi Jewish cystic fibrosis patients in Israel, with presentation of severe disease. *Am. J. Hum. Genet.* 50:222–28

146. Scriver CR. 1967. The phenotypic manifestations of hereditary tyrosinemia and tyrosyluria. *Can. Med. Assoc. J.* 97:1073–76

147. Scriver CR. 1970. Vitamin D dependency. *Pediatrics* 45:361–63

148. Scriver CR. 1971. Fondements biologiques de la sensibilité du rachitisme à la vitamine D. *Union Med. Can.* 100:462–74

149. Scriver CR. 1988. Cases are not incidence and vice versa. *Genet. Epidemiol.* 5:481–87

150. Scriver CR. 1988. Human genes: determinants of sick populations and sick patients. *Can. J. Public Health* 79:222–24

151. Scriver CR, Beaudet AL, Sly WS, Valle D, eds. 2001. *The Metabolic and Molecular Bases of Inherited Disease.* New York: McGraw-Hill. 8th ed.

152. Scriver CR, Byck S, Prevost L, Hoang L. 1996. The phenylalanine hydroxylase locus: a marker for the history of phenylketonuria and human genetic diversity. In *Variation in the Human Genome. Ciba Found. Symp. No. 197*, ed. KM Weiss, pp. 73–96. Chichester: Wiley

153. Scriver CR, Fujiwara TM. 1992. Invited editorial: cystic fibrosis genotypes and views on screening and both heterogeneous and population related. *Am. J. Hum. Genet.* 51:943–50

154. Scriver CR, John SMW, Rozen R, Eisensmith R, Woo SLC. 1993. Associations between populations, phenylketonuria mutations and RFLP haplotypes at the phenylalanine hydroxylase locus: an overview. *Dev. Brain Dysfunct.* 6:11–25

155. Scriver CR, Kaufman S. 2001. Hyper-phenylalaninemia: phenylalanine hydroxylase deficiency. See Ref. 151, pp. 1667–724

156. Scriver CR, Laberge C, Clow CL, Fraser FC. 1978. Genetics in medicine. An evolving relationship. *Science* 200:946–52

157. Scriver CR, Nowacki PM. 1999. Genomics, mutations and the internet: the naming and use of parts. *J. Inherit. Metab. Disord.* 22:519–30

158. Scriver CR, Tenenhouse HS. 1981. On the heritability of rickets, a common disease (Mendel, mammals and phosphate). *Johns Hopkins Med. J.* 149:179–87

159. Scriver CR, Waters PJ. 1999. Monogenic traits are not simple. Lessons from phenylketonuria. *Trends Genet.* 15:267–72

160. St Arnaud R, Messerlian S, Moir JM, Omdahl JL, Glorieux FH. 1997. The 25-hydroxyvitamin D 1 α-hydroxylase gene maps to the pseudovitamin D-deficiency rickets (PDDR) disease locus. *J. Bone Miner. Res.* 12:1552–59

161. Thompson EA, Neel JV. 1997. Allelic disequilibrium and allele frequency distribution as a function of social and demographic history. *Am. J. Hum. Genet.* 60:197–204

162. Treacy E, Byck S, Clow C, Scriver CR. 1993. Celtic phenylketonuria chromosomes found? Evidence in two regions of Quebec province. *Eur. J. Hum. Genet.* 1:220–28

163. van Amstel JKP, Bergman AJI, van Beurden EA, Roijers JFM, Peelen T, et al. 1996. Hereditary tyrosinemia type I: novel missense, nonsense and splice consensus mutations in the human *fumarylacetoacetate hydrolase* gene; variability of the genotype-phenotype relationship. *Hum. Genet.* 97:51–59

164. Vohl M-C, Moorjani S, Roy M, Gaudet D, Torres AL, et al. 1997. Geographic distribution of French Canadian low density lipoprotein receptor gene mutations in the province of Quebec. *Clin. Genet.* 52:1–6

165. Waite P. 1987. Between three oceans: challenges of a continental destiny, 1840–1900. See Ref. 22, pp. 279–374

166. Weatherall DJ, Clegg JB, Higgs DR, Wood WG. 2001. The hemoglobinopathies. See Ref. 151, pp. 4571–636

167. Weiss KM. 1996. Is there a paradigm shift in genetics? Lessons from the study of human diseases. *Mol. Phylogenet. Evol.* 5:259–65

168. Welsch MJ, Ramsey BW, Accuraso F, Cutting GR. 2001. Cystic fibrosis. See Ref. 151, pp. 5121–88

169. Wood S, Schertzer M, Hayden M, Ma Y. 1993. Support for founder effect for two lipoprotein lipase (LPL) gene mutations in French Canadians by analysis of GT microsatellites flanking the LPL gene. *Hum. Genet.* 91:312–16

170. Wynn G. 1987. On the margins of empire, 1760–1840. See Ref. 22, pp. 189–278

ADDENDUM

The HHH Syndrome (hyperornithinaemia-hyperammonaemia-hypercitrullinuria, OMIM 238970), an inborn error or urea cycle function, results from mutations in the *ORNTI* gene (locus 13q14) encoding the inner mitochondrial membrane ornithine transporter. The disease clusters in French Canadians where the F188 deletion mutation accounts for 19 of the 20 mutant alleles. Ancestry of probands maps to various regions in Southern Quebec, but not to the northeast region. Here is further evidence that Quebec province harbors genetically diverse subpopulations of French Canadians. (Camacho JA, Obie C, Biery B, Goodman BK, Hu C-A, et al. 1999. Hyperornithinaemia-hyperammonaemia-homocitrullinuria syndrome is caused by mutations in a gene encoding a mitochondrial ornithine transporter. *Nat. Genet.* 22:151–58.)

Annu. Rev. Genomics Hum. Genet. 2001. 2:103–28

HUMAN POPULATION GENETICS:
Lessons from Finland

Juha Kere
Finnish Genome Center, University of Helsinki, Helsinki 00014, Finland;
e-mail: juha.kere@helsinki.fi

Key Words disease gene, genetic mapping, genetic marker, recessive inheritance, common disease

■ **Abstract** A population of about 5 million at the northern corner of Europe is unlikely to arouse the attention of the human genetics community, unless it offers something useful for others to learn. A combination of coincidences has finally made this population one that, out of proportion for its size, has by example shaped research in human disease genetics. This chapter summarizes advances made in medical genetics that are based on research facilitated by Finland's population structure. The annotation of the human genome for its polymorphism and involvement in disease is not over; it is, therefore, of interest to assess whether genetic studies in populations such as the Finnish might help in the remaining tasks.

INTRODUCTION

Finland as a Model Population for Human Genetics

Finland is inhabited by a population that is readily distinguished from other European populations by its unusual language, which does not belong to the large family of Indoeuropean languages. Finnish and three other major languages, Hungarian, Estonian, and Saame, stem from the Uralic language family. This geographically marginal population belongs to the genetically tight cluster of European populations (5).

Although the relationship of Finns to the other European populations has been the subject of many population genetic studies, less emphasis is often put on the internal structure of the population, even though it has current practical importance. Isolated founder populations have become emphasized as fields for the study of common diseases and genetic susceptibility (17, 42, 92). I believe that it is important to regard such research from a broad population point of view that could help to avoid many pitfalls and, thus, arrive at more likely interpretations of the data. Therefore, in this review I concentrate primarily on medical genetics, its

1527-8204/01/0728-0103$14.00
103

lessons about recent genetic history, and the structure of the present population, with particular attention to genes in disease.

Geography and Climate

Some background information is useful to understand basic ideas on population structure. Finland spans from 59° to 70° northern latitude. This location puts Finland's capital, Helsinki, at the same latitude as Anchorage, Alaska. The climate of Finland is, however, much milder. In southern Finland, the average temperature in July is about 17°C, whereas the average winter temperatures remain below 0°C from December to March. Modern agriculture yields two crops of wheat and one crop of barlow, oats, and rye, but no corn or rice. During recorded history, poor years have caused famines several times, the last one in the 1860s.

Finland has land boundaries with Russia, Norway, and Sweden, and a seashore that spans about a third of its circumference. The area of Finland is 338,000 sq km (131,000 sq mi), ranking in size just after Alaska, Texas, California, and Montana, or between Italy or Poland on one hand and Germany on the other. Most of the area is covered by forest (68%) or wetland (11%), and a large fraction by numerous lakes (9.9%), leaving 11% for agriculture or built-up area. The average population density is presently about 17 inhabitants per sq km; however, this is not a relevant figure because the population is very unevenly spread. The northernmost half is inhabited by 13% of the population, whereas 27% of the population reside in the six largest cities (79).

Language as a Cultural Barrier

The Finnish language is completely unintelligible to all its neighbors except Estonians. Throughout historical times, the linguistic barrier has discouraged immigration, which has not exceeded the annual level of 0.2%, first reached in the 1960s. In addition to Finnish, native minority populations have Swedish (5.7%) or Saami (0.03%) as their mother tongues; other languages are spoken by 1.4% of the population (79).

Genetic admixture between the Swedish-speaking and Finnish-speaking populations has occurred, even though the linguistic cultures have remained distinct (88). The Saami population, residing presently in the northernmost part of Finland, represents a distinct population within Finland. It seems to have avoided admixture with the southern Finnish population in spite of the linguistic similarity and remains an outlier in the family tree of European populations (5, 35–37).

Early History

Although archaelogical signs of human activity date back more than 9000 years, the oldest archaelogical signs of agriculture are from the end of the Stone Age, 3300 to 4000 years ago (59). The southern and western coastal regions have had permanent settlement for at least 2000 years, whereas vast parts of the country were inhabited permanently only after A.D. 1550 (Figure 1).

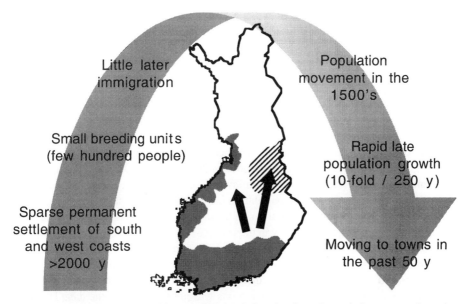

Figure 1   A brief population history of Finland. Starting from lower left corner and moving clockwise, the main features of population history are listed. The map shows the coastal region settled permanently before A.D. 1550 (dotted area), and the arrows indicate population movement in the sixteenth and seventeenth centuries. The province of Kainuu is highlighted (striped area).

A question of much local interest concerns the origin of Finns, but an embraced understanding has not been achieved among archaelogists, linguists, and geneticists. The problems are obvious: Although archaelogical findings provide indisputable evidence for settlement, it is often difficult to tell who the people were, i.e., to distinguish cultural transfer from population movement. A similar problem concerns linguistic interpretations. Genetic evidence pointing to a small relationship between the Finnish and Saami populations is not readily evident from their close linguistic relationship (74). Finally, the continuity of sparse populations in contrast to replacement by new populations is a difficult question to solve. Genetic evidence can seldom help with genes that were lost, thus biasing its view of history.

Summarizing much of the population genetic data from Europe, Cavalli-Sforza et al (5) place Finns closest to their geographical neighbors. Recent evidence from Y-chromosomal variation in Finland has indicated remarkably low diversity, suggesting a narrow bottleneck, whereas mitochondrial diversity is broader (29, 37, 75). A European founder effect may also have its origins in early Uralic populations. A single mutation has been found in the chemokine receptor 5 gene (CCR5) that confers protection against HIV-1 infections. In 18 European populations studied, the δ-CCR5 variant is embedded in a haplotype that is otherwise rare.

The highest frequencies of the variant were found in the Finnish and Mordvinian populations (16%), and the lowest in Sardinia (4%), suggesting a northeastern origin (47).

POPULATION STRUCTURE

Isolation by Density

Before modern times, inhabitation concentrated around waterways—sea, lakes, and rivers—that allowed easier communication and trade. A patchwork of forests and wetland was rich in game but uninviting for traffic, even after large areas were cleared for agriculture. The population has been estimated at about 250,000 until the 1600s. Even with consideration for the large parts of the country that were uninhabited, the population density in those times was very low. Walking and paddling in summer and skiing in winter were the most common means of communication, which must have had strong social and, thus, genetic consequences.

Indeed, mating probably occurred in units that consisted of hundreds, rather than thousands, of individuals; panmixis was not the word of the day. This concept has been called isolation by density, or dynamic isolation (56), and it refers to multiple small breeding units and little gene flow between the units. Small breeding units and little external gene flow allowed genetic drift to play with gene frequencies.

Population Expansion

Much of the population spreading occurred in the seventeenth century, when the previous wildmarks were settled permanently (Figure 1). Finland's population has grown at an unusually high rate, approximately 10-fold, during the past 250 years (Figure 2). The growth rates in Finland through most of the 1800s are comparable to those of the fastest-growing populations of Asia today. The oldest census records available show that from 1751–1760, the population averaged 457,000, with 20,500 live births and 13,300 deaths annually, yielding a growth figure of 1.5% per year. By comparison, from 1991–1995, the same figures were 5,064,000; 65,100; 49,500; and 0.45%, respectively. The proportion of children <15 years was 37% in 1751 and 19% in 1995 (79). Until World War II, the population remained mostly rural, and there is direct evidence that the breeding units remained small, continuing the concept of dynamic isolation with the increased population density that was, however, still low in absolute terms.

Some direct evidence for breeding unit size comes from studies by Nevanlinna (56) who studied population registries to assess the geographical origins of marriages. His study of marital habits in Hirvensalmi (a community in central southern Finland) during two periods in the 1800s showed that half of the marriages were between individuals from the same village (each village had 100–500 inhabitants). Based on these and other observations, Nevanlinna estimated the breeding unit size

Population of Finland 1750-1996

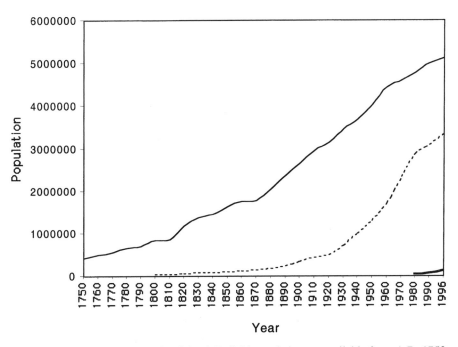

Figure 2   Population growth in Finland. Reliable statistics are available from A.D. 1750 onward. The solid line shows total population, and the dotted line represents urban population. The bold line (bottom right corner) depicts the population born outside Finland.

at 200 to 600 individuals. Subsequent studies have been published from other communities with different geographical structures (73). Cousin marriages, however, have been remarkably rare, indicating active avoidance as compared to random mating (18). There is reason to believe that these patterns prevailed through much of the period of rapid population growth.

Founder Effects and Drift

What should happen with genes in such circumstances? In a computer simulation using POPULUS software (V. Ollikainen & H. Mannila, unpublished), 50 founders were expanded to a population of 100,000 in 20 generations (corresponding to an annual growth rate of about 1.5%). The expanding population was allowed to mate randomly. The fate of each of the founding chromosomes was followed for one locus, and the effects of random drift (loss and enrichment of alleles) were recorded. The simulation was repeated 1000 times; the results are shown in Figure 3. Starting from a frequency of 1% for each allele (100 founding chromosomes), a rapid enrichment occurred for some alleles, reaching 6% for the most common allele

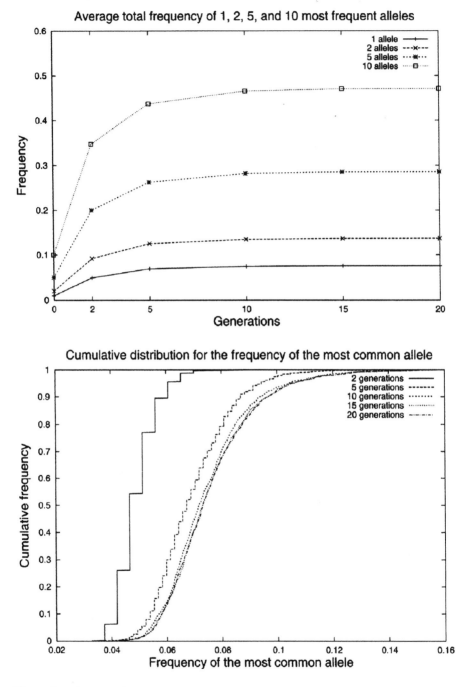

Figure 3 *(Continued)*

Cumulative distribution for the number of alleles remaining in the population

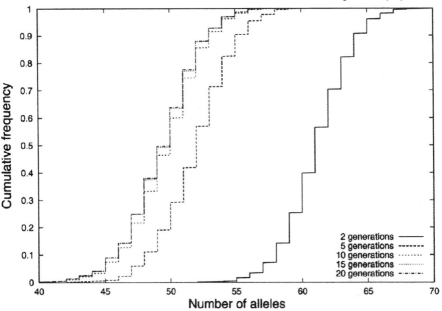

Figure 3 Effect of drift in a small expanding population. A simulation was performed using the POPULUS software (V. Ollikainen & H. Mannila, unpublished). 50 founders (100 founding alleles) expanded to 100,000 over 20 generations. Fate of alleles was recorded at 2, 5, 10, 15, and 20 generations (population size 107, 334, 2236, 14,954, and 100,000, respectively). The top panel shows increase in frequency for the most common alleles (average from 1000 simulations). The middle panel shows the frequency distribution of the most common allele in 1000 simulations. In all simulations, the most common allele was enriched almost 5-fold and, though rarely, up to 15-fold. The bottom panel shows the number of alleles remaining in the population in 1000 simulations. Only 40%–57% of the founder alleles survived in the population.

and levels of over 40% for the 10 most common alleles. The enrichment was ac-companied by the rapid loss of most alleles, so that 40% to 57% of alleles survived after 10 generations in different simulations (Figure 3). This simulation illustrates what strong effects random drift may have in small expanding populations.

The Example of Kainuu

The province of Kainuu has become one of the model populations for genetic studies. It is located at the narrowest middle part of Finland, extending roughly from the midline to the eastern border and spanning about 200 km vertically (Figure 1). Approximately 95% of the total area is used for forestry. Kainuu is sparsely inhabited, with only 4.1 inhabitants/sq km.

The population structure and regional history of Kainuu have been extensively studied (24, 27). First signs of permanent settlement, attained by pollen analysis, date back to the eighth century, but the region remained largely uninhabited until the sixteenth century. Settlements were totally destroyed during the Russian war, 1570–1595, and resettlement continued until the mid-seventeenth century, mainly from the Savo region (75%) but also from Ostrobothnia (25%). Population data from those dates come from account books, and population has been estimated at 1400 to 2700 persons, in about 240 households, before the 1650s. After that period, church records are also available for accurate figures. Over the period 1880–1963, the population grew faster than in the rest of Finland, but since that time, emigration has led to a decrease in the population.

Immigration was low during the entire period of rapid growth, consisting in large part of return migration by people born in the region. The region is further characterized by extremely stable ecclestical and administrative boundaries that date back to 1599 (Paltamo parish). Throughout its history, the region has belonged to the same bishopric, province, court of appeal and of lower appeal, and inferior court (21). The eastern boundary still follows the line drawn after the Russian War in 1595. In 1996, the population in the region was 94,000, including the town of Kajaani (population 37,000).

MOLECULAR GENETIC MAKEUP OF THE FINNISH POPULATION

Common and Rare Alleles

Nevanlinna (56) used common and rare blood group antigens to study genetic drift and assess the genetic makeup of different communities within Finland. He chose seven communities from four non-neighboring locations and measured allele frequencies for 10 loci in carefully ascertained samples. The results show unmistakably the effect of genetic drift: Gene frequencies in each of the seven communities varied up to over six standard deviations around the mean. At the same time, the gene frequencies in individual communities distributed symmetrically around the national frequency estimates. These data were interpreted to indicate that individual communities had a common seed population from which the subisolates were formed by drift. Nevanlinna estimated further that about 5% of subisolates had undergone gene enrichments greater than six standard deviations (57). Similar wide geographical variations for allele frequencies have been observed for HLA antigens in a large set of 10,000 subjects (78).

For common alleles, the overall frequencies in Finland are similar to other European countries. This is most easily seen now for the numerous microsatellite marker alleles that have been used in various gene mapping studies elsewhere and in Finland. However, a closer look at small subisolates reveals variations that may be significant even though they are not as numerically dramatic as those for rare genes. For example, the frequency of the common blood group allele A1, with an overall frequency of 0.213 in Finland, varied between 0.136 (−3 SD) and 0.253 (+2.5 SD) in the seven communities studied by Nevanlinna (56). At the province level, the gene frequencies for common alleles are already closer to the mean.

The small number of settlers and the small size of the breeding units caused a total loss of very rare alleles in most subisolates. Nevanlinna observed that the frequency of genes found in southern Finland at frequencies of 1/500 or less were completely absent from the parts of Finland that were settled later. On the other hand, genes with frequencies in the few percent range often showed higher than 10-fold differences in frequency among the seven different communities that he studied (56). The genes causing rare recessive Finnish diseases belong to this category and consequently show the typical uneven distributions for parental or grandparental birthplaces. In an ongoing study, we are using microsatellite markers and SNPs to assess the distribution of various rare alleles further (M.L. Savontaus, P. Lahermo, P. Sistonen, E. Salmela, & J. Kere, unpublished).

Finnish Disease Heritage

Finland was called the "promised land of rare hereditary traits" by Norio, Nevanlinna, and Perheentupa (58), but the idea was better expressed in the term "Finnish disease heritage," which is used to refer to the recessive diseases occurring in Finland at higher than usual frequencies (as compared to other European countries). The list of diseases now includes 35 entities (Table 1), with five of them added in the 1990s. Those include tibial muscular dystrophy (OMIM 253600), a form of infantile cerebellooptic atrophy (OMIM 260565), northern epilepsy (OMIM 600143), gonadal dysgenesis (OMIM 233300), and a lethal metabolic syndrome (OMIM 603358).

The delineation of these genetic disorders as new disease entities was made in all cases by observant clinicians who carefully excluded known causes and paid attention to the family histories of individual cases. These successes suggest that new diseases awaiting discovery are likely to also occur in other countries, especially when a subisolate structure has allowed the enrichment of recessive mutations. All but two of the disease genes have been mapped, and the gene and its common mutations are known for 22 diseases. The successes in positional cloning have been recently reviewed in more detail (6, 68, 69).

Some unusual incidences characterized the positional cloning of different disease genes, often progressing on neighboring benches or in neighboring institutes. As shown in Table 1, some genes for completely unrelated diseases mapped very close to each other (e.g., CSTB and AIRE; MUL and MKS1), promoting joint efforts in physical mapping. In another coincidence, genes for very phenotypically different disorders turned out to belong to the same gene family (SLC26A2, mutated in diastrophic dysplasia, and SLC26A3, mutated in congenital chloride diarrhea). Finally, for more than one disease, the mutated gene happened to carry

TABLE 1   Diseases commonly listed for Finnish disease heritage*

Disease	Gene, position	Main mutation (>75%)	OMIM #
Autosomal recessive, incidence >1:10,000			
Congenital nephrosis, Finnish type	NPHN, 19q31.1	2-BP DEL, 121CT	256300, 602716
Autosomal recessive, incidence 1:10,000–1:40,000			
Aspartyl-glucosaminuria	AGA, 4q32-q33	CYS163SER	208400
Infantile neuronal ceroid lipofuscinosis; Santavuori-Haltia disease	PPT1, 1p32	ARG122TRP	256730, 600722
Meckel syndrome, type 1	MKS1, 17q22-q23	?	249000
Hydrolethalus syndrome	11q23-q25	?	236680
Diastrophic dysplasia	SLC26A2, 5q32-33.1	−26, T-C, +2 (GT-to-GC transition in a splice donor site)	222600
Cartilage-hair hypoplasia	CHH, 9p13	?	250250
Ovarian dysgenesis; XX gonadal dysgenesis	FSHR, 2p21-p16	ALA189VAL	233300, 136435
Myoclonic epilepsy of Unverricht and Lundborg	CSTB, 21q22.3	12-mer expansion in promoter	254800, 601145
Lethal congenital contracture syndrome; Herva disease	LCCS, 9q34	?	253310
Autoimmune polyendocrinopathy syndrome, type I	AIRE, 21q22.3	ARG257TER	240300
Salla disease; sialuria, Finnish type	SLC17A5, 6q14-q15	ARG39CYS	604369, 604322
Congenital chloride diarrhea	SLC26A3, 7q22-q31.1	VAL317DEL	214700, 126650
Autosomal recessive, incidence <1:40000			
Mulibrey nanism	MUL, 17q22-q23	5-BP DEL, NT493–497	253250, 605073
Nonketotic hyperglycinemia	GLDC, 9p22	SER564ILE	238300
Peho syndrome	?	?	260565
Ornithine aminotransferase deficiency	OAT, 10q26	ARG180THR	258870
Lysinuric protein intolerance	SLC7A7, 14q11.2	1181, A-T, -2 (10-bp deletion beginning at 1181)	222700, 603593
Usher syndrome, type III	USH3, 3q21-q25	?	276902
Cohen syndrome	COH1, 8q22-q23	?	216550

TABLE 1 (*Continued*)

Disease	Gene, position	Main mutation (>75%)	OMIM #
Cornea plana type 2	KERA, 12q21.3-q22	ASN247SER	217300, 603288
Infantile-onset spinocerebellar ataxia	IOSCA, 10q24	?	271245
Tibial muscular dystrophy	TMD, 2q11	?	600334
Hereditary fructose intolerance	ALDOB, 9q22.3	Multiple	229600
Imerslund-Grasbeck syndrome; megaloblastic anemia type 1	CUBN, 10p12.1	PRO1297LEU	261100, 602997
Northern epilepsy; progressive epilepsy with mental retardation	CLN8, 8pter-p22	ARG24GLY	600143
Finnish lethal neonatal metabolic syndrome	FLNMS, 2q33-q37	?	603358
Muscle-eye-brain disease	MEB, 1p34-p32	?	253280
Polycystic lipomembranous osteodysplasia with sclerosing Leuko-encephalopathy; Nasu-Hakola syndrome	TYROBP, 19q31.1	EX1-4 DEL	221770, 604142
Neuronal ceroid lipofuscinosis, late infantile type, Finnish variant	CLN5, 13q21.1-q32	2467AT DEL, TER	256731
Lactase deficiency	2q21	?	223000
Rapadilino syndrome	?	?	266280
Autosomal dominant			
Finnish type amyloidosis; amyloidosis V	GSN, 9q34	ASP187ASN	105120, 137350
X-linked recessive			
Choroideremia	CHM, Xq21.2	Multiple	303100
X-linked juvenile retinoschisis 1	RS1, Xp22.2-p22.1	Multiple	312700

*The list includes 32 autosomal recessive diseases, 1 dominant, and 2 X chromosomal recessive. The autosomal recessive diseases are listed in approximate order of incidence; for the rarest diseases, the order is arbitrary. A single predominant founder mutation (>75% of chromosomes) is listed; for other entries, the mutation(s) remain unknown (?) or there are multiple mutations. For references on individual diseases, OMIM entry numbers are given (52).

an associated polymorphism (AGA, SLC26A3). Distinguishing the functional mutation from the polymorphism was finally accomplished by functional assays.

On the other hand, random sampling and the subsequent drift have produced a Finnish disease heritage with unexpectedly low frequencies of some genes that are much more common elsewhere. The most prominent examples include phenylketonuria, of which only four cases have ever been diagnosed in Finland (10), and cystic fibrosis, which shows regional founder effects with different mutations (28).

Other Recessive Diseases

All recessive diseases that occur in Finland, however, do not show as much remarkable homogeneity of mutations as many of the 35 Finnish diseases. There are already many diseases in Finland where multiple mutations have been identified. Some of these examples include diseases of the Finnish heritage: The first to be identified was ornithine aminotransferance deficiency (OMIM 258870) where the most prevalent Finnish mutation, Leu402Pro, accounted for only ∼85% of all mutations. At the time of its finding, this news was surprising; but investigators quickly understood that the relative enrichment of one mutation would serve to pick up all additional mutations in the population (as affected compound heterozygotes), including those with frequencies so low that homozygotes would be extremely rare. Another example of a Finnish disease with multiple mutations is retinoschisis (OMIM 312700). It shows both southern and northern geographic clusters in Finland, and its relative overrepresentation is caused by three widespread founder mutations (16).

The study of recessive diseases that are not particularly overrepresented in Finland in comparison to other European populations has often revealed multiple mutations. Each of them, however, shows distinct geographic clustering that is indicative of founder effects. Examples of these diseases include transglutaminase-1 (TGM1) mutations in autosomal recessive congenital ichthyoses and mapping of a new ichthyose locus with a founder location (38, 87), sulfonylurea receptor (SUR1) mutations in persistent hyperinsulinemic hypoglycemia of infancy (64), and coagulation factor XIII mutations in a rare bleeding disorder (53).

Especially illustrative are the results on steroid 21-hydroxylase gene (CYP21) mutations (45, 46). Homozygous CYP21 deficiency causes congenital adrenal hyperplasia and virilization in girls. The gene maps within the class III gene cluster in the HLA region in chromosome 6p21 and has a nearby pseudogene (CYP21P). This genomic structure favors unequal crossover events and frequent deletions as a result; indeed, the majority of CYP21 mutations are genomic rearrangements. The severe salt-wasting and virilizing forms are not uncommon and occur at a frequency of about 1/15,000 births in different populations (66). The location of the CYP21 gene within the HLA cluster makes it a highly interesting marker for population studies as well because the extreme polymorphism of HLA haplotypes and their associations with CYP21 mutations can be utilized. Although some CYP21 mutation-HLA haplotype correlations are enriched in some populations, in mixed populations CYP21 variants do not show significant linkage disequilibrium with HLA haplotypes. This is consistent with the high mutation rate and recurrent independent mutation events for CYP21. How does this rather mutation-prone gene behave in Finland?

Levo et al (45) studied the number of different mutations that were present in two thirds of all diagnosed congenital adrenal hyperplasia patients in Finland, representing 74 unrelated families. They identified a total of 19 different mutation-haplotype combinations, consistent with the expectedly high mutation rate. However, three mutations that occurred in otherwise very rare HLA haplotypes accounted for half of all mutations (46). Over 80% of all mutation-haplotype combinations were observed repeatedly, suggesting founder mutations. Indeed, plotted on a map of Finland based on grandparental birthplaces, each mutation-haplotype combination showed clustering consistent with founder effects. Some mutations were probably imported and old, others were more likely to have first arisen locally. Notably, a genealogical analysis revealed no consanguinity between the 74 families back to the grandparental level, but extended analysis allowed the construction of some multiply consanguineous pedigrees with common ancestors in the 1600s and 1700s (up to 11 generations back). For most families, common ancestors could not be identified, suggesting only remote links. These results again support the picture of Finland as a country of multiple subisolates. The process of isolate formation and their expansion has been so recent and rapid that it also predominates over the diversifying effect of new mutations for this relatively mutable gene.

Dominant Diseases

For dominant genes, the founder effects are equally visible. One dominant disorder, amyloidosis V or Finnish-type amyloidosis, is listed among the 35 Finnish diseases, and it has a single founder mutation in the gelsolin gene (Table 1). In long QT syndrome, a familial cardiac arrhythmia syndrome, a founder mutation was recently identified in Finland (72). Among more common dominant diseases, acute intermittent porphyria is illustrative. It is caused by mutations in the porphobilinogen deaminase gene. Forty known families in Finland possess a total of 26 mutations; the most common of them form founder clusters when plotted on a map (26, 54, 55).

Distinct founder effects have been observed in familial hypercholesterolemia (FH), inherited nonpolyposis colon cancer, and breast cancer. One specific mutation in the low-density lipoprotein lipase receptor gene (LDLR), called FH-North Karelia, accounts for almost 90% of cases in an eastern subpopulation of about 180,000, with at least 340 carriers identified (89). In all of Finland, four mutations (FH-North Karelia, FH-Helsinki, FH-Turku, and FH-Pori) account for about three quarters of all patients; FH-North Karelia alone is responsible for two thirds. The distribution of mutations varies in different parts of the country (30, 31). Interestingly, these mutations are not common for other Nordic countries: A survey among Swedish FH patients revealed only 5.5% (10 cases) of FH-Helsinki and a single case of FH-North Karelia (48). This observation is consistent with the expected consequences of rapid recent population growth in a subisolate structure and little admixture with neighbors.

Two founder mutations in the MLH1 gene are responsible for about two thirds of hereditary nonpolyposis colorectal cancer families in Finland (60–62), and mutations in the BRCA1 and BRCA2 genes have given rise to multiple founder effects in inherited breast cancer (76).

LESSONS LEARNED

Utility of Linkage Disequilibrium

What have we learned from the study of rare recessive diseases? Perhaps the most notable general lessons have been associated with the repeated successes in making use of the genetic founder effects, also called linkage disequilibrium mapping. This is based on the expectation that a single mutation has been increased in frequency by founder effect and drift for most recessive diseases that have a nonuniform geographic distribution. Introduced originally in a single copy, the allelic composition of that chromosome has eroded over time by recombinations, but it is still preserved near the disease gene. When a chromosome that carries a disease gene (found homozygous in affected individuals) is studied with polymorphic microsatellite markers, certain alleles for each marker show remarkable overrepresentation when compared to the general allele frequencies in the population (usually calculated on the basis of the untransmitted parental chromosomes). The alleles form conserved haplotypes that are repeatedly observed in patients from seemingly unrelated families. Linkage disequilibrium in this context refers to the association of certain alleles and haplotypes to the presence of the disease gene, and its significance is assessed by contingency tables and χ-square tests comparing marker allele frequencies in patient chromosomes versus control chromosomes. One then looks for markers that are monomorphic in patient chromosomes or show the strongest allelic association: They are most likely to be closest to the disease gene.

A common misconception is that, given the relatively young age of the population subisolates, the conserved chromosomal segments are too long to provide sufficient resolution for positional cloning. This has, however, not been a practical problem. Although it is true that the conserved segments often span several cM and marker alleles of up to 5 cM away may show statistically significant associations, the conserved segments do not fully overlap between different families. The chromosomes that carry disease genes have undergone different historical recombinations in varying branches of the superpedigree that describe the descent of the common mutation. Thus, by compiling the haplotypes and comparing the likely positions of historical recombinations, one can infer the position of the disease gene very accurately.

This is exemplified in Figure 4, which shows haplotype data for congenital chloride diarrhea (14). Compilation of the haplotypes from only 24 core families (most of them with just one affected child) mapped the gene exactly between two markers, even though marker allele associations were statistically significant as far as 5–6 cM away from the disease gene.

Further accuracy for the predictions of gene localizations was obtained from the application of Luria & Delbrück's equation, originally introduced to assess mutations in bacterial cultures (49). The reinvention of the equation for the use of estimating gene position in isolated population settings was made by E. Lander, working with J. Hästbacka & A. de la Chapelle on the positional cloning of the diastrophic dysplasia gene (11, 12). The formula was simplified (43) to

$$p_{excess} = \alpha(1 - \Theta)^g \approx \alpha\, e^{-\Theta g}, \text{ where } p_{excess} = (p_{affected} - p_{control})/(1 - p_{control}).$$

In this equation, α denotes the fraction of chromosomes that are expected to have a common mutation; g, the number of generations; Θ, recombination distance from marker to mutation; and $p_{affected}$ and $p_{control}$, the overrepresented allele frequency for the marker in disease gene–carrying and control chromosomes, respectively.

The accuracy and limits of the estimates became the subject of statistical debate (22, 23). Later, de la Chapelle & Wright (5) considered sources of error for the estimates and introduced a corrected standard error estimate for the Luria-Delbrück method.

In spite of confidence limits that were too narrow and the need to guess the number of generations since the founding of the branching mutation pedigree,

4	6	6	2	4	2	3	4	6	4	6	6	4	5	4	4	3	4	6	6	6	4	6	D7S658
6	6	6	6	6	6	6	2	8	8	8	1	6	6	6	2	6	6	6	6	6	6	2	D7S501
6	6	6	6	6	6	6	6	1	6	6	6	6	6	6	6	6	6	6	6	6	6	1	D7S496
4	4	4	4	4	4	4	1	4	4	4	4	4	4	4	4	4	2	3	3	1	4	4	D7S692
8	8	8	8	3	8	8	2	8	8	8	8	8	8	8	8	5	1	2	5	2	7	3	D7S799
3	3	4	3	6	4	3	2	3	3	4	3	4	3	5	3	4	4	2	4	6	6	4	D7S523
8	8	5	3	2	2	1	2	2	1	1	1	1	1	1	1	1	1	1	1	1	1	1	Nr of chr

Figure 4 Six marker haplotypes in chromosomes from patients with congenital chloride diarrhea [adapted from (14)]. The microsatellite markers, spanning 12 cM, are indicated on the right. Each haplotype observed in patients is depicted as a vertical column of six alleles, with the number of times it was encountered (bottom row). In 16 chromosomes, the haplotype was 4 or 6-6-6-4-8-3, suggesting the founder haplotype with an early recombination at D7S658 (indicated as 4/6 variation for that marker). The completely conserved haplotype segments in the remaining chromosomes are highlighted with gray. All chromosome segments overlap only between D7S496 and D7S692, suggesting a likely position for the gene. The gene is located 290 kb from D7S496 toward D7S692 (13).

application of the Luria-Delbrück estimations proved highly useful for practical positional cloning purposes. Refined mapping of the disease gene by the haplotype compilation or Luria-Delbrück methods was presented as an interlude to cloning for most of the diseases listed in Table 1. The endgame of positional cloning was played in some instances by isolating entirely new genes; but in later cases, the presence of positional candidate genes facilitated direct mutation analyses.

After genes became identified, the Luria-Delbrück equation was also used to estimate the age of mutation since its introduction to the expanding population, even though the statistical accuracy of this estimate might well be questioned. From the equation above, g was solved for several markers based on p_{excess} in mutation-carrying chromosomes and the distances from markers to the mutation, and the estimate was obtained by averaging overmarker-specific g's (13). For example, the founder mutation in congenital chloride diarrhea was estimated to have spread for about 19 generations, in good accordance with the age of the eastern Finnish expansion (400 years).

Study Design and Sampling

A population structure with small breeding units, large local variations in allele frequencies, and distinct subpopulations should make one cautious about extrapolations and predictions based on simple population models and analytical calculations (91). The unorthodox population structure should be already considered when applying the Hardy-Weinberg equation for estimating carrier frequencies from disease incidence. Also, predictions of levels of linkage disequilibrium based on a global view of Finland as a panmictic population (33) yield values that are too small in comparison to those observed for many rare genes. More complex models are clearly needed to aid in designing studies. In ongoing work, we are attempting to integrate pedigree and meiosis simulation with geographical frameworks to yield population models that might more accurately reflect the genetic structure of a population (V. Ollikainen, H. Mannila & J. Kere, unpublished).

In genetic studies, sampling is important, all the more so in a population that has a distinct substructure caused by founder effects and genetic drift, as is the case in rural Finland. In disease gene studies, cases and controls need to be genetically matched. One of the best ways to achieve this is to use family-based controls (84). If such a strategy cannot be adopted, then at least the origin of cases and controls should be controlled and matched geographically, considering potential pitfalls of population substructure.

These considerations are even more important when common diseases become subjects for genetic study. Relatively large variations for disease frequencies have been observed in epidemiological studies among different parts of the country, and these differences have been attributed in part to genetic variances (20, 81). Closer analysis shows that clinical and laboratory parameters also seem to profile patients differently in varying parts of the country, which further emphasizes the need to pay careful attention to sampling (4, 44, 80).

Especially when large numbers of coding SNPs become available, studies that remain inconclusive are all too easy to plan and execute. A tempting outline for a study might be, for example, the following: One could start by looking for geographic differences in the frequency of a common disease and sample families from relative high-frequency and low-frequency areas. For the next step, one could look at differences in the frequencies of candidate gene SNPs in both sample sets. If one would find a certain SNP at a significantly higher frequency in the high disease frequency area compared to the low frequency area, that might be interpreted as providing useful information about genes that contribute to the disease. This conclusion, however, would most likely be wrong. As shown already for both rare and common neutral alleles (e.g., the blood group markers), significant differences in their frequencies exist between various communities within Finland. Such differences affect all genes and are caused purely by the random effects of the low numbers of founders and the drift in small breeding units. Thus, almost any gene may have variances in frequency within two communities if they are small enough, and these variances may even be significant for a large number of genes. The phenomenon of different disease frequency in two communities may well be caused by founder events and drift, but one cannot tell by simple association which of the enriched genes contributes to the disease pathogenesis and which was enriched coincidentally.

FOCUS ON MULTIFACTORIAL DISEASES

First Experiences from Linkage Approaches

First experiences from mostly linkage-based projects aimed at identifying susceptibility genes in common diseases have started to accumulate from the Finnish population. The spectrum of diseases is already wide, including type 2 diabetes (9, 50, 90), multiple sclerosis (34), asthma (38a, 39), myocardial infarction (67), familial combined hyperlipidemia (65), hypertension (70), obesity (63), schizophrenia (8, 15), psoriasis (1), and systemic lupus (32). Several other projects are in progress. It is too early to draw conclusions based on this experience, but some first observations are possible.

The geographic clustering noted for some phenotypes has obviously not resulted from the simple enrichment of a single mutation. Three notable examples include type 2 diabetes, with high incidence on west coastal regions (50); multiple sclerosis, which is also enriched in a western subpopulation (34); and schizophrenia (8, 15). In schizophrenia, one genome scan was performed in the young subpopulation of Kuusamo (a high-incidence region), and another on sibpairs from other parts of Finland. Both studies suggest the presence of multiple loci (8, 15), and convincing haplotype association was not found. On the other hand, a genome scan on hypertension identified one locus as the most significant in Finland (70), and a genome-wide scan for asthma susceptibility genes in the Kainuu subpopulation resulted in significant linkage mapping of one locus, suggesting higher

homogeneity than in general populations (38a). As long as individual susceptibility genes and their disease-associated polymorphisms remain unidentified, not
much more can be said about the success of these lines of work.

A strong interest in haplotype association as a tool to extract more information
from the isolated population settings has sparked the development of statistical and
computing methods (17, 41, 51, 85). At least in simulated settings, these methods
have shown promise (77).

Linkage Disequilibrium and Common Markers

The prospect of utilizing linkage disequilibrium to map common disease genes
has sparked much recent interest. Several studies have recently addressed the extent of linkage disequilibrium in the population of Finland as well as elsewhere
(7, 19, 71, 82, 86). The results have uniformly indicated that levels of linkage disequilibrium between nearby markers in Finland do not differ from levels observed
in other European populations. Some authors have interpreted this to suggest that
for the mapping of common disease genes studies on populations, such as those
in Finland, might offer no specific advances.

The results are not unexpected, but the interpretation may be too generalized
(2, 17, 82). Studies on overall levels of linkage disequilibrium have used samples collected from healthy individuals. As genetic markers, the studies have employed either microsatellites or SNPs, and they have especially considered the most
common alleles for both types of markers. The common alleles have obviously
been imported in the form of thousands of copies with the first and later settlers
to Finland, and they even have been spread to the subisolates in tens or hundreds
of copies, depending on each allele's frequency, and undergone recombinations at
rates common to all populations. Thus, the overall measures of linkage disequilibrium for common alleles should be similar in Finland and elsewhere, especially
when sampling is made on the "general" Finnish population.

Taillon-Miller et al (82) considered linkage disequilibrium for common SNP
alleles in the Kainuu subpopulation. The levels of linkage disequilibrium were
indistinguishable from, for example, the levels observed in CEPH samples. These
data are consistent with the fact that Kainuu was founded by more than a few individuals. Interestingly, the haplotype patterns for two tight X-chromosomal clusters
of markers were remarkably different between Kainuu and Sardinia. This result
illustrates simply that the two isolates have different population histories and that
different haplotypes have been traded differently by drift. Overall, these data are
consistent with the previous finding that in local isolates in Finland even common
blood group alleles may show significant deviations from the population mean.
Similar results were obtained for the Kuusamo subisolate (86). Some linkage disequilibrium was observed between X-chromosomal markers in Kuusamo but very
little between autosomal markers in both Kuusamo and a general Finnish sample
set. Do these results invalidate linkage disequilibrium mapping for susceptibility
genes in common multifactorial diseases?

Linkage Disequilibrium in Disease Gene Studies

Whenever patients are collected, the sample set will become enriched for disease or susceptibility alleles at the same time as the sample represents a random sampling of alleles that are irrelevant for disease pathogenesis (or at least that is what investigators hope to achieve). It is important to note that this is true even for SNPs within one gene: A disease gene may contain both disease-causing SNPs and neutral SNPs, and depending on the history of each polymorphism, the latter may or may not be associated to the disease. Thus, a random SNP in a susceptibility gene may be totally useless for linkage disequilibrium mapping, which should be considered when designing SNP association studies. For example, a polymorphic gene (HCR) in the psoriasis susceptibility region PSORS1 contained SNPs that were tightly associated with psoriasis, as expected, but also a large number of presumably neutral unassociated SNPs (1).

The level of enrichment for any disease allele is dependent on both the biology and pathogenesis of the disease (how well the disease status predicts a defect in a particular gene) and on the genetic makeup of the population from which the samples were collected (how many different disease-causing alleles there are to choose from). The significance of disequilibrium observed around a susceptibility gene will then depend on the frequencies of the surrounding alleles and haplotypes in patient and control chromosomes.

In an extreme case, a disease is caused by a single mutation that occurs homozygous in all patients. That extreme case is what has been observed for many recessive monogenic diseases in Finland, and such single mutations are likely to be in strong linkage disequilibrium with the surrounding alleles and haplotype. On the other hand, any fully penetrant dominant gene is likely to be found in only half of all patient chromosomes, and linkage disequilibrium will thus be smaller when markers on all chromosomes are considered. With decreasing levels of penetrance, increasing proportion of phenocopies, and increasing levels of allelic heterogeneity behind the disease in the particular population that was sampled, linkage disequilibrium will diminish, ultimately becoming undetectable. If there is no enrichment for any particular allele, as is the case when normal individuals are sampled, one will observe only "background" disequilibrium. But by selecting for disease (i.e., introducing an intended bias), the expected level of linkage disequilibrium in the sample cannot be predicted from background disequilibrium.

Power Considerations

It may be useful to consider issues of power when linkage disequilibrium analyses are used as part of the mapping strategy for susceptibility genes in common diseases. For example, Laitinen et al (39) recruited asthma patients and their families from the Kainuu region for susceptibility gene studies. The origin of the subjects was verified from population registries, and several candidate gene regions were considered (25, 39, 40). Haplotype association analysis was employed in addition to genetic linkage analysis to increase power, and the investigators estimated that the study material would have allowed a good chance to identify an associated haplotype present in 15% of all patient chromosomes. Indeed, moderate but statistically significant haplotype associations were observed for two candidate gene regions, around the FCER2 gene in chromosome 19p and around the IL9R gene in the Xq pseudoautosomal region (25, 40).

How should such associations be interpreted? Of course, a false positive association remains a possibility. On the other hand, the population history of Kainuu may be one that is made up of founders who brought along a small number of susceptibility alleles for these genes that are neither necessary nor sufficient to induce the development of asthma in their carriers. Drift modified the allele frequencies so that perhaps one of them became more common than the others. When present-day patients were sampled, those specific susceptibility alleles with their surrounding haplotypes became further enriched by the process of sampling so that the difference between patient and control chromosomes became detectable. Recent considerations on the nature and frequency of disease-causing alleles that may well be relatively rare in general populations (93) would be compatible with this hypothesis. Clearly, until functional differences that are specific for associated alleles and compatible with disease pathogenesis have been demonstrated, the results only suggest a path for further study.

A MODEL POPULATION: PROMISES AND LIMITATIONS

The allelic diversity of a disease in a particular population is a most important issue when susceptibility gene studies are planned. As is evident from marker data, allelic diversity in any founder population that has grown in isolation is different from other populations, and the differences will be most prominent for relatively rare alleles. The varying frequencies of alleles in small subpopulations that are caused by drift and simplified allelic diversity might aid genetic studies when SNP association becomes a major approach (3). For example, the finding of a corneodesmosin gene allele that was previously suggested as a psoriasis susceptibility gene at a very high frequency in Kainuu helped to exclude it as a susceptibility allele (1).

The rarer an allele is, the more likely it is to get lost because of drift, but occasional rare alleles survive at unusually high frequencies (Figure 3). Many disease susceptibility alleles may turn out to have low general population frequencies and high allelic diversity (93). For a disease gene with a large number of different mutations, this may mean the loss of most mutated alleles but the preservation of one or a few major mutations in the population isolate, accompanied with associated haplotypes. Unfortunately, predicting how individual alleles have fared is not possible.

On the other hand, if the disease alleles are very common and confer only a slight increase in disease risk (have a low penetrance), their identification in an isolated population will become difficult. The main limiting problem is likely to be practical: The population size in an isolated population may not be large enough to allow for the collection of a sufficient number of patients that is needed to show a small increase in susceptibility. This practical issue also becomes a problem with some suggested solutions to the dilemma of identifying susceptibility genes. The Saami population is characterized genetically as one without signs of expansion, but rather long-time isolation with constant population size; and such a situation might help to map some old and common mutations (35, 83). However, with less than 2000 individuals announcing Saami as their mother tongue (79), the population appears too small for the study of even the most common multifactorial diseases.

During the past decade, most of the work that was needed to dissect the molecular genetics of the Finnish disease heritage has been accomplished. This decade is devoted more to the study of multifactorial diseases in Finland as well as elsewhere. Although the outcome from these studies is still difficult to predict, positive signs have started to emerge (38a). Regardless, we should expect to learn much more about how different populations yield information that can be applied to the annotation of the human morbidity map.

ACKNOWLEDGMENTS

I thank Vesa Ollikainen for the computer simulation presented in this article. Members of our research group and staff of the Finnish Genome Center have contributed via numerous discussions that were essential for the contents of this review. Our work is supported by the Academy of Finland, Sigrid Jusélius Foundation, Finnish Technology Fund Tekes, Ulla Hjelt Fund, and Helsinki University Hospital research funds. JK is a member of Biocentrum Helsinki and the Center of Excellence for Disease Genetics, University of Helsinki.

Visit the Annual Reviews home page at www.AnnualReviews.org

LITERATURE CITED

1. Asumalahti K, Laitinen T, Itkonen-Vatjus R, Lokki ML, Suomela S, et al. 2000. A candidate gene for psoriasis near HLA-C, HCR (Pg8), is highly polymorphic with a disease-associated susceptibility allele. *Hum. Mol. Genet.* 9:1533–42
2. Boehnke M. 2000. A look at linkage disequilibrium. *Nat. Genet.* 25:246–47
3. Cargill M, Altshuler D, Ireland J, Sklar P, Ardlie K, et al. 1999. Characterization of single-nucleotide polymorphisms in coding regions of human genes. *Nat. Genet.* 22:231–38

4. Carmelli D, Williams RR, Rissanen A. 1982. Contrasting patterns of familiality for cholesterol and triglyceride in Finland according to type of coronary manifestations and locations. *Am. J. Epidemiol.* 116:617–21
5. Cavalli-Sforza LL, Menozzi P, Piazza A. 1994. *The History and Geography of Human Genes.* Princeton, NJ: Princeton Univ. Press
6. de la Chapelle A, Wright FA. 1998. Linkage disequilibrium mapping in isolated populations: the example of Finland

revisited. *Proc. Natl. Acad. Sci. USA* 95:12416–23

7. Eaves IA, Merriman TR, Barber RA, Nutland S, Tuomilehto-Wolf E, et al. 2000. The genetically isolated populations of Finland and Sardinia may not be a panacea for linkage disequilibrium mapping of common disease genes. *Nat. Genet.* 25:320–23

8. Ekelund J, Lichtermann D, Hovatta I, Ellonen P, Suvisaari J, et al. 2000. Genomewide scan for schizophrenia in the Finnish population: evidence for a locus on chromosome 7q22. *Hum. Mol. Genet.* 9:1049–57

9. Ghosh S, Watanabe RM, Valle TT, Hauser ER, Magnuson VL, et al. 2000. The Finland–United States investigation of non-insulin-dependent diabetes mellitus genetics (FUSION) study. I. An autosomal genome scan for genes that predispose to type 2 diabetes. *Am. J. Hum. Genet.* 67:1174–85

10. Guldberg P, Henriksen KF, Sipilä I, Guttler F, de la Chapelle A. 1995. Phenylketonuria in a low incidence population: molecular characterisation of mutations in Finland. *J. Med. Genet.* 32:976–78

11. Hästbacka J, de la Chapelle A, Kaitila I, Sistonen P, Weaver A, Lander E. 1992. Linkage disequilibrium mapping in isolated founder populations: diastrophic dysplasia in Finland. *Nat. Genet.* 2:204–11

12. Hästbacka J, de la Chapelle A, Mahtani MM, Clines G, Reeve-Daly MP, et al. 1994. The diastrophic dysplasia gene encodes a novel sulfate transporter: positional cloning by fine-structure linkage disequilibrium mapping. *Cell* 78:1073–87

13. Höglund P, Haila S, Socha J, Tomaszewski L, Saarialho-Kere U, et al. 1996. Mutations in the down-regulated in adenoma (DRA) gene cause congenital chloride diarrhoea. *Nat. Genet.* 14:316–19

14. Höglund P, Sistonen P, Norio R, Holmberg C, Dimberg A, et al. 1995. Fine mapping of the congenital chloride diar-

rhea gene by linkage disequilibrium. *Am. J. Hum. Genet.* 57:95–102

15. Hovatta I, Varilo T, Suvisaari J, Terwilliger JD, Ollikainen V, et al. 1999. A genomewide screen for schizophrenia genes in an isolated Finnish subpopulation, suggesting multiple susceptibility loci. *Am. J. Hum. Genet.* 65:1114–24

16. Huopaniemi L, Rantala A, Forsius H, Somer M, de la Chapelle A, Alitalo T. 1999. Three widespread founder mutations contribute to high incidence of X-linked juvenile retinoschisis in Finland. *Eur. J. Hum. Genet.* 7:368–76

17. Jorde LB. 2000. Linkage disequilibrium and the search for complex disease genes. *Genome Res.* 10:1435–44

18. Jorde LB, Pitkänen KJ. 1991. Inbreeding in Finland. *Am. J. Phys. Anthropol.* 84:127–39

19. Jorde LB, Watkins WS, Kere J, Nyman D, Eriksson AW. 2000. Gene mapping in isolated populations: new roles for old friends? *Hum. Hered.* 50:57–65

20. Jousilahti P, Vartiainen E, Tuomilehto J, Pekkanen J, Puska P. 1998. Role of known risk factors in explaining the difference in the risk of coronary heart disease between eastern and southwestern Finland. *Ann. Med.* 30:481–87

21. Jutikkala E. 1959. *Suomen Historian Kartasto (Atlas of Finnish History)*. Porvoo: Finnish Acad. Sci. 2nd ed.

22. Kaplan NL, Hill WG, Weir BS. 1995. Likelihood methods for locating disease genes in nonequilibrium populations. *Am. J. Hum. Genet.* 56:18–32

23. Kaplan NL, Weir BS. 1995. Are moment bounds on the recombination fraction between a marker and a disease locus too good to be true? Allelic association mapping revisited for simple genetic diseases in the Finnish population. *Am. J. Hum. Genet.* 57:1486–98

24. Karjalainen E. 1989. Migration and regional development in the rural communes of Kainuu, Finland in 1980–1985. *Nordia* 23:1–89

25. Kauppi P, Laitinen T, Ollikainen V, Mannila H, Laitinen LA, Kere J. 2000. The IL9R region contribution in asthma is supported by genetic association in an isolated population. *Eur. J. Hum. Genet.* 8:788–92

26. Kauppinen R, Mustajoki S, Pihlaja H, Peltonen L, Mustajoki P. 1995. Acute intermittent porphyria in Finland: 19 mutations in the porphobilinogen deaminase gene. *Hum. Mol. Genet.* 4:215–22

27. Keränen J. 1984. Kainuun asuttaminen (the settling of Kainuu). *Studia Historica Jyväskyläensia*, Vol. 28. Jyväskylä: Univ. Jyväskylä. 282 pp.

28. Kere J, Estivill X, Chillon M, Morral N, Nunes V, et al. 1994. Cystic fibrosis in a low-incidence population: two major mutations in Finland. *Hum. Genet.* 93:162–66

29. Kittles RA, Bergen AW, Urbanek M, Virkkunen M, Linnoila M, et al. 1999. Autosomal, mitochondrial, and Y chromosome DNA variation in Finland: evidence for a male-specific bottleneck. *Am. J. Phys. Anthropol.* 108:381–99

30. Koivisto UM, Viikari J S, Kontula K. 1995. Molecular characterization of minor gene rearrangements in Finnish patients with heterozygous familial hypercholesterolemia: identification of two common missense mutations (Gly823-to-Asp and Leu380-to-His) and eight rare mutations of the LDL receptor gene. *Am. J. Hum. Genet.* 57:789–97

31. Kontula K, Koivisto UM, Koivisto P, Turtola H. 1992. Molecular genetics of familial hypercholesterolaemia: common and rare mutations of the low density lipoprotein receptor gene. *Ann. Med.* 24:363–67

32. Koskenmies S, Widén E, Kere J, Julkunen H. 2001. Familial systemic lupus erythematosus in Finland. *J. Rheumatol.* 28:758–60

33. Kruglyak L. 1999. Prospects for whole-genome linkage disequilibrium mapping of common disease genes. *Nat. Genet.* 22:139–44

34. Kuokkanen S, Gschwend M, Rioux JD, Daly MJ, Terwilliger JD, et al. 1997. Genomewide scan of multiple sclerosis in Finnish multiplex families. *Am. J. Hum. Genet.* 61:1379–87

35. Laan M, Pääbo S. 1997. Demographic history and linkage disequilibrium in human populations. *Nat. Genet.* 17:435–38

36. Lahermo P, Sajantila A, Sistonen P, Lukka M, Aula P, et al. 1996. The genetic relationship between the Finns and the Finnish Saami (Lapps): analysis of nuclear DNA and mtDNA. *Am. J. Hum. Genet.* 58:1309–22

37. Lahermo P, Savontaus ML, Sistonen P, Beres J, de Knijff P, et al. 1999. Y chromosomal polymorphisms reveal founding lineages in the Finns and the Saami. *Eur. J. Hum. Genet.* 7:447–58

38. Laiho E, Ignatius J, Mikkola H, Yee VC, Teller DC, et al. 1997. Transglutaminase 1 mutations in autosomal recessive congenital ichthyosis: private and recurrent mutations in an isolated population. *Am. J. Hum. Genet.* 61:529–38

38a. Laitinen T, Daly MJ, Rioux JD, Kauppi P, Laprise C, et al. 2001. A susceptibility locus for asthma-related traits on chromosome 7 revealed by genome-wide scan in a founder population. *Nat. Genet.* 28:87–91

39. Laitinen T, Kauppi P, Ignatius J, Ruotsalainen T, Daly MJ, et al. 1997. Genetic control of serum IgE levels and asthma: linkage and linkage disequilibrium studies in an isolated population. *Hum. Mol. Genet.* 6:2069–76

40. Laitinen T, Ollikainen V, Lazaro C, Kauppi P, de Cid R, et al. 2000. Association study of the chromosomal region containing the FCER2 gene suggests it has a regulatory role in atopic disorders. *Am. J. Respir. Crit. Care Med.* 161:700–6

41. Lake SL, Blacker D, Laird NM. 2000. Family-based tests of association in the presence of linkage. *Am. J. Hum. Genet.* 67:1515–25

42. Lander ES, Schork NJ. 1994. Genetic

dissection of complex traits. *Science* 265:2037–48

43. Lehesjoki AE, Koskiniemi M, Norio R, Tirrito S, Sistonen P, et al. 1993. Localization of the EPM1 gene for progressive myoclonus epilepsy on chromosome 21: linkage disequilibrium allows high resolution mapping. *Hum. Mol. Genet.* 2:1229–34

44. Lehtinen S, Luoma P, Nayha S, Hassi J, Ehnholm C, et al. 1998. Apolipoprotein A-IV polymorphism in Saami and Finns: frequency and effect on serum lipid levels. *Ann. Med.* 30:218–23

45. Levo A, Jääskeläinen J, Sistonen P, Siren MK, Voutilainen R, Partanen J. 1999. Tracing past population migrations: genealogy of steroid 21-hydroxylase (CYP21) gene mutations in Finland. *Eur. J. Hum. Genet.* 7:188–96

46. Levo A, Partanen J. 1997. Mutation-haplotype analysis of steroid 21-hydroxylase (CYP21) deficiency in Finland. Implications for the population history of defective alleles. *Hum. Genet.* 99:488–97

47. Libert F, Cochaux P, Beckman G, Samson M, Aksenova M, et al. 1998. The delta-ccr5 mutation conferring protection against HIV-1 in Caucasian populations has a single and recent origin in northeastern Europe. *Hum. Mol. Genet.* 7:399–406

48. Lind S, Eriksson M, Rystedt E, Wiklund O, Angelin B, Eggertsen G. 1998. Low frequency of the common Norwegian and Finnish LDL-receptor mutations in Swedish patients with familial hypercholesterolaemia. *J. Intern. Med.* 244:19–25

49. Luria SE, Delbrück M. 1943. Mutations of bacteria from virus sensitivity to virus resistance. *Genetics* 28:491–511

50. Mahtani MM, Widen E, Lehto M, Thomas J, McCarthy M, et al. 1996. Mapping of a gene for type 2 diabetes associated with an insulin secretion defect by a genome scan in Finnish families. *Nat. Genet.* 14:90–94

51. McIntyre LM, Martin ER, Simonsen KL, Kaplan NL. 2000. Circumventing multiple testing: a multilocus Monte Carlo approach to testing for association. *Genet. Epidemiol.* 19:18–29

52. McKusick VA. 1998. *Mendelian Inheritance in Man.* Baltimore: Johns Hopkins Univ. Press. 12th ed. http://www.ncbi.nlm.nih.gov/omim

53. Mikkola H, Syrjälä M, Rasi V, Vahtera E, Hämäläinen E, et al. 1994. Deficiency in the A-subunit of coagulation factor XIII: two novel point mutations demonstrate different effects on transcript levels. *Blood* 84:517–25

54. Mustajoki P, Desnick RJ. 1985. Genetic heterogeneity in acute intermittent porphyria: characterisation and frequency of porphobilinogen deaminase mutations in Finland. *Br. Med. J.* 291:505–9

55. Mustajoki S, Pihlaja H, Ahola H, Petersen NE, Mustajoki P, Kauppinen R. 1998. Three splicing defects, an insertion, and two missense mutations responsible for acute intermittent porphyria. *Hum. Genet.* 102:541–48

56. Nevanlinna HR. 1972. The Finnish population structure. A genetic and genealogical study. *Hereditas* 71:195–236

57. Nevanlinna HR. 1980. Rare hereditary diseases and markers in Finland: an introduction. In *Population Structure and Genetic Disorders*, ed. A Eriksson. London: Academic

58. Norio R, Nevanlinna HR, Perheentupa J. 1973. Hereditary diseases in Finland; rare flora in rare soil. *Ann. Clin. Res.* 5:109–41

59. Nunez MG. 1987. A model for the early settlement of Finland. *Fennosc. Archaeol.* 4:3–18

60. Nyström-Lahti M, Kristo P, Nicolaides NC, Chang SY, Aaltonen LA, et al. 1995. Founding mutations and Alu-mediated recombination in hereditary colon cancer. *Nat. Med.* 1:1203–6

61. Nyström-Lahti M, Sistonen P, Mecklin JP, Pylkkänen L, Aaltonen LA, et al. 1994. Close linkage to chromosome 3p

and conservation of ancestral founding haplotype in hereditary nonpolyposis colorectal cancer families. *Proc. Natl. Acad. Sci. USA* 91:6054–58

62. Nyström-Lahti M, Wu Y, Moisio AL, Hofstra RM, Osinga J, et al. 1996. DNA mismatch repair gene mutations in 55 kindreds with verified or putative hereditary non-polyposis colorectal cancer. *Hum. Mol. Genet.* 5:763–69

63. Ohman M, Oksanen L, Kaprio J, Koskenvuo M, Mustajoki P, et al. 2000. Genomewide scan of obesity in Finnish sibpairs reveals linkage to chromosome Xq24. *J. Clin. Endocrinol. Metab.* 85:3183–90

64. Otonkoski T, Ämmälä C, Huopio H, Cote GJ, Chapman J, et al. 1999. A point mutation inactivating the sulfonylurea receptor causes the severe form of persistent hyperinsulinemic hypoglycemia of infancy in Finland. *Diabetes* 48:408–15

65. Pajukanta P, Terwilliger JD, Perola M, Hiekkalinna T, Nuotio I, et al. 1999. Genomewide scan for familial combined hyperlipidemia genes in Finnish families, suggesting multiple susceptibility loci influencing triglyceride, cholesterol, and apolipoprotein B levels. *Am. J. Hum. Genet.* 64:1453–63

66. Pang SY, Wallace MA, Hofman L, Thuline HC, Dorche C, et al. 1988. Worldwide experience in newborn screening for classical congenital adrenal hyperplasia due to 21-hydroxylase deficiency. *Pediatrics* 81:866–74

67. Pastinen T, Perola M, Niini P, Terwilliger J, Salomaa V, et al. 1998. Array-based multiplex analysis of candidate genes reveals two independent and additive genetic risk factors for myocardial infarction in the Finnish population. *Hum. Mol. Genet.* 7:1453–62

68. Peltonen L. 2000. Positional cloning of disease genes: advantages of genetic isolates. *Hum. Hered.* 50:66–75

69. Peltonen L, Jalanko A, Varilo T. 1999. Molecular genetics of the Finnish disease heritage. *Hum. Mol. Genet.* 8:1913–23

70. Perola M, Kainulainen K, Pajukanta P, Terwilliger JD, Hiekkalinna T, et al. 2000. Genome-wide scan of predisposing loci for increased diastolic blood pressure in Finnish siblings. *J. Hypertens.* 18:1579–85

71. Peterson AC, Di Rienzo A, Lehesjoki AE, de la Chapelle A, Slatkin M, Freimer NB. 1995. The distribution of linkage disequilibrium over anonymous genome regions. *Hum. Mol. Genet.* 4:887–94

72. Piippo K, Laitinen P, Swan H, Toivonen L, Viitasalo M, et al. 2000. Homozygosity for a HERG potassium channel mutation causes a severe form of long QT syndrome: identification of an apparent founder mutation in the Finns. *J. Am. Coll. Cardiol.* 35:1919–25

73. Pitkänen K, Jorde LB, Mielke JH, Fellman JO, Eriksson AW. 1988. Marital migration and genetic structure in Kitee, Finland. *Ann. Hum. Biol.* 15:23–33

74. Sajantila A, Lahermo P, Anttinen T, Lukka M, Sistonen P, et al. 1995. Genes and languages in Europe: an analysis of mitochondrial lineages. *Genome Res.* 5:42–52

75. Sajantila A, Salem AH, Savolainen P, Bauer K, Gierig C, Pääbo S. 1996. Paternal and maternal DNA lineages reveal a bottleneck in the founding of the Finnish population. *Proc. Natl. Acad. Sci. USA* 93:12035–39

76. Sarantaus L, Huusko P, Eerola H, Launonen V, Vehmanen P, et al. 2000. Multiple founder effects and geographical clustering of BRCA1 and BRCA2 families in Finland. *Eur. J. Hum. Genet.* 8:757–63

77. Sevon P, Ollikainen V, Onkamo P, Toivonen HTT, Mannila H, Kere J. 2001. Mining associations between genetic markers, phenotypes and covariates. *Genet. Epidemiol.* (Suppl.) In press

78. Siren MK, Sareneva H, Lokki ML, Koskimies S. 1996. Unique HLA antigen frequencies in the Finnish population. *Tissue Antigens* 48:703–7

79. Statistics Finland. 1997. *Statistical Yearbook of Finland*, Vol. 92. Hämeenlinna: Statistics Finland

80. Stengård JH, Kardia SL, Tervahauta M, Ehnholm C, Nissinen A, Sing CF. 1999. Utility of the predictors of coronary heart disease mortality in a longitudinal study of elderly Finnish men aged 65 to 84 years is dependent on context defined by Apo E genotype and area of residence. *Clin. Genet.* 56:367–77

81. Suhonen O, Reunanen A, Aromaa A, Knekt P, Pyörälä K. 1985. Four-year incidence of myocardial infarction and sudden coronary death in twelve Finnish population cohorts. *Acta Med. Scand.* 217:457–64

82. Taillon-Miller P, Bauer-Sardiña I, Saccone NL, Putzel J, Laitinen T, et al. 2000. Juxtaposed regions of extensive and minimal linkage disequilibrium in human Xq25 and Xq28. *Nat. Genet.* 25:324–28

83. Terwilliger JD, Zollner S, Laan M, Pääbo S. 1998. Mapping genes through the use of linkage disequilibrium generated by genetic drift: "drift mapping" in small populations with no demographic expansion. *Hum. Hered.* 48:138–54

84. Thomson G. 1995. Mapping disease genes: family-based association studies. *Am. J. Hum. Genet.* 57:487–98

85. Toivonen HTT, Onkamo P, Vasko K, Ollikainen V, Sevon P, et al. 2000. Data mining applied to linkage disequilibrium mapping. *Am. J. Hum. Genet.* 67:133–45

86. Varilo T, Laan M, Hovatta I, Wiebe V, Terwilliger JD, Peltonen L. 2000. Linkage disequilibrium in isolated populations: Finland and a young sub-population of Kuusamo. *Eur. J. Hum. Genet.* 8:604–12

87. Virolainen E, Wessman M, Hovatta I, Niemi K-M, Ignatius J, et al. 2000. Assignment of a novel locus for autosomal recessive congenital ichthyosis to chromosome 19p13.1–p13.2. *Am. J. Hum. Genet.* 66:1132–37

88. Virtaranta-Knowles K, Sistonen P, Nevanlinna HR. 1991. A population genetic study in Finland: comparison of the Finnish- and Swedish-speaking populations. *Hum. Hered.* 41:248–64

89. Vuorio AF, Turtola H, Piilahti KM, Repo P, Kanninen T, Kontula K. 1997. Familial hypercholesterolemia in the Finnish north Karelia. A molecular, clinical, and genealogical study. *Arterioscler. Thromb. Vasc. Biol.* 17:3127–38

90. Watanabe RM, Ghosh S, Langefeld CD, Valle TT, Hauser ER, et al. 2000. The Finland–United States investigation of non-insulin-dependent diabetes mellitus genetics (FUSION) study. II. An autosomal genome scan for diabetes-related quantitative-trait loci. *Am. J. Hum. Genet.* 67:1186–200

91. Workman PL, Mielke JH, Nevanlinna HR. 1976. The genetic structure of Finland. *Am. J. Phys. Anthropol.* 44:341–67

92. Wright AF, Carothers AD, Pirastu M. 1999. Population choice in mapping genes for complex diseases. *Nat. Genet.* 23:397–404

93. Zwick ME, Cutler DJ, Chakravarti A. 2000. Patterns of genetic variation in Mendelian and complex traits. *Annu. Rev. Genomics Hum. Genet.* 1:387–407

Annu. Rev. Genomics Hum. Genet. 2001. 2:129–51

CONGENITAL DISORDERS OF GLYCOSYLATION

Jaak Jaeken[1] and Gert Matthijs[2]

[1]Department of Paediatrics, Centre for Metabolic Disease, University of Leuven, Leuven,
Belgium; e-mail: jaak.jaeken@uz.kuleuven.ac.be
[2]Centre for Human Genetics, University of Leuven, Leuven, Belgium;
e-mail: Gert.Matthijs@med.kuleuven.ac.be

Key Words CDG, CDG-x, N-glycan, O-glycan, transferrin

■ **Abstract** Congenital disorders of glycosylation (CDG) are a rapidly growing
group of genetic diseases that are due to defects in the synthesis of glycans and in
the attachment of glycans to other compounds. Most CDG are multisystem diseases
that include severe brain involvement. The CDG causing sialic acid deficiency of N-
glycans can be diagnosed by isoelectrofocusing of serum sialotransferrins. An efficient
treatment, namely oral D-mannose, is available for only one CDG (CDG-Ib). In many
patients with CDG, the basic defect is unknown (CDG-x). Glycan structural analysis, yeast genetics, and knockout animal models are essential tools in the elucidation
of novel CDG. Eleven primary genetic glycosylation diseases have been discovered
and their basic defects identified: six in the N-glycan assembly, three in the N-glycan
processing, and two in the O-glycan (glycosaminoglycan) assembly. This review summarizes their clinical, biochemical, and genetic characteristics and speculates on further
developments in this field.

INTRODUCTION

Congenital defects of glycosylation (CDG), formerly called carbohydrate deficient glycoprotein syndromes (CDGS) (45a), are genetic diseases caused by deficient glycosylation of glycoconjugates, such as glycoproteins and glycolipids.
The glycans on proteins are either N-linked (to the amide group of asparagine
via an N-acetylglucosamine residue) or O-linked (to the hydroxyl group of serine or threonine via an N-acetylgalactosamine or a xylose residue). In humans,
most known CDG are N-glycosylation defects; only a few defects are known in
O-glycosylation, specifically in glycosaminoglycan synthesis.

The synthesis of the N-glycans encompasses a much longer pathway than that
of the O-glycans (5, 21, 91, 129) because N-glycosylation, apart from an assembly
pathway, also comprises a processing pathway that is lacking in O-glycosylation.
N-glycosylation encompasses three cellular compartments: the cytosol, the endoplasmic reticulum (ER), and the Golgi. In the cytosol, the mannose donor,
GDP-mannose, is synthesized from fructose 6-phosphate, an intermediate of the

glycolytic pathway. N-glycan assembly is carried out first on the cytoplasmic and then on the lumenal side of the ER. This results in the formation of the lipid-linked oligosaccharide dolichylpyrophosphate-$GlcNAc_2Man_9Glc_3$ (Figure 1, see color insert). The oligosaccharide moiety of this compound is then transferred to selected asparagines of the nascent proteins. Finally, the glycan of the newly formed glycoprotein is processed first in the ER, by trimming off the three glucoses, and further in the Golgi, by trimming off six mannoses and replacing these with two residues each of N-acetylglucosamine, galactose, and sialic acid.

O-glycan biosynthesis starts in the Golgi, and the initiating event is the addition of GalNAc (or xylose in the case of glycosaminoglycans) to serine or threonine residues. O-glycans are less branched than most N-glycans. They are generally found on mucins, but some proteins contain O-glycans either as short chains or as elongated bi-antennary structures (99).

Most glycosylation disorders are severe, multisystem diseases, underlining the extremely important biological roles of glycans (130, 155). The first patients with a CDG were reported in 1980 (78). Since then, investigators have discovered 11 CDG: 6 in the N-glycan assembly: phosphomannomutase deficiency (CDG-Ia), phosphomannose isomerase deficiency (CDG-Ib), glucosyltransferase I deficiency (CDG-Ic), mannosyltransferase VI deficiency (CDG-Id), dolichol-phosphate-mannose synthase-1 deficiency (CDG-Ie), and CDG-If; 3 in the N-glycan processing: N-acetylglucosaminyltransferase II deficiency (CDG-IIa), glucosidase I deficiency (CDG-IIb), and GDP-fucose transporter deficiency; and 2 in the O-glycan assembly: galactosyltransferase I deficiency and EXT1/EXT2 complex deficiency (66–68, 159).

DISORDERS OF N-GLYCAN ASSEMBLY

Phosphomannomutase Deficiency (CDG-Ia)

PHENOTYPE The phosphomannomutase (PMM) deficiency (MIM 212065, 601785) is the longest and, thus, best known CDG. An estimated 300 patients are known worldwide. It has been found in more than twenty different populations from Asia, Australia, Europe, and North and South America (3, 7, 14, 19, 26, 37, 40, 64, 69, 71, 97, 109, 112, 121, 126). The clinical picture comprises mild to severe neurological disease, mild to pronounced dysmorphy, and variable involvement of many organs. In the neonatal period and infancy, common striking features are abnormal eye movements, alternating internal strabism, and axial hypotonia. Variable feeding problems occur (anorexia, vomiting, diarrhea). Psychomotor retardation is constant and generally marked. Later, retinitis pigmentosa develops, and though less frequent, stroke-like episodes, epilepsy, and joint contractures may also develop. Walking without support is seldom achieved, but as a rule there is no psychomotor regression. Dysmorphy ranges from mild and aspecific to rather characteristic, particularly as an abnormal subcutaneous adipose tissue distribution with fat pads and nipple retraction. A minority of infants develop severe and mostly fatal organ disease, such as acute cerebral hemorrhage, cardiomyopathy with or without pericardial effusion, liver failure, or nephrotic syndrome (11, 14, 24, 25, 54, 59, 76, 77, 82, 92, 93, 149, 153). Nearly all organs show structural/morphological abnormalities, as demonstrated by radiological and histological techniques. The most important of these are olivopontocerebellar hypoplasia, decreased myelin, multivacuolar inclusions in the Schwann cells of peripheral nerves, renal cysts, liver fibrosis, lamellar inclusions in the lysosomes of the hepatocytes (but not of the Kupffer cells), and osteopenia (32, 46, 48, 98, 111, 144, 145).

Mortality is about 20% in the first years of life. Adults have a stable mental retardation, variable peripheral neuropathy, and sometimes severe kyphoscoliosis and premature aging (137). Hypogonadism also occurs in females. In spite of all these handicaps, most patients have an extrovert and cheerful personality. More recently, patients with a very mild presentation have been identified. As a result, the clinical criteria have been redefined: All CDG-Ia patients showed mental retardation, hypotonia, cerebellar hypoplasia, and strabismus, but the classical hallmarks, i.e., the inverted nipples and fat pads, are not always present (50).

The glycosylation defect causes abnormalities in a large number of glycoproteins. This has been best documented with regard to serum glycoproteins (10, 13, 39, 51, 52, 55, 57, 60, 93, 134, 139, 140, 146, 151). Most glycoprotein concentrations or enzyme activities in serum are decreased (striking examples are clotting factor XI and cholinesterase); the others are increased, such as lysosomal enzyme activities (e.g., arylsulphatase A and β-glucuronidase), or normal. Isoelectrofocusing (IEF) of serum glycoproteins shows a cathodal shift due to the deficiency of sialic acid, a negatively charged monosaccharide. This feature has been applied for diagnostic purposes (79, 135), and IEF of serum transferrin is still the most widely used screening test for CDG. In PMM deficiency and in the other defects of N-glycan assembly, a so-called type 1 pattern is obtained, characterized by a decrease of anodal fractions and an increase of disialo- and asialotransferrin (Figure 2). The reduction in GDP-mannose level reduces the dolichol pyrophosphate-linked oligosaccharide pool. Entire sugar chains are missing from the glycoproteins, leaving glycosylation sites unoccupied and thus resulting in a general hypoglycosylation. This pattern is opposed to the type 2 pattern that also shows an increase of the threesialo- and monosialofractions, most likely because of the incorporation of truncated or monoantennary sugar chains. It is seen in defects of the N-glycan processing (see below). It has to be noted that the transferrin IEF test can be normal in well documented PMM deficiency (47). On the other hand, an abnormal pattern does not always imply CDG; a transferrin protein variant or an artifact must always be considered first (Figure 2). Hypoglycosylation of a brain glycoprotein (β-trace protein) has also been reported (124). Remarkably, a normal glycosylation pattern is found in the fetus (28).

In terms of the biochemical defect, a decisive contribution to its elucidation was the finding that serum transferrin from CDG-Ia patients lacks one or both of the

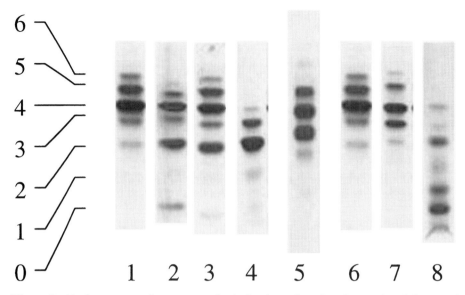

Figure 2   Basic patterns of serum transferrin isoelectrofocusing. Lanes 1 and 6, normal pattern; lanes 2 and 3, type 1 pattern; lane 4, type 2 pattern; lane 5, CDG-x; lane 7, transferrin protein variant; lane 8, the artifactual pattern obtained when using normal EDTA plasma. Figures on the left indicate sialotransferrin fractions 0 to 6.

two glycans (156, 163). This suggested a defect in an early glycosylation step and eventually led to the discovery of the enzymatic defect 15 years after the clinical report of the index patients (31, 65, 154).

In affected patients, the PMM defect is associated with a decrease of mannose 1-phosphate, GDP-mannose, GDP-fucose, and dolichyl-phosphomannose in their fibroblasts and a decrease of mannose levels in their serum (42, 43). However, the substrate mannose 6-phosphate is not increased. It was subsequently shown that two different PMM isozymes exist in human tissues and that only PMM2 is deficient in these patients (105, 106). In most patients, very low activities (<5% of mean normal activity) have been found in liver, leukocytes, fibroblasts, or lymphoblasts. Remarkably, in some patients significantly higher activities (up to 25% of the normal activity) have been recorded in fibroblasts or lymphoblasts versus leukocytes (50). PMM1 and PMM2 have different expression patterns in human tissues. It is noteworthy that the latter is only weakly expressed in brain, which is one of the most severely affected organs in CDG-Ia (122).

The three-dimensional structure of PMM2 is still unknown, but the protein belongs to the haloacid dehalogenase superfamily of proteins, which are characterized by the conservation of three different motifs that are probably involved in the catalytic activity (6). The reaction mechanism of the enzyme involves the phosphorylation of the first aspartate in an extremely conserved DXDXT/V sequence

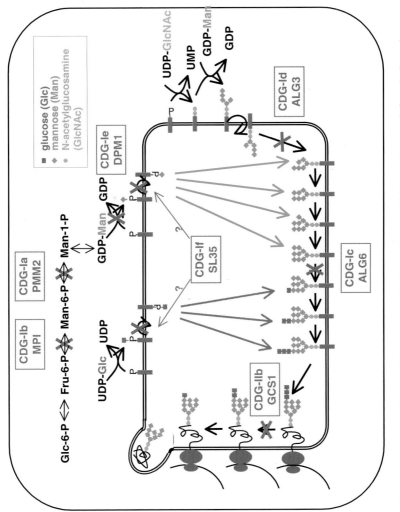

Figure 1 Scheme of the cytosolic and ER part of the N-glycosylation pathway, with known human defects indicated.

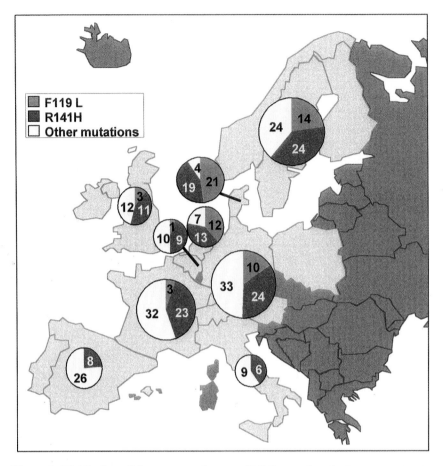

Figure 3 Distribution of the two most frequent *PMM2* mutations in Europe. R141H is found in every population, except Japan (not shown). The F119L is the "Viking" mutation, which shows a clear gradient from North to South (16, 17). The size of the pie corresponds roughly to the number of patients that are known in the different countries. No patients are known in the darker gray countries. Data for Spain and Portugal; the UK and Ireland; Germany, Austria, Switzerland, and Poland; and Sweden, Finland, and Norway were pooled. The number of other mutations ranges from 4 in Denmark to 13 in Spain and Portugal (see 102).

that is close to the amino-terminus of the enzyme (Asp-12) and is also found in a series of other phosphatases and phosphomutases (31). No efficient treatment is known for this disease (97), although correction of the glycosylation defect in vitro by supplementation of mannose or deprivation of glucose has been reported (90, 115–117).

GENOTYPE   The PMM2 protein is encoded by the *PMM2* gene, located on chromosome 16p13 (105). The gene is relatively small and the coding region is composed of 8 exons. The open reading frame (ORF) of 738 nucleotides predicts a protein of 246 amino acids. The paralogous *PMM1* gene on chromosome 22q13 is not implicated in disease (106, 133). Both the human *PMM1* and the human *PMM2* gene were identified on the basis of the similarity of EST sequences from the IMAGE consortium with the sequence of yeast Sec53 (104). Mutation analysis of PMM2 in CDG-Ia patients revealed a plethora of mainly missense mutations (102). A thorough screening of the exons and flanking intron sequences of the *PMM2* gene essentially detected all mutations in CDG-Ia cases (15, 64, 102, 107). The mutational spectrum comprises one particularly frequent mutation R141H, which has been identified in all Caucasian populations, and a few mutations that show a founder effect in distinct populations. Figure 3 (see color insert) presents the frequency and regional distribution of these mutations. In Scandinavian countries, R141H and F119L together make up 72% of all mutations (15, 85, 86). As a result, the most common genotype is F119L/R141H. In contrast, the genotypes are much more heterogenous in other European countries, e.g., Spain, Portugal, France, or Italy (102).

There is a clear discrepancy between the frequency of the most prevalent *PMM2* mutation, R141H, and its occurrence in CDG-Ia. On the basis of the observed carrier frequency, approximately 1/20000 homozygotes for R141H are expected under Hardy-Weinberg equilibrium (132). Thus, one would expect to find the homozygous R141H/R141H genotype in 45% to 60% of the CDG-Ia patients. Not a single has been found and this is statistically significant. The lack of homozygotes for R141H cannot be explained by genetic drift or nonrandom mating, but it is easily explained by the severity of the mutation: The enzymatic activity of recombinant R141H protein is virtually zero (86, 123). Therefore, homozygosity for R141H is probably incompatible with life. The disease frequency is given by the frequency of compound heterozygotes for R141H and another mutation as well as the occurrence of combinations of the other mutations. The frequency of the other mutations is estimated to be between 1/300 and 1/400, depending on the population. The combined data allowed us to give a rough estimate of the frequency of the disease: It could be as high as 1/20000.

Because the R141H mutation is genetically lethal, it should disappear unless the loss of alleles is compensated by new mutations or by a heterozygous advantage. Remarkably, the common R141H mutation is associated with the same haplotype in most cases (132). The mutational event must have occurred at least 200 generations ago (132). This implies a heterozygous advantage for this mutation, whereby the coefficient for fitness is 1.06 in favor of the mutated allele. Because this effect is small, pinpointing the selective pressure that maintains this mutation in the gene pool will be difficult.

Phosphomannose Isomerase Deficiency (CDG-Ib)

PHENOTYPE   Although the basic defect is in the enzyme just preceding phosphomannomutase, the clinical picture of CDG-Ib (MIM 154550, 602579) is markedly different from that of CDG-Ia. At least 12 patients, belonging to nine families, have been diagnosed. The majority showed a hepatic-intestinal disease with liver fibrosis and protein-losing enteropathy, which in some was associated with coagulation disturbances and/or hyperinsulinemic hypoglycemia (2, 9, 36, 73, 110, 150). In one family with three affected siblings, the main feature was prolonged episodic vomiting, sometimes associated with diarrhea (34, 35). One patient had only transient liver disease (73), and another subject was a clinically normal woman whose affected sibling died at age 5 (87, 119). Symptoms started between 2 and 12 months, and there was no, or only transient, neurological involvement. Three patients have died. There is a report from 1986 on four infants who most probably suffered from the same disease and who died between 4 and 21 months (120). We speculate that many patients with this disorder are not recognized as such. What led to the diagnosis of CDG in these patients? The observation of a profound deficiency of antithrombin III, a common feature of CDG, was the primary reason that we, and other investigators, performed the transferrin assay and diagnosed CDG-Ib. The blood biochemical abnormalities are similar to those found in PMM2 deficiency. Contrary to PMM2 deficiency, incorporation of $[2-^3H]$-mannose in fibroblasts into newly synthesized glycoproteins is increased (73, 110). Phosphomannose isomerase (PMI) deficiency is noted in leukocytes, fibroblasts, and liver. PMI is widely distributed in human tissues, and the molecule contains zinc. The use of metal chelators should thus be avoided in sample preparation and measurement of the enzyme (72).

The clinical differences between PMI and PMM2 deficiency, in particular with regard to the neurological manifestations, are intriguing. Several explanations have been provided (72). One of these is that brain hexokinase, which can convert mannose to mannose 6-phosphate thus bypassing PMI, has a rather high affinity for mannose, whereas glucokinase, the major hexokinase present in hepatocytes, has a very low affinity for mannose and is thus less efficient in phosphorylating this substrate.

PMI deficiency is the only known CDG for which an efficient treatment is available, namely oral D-mannose administration. The ability to directly convert mannose to mannose 6-phosphate by hexokinases made this possible. Doses of 100–150 mg/kg three to six times per day are effective, but the biochemical parameters normalize only after several months of treatment (9, 36, 110). Screening for this potentially lethal disorder is worthwhile because this treatment exists.

GENOTYPE   Based on its similarity to yeast, the human mannose 6-phosphate isomerase (MPI), or PMI, cDNA sequence was originally cloned by Proudfoot et al (125). The gene is composed of eight exons, spans only 5 kb (131), and is located on chromosome 15 (152). The protein contains 423 amino acids, and the structure (of the *Candida albicans ortholog*) of the protein has been published (29).

To date, 12 different mutations have been found in nine families (36, 73, 110, 131). Ten of the 12 mutations are missense mutations, mostly affecting amino acids conserved among human, mouse, and *C. albicans* PMI. These amino acids may be important for the structural integrity or catalytic activity of the enzyme. Only mutation Y255C (36) affects a semiconserved position. M51, D131, M138, and S102 are structurally close to the active site of the enzyme. R152, R219, G250, and Y255 are positioned in the helical domain, and I398 and R418 are in the C-terminal domain.

In the first patient with CDG-Ib, as described by Niehues et al (110), a missense mutation, R219Q, was originally identified at the cDNA level. The second mutation, an insertion of C in exon 3 (c. 166-167insC), escaped identification until the patient's DNA was sequenced at the genomic level (131). This frameshift mutation results in a premature stop after codon 62 and therefore must result in an inactive protein. Apparently, the mRNA is unstable. In another patient who was compound heterozygous for the IVS5-1G>C splice mutation and had a missense mutation (R418H), the second allele was also overrepresented at the mRNA level (131). Mutation analysis on genomic DNA is clearly an asset for the identification of mutations that affect the stability or splicing of the mRNA.

In CDG-Ib, three patients were homozygous for the mutations M51T, D131N, and R152Q, respectively (131). The presence of homozygous patients is expected for a rare recessive disease.

Glucosyltransferase I Deficiency (CDG-Ic)

PHENOTYPE   About 30 patients, only half of which have been reported, with CDG-Ic (MIM 603147, 604566) are known to the authors (22, 49, 53, 61, 62, 88). The clinical presentation is mainly neurological but milder than in PMM2 deficiency. It comprises moderate psychomotor retardation, pronounced axial hypotonia, strabismus, epilepsy, and in a few patients, ataxia. There is no important cerebellar hypoplasia, no peripheral neuropathy, and inconstant retinitis pigmentosa. Serum transaminases can be increased during infections. Hypoalbuminemia and proteinuria, features of CDG-Ia, are absent. A number of serum glycoproteins are decreased, and levels of factor XI and apolipoprotein B are particularly low. IEF of serum transferrin shows a type 1 pattern. Lipid-linked oligosaccharide analysis in fibroblasts reveals an accumulation of dolichylpyrophosphate-linked $Man_9GlcNAc_2$, owing to a deficiency of dolichyl-P-Glc: $Man_9GlcNAc_2$-pyrophosphate-dolichyl glucosyltransferase (ALG6).

This is the first CDG whose basic defect was elucidated on the basis of yeast genetics (22, 88). Noteworthy, the first CDG-Ic patients were also compiled as a group, preselected based on clinical examination (49). However, it is now clear that the clinical presentation is less homogeneous than originally described (22).

GENOTYPE   The ALG6 gene is much larger than the PMM2 or MPI genes, and this hampers molecular diagnostics. The gene has 14 exons and spans 55 kb. It is located on chromosome 1p22.3 (62). The ORF of 1521 bp predicts a protein of 507 amino acids, with an identity of 32% to *Saccharomyces cerevisiae* alg6p (61). In the original publication, all four patients from a related family were homozygous for the A333V mutation (61). This mutation has also been identified in other patients, and haplotype analysis revealed a founder effect for this mutation. In seven additional patients from six families, apart from A333V, two other mutant alleles were identified: IVS3+5 G>A and the combination of F304S and S478P (62). The detrimental effect of the A333V, the splice mutant, and the S478P mutations was confirmed by the lack of complementation of alg6 yeast mutants. Interestingly, none of the missense mutations completely abolished the activity, indicating the important function of ALG6 for N-linked glycosylation of proteins. However, the clinical presentation in another CDG-Ic patient (originally described as CDG type V) is very similar to the above, even though the patient was homozygous for a 3 bp deletion (del1299) in combination with the F304S variant (53, 88). The del1 299 mutation was also observed in other patient (157). F304S is most likely a frequent polymorphism.

Again, the detection of a mutation in a splicing donor site underlined the importance of analyzing the gene sequence at the genomic level.

Mannosyltransferase VI Deficiency (CDG-Id)

PHENOTYPE   In 1995, Stibler et al reported on an infant with nearly absent psychomotor development, hypsarrhythmia, postnatal microcephaly, optic atrophy, iriscoloboma, and brain and corpus callosum atrophy (141). IEF of serum transferrin showed a type 1 pattern but without increase of asialotransferrin. In 1999, the lipid-linked oligosaccharide analysis revealed an accumulation of the dolichylpyrophosphate-$Man_5GlcNAc_2$ precursor in fibroblasts of this patient. Subsequently, the basic defect in this child was identified as a deficiency of the ER mannosyltransferase that attaches a mannosylresidue from dolicholphosphomannose to $Man_5GlcNAc_2$-pyrophosphate-dolichol(dolichyl-P-Man:$Man_5GlcNAc_2$-pyrophosphate-dolichylmannosyltransferase) (MIM 601110) (89). The patient was initially described as CDG type IV, but the disease was renamed CDG-Id, according to the novel nomenclature (45a).

GENOTYPE   The cDNA of the ALG3, or the human Not 56-like protein gene, predicts a protein of 438 amino acids with 30% identity to yeast alg3p. A search into the publicly available databases revealed that the gene has nine exons and spans at least 5 kb. To date, only one patient has been described who is homozygous for the G118D mutation in ALG3 (89). The human Not 56–like protein can indeed substitute for yeast alg3p, and the mutant cannot correct the size of the LLO in the deficient yeast strain, which leads to an accumulation of dolichylpyrophosphate-$Man_5GlcNAc_2$.

Dolichol-Phosphate-Mannose Synthase-1 Deficiency (CDG-Ie)

PHENOTYPE   This CDG (MIM 603503) was identified in four children (from three families) with pronounced psychomotor retardation, severe epilepsy, hypotonia, failure to thrive, and mild dysmorphy. Serum transaminases and creatine phosphokinase were mildly to moderately elevated. IEF of serum transferrin showed a type 1 pattern with only little or no increase of asialotransferrin. Analysis of the lipid-linked oligosaccharides in the ER of fibroblasts revealed an accumulation of dolichylpyrophosphate-$Man_5GlcNAc_2$ instead of $Glc_3Man_9GlcNAc_2$, due to a reduced activity of dolichol-phosphate-mannose synthase (63, 83, 114).

GENOTYPE   The chromosomal localization and genomic organization of the human DPM1 gene were determined in silico: A contig sequence in the public database assigned to chromosome 20q13 contains the entire gene. It spans 23.6 kbp, consists of nine exons, and encodes a protein of 260 amino acids (63). Only three patients have been analyzed at the molecular level (63, 83). One patient was homozygous for R92G, whereas the two other patients were compound heterozygous for R92G with either a 13 bp deletion (331-343del13) in exon 4 or a single bp deletion (628 delC) in exon 8.

Comparison of DPM1 protein sequences from several species indicated that the arginine residue at position 92 was semiconserved.

CDG-If

This novel type of CDG has been identified in two siblings (70). One of them is now 18 years old and shows dwarfism, congenital ichthyosis, psychomotor retardation, and retinopathy. The other sibling died at 3 months with a similar syndrome. The proposita showed a transient growth hormone deficiency and decreased values of serum LDL-cholesterol, cholinesterase, progesterone, and oestradiol as well as blood factor XI, antithrombin, protein C, and protein S. Electronmicroscopy of a liver biopsy revealed lamellar lysosomal inclusions. Patient's fibroblasts accumulated incomplete oligosaccharides in the ER and transferred complete as well as truncated $Glc_3Man_9GlcNAc_2$ oligosaccharides to protein. Surface expression of CD59, a GPI-anchored protein was reduced; however, the dolichylphosphate (Dol-P) level in the cells was normal. Sequence analysis of the *SL15* (*Lec35*) gene, which is involved in the utilization of Dol-P, revealed mutations in the index patient and in two other patients (B. Schenk, T. Imbach, C.G. Frank, C.E. Grubenmann,

G.V. Raymond, H. Hurvitz, J. Jaeken, E.G. Berger, G. Matthijs, T. Hennet, & M. Aebi, manuscript in preparation).

DISORDERS OF N-GLYCAN PROCESSING

N-acetylglucosaminyltransferase II Deficiency (CDG-IIa)

PHENOTYPE   Three patients have been reported with this severe mental retardation dysmorphy syndrome (MIM 212066, 602616) (27, 74, 75, 128). They had epilepsy and a striking stereotypic behavior. Laboratory investigations revealed lowered serum values of a number of glycoproteins, similar to those with PMM deficiency. There was an increase of serum glutamic-oxaloacetic, but not of glutamic-pyruvic transaminase. Isoelectrofocusing of serum transferrin showed a type 2 pattern (see PMM Deficiency) but with nearly absent tetrasialotransferrin. Contrary to the defects in the CDG-I group, liver biopsy showed no lysosomal inclusions. Fine structure analysis of the glycans on serum transferrin revealed that some of the normal, disialo-biantennary N-glycans are replaced by truncated, monosialo-monoantennary N-glycans and pinpointed the defect to the GlcNAc-transferase II (GnT II) in the Golgi [for details, see (74)]. GnT II activity was reduced by over 98% in fibroblasts and mononuclear blood cells from patients (75).

GENOTYPE   The human UDP-N-acetylglucosamine:alpha-6-D-mannoside-beta-1,2-N-acetylglucosaminyltransferase II gene (MGAT2) is located on chromosome 14q21. The coding region contains only one exon, with an open reading frame of 1341 bp that encodes a protein of 447 amino acids (147).

Mutation analysis of the GnT II coding sequence (*MGAT2* gene) in the two patients revealed that they were homozygous for the S290F and H262R mutations, respectively (148). Both mutations occur in the C-terminal catalytic domain, locations that are conserved between rat and human GnT II and inactivate the enzyme (148). Since the original description and molecular characterization of CDG-IIa, only one other case has been identified (33). This patient was compound heterozygous for a missense (N318D) and a nonsense (C339X) mutation. One probable explanation for the paucity of diagnosed cases is that the GnT II deficiency is a very severe disorder. This is illustrated by the knockout mouse model. Homozygous knockout (*Mgat2*−/−) mice survive to term, but they are born stunted with various congenital abnormalities and die early in the neonatal phase (23).

Glucosidase-I Deficiency (CDG-IIb)

PHENOTYPE   A defect in the first step of the N-glycan processing was reported in a neonate with dysmorphic features and generalized hypotonia (38) (MIM 601336). The clinical course was characterized by hypoventilation, seizures, feeding problems, hepatomegaly, and fatal outcome at 2.5 months. Interestingly, isoelectrofocusing of serum transferrin was normal, whereas that of serum hexosaminidase showed a slight cathodal shift. Thin layer chromatography of oligosaccharides in urine revealed an abnormal band that was identified as the tetrasaccharide Glc(α1-2)Glc(α1-3)Glc(α1-3)Man. Electronmicroscopy showed lamellar inclusions in lysosomes of liver parenchymal cells and macrophages as well as numerous empty, membrane-bound vacuoles in neurons of the frontal and occcipital brain lobes. Severe glucosidase I deficiency was found in the liver and cultured skin fibroblasts (<3% residual activity). This observation shows that, in neonates with a possible diagnosis of a glycosylation defect, analysis of oligosaccharides in the urine might be a worthwhile test.

GENOTYPE   The full-length human GCS1 cDNA contains 2881 bp and encodes a protein of 834 amino acids, which corresponds to a molecular mass of $\sim$92 kDa (80). GCS1 is located on chromosome 2p12-p13 (81). The only known patient, described above, is compound heterozygous for two missense mutations in the GCS1 gene, R486T and F652L (38).

GDP-fucose Transporter Defect

The three reported patients had severe developmental retardation, microcephaly, hypotonia, retarded growth, and craniofacial dysmorphy (44, 95). They also suffered from recurrent infections, with marked leukocytosis even during and between the infections. The biochemical hallmark of this disease is a generalized hypofucosylation of N- and O-glycoproteins. The neutrophils of these patients lack sialyl-Lewis X, a fucose-containing carbohydrate ligand of the selectin family of cell adhesion molecules that is required for recruitment of neutrophils to sites of inflammation (hence the name, leukocyte adhesion deficiency II syndrome). The isoelectrofocusing pattern of serum sialotransferrins is normal because this glycosylation disorder is not accompanied by a deficiency of glycoprotein sialic acid.

Transport of GDP-fucose into the Golgi by a newly described, specific GDP-fucose transporter is deficient (95; K. von Figura, personal communication).

DISORDERS OF O-GLYCAN ASSEMBLY

Galactosyltransferase I Deficiency

PHENOTYPE   One patient has been reported with this so-called progeroid variant of Ehlers-Danlos syndrome (127) (MIM 130070, 604327). This child had the appearance of an old man and showed psychomotor retardation, macrocephaly, hyperlaxity of the joints, and loose, elastic skin. A defect was found in the common linkage region of glycosaminoglycans (O-linked xylose-galactose-galactose-glucuronic acid) and, more specifically, in the attachment of the first galactose to xylose. Galactosyltransferase I activity was less than 5% of normal in the patient's fibroblasts.

GENOTYPE   The cDNA for the galactosyltransferase (XGALT-1 or B4GALT7) was only recently isolated by searching an EST database (113). The gene shows high homology (38%) to the sqv3 gene of *Caenorhabditis elegans*, and its product shows specific activity of galactosyltransferase on p-nitro-phenyl-beta-D-xylopyranoside with beta-1,4 linkage. The patient was compound heterozygous for A186D and L206P mutations in the gene. The former is a mild mutation, and the mutant protein retains approximatively 50% of activity. However, the latter is a severe mutation with no residual activity of the mutant protein (113).

Golgi Localized EXT1/EXT2 Complex Deficiency

PHENOTYPE   This defect has been identified in patients with the multiple exostoses syndrome, which has an autosomal dominant inheritance (158) (MIM 133700, 133701). More than 1000 patients have been reported with this syndrome. Its estimated prevalence is 1/50000. It is characterized by diaphyseal juxtaepiphyseal, cartilage capped outgrowths that cause deformity. They are often present at birth but usually not diagnosed until early childhood. Their growth slows at adolescence and stops in adulthood. They are also most prominent at the ends of long bones. There is a 3% incidence of sarcoma from these lesions. Complications may arise from compression of peripheral nerves and blood vessels (158). The basic defect is in the Golgi-localized EXT1/EXT2 complex that has glucuronyltransferase as well as N-acetyl-D-glucosaminyltransferase activities and that catalyses the polymerization of heparan sulfate (84, 108).

GENOTYPE   The EXT1 and EXT2 loci were localized using linkage analysis in large families with the autosomal dominant hereditary multiple exostoses, and the genes were identified by positional cloning. EXT1 is located on chromosome 8q23-q24, and EXT2 is located on 11p11-p12. A third EXT locus, which is probably only involved in rare instances, (EXT3) lies on chromosome 19p (162). The availability of the genes did not lead investigators to the identification of the biological function because the sequence did not show any recognized functional domains or structural motifs. Recent functional data revealed that both proteins are in fact glycosyltransferases, which are involved in heparan sulphate biosynthesis (94, 108). EXT1 and EXT2 encode homologous proteins of 746 and 718 amino acids, respectively (4, 143, 161). The overall identity is 31%. These genes are large: EXT1 contains 11 exons and spans more than 250 kb; EXT2 contains 16 exons and spans more than 100 kb (30, 96). The *Drosophila* homologue of EXT1 is the tout-velu gene, which has a role in hedgehog (hh) signaling (12). Indian hh (Ihh) is one of the mammalian homologues of hh, and it regulates cartilage differentiation. The current model for EXT involvement in the development of exostoses states that the mutation in the glycosyltransferase(s) impairs the synthesis of a glycosaminoglycan that normally regulates the diffusion of Ihh and thus disrupts the negative feedback loop of chondrocyte differentiation. Interestingly, both the EXT1 and EXT2 regions show loss of heterozygosity (LOH) in EXT-related and non-EXT-related chondrosarcomas (18).

Almost 50 different *EXT1* and 25 different *EXT2* mutations have been reported (162). In the case of *EXT1*, 73% of all mutations cause an interruption of the ORF and premature termination of the EXT1 protein. These are nonsense and splice site mutations as well as small deletions and insertions. The other mutations are missense mutations or in-frame deletions and insertions. Only one mutation has been observed in more than one family, and it may be a recurrent mutation. For *EXT2*, a comparable mutation spectrum has been observed, also with a majority of nonsense, splice, and frameshift mutations as well as only a few known missense mutations and no recurrent mutations.

CONCLUSIONS AND OUTLOOK

From the large number of enzymes and transporters involved in glycosylation, it is clear that only a small "tip of the iceberg" regarding glycosylation disorders is known.

The number of patients with an unidentified glycosylation defect is steadily growing (CDG-x). They have known phenotypes (41, 138, 142) but also novel phenotypes such as hemolytic anemia, profound thrombocytopenia, corneal dystrophy, Budd-Chiari syndrome, etc. (1, 20, 45, 58). Their identification will be greatly facilitated by the availability of yeast mutants (136) and knockout mice (8, 23, 56). A screening for CDG is recommended, although investigators should know that a normal result does not exclude a CDG (47). Screening is accomplished by performing IEF of serum sialotransferrins in any unexplained clinical disorder. A major challenge is to devise an efficient treatment, particularly for CDG-Ia, which is by far the most frequent CDG (160).

APPENDIX

A public website on CDG and a mutation database is currently maintained at http://www.kuleuven.ac.be/med/cdg.

Visit the Annual Reviews home page at www.AnnualReviews.org

LITERATURE CITED

1. Acarregui M, George TN, Rhead WJ. 1998. Carbohydrate-deficient glycoprotein syndrome type 1 with profound thrombocytopenia and normal phosphomannomutase and phosphomannose isomerase activities. *J. Pediatr.* 133:697–700
2. Adamowicz M, Matthijs G, Van Schaftingen E, Jaeken J, Rokicki D, et al. 2000. New case of phosphomannose isomerase deficiency. *J. Inherit. Metab. Dis.* 23 (Suppl. 1):184
3. Agamanolis DP, Potter JL, Naito HK, Robinson HB Jr, Kulasekaran T. 1986. Lipoprotein disorder, cirrhosis, and olivopontocerebellar degeneration in two siblings. *Neurology* 36:674–81

4. Ahn J, Ludecke HJ, Lindow S, Horton WA, Lee B, et al. 1995. Cloning of the putative tumour suppressor gene for hereditary multiple exostoses (EXT1). *Nat. Genet.* 11:137–43
5. Alton G, Hasilik M, Niehues R, Panneerselvam K, Etchison JR, et al. 1998. Direct utilization of mannose for mammalian glycoprotein biosynthesis. *Glycobiology* 8:285–95
6. Aravind L, Galperin MY, Koonin EV. 1998. The catalytic domain of the P-type ATPase has the haloacid dehalogenase fold. *Trends Biochem. Sci.* 23:127–29
7. Artigas J, Cardo E, Pineda M, Nosas R, Jaeken J. 1998. Phosphomannomutase deficiency and normal pubertal development. *J. Inherit. Metab. Dis.* 21:78–79
8. Asano M, Furukawa K, Kido M, Matsumoto S, Umesaki Y, et al. 1997. Growth retardation and early death of beta-1,4-galactosyltransferase knockout mice with augmented proliferation and abnormal differentiation of epithelial cells. *EMBO J.* 16:1850–57
9. Babovic-Vuksanovic D, Patterson MC, Schwenk WF, O'Brien JF, Vockley J, et al. 1999. Severe hypoglycemia as a presenting symptom of carbohydrate-deficient glycoprotein syndrome. *J. Pediatr.* 135:775–81
10. Barone R, Carchon H, Jansen E, Pavone L, Fiumara A, et al. 1998. Lysosomal enzyme activities in serum and leukocytes from patients with carbohydrate-deficient glycoprotein syndrome type IA (phosphomannomutase deficiency). *J. Inherit. Metab. Dis.* 21:167–72
11. Barone R, Pavone L, Fiumara A, Bianchini R, Jaeken J. 1999. Developmental patterns and neuropsychological assessment in patients with carbohydrate-deficient glycoconjugate syndrome type IA (phosphomannomutase deficiency). *Brain Dev.* 21:260–63
12. Bellaiche Y, The I, Perrimon N. 1998. Tout-velu is a Drosophila homologue of the putative tumour suppressor EXT-1 and is needed for Hh diffusion. *Nature* 394:85–88
13. Bergmann M, Gross HJ, Abdelatty F, Möller P, Jaeken J, Schwartz-Albiez R. 1998. Abnormal surface expression of sialoglycans on B lymphocyte cell lines from patients with carbohydrate deficient glycoprotein syndrome type IA (CDGS IA). *Glycobiology* 8:963–72
14. Billette de Villemeur T, Poggi-Travert F, Laurent J, Jaeken J, Saudubray JM. 1995. Le syndrome d'hypoglycosylation des protéines: un nouveau groupe de maladies héréditaires à expression multisystémique. In *Journées Parisiennes de Pédiatrie*, ed. M Arthuis, F Beaufils, B Caille, pp. 119–24. Paris: Flammarion. 365 pp.
15. Bjursell C, Erlandson A, Nording M, Nilsson S, Wahlström J, et al. 2000. PMM2 mutation spectrum, including 10 novel mutations, in a large CDG type 1A family material with a focus on Scandinavian families. *Hum. Mutat.* 16:395–400
16. Bjursell C, Stibler H, Wahlstrom J, Kristiansson B, Skovby F. et al. 1997. Fine mapping of the gene for carbohydrate-deficient glycoprotein syndrome, type I (CDG1): linkage disequilibrium and founder effect in Scandinavian families. *Genomics* 39:247–53
17. Bjursell C, Wahlstrom J, Berg K, Stibler H, Kristiansson B, et al. 1998. Detailed mapping of the phosphomannomutase 2 (PMM2) gene and mutation detection enable improved analysis for Scandinavian CDG type I families. *Eur. J. Hum. Genet.* 6:603–11
18. Bovee JV, Cleton-Jansen AM, Wuyts W, Caethoven G, Taminiau AH, et al. 1999. EXT-mutation analysis and loss of heterozygosity in sporadic and hereditary osteochondromas and secondary chondrosarcomas. *Am. J. Hum. Genet.* 65:689–98
19. Briones P, Vilaseca MA, Garcia-Silva MT, Pineda M, Colomer J, et al. 2000. Congenital disorders of glycosylation

(CDG) may be underdiagnosed when mimicking mitochondrial disease. *J. Inherit. Metab. Dis.* 23 (Suppl. 1):180

20. Buist N, Grompe M, Steiner R, Miura Y, O'Brien J, Freeze H. 2000. Yet another congenital disorder of glycosylation! (CDGS). *J. Inherit. Metab. Dis.* 23 (Suppl. 1):187

21. Burda P, Aebi M. 1998. The dolichol pathway of *N*-linked glycosylation. *Biochim. Biophys. Acta* 1426:239–57

22. Burda P, Borsig L, de Rijk-van Andel J, Wevers R, Jaeken J, et al. 1998. A novel carbohydrate-deficient glycoprotein syndrome characterized by a deficiency in glucosylation of the dolichol-linked oligosaccharide. *J. Clin. Invest.* 102:647–52

23. Campbell R, Tan J, Schachter H, Bendiak B, Marth J. 1997. Targeted inactivation of the murine UDP-GlcNAc:α-6-D-mannoside β-1-2-N-acetylglucosaminyltransferase II gene. *Glycobiology* 7:1050

24. Carchon H, Van Schaftingen E, Matthijs G, Jaeken J. 1999. Carbohydrate-deficient glycoprotein syndrome type 1A (phosphomannomutase-deficiency). *Biochim. Biophys. Acta* 1455:155–65

25. Casteels I, Spileers W, Leys A, Lagae L, Jaeken J. 1996. Evolution of ophthalmic and electrophysiological findings in identical twin sisters with the carbohydrate deficient glycoprotein syndrome type 1 over a period of 14 years. *Br. J. Ophthalmol.* 80:900–2

26. Charlwood J, Clayton P, Keir G, Mian N, Young E, Winchester B. 1998. Prenatal diagnosis of the carbohydrate-deficient glycoprotein syndrome type 1A (CDG1A) by a combination of enzymology and genetic linkage analysis after amniocentesis or chorionic villus sampling. *Prenat. Diagn.* 18:693–99

27. Charuk JHM, Tan J, Bernardini M, Haddad S, Reithmeier RAF, et al. 1995. Carbohydrate-deficient glycoprotein syndrome type II: an autosomal recessive N-acetylglucosaminyltransferase II deficiency different from typical hereditary erythroblastic multinuclearity, with a positive acidified-serum lysis test (HEMPAS). *Eur. J. Biochem.* 230:797–805

28. Clayton P, Winchester B, di Tomaso E, Young E, Keir G, Rodeck C. 1993. Carbohydrate-deficient glycoprotein syndrome: normal glycosylation in the fetus. *Lancet* 341:956

29. Cleasby A, Wonacott A, Skarzynski T, Hubbard RE, Davies GJ, et al. 1996. The X-ray crystal structure of phosphomannose isomerase from Candida albicans at 1.7 Å resolution. *Nat. Struct. Biol.* 3:470–79

30. Clines GA, Ashley JA, Shah S, Lovett M. 1997. The structure of the human multiple exostoses 2 gene and characterization of homologs in mouse and *Caenorhabditis elegans*. *Genome Res.* 7:359–67

31. Collet JF, Stroobant V, Pirard M, Delpierre G, Van Schaftingen E. 1998. A new class of phosphotransferases phosphorylated on an aspartate residue in a DXDXT/V motif. *J. Biol. Chem.* 273:14107–12

32. Conradi N, De Vos R, Jaeken J, Lundin P, Kristiansson B. 1991. Liver pathology in the carbohydrate-deficient glycoprotein syndrome. *Acta Paediatr. Scand. Suppl.* 375:50–54

33. Cormier-Daire V, Amiel J, Vuillaumier-Barrot S, Tan J, Durand G, et al. 2000. Congenital disorders of glycosylation IIa cause growth retardation, mental retardation, and facial dysmorphism. *J. Med. Genet.* 37:875–76

34. de Koning TJ, Dorland L, van Diggelen OP, Boonman AMC, de Jong GJ, et al. 1998. A novel disorder of N-glycosylation due to phosphomannose isomerase deficiency. *Biochem. Biophys. Res. Commun.* 245:38–42

35. de Koning TJ, Nikkels PGJ, Dorland L, Bekhof J, De Schrijver JEAR, et al. 2000. Congenital hepatic fibrosis in 3 siblings

with phosphomannose isomerase deficiency. *Virchows Arch.* 437:101–5

36. de Lonlay P, Cuer M, Vuillaumier-Barrot S, Beaune G, Castelnau P, et al. 1999. Hyperinsulinemic hypoglycemia as a presenting sign in phosphomannose isomerase deficiency: a new manifestation of carbohydrate deficient glycoprotein syndrome treatable with mannose. *J. Pediatr.* 135:379–83

37. de Michelena MI, Franchi LM, Summers PG, De La Fuente C, Campos PJ, Jaeken J. 1999. Carbohydrate-deficient glycoprotein syndrome due to phosphomannomutase deficiency: the first reported cases from latin America. *Am. J. Med. Genet.* 84:481–83

38. De Praeter CM, Gerwig GJ, Bause E, Nuytinck LK, Vliegenthart JFG, et al. 2000. A novel disorder caused by defective biosynthesis of N-linked oligosaccharides due to glucosidase I deficiency. *Am. J. Hum. Genet.* 66:1744–56

39. de Zegher F, Jaeken J. 1995. Endocrinology of the carbohydrate-deficient glycoprotein syndrome type 1 from birth through adolescence. *Pediatr. Res.* 37: 395–401

40. Di Rocco M, Barone R, Adam A, Burlina A, Carrozzi M, et al. 2000. Carbohydrate-deficient glycoprotein syndromes: the Italian experience. *J. Inherit. Metab. Dis.* 23:391–95

41. Dorland L, de Koning TJ, Toet M, de Vries LS, van den Berg IET, Poll-The BT. 1997. Recurrent non-immune hydrops fetalis associated with carbohydrate-deficient glycoprotein syndrome. *J. Inherit. Metab. Dis.* 20 (Suppl.):88

42. Dupre T, Ogier-Denis E, Moore SEH, Cormier-Daire V, Dehoux M, et al. 1999. Alteration of mannose transport in fibroblasts from type I carbohydrate deficient glycoprotein syndrome patients. *Biochim. Biophys. Acta* 1453:369–77

43. Etchison JR, Freeze HH. 1997. Enzymatic assay of D-mannose in serum. *Clin. Chem.* 43:533–38

44. Etzioni A, Frydman M, Pollack S, Avidor I, Phillips ML, et al. 1992. Recurrent severe infections caused by a novel leukocyte adhesion deficiency. *N. Engl. J. Med.* 327:1789–92

45. Eyskens F, Ceuterick C, Martin JJ, Janssens G, Jaeken J. 1994. Carbohydrate-deficient glycoprotein syndrome with previously unreported features. *Acta Paediatr.* 83:892–96

45a. First Int. Workshop CDGS, Leuven, Belgium, Nov. 12–13. 1999. Carbohydrate-deficient glycoprotein syndromes become congenital disorders of glycosylation: an updated nomenclature for CDG. *Glycoconj. J.* 16:669–71

46. Fiumara A, Barone R, Nigro F, Sorge G, Pavone L. 1996. Familial Dandy-Walker variant in CDG syndrome. *Am. J. Med. Genet.* 63:412

47. Fletcher JM, Matthijs G, Jaeken J, Van Schaftingen E, Nelson PV. 2000. Carbohydrate-deficient glycoprotein syndrome: beyond the screen. *J. Inherit. Metab. Dis.* 23:396–98

48. Garel C, Baumann C, Besnard M, Ogier H, Jaeken J, Hassan M. 1998. Carbohydrate-deficient glycoprotein syndrome type 1: a new cause of dysostosis multiplex. *Skelet. Radiol.* 27:43–45

49. Grünewald S, Imbach T, Huijben K, Rubio-Gozalbo ME, Verrips A, et al. 2000. Clinical and biochemical characteristics of congenital disorder of glycosylation type Ic, the first recognized endoplasmic reticulum defect in N-glycan synthesis. *Ann. Neurol.* 47:776–81

50. Grünewald S, Schollen E, Van Schaftingen E, Jaeken J, Matthijs G. 2001. High residual activity of PMM2 in patients' fibroblasts: possible pitfall in the diagnosis of CDG-Ia (phosphomannomutase deficiency). *Am. J. Hum. Genet.* 68:347–54

51. Gu J, Kondo A, Okamoto N, Wada Y. 1994. Oligosaccharide structures of immunoglobulin G from two patients with carbohydrate-deficient glycoprotein syndrome. *Glycosylation Dis.* 1:247–52

52. Gu J, Wada Y. 1995. Aberrant expressions of decorin and biglycan genes in the carbohydrate-deficient glycoprotein syndrome. *J. Biochem.* 117:1276–79

53. Hanefeld F, Körner C, Holzbach-Eberle U, von Figura K. 2000. Congenital disorder of glycosylation-Ic: case report and genetic defect. *Neuropediatrics* 31:60–62

54. Harding B, Dunger DB, Grant DB, Erdohazi M. 1988. Familial olivopontocerebellar atrophy with neonatal onset: a recessively inherited syndrome with systemic and biochemical abnormalities. *J. Neurol. Neurosurg. Psychiatry* 51:385–90

55. Harrison HH, Miller KL, Harbison MD, Slonim AE. 1992. Multiple serum protein abnormalities in carbohydrate-deficient glycoprotein syndrome: pathognomonic finding of two-dimensional electrophoresis? *Clin. Chem.* 38:1390–92

56. Hennet T, Chui D, Paulson JC, Marth JD. 1998. Immune regulation by the ST6Gal sialyltransferase. *Proc. Natl. Acad. Sci. USA* 95:4504–9

57. Henry H, Tissot J-D, Messerli B, Markert M, Muntau A, et al. 1997. Microheterogeneity of serum glycoproteins and their liver precursors in patients with carbohydrate-deficient glycoprotein syndrome type I: apparent deficiencies in clusterin and serum amyloid P. *J. Lab. Clin. Med.* 129:412–21

58. Huemer M, Huber WD, Schima W, Moeslinger D, Holzbach U, et al. 2000. Budd-Chiari syndrome associated with coagulation abnormalities in a child with carbohydrate deficient glycoprotein syndrome type Ix. *J. Pediatr.* 136:691–95

59. Hutchesson ACJ, Gray RGF, Spencer DA, Keir G. 1995. Carbohydrate deficient glycoprotein syndrome: multiple abnormalities and diagnostic delay. *Arch. Dis. Child* 72:445–46

60. Iijima K, Murakami F, Nakamura K, Ikawa S, Yuasa I, et al. 1994. Hemostatic studies in patients with carbohydrate-deficient glycoprotein syndrome. *Thromb. Res.* 76:193–98

61. Imbach T, Burda P, Kuhnert P, Wevers RA, Aebi M, et al. 1999. A mutation in the human ortholog of the *Saccharomyces cerevisiae ALG6* gene causes carbohydrate-deficient glycoprotein syndrome type-Ic. *Proc. Natl. Acad. Sci. USA* 96:6982–87

62. Imbach T, Grünewald S, Schenk B, Burda P, Schollen E, et al. 2000. Multiallelic origin of congenital disorder of glycosylation (CDG)-Ic. *Hum. Genet.* 106:538–45

63. Imbach T, Schenk B, Schollen E, Burda P, Stutz A, et al. 2000. Deficiency of dolichol-phosphate-mannose synthase-1 causes congenital disorder of glycosylation type Ie. *J. Clin. Invest.* 105:233–39

64. Imtiaz F, Worthington V, Champion M, Beesley C, Charlwood J, et al. 2000. Genotypes and phenotypes of patients in the UK with carbohydrate-deficient glycoprotein syndrome type 1. *J. Inherit. Metab. Dis.* 23:162–74

65. Jaeken J, Artigas J, Barone R, Fiumara A, de Koning TJ, et al. 1997. Phosphomannomutase deficiency is the main cause of carbohydrate-deficient glycoprotein syndrome with type I isoelectrofocusing pattern of serum sialotransferrins. *J. Inherit. Metab. Dis.* 20:447–49

66. Jaeken J, Carchon H. 2000. What's new in congenital disorders of glycosylation? *Eur. J. Paediatr. Neurol.* 4:163–67

67. Jaeken J, Carchon H, Stibler H. 1993. The carbohydrate-deficient glycoprotein syndromes: pre-Golgi and Golgi disorders? *Glycobiology* 3:423–28

68. Jaeken J, Casaer P. 1997. Carbohydrate-deficient glycoconjugate (CDG) syndromes: a new chapter of neuropaediatrics. *Eur. J. Paediatr. Neurol.* 2/3:61–66

69. Jaeken J, Hagberg B, Strømme P. 1991. Clinical presentation and natural course of the carbohydrate-deficient glycoprotein syndrome. *Acta Paediatr. Scand. Suppl.* 375:6–13

70. Jaeken J, Imbach T, Schenk B, Smeets E, Carchon H, et al. 2000. A newly

recognized glycosylation defect with psychomotor retardation, ichthyosis and dwarfism. *J. Inherit. Metab. Dis.* 23 (Suppl. 1):186

71. Jaeken J, Matthijs G, Barone R, Carchon H. 1997. Carbohydrate deficient glycoprotein (CDG) syndrome type I. *J. Med. Genet.* 34:73–76

72. Jaeken J, Matthijs G, Carchon H, Van Schaftingen E. 2001. Defects of N-glycan synthesis. In *The Metabolic and Molecular Bases of Inherited Disease*, ed. CR Scriver, AL Beaudet, WS Sly, D Valle, pp. 1601–21. New York: McGraw-Hill

73. Jaeken J, Matthijs G, Saudubray JM, Dionisi-Vici C, Bertini E, et al. 1998. Phosphomannose isomerase deficiency: a carbohydrate-deficient glycoprotein syndrome with hepatic-intestinal presentation. *Am. J. Hum. Genet.* 62:1535–39

74. Jaeken J, Schachter H, Carchon H, De Cock P, Coddeville B, Spik G. 1994. Carbohydrate deficient glycoprotein syndrome type II: a deficiency in Golgi localised N-acetyl-glucosaminyltransferase II. *Arch. Dis. Child.* 71:123–27

75. Jaeken J, Spik G, Schachter H. 1996. Carbohydrate-deficient glycoprotein syndrome Type II: an autosomal recessive disease due to mutations in the N-acetylglucosaminyltransferase II gene. In *Glycoproteins and Disease*, ed. J Montreuil, JFG Vliegenthart, H Schachter, 30: 457–67. Amsterdam: Elsevier Sci. 486 pp.

76. Jaeken J, Stibler H. 1989. A newly recognized inherited neurological disease with carbohydrate-deficient secretory glycoproteins. In *Genetics of Neuropsychiatric Diseases*, ed. L Wetterberg, 51:69–80. New York: Macmillan. 363 pp.

77. Jaeken J, Stibler H, Hagberg B. 1991. The carbohydrate-deficient glycoprotein syndrome: a new inherited multisystemic disease with severe nervous system involvement. *Acta Paediatr. Scand. Suppl.* 375:71 pp.

78. Jaeken J, Vanderschueren-Lodeweyckx M, Casaer P, Snoeck L, Corbeel L,

et al. 1980. Familial psychomotor retardation with markedly fluctuating serum proteins, FSH and GH levels, partial TBG-deficiency, increased serum arylsulphatase A and increased CSF protein: a new syndrome? *Pediatr. Res.* 14:179

79. Jaeken J, van Eijk HG, van der Heul C, Corbeel L, Eeckels R, Eggermont E. 1984. Sialic acid-deficient serum and cerebrospinal fluid transferrin in a newly recognized genetic syndrome. *Clin. Chim. Acta* 144:245–47

80. Kalz-Fuller B, Bieberich E, Bause E. 1995. Cloning and expression of glucosidase I from human hippocampus. *Eur. J. Biochem.* 231:344–51

81. Kalz-Fuller B, Heidrich-Kaul C, Nothen M, Bause E, Schwanitz G. 1996. Localization of the human glucosidase I gene to chromosome 2p12-p13 by fluorescence in situ hybridization and PCR analysis of somatic cell hybrids. *Genomics* 34:442–43

82. Keir G, Winchester BG, Clayton P. 1999. Carbohydrate-deficient glycoprotein syndromes: inborn errors of protein glycosylation. *Ann. Clin. Biochem.* 36:20–36

83. Kim S, Westphal V, Srikrishna G, Mehta DP, Peterson S, et al. 2000. Dolichol phosphate mannose synthase (DPM1) mutations define congenital disorder of glycosylation Ie (CDG-Ie). *J. Clin. Invest.* 105:191–98

84. Kitagawa H, Shimakawa H, Sugahara K. 1999. The tumor suppressor EXT-like gene *EXTL2* encodes an α1, 4-N-acetyl-hexosaminyltransferase that transfers N-acetylgalactosamine and N-acetylglucosamine to the common glycosaminoglycan-protein linkage region. *J. Biol. Chem.* 274:13933–37

85. Kjaergaard S, Skovby F, Schwartz M. 1998. Absence of homozygosity for predominant mutations in PMM2 in Danish patients with carbohydrate-deficient glycoprotein syndrome type 1. *Eur. J. Hum. Genet.* 6:331–36

86. Kjaergaard S, Skovby F, Schwartz M.

1999. Carbohydrate-deficient glycoprotein syndrome type 1A: expression and characterisation of wild type and mutant PMM2 in *E. coli. Eur. J. Hum. Genet.* 7:884–88

87. Kjaergaard S, Westphal V, Davis JA, Peterson SM, Freeze HH, Skovby F. 2000. Variable outcome and the effect of mannose in congenital disorder of glycosylation type Ib (CDG-Ib). *J. Inherit. Metab. Dis.* 23 (Suppl. 1):184

88. Körner C, Knauer R, Holzbach U, Hanefeld F, Lehle L, von Figura K. 1998. Carbohydrate-deficient glycoprotein syndrome type V: deficiency of dolichyl-P-Glc:Man$_9$GlcNAc$_2$–PP-dolichyl glucosyltransferase. *Proc. Natl. Acad. Sci. USA* 95:13200–5

89. Körner C, Knauer R, Stephani U, Marquardt T, Lehle L, von Figura K. 1999. Carbohydrate deficient glycoprotein syndrome type IV: deficiency of dolichyl-P-Man:Man$_5$GlcNAc$_2$-PP-dolichyl mannosyltransferase. *EMBO J.* 18:6816–22

90. Körner C, Lehle L, von Figura K. 1998. Carbohydrate-deficient glycoprotein syndrome type 1: correction of the glycosylation defect by deprivation of glucose or supplementation of mannose. *Glycoconj. J.* 15:499–505

91. Kornfeld R, Kornfeld S. 1985. Assembly of asparagine-linked oligosaccharides. *Annu. Rev. Biochem.* 54:631–64

92. Kristiansson B, Stibler H, Conradi N, Eriksson BO, Ryd W. 1998. The heart and pericardial effusions in CDGS-I (carbohydrate-deficient glycoprotein syndrome type I). *J. Inherit. Metab. Dis.* 21:112–24

93. Kristiansson B, Stibler H, Wide L. 1995. Gonadal function and glycoprotein hormones in the carbohydrate-deficient glycoprotein (CDG) syndrome. *Acta Paediatr.* 84:655–60

94. Lind T, Tufaro F, McCormick C, Lindahl U, Lidholt K. 1998. The putative tumor suppressors EXT1 and EXT2 are glycosyltransferases required for the biosynthesis of heparan sulfate. *J. Biol. Chem.* 273:26265–68

95. Lübke T, Marquardt T, von Figura K, Körner C. 1999. A new type of carbohydrate-deficient glycoprotein syndrome due to a decreased import of GDP-fucose into the Golgi. *J. Biol. Chem.* 274:25986–89

96. Ludecke HJ, Ahn J, Lin X, Hill A, Wagner MJ, et al. 1997. Genomic organization and promoter structure of the human *EXT1* gene. *Genomics* 40:351–54

97. Marquardt T, Hasilik M, Niehues R, Herting M, Muntau A, et al. 1997. Mannose therapy in carbohydrate-deficient glycoprotein syndrome type 1: first results of the German multicenter study. *Amino Acids* 12:389

98. Marquardt T, Ullrich K, Zimmer P, Hasilik A, Deufel T, et al. 1995. Carbohydrate-deficient glycoprotein syndrome (CDGS): glycosylation, folding and intracellular transport of newly synthesized glycoproteins. *Eur. J. Cell Biol.* 66:268–73

99. Marth JD. 1999. O-glycans. In *Essentials of Glycobiology*, ed. A Varki, R Cummings, J Esko, H Freeze, G Hart, J Marth, pp. 101–13. New-York: Cold Spring Harbor Lab. Press. 653 pp.

100. Martinsson T, Bjursell C, Stibler H, Kristiansson B, Skovby F, et al. 1994. Linkage of a locus for carbohydrate-deficient glycoprotein syndrome type I (CDG1) to chromosome 16p, and linkage disequilibrium to microsatellite marker D16S406. *Hum. Mol. Genet.* 3:2037–42

101. Matthijs G, Legius E, Schollen E, Vandenberk P, Jaeken J, et al. 1996. Evidence for genetic heterogeneity in the carbohydrate-deficient glycoprotein syndrome type I (CDG1). *Genomics* 35:597–99

102. Matthijs G, Schollen E, Bjursell C, Erlandson A, Freeze H, et al. 2000. Mutations update: mutations in PMM2 that cause congenital disorders of glycosylation, type Ia (CDG-Ia). *Hum. Mutat.* 16:386–94

103. Matthijs G, Schollen E, Cassiman JJ, Cormier-Daire V, Jaeken J, et al. 1998. Prenatal diagnosis in CDG1 families: beware of heterogeneity. *Eur. J. Hum. Genet.* 6:99–104

104. Matthijs G, Schollen E, Heykants L, Grünewald S. 1999. Phosphomannomutase deficiency: the molecular basis of the classical Jaeken syndrome (CDGS type Ia). *Mol. Genet. Metab.* 68:220–26

105. Matthijs G, Schollen E, Pardon E, Veiga-Da-Cunha M, Jaeken J, et al. 1997. Mutations in PMM2, a phosphomannomutase gene on chromosome 16p13, in carbohydrate-deficient glycoprotein type I syndrome (Jaeken syndrome). *Nat. Genet.* 16:88–92

106. Matthijs G, Schollen E, Pirard M, Budarf ML, Van Schaftingen E, Cassiman JJ. 1997. PMM (PMM1), the human homologue of SEC53 or yeast phosphomannomutase, is localized on chromosome 22q13. *Genomics* 40:41–47

107. Matthijs G, Schollen E, Van Schaftingen E, Cassiman JJ, Jaeken J. 1998. Lack of homozygotes for the most frequent disease allele in carbohydrate-deficient glycoprotein syndrome type 1A. *Am. J. Hum. Genet.* 62:542–50

108. McCormick C, Duncan G, Goutsos T, Tufaro F. 2000. The putative tumor suppressors EXT1 and EXT2 form a stable complex that accumulates in the Golgi apparatus and catalyzes the synthesis of heparan sulfate. *Proc. Natl. Acad. Sci. USA* 97:668–73

109. Midro AT, Hanefeld F, Zadrozna Tolwińska B, Stibler H, Olchowik B, Stasiewicz-Jarocka B. 1996. Jaeken's (CDG) syndrome in siblings. *Pediatr. Pol.* 71: 621–28

110. Niehues R, Hasilik M, Alton G, Körner C, Schiebe-Sukumar M, et al. 1998. Carbohydrate-deficient glycoprotein syndrome type Ib: phosphomannose isomerase deficiency and mannose therapy. *J. Clin. Invest.* 101:1414–20

111. Nordborg C, Hagberg B, Kistiansson B. 1991. Sural nerve pathology in the carbohydrate-deficient glycoprotein syndrome. *Acta Paediatr. Scand. Suppl.* 375:39–49

112. Ohno K, Yuasa I, Akaboshi S, Itoh M, Yoshida K, et al. 1992. The carbohydrate deficient glycoprotein syndrome in three Japanese children. *Brain Dev.* 14:30–35

113. Okajima T, Fukumoto S, Furukawa K, Urano T, Furukawa K. 1999. Molecular basis for the progeroid variant of Ehlers-Danlos syndrome. *J. Biol. Chem.* 274:28841–44

114. Orlean P. 2000. Congenital disorders of glycosylation caused by defects in mannose addition during N-linked oligosaccharide assembly. *J. Clin. Invest.* 105:131–32

115. Panneerselvam K, Etchinson JR, Freeze HH. 1997. Human fibroblasts prefer mannose over glucose as a source of mannose for N-glycosylation: evidence for the functional importance of transported mannose. *J. Biol. Chem.* 272:23123–29

116. Panneerselvam K, Freeze HH. 1996. Mannose corrects altered N-glycosylation in carbohydrate-deficient glycoprotein syndrome fibroblasts. *J. Clin. Invest.* 97:1478–87

117. Panneerselvam K, Freeze HH. 1996. Mannose enters mammalian cells using a specific transporter that is insensitive to glucose. *J. Biol. Chem.* 271:9417–21

118. Deleted in proof

119. Pedersen PS, Tygstrup I. 1980. Congenital hepatic fibrosis combined with protein-losing enteropathy and recurrent thrombosis. *Acta Paediatr. Scand.* 69:571–74

120. Pelletier VA, Galeano N, Brochu P, Morin CL, Weber AM, Roy CC. 1986. Secretory diarrhea with protein-losing enteropathy, enterocolitis cystica superficialis, intestinal lymphangiectasia, and congenital hepatic fibrosis: a new syndrome. *J. Pediatr.* 108:61–65

121. Pineda M, Pavia C, Vilaseca MA, Ferrer I, Temudo T, et al. 1996. Normal pubertal

development in a female with carbohydrate deficient glycoprotein syndrome. *Arch. Dis. Child.* 74:242–43

122. Pirard M, Achouri Y, Collet JF, Schollen E, Matthijs G, Van Schaftingen E. 1999. Kinetic properties and tissular distribution of mammalian phosphomannomutase isozymes. *Biochem. J.* 339:201–7

123. Pirard M, Matthijs G, Heykants L, Schollen E, Grünewald S, et al. 1999. Effect of mutations found in carbohydrate-deficient glycoprotein syndrome type IA on the activity of phosphomannomutase 2. *FEBS Lett.* 452:319–22

124. Pohl S, Hoffmann A, Rüdiger A, Nimtz M, Jaeken J, Conradt HS. 1997. Hypoglycosylation of a brain glycoprotein (β-trace protein) in CDG syndromes due to phosphomannomutase deficiency and N-acetylglucosaminyl-transferase II deficiency. *Glycobiology* 7:1077–84

125. Proudfoot AE, Turcatti G, Wells TN, Payton MA, Smith DJ. 1994. Purification, cDNA cloning and heterologous expression of human phosphomannose isomerase. *Eur. J. Biochem.* 219:415–23

126. Quelhas D, Vilarinho L, Jaeken J. 2000. Carbohydrate deficient glycoprotein syndrome type I. Portuguese experience. *J. Inherit. Metab. Dis.* 23 (Suppl. 1):183

127. Quentin E, Gladen A, Roden L, Kresse H. 1990. A genetic defect in the biosynthesis of dermatan sulfate proteoglycan: galactosyltransferase I deficiency in fibroblasts from a patient with a progeroid syndrome. *Proc. Natl. Acad. Sci. USA* 87:1342–46

128. Ramaekers VT, Stibler H, Kint J, Jaeken J. 1991. A new variant of the carbohydrate deficient glycoproteins syndrome. *J. Inherit. Metab. Dis.* 14:385–88

129. Schachter H. 1991. The 'yellow brick road' to branched complex N-glycans. *Glycobiology* 1:453–61

130. Schachter H, Tan J, Sarkar M, Yip B, Chen S, et al. 1998. Defective glycosyltransferases are not good for your health. *Adv. Exp. Med. Biol.* 435:9–27

131. Schollen E, Dorland L, de Koning TJ, Van Diggelen OP, Huijmans JGM, et al. 2000. Genomic organization of the human phosphomannose isomerase (MPI) gene and mutation analysis in patients with congenital disorders of glycosylation type Ib (CDG-Ib). *Hum. Mutat.* 16:247–52

132. Schollen E, Kjaergaard S, Legius E, Schwartz M, Matthijs G. 2000. Lack of Hardy-Weinberg equilibrium for the most prevalent *PMM2* mutation in CDG-Ia (congenital disorders of glycosylation type Ia). *Eur. J. Hum. Genet.* 8:367–71

133. Schollen E, Pardon E, Heykants L, Renard J, Doggett NA, et al. 1998. Comparative analysis of the phosphomannomutase genes PMM1, PMM2 and PMM2-psi: the sequence variation in the processed pseudogene is a reflection of the mutations found in the functional gene. *Hum. Mol. Genet.* 7:157–64

134. Seta N, Barnier A, Hochedez F, Besnard MA, Durand G. 1996. Diagnostic value of Western blotting in carbohydrate-deficient glycoprotein syndrome. *Clin. Chim. Acta* 254:131–40

135. Spaapen LJM, Bakker JA, Van der Meer SB, Velmans MH, van Pelt J. 1995. Sialotransferrin patterns and carbohydrate-deficient transferrin values in neonatal and umbilical cord blood. *33rd SSIEM Annu. Symp.*, Vol. 71. Lancaster: Kluwer Acad. (Abstr.)

136. Stanley P, Ioffe E. 1995. Glycosyltransferase mutants: key to new insights in glycobiology. *FASEB J.* 9:1436–44

137. Stibler H, Blennow G, Kristiansson B, Lindehammer H, Hagberg B. 1994. Carbohydrate-deficient glycoprotein syndrome: clinical expression in adults with a new metabolic disease. *J. Neurol. Neurosurg. Psychiatry* 57:552–56

138. Stibler H, Gylje H, Uller A. 1999. A neurodystrophic syndrome resembling carbohydrate-deficient glycoprotein syndrome type III. *Neuropediatrics* 30:90–92

139. Stibler H, Holzbach U, Tengborn L,

Kristiansson B. 1996. Complex functional and structural coagulation abnormalities in the carbohydrate-deficient glycoprotein syndrome type I. *Blood Coagul. Fibrinolysis* 7:118–26

140. Stibler H, Jaeken J, Kristiansson B. 1991. Biochemical characteristics and diagnosis of the carbohydrate-deficient glycoprotein syndrome. *Acta Paediatr. Scand. Suppl.* 375:21–31

141. Stibler H, Stephani U, Kutsch U. 1995. Carbohydrate-deficient glycoprotein syndrome: a fourth subtype. *Neuropediatrics* 26:235–37

142. Stibler H, Westerberg B, Hanefeld F, Hagberg B. 1993. Carbohydrate-deficient glycoprotein (CDG) syndrome: a new variant, type III. *Neuropediatrics* 24:51–52

143. Stickens D, Clines G, Burbee D, Ramos P, Thomas S, et al. 1996. The EXT2 multiple exostoses gene defines a family of putative tumour suppressor genes. *Nat. Genet.* 14:25–32

144. Strom EH, Stromme P, Westvik J, Pedersen SJ. 1993. Renal cysts in the carbohydrate-deficient glycoprotein syndrome. *Pediatr. Nephrol.* 7:253–55

145. Stromme P, Maehlen J, Strom EH, Torvik A. 1991. Postmortem findings in two patients with the carbohydrate-deficient glycoprotein syndrome. *Acta Paediatr. Scand. Suppl.* 375:55–62

146. Subhedar NV, Isherwood DM, Davidson DC. 1996. Hyperglycinaemia in a child with carbohydrate-deficient glycoprotein syndrome type I. *J. Inherit. Metab. Dis.* 19:796–97

147. Tan J, D'Agostaro GAF, Bendiak B, Reck F, Sarkar M, et al. 1995. The human UDP-N-acetylglucosamine: alpha-6-D-mannoside-beta-1,2-N-acetylglucosaminyltransferase II gene (MGAT2): cloning of genomic DNA, localization to chromosome 14q21, expression in insect cells and purification of the recombinant protein. *Eur. J. Biochem.* 231:317–28

148. Tan J, Dunn J, Jaeken J, Schachter H. 1996. Mutations in the MGAT2 gene controlling complex N-glycan synthesis cause carbohydrate-deficient glycoprotein syndrome type II, an autosomal recessive disease with defective brain development. *Am. J. Hum. Genet.* 59:810–17

149. van der Knaap MS, Wevers RA, Monnens L, Jakobs C, Jaeken J, van Wijk JAE. 1996. Congenital nephrotic syndrome: a novel phenotype of type I carbohydrate-deficient glycoprotein syndrome. *J. Inherit. Metab. Dis.* 19:787–91

150. van Diggelen OP, Maat-Kievit JA, de Klerk JBC, Boonman AMC, van Noort WL, et al. 1998. Two more Dutch cases of CDG syndrome Ib: phosphomannose isomerase deficiency. *J. Inherit. Metab. Dis.* 21 (Suppl. 2):97

151. Van Geet C, Jaeken J. 1993. A unique pattern of coagulation abnormalities in carbohydrate-deficient glycoprotein syndrome. *Pediatr. Res.* 33:540–41

152. van Heyningen V, Bobrow M, Bodmer WF, Gardiner SE, Povey S, Hopkinson DA. 1975. Chromosome assignment of some human enzyme loci: mitochondrial malate dehydrogenase to 7, mannosephosphate isomerase and pyruvate kinase to 15 and probably, esterase D to 13. *Ann. Hum. Genet.* 38:295–303

153. van Ommen CH, Peters M, Barth PG, Vreken P, Wanders RJA, Jaeken J. 2000. Carbohydrate-deficient glycoprotein syndrome type 1a: a variant phenotype with borderline cognitive dysfunction, cerebellar hypoplasia, and coagulation disturbances. *J. Pediatr.* 136:400–3

154. Van Schaftingen E, Jaeken J. 1995. Phosphomannomutase deficiency is a cause of carbohydrate-deficient glycoprotein syndrome type I. *FEBS Lett.* 377:318–20

155. Varki A. 1993. Biological roles of oligosaccharides: all of the theories are correct. *Glycobiology* 3:97–130

156. Wada Y, Nishikawa A, Okamoto N, Inui K, Tsukamoto H, et al. 1992.

Structure of serum transferrin in carbohydrate-deficient glycoprotein syndrome. *Biochem. Biophys. Res. Commun.* 189:832–36

157. Westphal V, Schottstädt C, Marquardt T, Freeze HH. 2000. Analysis of multiple mutations in the h*ALG6* gene in a patient with congenital disorder of glycosylaton Ic. *Mol. Genet. Metab.* 70:219–23

158. Wicklund CL, Pauli RM, Johnston D, Hecht JT. 1995. Natural history study of hereditary multiple exostoses. *Am. J. Med. Genet.* 55:43–46

159. Winchester B, Clayton P, Mian N, di Tomaso E, Dell A, et al. 1995. The carbohydrate-deficient glycoprotein syndrome: an experiment of nature in glycosylation. *Biochem. Soc. Trans.* 23:185–88

160. Winchester B, Fleet GWJ. 2000. Modification of glycosylation as a therapeutic strategy. *J. Carbohydr. Chem.* 19:471–83

161. Wuyts W, Van Hul W, Wauters J, Nemtsova M, Reyniers E, et al. 1996. Positional cloning of a gene involved in hereditary multiple exostoses. *Hum. Mol. Genet.* 5:1547–57

162. Wuyts W, Van Hul W. 2000. Molecular basis of multiple exostoses: mutations in the EXT1 and EXT2 genes. *Hum. Mutat.* 15:220–27

163. Yamashita K, Ohkura T, Ideo H, Ohno K, Kanai M. 1993. Electrospray ionization-mass spectrometric analysis of serum transferrin isoforms in patients with carbohydrate-deficient glycoprotein syndrome. *J. Biochem.* 114:766–69

Annu. Rev. Genomics Hum. Genet. 2001. 2:153–75

Genome Organization, Function, and Imprinting in Prader-Willi and Angelman Syndromes

Robert D. Nicholls and Jessica L. Knepper

Center for Neurobiology and Behavior, Department of Psychiatry, and Department of Genetics, University of Pennsylvania, CRB 528, 415 Curie Blvd., Philadelphia, Pennsylvania 19104-6140; e-mail: robertn@mail.med.upenn.edu, knepper@mail.med.upenn.edu

Key Words chromatin, DNA methylation, duplicon, epigenetic, polycistronic RNA

■ **Abstract** The chromosomal region, 15q11-q13, involved in Prader-Willi and Angelman syndromes (PWS and AS) represents a paradigm for understanding the relationships between genome structure, epigenetics, evolution, and function. The PWS/AS region is conserved in organization and function with the homologous mouse chromosome 7C region. However, the primate 4 Mb PWS/AS region is bounded by duplicons derived from an ancestral *HERC2* gene and other sequences that may predispose to chromosome rearrangements. Within a 2 Mb imprinted domain, gene function depends on parental origin. Genetic evidence suggests that PWS arises from functional loss of several paternally expressed genes, including those that function as RNAs, and that AS results from loss of maternal *UBE3A* brain-specific expression. Imprinted expression is coordinately controlled in *cis* by an imprinting center (IC), a genetic element functional in germline and/or early postzygotic development that regulates the establishment of parental specific allelic differences in replication timing, DNA methylation, and chromatin structure.

INTRODUCTION

Prader-Willi and Angelman syndromes (PWS and AS, respectively) are clinical disorders linked to abnormalities in inheritance of chromosome 15q11-q13. PWS is characterized by hypotonia, respiratory distress, and a failure to thrive in the postnatal period, with hyperphagia in early childhood resulting in obesity (Figure 1*a*, see color insert), as well as short stature, small hands and feet, hypogonadism, mild-moderate mental retardation, temper tantrums, and obsessive-compulsive mannerisms (10). Major characteristics of AS include developmental delay, severe mental retardation with a lack of speech, movement ataxia, hyperactivity, seizures, aggressive behavior, and excessive inappropriate laughter (Figure 1*b*, see color insert) (42, 60). At the molecular level, 15q11-q13 represents a very

1527-8204/01/0728-0153$14.00

153

complex chromosomal region and study of its scientifically fascinating disorders has defined or been illustrative of many of the novel genetic concepts identified in modern human genetics.

In particular, genes within a 2 Mb domain spanning half of 15q11-q13 are imprinted, with expression of these genes dependent on the sex of the parent-of-origin. Imprinting is a still relatively poorly understood epigenetic phenomenon, controlled in *cis* by a multitude of parent-of-origin chromatin events beginning in the male and female germlines and early postfertilization, and maturing as development proceeds (3, 54, 60, 64, 80). With the emergence of the mouse in recent years as a model for PWS and AS, as the genetic organization in the homologous region of chromosome 7C is almost identical to that in human, PWS/AS studies will contribute significantly to an understanding of epigenetic and genetic phenomena. Here, we provide a descriptive framework for understanding the complexities of gene organization and function within the homologous human 15q11-q13 and mouse 7C PWS/AS genomic domains.

INHERITANCE IN PWS AND AS

The inheritance of genetic abnormalities in the region of 15q11-q13 is very complex (Figure 2, see color insert). Nevertheless, in each mechanism, the end result is that PWS is associated primarily with loss of expression of paternally derived alleles, and the major features of AS are associated with loss of expression of the maternal allele. The majority (65%–75%) of PWS and AS cases are due to de novo deletions spanning 4 to 4.5 Mb of 15q11-q13. There are two classes of common PWS/AS deletions, one spanning from breakpoint 1 (BP1) to breakpoint 3 (BP3) and the other from breakpoint 2 (BP2) to BP3 (Figure 3a, see color insert), and these deletions are thought to occur by recombination between duplicated sequences (termed duplicons) at these three locations (discussed below).

The next most common PWS genetic abnormality (20%–30%) is maternal uniparental disomy (matUPD), which arises in most or all cases from maternal meiotic nondisjunction that is followed by mitotic loss of the single paternal chromosome 15 postzygotically (65). Most matUPD likely originate from meiosis I errors because recombination occurs prior to the segregation defect, resulting in heterodisomic and isodisomic chromosome regions. Similarly, paternal UPD occurs in up to 5% of AS cases, but as almost all cases are isodisomic, the likely origin is maternal nondisjunction with postzygotic duplication of the single sperm-derived chromosome (65).

Other less common mutations in PWS and AS include two types of imprinting defects (ID), accounting for up to 5% of cases in each disorder, those with submicroscopic deletions of a genetic element termed the imprinting center (IC) and those with solely an abnormal imprint (or epigenotype) and no detectable mutation (see Regulation in *cis* Based on Inherited Imprinting Mutations) (7, 20, 60–62). In PWS ID patients, the paternally derived chromosome has either an IC mutation or an "epimutation" resulting in a maternal epigenotype, whereas AS ID patients similarly have an incorrect paternal epigenotype on the maternal chromosome (Figure 2); the resultant inheritance or acquisition of uniparental imprints results in the imprinted disorder.

Balanced translocations occur but are rare (~0.1% of cases) in PWS (Figure 2). Thus far, five paternally derived translocations have been reported, all of which disrupt the *SNURF-SNRPN* gene in either of two locations (Figure 3*a*), although the exact phenotype and downstream regulation may differ in these cases (see Imprinted PWS/AS Domain as well as snoRNAs and PWS) (59, 85). To date, all PWS patients with a classical phenotype have a deletion, matUPD, ID, or balanced translocation, with no single gene mutations causing PWS or a PWS-like disorder. Therefore, PWS may be a multigenic disorder involving the loss of function of several paternally expressed genes, although single gene mutations may occur in patients with phenotypic subclasses.

In contrast to PWS, single gene mutations occur in ~10% of AS cases (Figure 2) in the *UBE3A* gene that encodes an E3 ubiquitin ligase involved in the ubiquination pathway of protein degradation (42, 57). All 50 independent mutations (Figure 4, see color insert) occurred de novo or were inherited on the maternal allele, with no phenotype on paternal inheritance for familial cases (47, 51a, 67). Two thirds of *UBE3A* mutations are frameshift (deletion or insertion/duplication), whereas 20% are nonsense mutations (Figure 4), creating null alleles. At least 8 of 21 deletions and 5 of 13 insertions (those that are direct duplications) identified to date in *UBE3A* likely result from "slipped mispairing" replication errors (3 additional deletions and 5 duplications of a single nucleotide may also occur by this mechanism), whereas 6/16 point mutations are C $\rightarrow$ T transitions likely caused by deamination of 5-methylcytosine in CpG dinucleotide residues (Figure 4; data not shown). All but one of the independently recurring *UBE3A* mutations occurs by these two mechanisms, identifying potential mutation hotspots.

A substantial fraction of *UBE3A* mutations in AS, as well as IC microdeletions in PWS and AS ID patients that are usually familial, can transmit silently through one sex (fathers in AS pedigrees, mothers in PWS pedigrees), although resulting in a 50% recurrence risk when transmitted through a parent of the opposite sex (maternal for AS, paternal for PWS), composing an important genetic counseling issue (61). Finally, 10%–15% of AS cases have an unknown etiology (Figure 2) and may arise as a consequence of epigenetic or genetic events affecting *UBE3A* or unlinked genes (42, 51a).

Genotype-phenotype analyses of the five subclasses of AS have shown that, while all share the cardinal features of AS (developmental delay, speech impairment, movement and balance disorder, and characteristic behavior), class I deletion patients have a more classical and severe phenotype with the highest incidence and earliest age of onset of seizures (51a). Haploinsufficiency for codeleted genes that are not imprinted may contribute to the severe AS phenotype, and *GABRB3* may have a role in the susceptibility to severe seizures in conjunction with *UBE3A* deficiency (17, 42, 57). UPD and ID patients are significantly less severe in most features and have high body mass index (BMI) values, indicating obesity (51a).

Class IV (*UBE3A* mutation) patients fall in the middle of these other groups and indicate that deficiency for maternally expressed *UBE3A* is sufficient to account for the classical clinical features of AS. In addition to the C-terminal ubiquitin ligase domain, *UBE3A* interacts with and coactivates the transcriptional activity of members of the nuclear hormone receptor superfamily (Figure 4) (56). Several AS mutations were tested in vitro and found to abolish ubiquination of a substrate (HHR23A; other known substrates of *UBE3A* include itself, p53, the Src family kinase Blk, and the MCM-7 replication protein) (see 37) but not the coactivator function (56). This observation, coupled with specific AS mutations in the C-terminal ubiquitin ligase domain, make it clear that AS neurological features result from a defect in proteasome-mediated protein degradation.

GENOMIC STRUCTURE AND FUNCTION OF HUMAN 15q11-q13 AND MOUSE 7C

Imprinted PWS/AS Domain

The 15q11-q13 and conserved mouse 7C regions are highlighted by a 2 Mb region that contains multiple paternally expressed, imprinted genes associated with the PWS phenotype and a single maternally expressed gene associated with AS (Figure 3a). In contrast to the paternally expressed genes that maintain the imprint in most or all tissues, maternal-specific expression of *UBE3A* is spatially restricted to brain in the human and mouse. Studies in mouse have shown imprinted *Ube3a* expression in cerebellar Purkinje cells, hippocampal neurons, and olfactory mitral cells (see 42). Therefore, brain-restricted imprinted gene expression defines the neurological basis of the disorder. Maternal (m) inheritance of an *Ube3a*-null allele (m^-/p^+) results in fertile animals that have a mild neurobehavioral phenotype, representing a mouse model of AS (41). The AS-like features include motor and coordination deficits, electrencephalogram (EEG) abnormalities and inducible seizures, and an impairment of hippocampal long-term synaptic potentiation (LTP) associated with a context-dependent learning deficit (41, 57). In addition to the *Ube3a* (m^-/p^+) model of AS, other mouse models of PWS and AS have been developed, including those based on the major subclasses of deletion (Figure 1c, d, see color insert) (28), UPD (11, 12), and ID in a PWS mouse model (see Imprinting in PWS/AS) (4, 86). These animal models have been and will be crucial for characterization of the molecular pathogenesis of both disorders.

In contrast to AS, the presence of multiple paternally expressed genes has delayed identification of the critical loci for PWS because all such genes in the 1.5 Mb domain are positional candidates (58, 60). Genetic evidence based on microdeletions in ID in the human and balanced translocation patients has implicated the central *SNURF-SNRPN* locus as a strong candidate for a major role in PWS (Figure 3a). This exceedingly complex locus encodes at least four functions, including elements of the *cis*-acting regulatory region termed the imprinting center

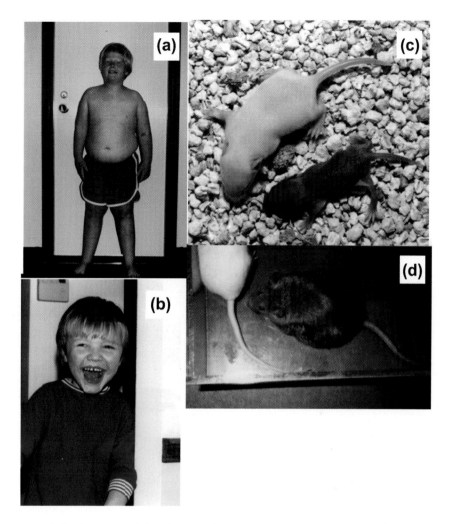

Figure 1 Prader-Willi and Angelman syndrome (PWS and AS) phenotypes in human and mouse. (*a*) A PWS child illustrates the obesity, short stature, and small hands and feet typical of PWS [reproduced with permission, (10)]. The hypopigmentation of the skin, hair, and eyes is associated with a 15q11-q13 deletion. (*b*) An individual with AS (due to an imprinting defect) exhibiting the typical excessive inappropriate laughter [reproduced with permission, (68)]. (*c*) The PWS deletion mouse model demonstrates a growth retardation resulting from failure to thrive and hypotonia that manifests soon after birth, compared to a wild-type littermate [reproduced with permission, (28)]. These features are similar to the human neonatal course. (*d*) The AS mouse model is associated with a late-onset obesity, compared to an age-matched wild-type littermate.

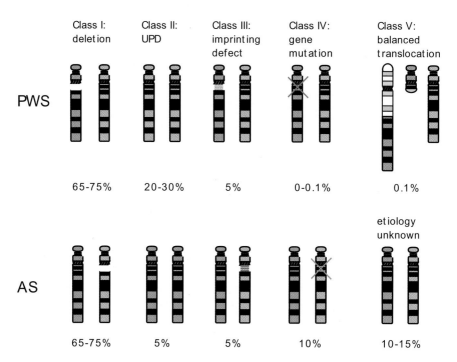

Figure 2 Complex genetic inheritance in PWS and AS. Shown are the five genetic classes of PWS and AS, with the parental origin (blue, paternal; pink, maternal) and frequencies observed for each. For PWS and AS, respectively, these are chromosome 15q11-q13 deletions of paternal or maternal origin, maternal or paternal uniparental disomy (UPD), and imprinting defects generating a maternal imprint on the paternally derived chromosome (PWS) or a paternal imprint on the maternal allele (AS). Imprinting defects have two subclasses, mutation and epimutation, where only the former have a detectable mutation in the imprinting center. Single gene mutations on the maternal allele occur in AS but are still not found in PWS. Balanced translocations are a rare cause of PWS, whereas a large class of AS patients have no known molecular basis.

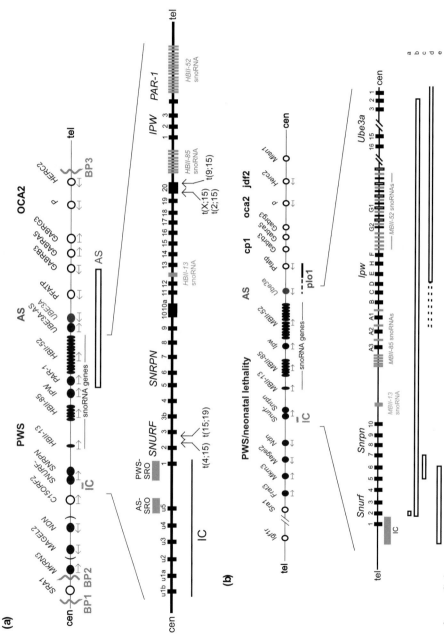

See legend page C-4

Figure 3 (see figure page C-3) Human chromosome 15q11-q13 and mouse chromosome
7C genomic structure. The parental origin (blue, paternal; red, maternal; open circles,
biparental) and transcriptional orientation (arrows) of genes are indicated, as well as the
phenotypes caused by loss of function of a particular segment or gene of the specified
parental origin (black, recessive). (*a*) The human 4 Mb 15q11-q13 region flanked by
PWS/AS common deletion breakpoints (BP, zigzag lines). The open bar represents a 1 Mb
deletion causing AS only on maternal inheritance (70). The *HBII-85* and *HBII-52*
snoRNAs are repeated 24 and 47 times, respectively (13). Parentheses indicate an
unknown relative orientation with respect to the centromere (cen) and telomere (tel) (IC,
imprinting center; OCA2, oculocutaneous albinism type II). The expansion shows the *IC-
SNURF-SNRPN-HBII-13* polycistronic locus, the flanking snoRNA repeats, and balanced
translocation breakpoints (arrows) found in PWS and PWS-like patients. The *PWCR1*
locus (18) spans two copies of the *HBII-85* snoRNA and intervening DNA (SRO, shortest
region of microdeletion overlap). (*b*) Mouse 7C gene map and expansion in the IC through
Ube3a segment. Copies of the *MBII-85* and *MBII-52* snoRNAs are interspersed within 5′
and 3′ exons of the *Ipw* gene, respectively (13, 18, 84). The bars under the enlargement of
the *Snurf-Snrpn* region of chromosome 15 indicate targeted deletions [a, b, (82); c, e, (4,
84); d, (43) and D.K. Johnson, personal communication] (plo1, p-locus-associated obesity; cp1, cleft palate 1; jdf2, juvenile development and fertility 2).

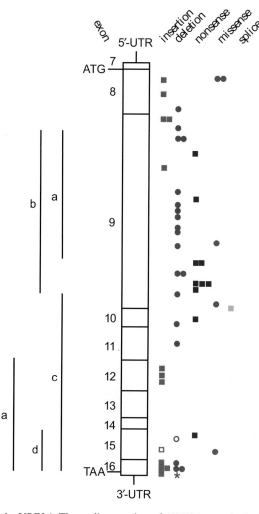

Figure 4 Mutations in *UBE3A*. The coding portion of *UBE3A* gene is depicted (exons 7–16), with the position of the start and stop codons. The color symbols represent the locations and molecular basis of mutations in sporadic and familial AS patients [(47, 67); (51a)]. Pink squares, insertion mutations (filled or open symbols represent frameshift or in-frame mutations, respectively); red circles, deletions (filled or open symbols, as above; star, 15-bp deletion of the stop codon, generating an abnormal C-terminus); blue squares or circles, nonsense or missense mutations, respectively; green square, splice mutation. The letters indicate various protein domains: a, progesterone receptor (PR) interaction domain; b, transactivation domain; c, HECT domain; d, E3 ubiquitin ligase domain.

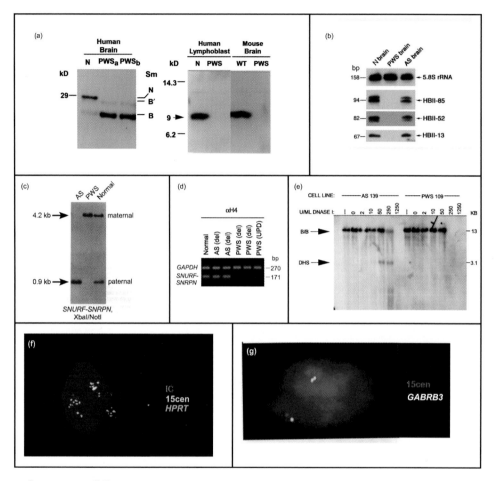

See text page C-7

Figure 5 (see figure page C-6) Imprinting in PWS/AS. (*a*) Imprinted expression of protein. Western analysis shows expression of SmN in normal (N) but not PWS brain samples, indicating translation from the paternal allele only (*left*). A significant upregulation of the closely related SmB and SmB´ proteins, two isoforms generated by alternative splicing, compensate for the loss of SmN in the brain [reproduced with permission, (33)]. Similarly, immunoprecipitation (IP)-Western analysis shows SNURF protein in normal or wild-type (WT) cells, but not PWS cells, in the human and mouse [*right*; reproduced with permission, (32)]. (*b*) Imprinted RNA expression. Paternal expression for 3 classes of snoRNAs is identified by Northern analysis of normal, AS, and PWS brain samples [reproduced with permission, (13). (*c*) Differential DNA methylation. Southern analysis with the methylation-sensitive enzyme *Not*I (and insensitive *Xba*I) identifies the methylated, maternal allele of the *SNURF-SNRPN* promoter in a PWS matUPD patient and the unmethylated, paternal allele in an AS patUPD patient, along with the normal biparental imprint [reproduced with permission, (68)]. (*d*) Histone acetylation. Chromatin-immunoprecipitation (ChIP) analysis identifies acetylated histone H4 on the paternal allele in normal and AS deletion (del) cells, indicative of an open chromatin structure and transcriptional activity, whereas PWS patients (maternal allele only) show no acetylation of histone H4, indicative of transcriptional silencing [reproduced with permission, (71)]. (*e*) Chromatin structure. A DNase I hypersensitive site is present in the *SNURF-SNRPN* promoter in the AS but not the PWS deletion cell lines, suggesting an open chromatin conformation and *trans*-factor occupancy exclusively on the paternal allele at this site [reproduced with permission, (62)]. (*f*) Matrix attachment regions. Both copies of the chromosome 15 α-satellite repeat probe (15cen, yellow) are associated with the nuclear matrix (blue) in halo nuclei from a normal female, whereas one allele of each of the *IC-SNURF-SNRPN* (red) and *HPRT* (green) probes is confined to the nuclear matrix with the homologue associated, at least in part, with the surrounding DNA halo [reproduced with permission, (34)]. Similar studies on cells from a female PWS deletion patient indicate that it is the maternal allele of the IC that is matrix associated (34). (*g*) Asynchronous replication timing. Asynchrony is shown by the single-double replication pattern of the *GABRB3* probe (white), located close to the polymorphic α-satellite probe (15cen, red). The large 15s$^+$ satellite locus identifies the paternal origin of the chromosome, with replication occurring earlier than on the maternal allele [reproduced with permission, (45)].

(a)

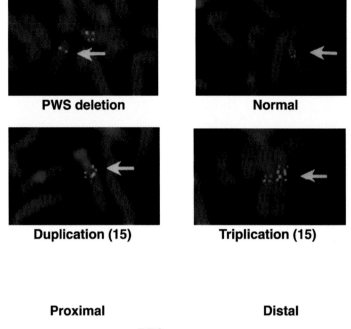

(b)

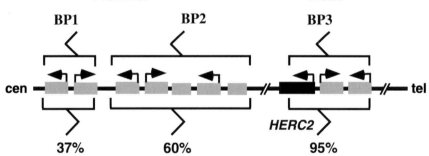

Figure 6 Duplicon-mediated instability in 15q11-q13. (*a*) *HERC2*-duplicons at 15q11 (BP1/BP2) and 15q13 (BP3) are detected on normal and abnormal chromosome 15 by fluorescent in situ hybridization (FISH) using the λ11A1 probe [reproduced with permission (1)]. One, two, three, or four sets of signals (arrows) demarcate zero, one, two, or three copies of the 15q11-q13 PWS/AS region, respectively. (*b*) Structural model for the arrangement of *HERC2* duplicons (rectangles) at BP1-3 in 15q11 and 15q13. The ancestral *HERC2* gene position and transcriptional orientation (arrow) is known, whereas the exact order and orientation of the duplicated sequences is unknown. The frequency of breakpoints at each site in PWS and AS deletions is also shown.

(IC) (see Imprinting in PWS/AS), two independent proteins, and a small nucleolar RNA (snoRNA) (13, 32, 60, 85). The encoded SmN and SNURF polypeptides (Figure 5a, see color insert) and *HBII-13* snoRNA (Figures 3a, 5b) are expressed in vivo in human and mouse but are absent in PWS and PWS mouse models, demonstrating that this is an imprinted polycistronic locus (13, 32). This novel genomic structure, more typical of prokaryotic operons, raises the question of how their expression is regulated and the possibility of these products functioning in a common biochemical pathway.

SmN is encoded by exons 4–10 of *SNRPN* and is a core spliceosomal protein involved in mRNA splicing in the brain where it replaces the otherwise consti- tutively expressed SmB'/B proteins. The latter represent two alternatively spliced isoforms encoded by an independent, ancestral chromosomal locus (33). Spe- cific gene ablation of the SmN-coding sequence [(Figure 3b, see color insert), note deletion c] yielded apparently normal fertile mice (86), although behavioral studies were not performed. However, in the absence of SmN protein in PWS mice (86) and in PWS patients (33) the SmB'/B isoforms are significantly upreg- ulated (Figure 5a). This maintains an overall constant level of SmN/B/B' in the brain, likely to prevent imbalance in the stoichiometry of the 7-member helical spliceosome structure (33). Thus, these loci are subject to a feedback regula- tion mechanism that senses and regulates the level of SmN and SmB/B'. This mechanism also serves as a natural gene therapy in which SmB'/B upregula- tion likely ameliorates the potential severity of loss of SmN gene function in PWS.

SNURF (*SNRPN* Upstream Reading Frame) encodes a 71 amino acid protein that is enriched in arginine (R) residues and could bind RNA, has a C-terminal RGG motif similar to ubiquitin, and is localized in the nucleus (32). Potential func- tions for SNURF include regulation of SmN or of the imprinting process, the latter as a consequence of SNURF being the only protein-coding locus to overlap the IC (see Imprinting in PWS/AS). Despite translation in vivo (Figure 5a) and high conservation in mammals (32), ablation of the *Snurf* coding sequence [(Figure 3b), note deletion a] in the absence of an IC defect has not revealed a phenotypic effect (82). Nevertheless, because *SNURF* is uniquely disrupted by balanced transloca- tions in the only two such patients with a classical PWS phenotype (Figure 3a) (59, 85), the role remains enigmatic.

Work by Horsthemke and colleagues has defined an extraordinarily complex pattern of 5' (see Imprinting in PWS/AS) (20, 23) and 3' (85) alternative splicing for the *SNURF-SNRPN* locus (Figure 3a). The majority of *SNURF-SNRPN* transcripts in human and mouse initiate at the major CpG-island associated with exon 1 and are 3' processed at a strong polyadenylation site in exon 10 [human exon 3b encodes a minor *SNURF*-only polyadenylation signal in muscle; (32)]. However, transcription continues beyond exon 10 and variably includes up to an additional eleven 3' noncoding exons [exons 10a–20; (Figure 3a)]. As the *HBII-13* snoRNA gene (see snoRNA's and PWS) (13) is encoded within intron 12 (85), this suggests a likely function for these 3' transcripts.

Several additional paternally expressed imprinted genes that encode proteins have been identified as a cluster of intronless genes in 15q11-q13 and mouse 7C (Figure 3*a, b*), including two MAGE protein family members, *NDN* and *MAGEL2* (discussed below), and *MKRN3* (formerly *ZNF127* in human, *Zfp127* in mouse). *MKRN3* encodes a polypeptide with a RING zinc-finger and multiple C_3H zinc-finger motifs, suggesting an unknown function as a ribonucleoprotein (44). *MKRN3* arose by retrotransposition from the ancestral *Mkrn1* locus ~80–120 million years ago (mya) (31), so although a mouse *Mkrn3* knockout is viable and fertile (see 44), maintenance over this evolutionary period strongly argues for a still unknown function in mammals. Another intronless gene present only in the mouse cluster, *Frat3* (Figure 3*b*), represents a recently evolved retrotransposon that is subject to the imprinting mechanism (J.H. Chai, D.P. Locke, T. Ohta, J.M. Greally, & R.D. Nicholls, submitted). This observation suggests that the imprint mechanism is transmitted solely through chromatin structure and/or the presence of a 5' CpG-island (see Imprinting in PWS/AS).

One intronless gene centromeric of the IC in the human and not present in the mouse, *C15ORF2* (Figure 3*a*), shows biparental expression in testis, but this is the only tissue in which it is expressed (24). As it is unmethylated only in sperm, if expression is in postmeiotic germ cells after erasure of the maternal imprint, then the gene would be imprinted during spermatogenesis (24). Furthermore, as a silent, highly methylated gene in somatic tissues, its presence does not violate the paternal-only expression domain in the PWS region. Adjacent to the imprinted, intronless genes, the *SRA1* gene (D.P. Locke, J.M. Greally, T. Ohta, J. Dunai, A. Yavor, E.E. Eichler, J.H.M. Knoll, & R.D. Nicholls, submitted) and two flanking genes (J.H. Chai, D.P. Locke, & R.D. Nicholls, unpublished data) are nonimprinted in human and mouse, and this observation, therefore, defines a boundary to the imprinted domain (Figure 3*a, b*). Given that the imprint signal generated from the IC transmits bidirectionally over 0.5–1 Mb in either direction (see Imprinting in PWS/AS), identifying the nature of the DNA or chromatin element that stops the IC-imprint signal from spreading further will be of interest.

The *NDN* Gene in PWS

Among the imprinted candidate genes for PWS, the gene *NDN* encoding the MAGE family NECDIN protein has been a particular focus of study (see 29, 55, 58, 79). Mouse *Ndn* mRNA is expressed predominantly in a subset of postmitotic neurons, with highest levels in the hypothalamus and several other brain regions at late embryonic and early postnatal stages, as well as other tissues (29, 55). Similarly, a second imprinted MAGE member, *MAGEL2*, is located adjacent to *NDN* in the human and mouse, with highest expression in mouse at late developmental stages and in the hypothalamus and other brain regions (6, 49). However, functional analyses of *Magel2* remain unperformed. In vitro studies have identified that NECDIN overexpression results in suppression of cell proliferation and that the protein can bind to transcription factors p53, E1A, E2F1, and the SV40 large T antigen, as well as to the Ca^{2+} binding protein NEFA (see 79); however, the in vivo relevance of these observations remains untested.

Three different *Ndn*-deficient mouse models have been developed (29, 55, 58, 81). Although one line had no apparent phenotype (81), the others had an early postnatal lethality with variable penetrance dependent on the strain genetic background (29, 55). There was no lethality on breeding to a hybrid background (29), indicating that the *Ndn*-deficiency phenotype was dependent on genetic modifiers (58). The lethality arose from respiratory distress associated with failure to properly inflate the lungs, and *Ndn* was highly expressed in the respiratory center in the pons and medulla of wild-type animals (29). Surviving mice in all three studies showed no overt phenotype and were fertile and nonobese. As has been suggested in PWS, these mice showed a reduction in hypothalamic oxytocin–producing and luteinizing hormone–releasing hormone-producing neurons (55), although no specific phenotypic consequence has been identified. Interestingly, *Ndn*-deficient males (females were not tested) had improved spatial learning and memory compared to wild-type littermates in a Morris water maze test, reminiscent of PWS patient strengths in some cognitive functions (55). *Ndn*-mutants also displayed increased skin scraping in an open field test, possibly mimicking the skin picking typical of PWS (55). Nevertheless, because these behavioral studies used animals from the F_2 generation and did not test animals from the 129/Sv strain from which the mutant ES cells were derived (55), the entire linked PWS/AS imprinted domain on the *Ndn*-mutation allele may be 129/Sv derived [compared to an origin of C57BL6/J for controls; see (1a)]. Thus, it remains possible that the observed phenotypes are a result of genetic variation in another PWS candidate gene. More complete studies of the basis for respiratory distress and behavior in *Ndn*-mutant mice are important, as is identification of the biochemical function of NECDIN and a search for *NDN* mutations in patients with a subset of the PWS phenotype as identified from the mouse models.

snoRNAs and PWS

Three families of C/D box snoRNA genes have recently been mapped between *SNURF-SNRPN* and *UBE3A* in human chromosome 15q11-q13 (Figure 3*a*) and mouse chromosome 7C (Figure 3*b*) (13, 18). All the 15q11-q13/7C snoRNA genes are intronic and are processed from primary transcripts in this region, as are other snoRNA genes (25). However, the 15q11-q13/7C snoRNAs are brain-specific RNAs that are imprinted and expressed from the paternal allele only (Figure 5*b*, see color insert) (13, 18). Single copy *HBII-13* and *MBII-13* (H, human; M, mouse; B, brain) snoRNAs map just 3′ to *SNURF-SNRPN* in each species, and as noted above, *HBII-13* is processed from intron 12 of the 3′ transcripts from this locus (13, 85). The other two snoRNAs, *HBII-85/MBII-85* and *HBII-52/MBII-52*, are each repeated tandemly [24 copies of *HBII-85* over 99 kb and 47 copies of *HBII-52* over 55 kb; (13)]. The former cluster maps between *SNURF-SNRPN* and *IPW* and the latter cluster maps between *IPW* and *UBE3A* in human (Figure 3*a*).

However, in mouse where the exon structure of the noncoding *Ipw* gene differs from human (84), the tandem repeats include a copy of *MBII-85* localized adjacent to each A exon of *Ipw* (18), whereas a copy of *MBII-52* is localized within an intron splitting each G exon of *Ipw* into exons G1 and G2 (13) (Figure 3*b*). These observations suggest that the loci characterized as *Ipw* in mouse and likely *IPW* in human may represent a part of a long transcript serving as the host gene for both families of tandemly repeated, imprinted snoRNA genes. The putative intronless *PWCR1* cDNA has one copy of *HBII-85* exactly at each of the 5′ and 3′ ends (18), suggesting that it represents a stable processing intermediate of this host gene rather than an independent gene. Consistent with the host gene model, *PWCR1* probes detect an abundant high molecular weight smear on Northern analysis (18). Numerous other imprinted noncoding transcripts previously identified in the *SNRPN-UBE3A* interval (50, 78) are also likely to originate from processing intermediates of the host gene(s) (13). If the snoRNA host gene(s) derive from extended *SNURF-SNRPN* transcripts beyond exon 20 or if an independent RNA polymerase II promoter exists is still unknown.

Several structural features of the *HBII-85*/*MBII-85* and *HBII-52*/*MBII-52* snoRNA genes warrant further discussion. The *HBII-52* snoRNA coding sequences show only 3% sequence divergence between copies, but the 1.9-kb repeat units diverge an average of 22% from the consensus and share no homology with the *MBII-52* repeats, other than the highly conserved snoRNA coding sequence (13). In contrast, the *HBII-85* snoRNA coding sequences each differ on average by 20%, and the repeat units are less evenly spaced than for *HBII-52* genes (13). The mouse *MBII-85* genes have a more homogenous and highly conserved structure (18), with only 2%–4% divergence between copies (because the genome sequence is incomplete, copy number is uncertain but may be over 50). Therefore, the tandemly arrayed snoRNA genes are undergoing concerted evolution within each cluster and species. The striking lack of repetitive elements (eg., *Alu*, L1, etc.) within the *HBII-85* and *HBII-52* clusters (13) is consistent with either their removal by sequence homogenization or phenotypic selection against disruption of the arrays. To speculate that the structural potential for recombination within tandemly repeated *HBII-85* or *HBII-52* snoRNA genes might lead to instability and localized deletions in rare patients with PWS-like phenotypes is tempting.

Is there any evidence supporting a role for these snoRNA genes in the PWS phenotype, and what are their cellular functions? All three families of snoRNAs lack expression in PWS tissues from human or from a PWS mouse model with an imprinting defect (13, 18). Paternally inherited deletions in the mouse from *Snurf* to *Ube3a* [(Figure 3*b*), note deletion b] generate mice with growth retardation, hypotonia, and ~80% lethality up to weaning (82), and this deletion includes all three classes of imprinted snoRNAs. Failure-to-thrive with lethality is 100% penetrant for the deletion, matUPD, and imprinting mutation PWS mouse models (12, 28, 86). The fully penetrant phenotype may be accounted for by the additive contributions of an early postnatal lethality in *Ndn*-mutant mice (see previous section) and the cumulative neonatal lethality in *Snurf-Ube3a* deletion mice. Thus, the loss of one or more of the *MBII-13*, *MBII-85*, or *MBII-52* snoRNA genes may be involved in the hypotonia and neonatal failure-to-thrive PWS phenotype.

Most C/D box snoRNAs are ubiquitous and function as guide RNAs to direct 2′-*O*-ribose methylation in rRNA [the C and D boxes are conserved sequence elements; (25)]. In contrast, the brain-specific 15q11-q13/7C snoRNAs do not have a region complementary to rRNA, and they may be involved in mRNA modification (13). For *HBII-52/MBII-52*, a guide region exists with an 18-nt complementarity to the serotonin receptor 2C mRNA in a region involved in adenosine-to-inosine editing and close to a site for alternative splicing (13). However, the *HBII-52* tandem array may map within a 1 Mb deletion (Figure 3*a*) that causes AS on maternal inheritance but has no phenotypic effect when transmitted through a male (70). Similarly, in the mouse, the p^{30PUb} radiation-induced deletion includes *Ube3a* and *Ipw* [(Figure 3*b*), note deletion d] and can be transmitted without phenotypic effect through males (43). This deletion should include at least the *MBII-52* snoRNA family (Figure 3*b*), and if so, *MBII-52* snoRNAs would not contribute to the PWS phenotype.

The potential RNA targets of the other PWS snoRNAs are unknown. However, the *HBII-85* snoRNA gene cluster maps distal of a cluster of three balanced translocation breakpoints within the noncoding exon 20 of the *SNURF-SNRPN* gene in PWS-like patients (Figure 3*a*) and is not expressed in these patients, supporting a role for the *HBII-85* snoRNA genes in PWS (85). Because *IPW* is expressed in cells from the two classical PWS patients with translocations within *SNURF* (see Imprinted PWS/AS Domain) (59, 85), assessing the snoRNA expression pattern in these patients will be important. Definitive evidence for snoRNA involvement in PWS will await further studies in patients and mouse models (knockouts or phenotypic rescue by transgenes), but the work to date is an exciting development.

IMPRINTING IN PWS/AS

An understanding of the parental imprinting process in PWS and AS has come from a descriptive characterization of the allelic differences in chromatin marking and organization, as well as experimental observations that are dissecting the genetic and molecular basis for establishment and maintenance of paternal and maternal gene expression. The organization of the following two sections reflects this progression and highlights the epigenetic regulation of *SNURF-SNRPN*, which has received the most attention in 15q11-q13/7C as a consequence of the localization of the IC to the 5′ end of this locus (see Regulation in *cis* Based on Inherited Imprinting Mutations).

Imprinted Features of Expression and Chromatin Regulation

DNA methylation was the first epigenetic modification identified in the imprinting process and is found in most imprinted genes (64, 80). Similarly, the CpG-islands associated with the 5′ promoters of paternally expressed imprinted genes in

15q11-q13/7C are methylated on the silent, maternal allele and unmethylated on the active, paternal allele [(Figure 5c, see color insert); (6, 27, 30, 44, 49, 60, 74, 86, 87); J.H. Chai, D.P. Locke, T. Ohta, J.M. Greally, & R.D. Nicholls, submitted]. The paternal allele of the 5' CpG-island of *SNURF-SNRPN* is specifically associated with acetylated histones H4 and H3, whereas the maternal allele is hypoacetylated (Figure 5d, see color insert), in keeping with the general roles of DNA methylation and histone deacetylation to cooperate in transcriptional silencing (71). Interestingly, the use of a DNA methyltransferase inhibitor results in a low level of demethylation and histone H4 (but not H3) acetylation coupled with reactivation of *SNURF-SNRPN* expression on the maternal allele (71). These results indicate that DNA methylation and histone H4 deacetylation may be sufficient to maintain silencing of the imprinted gene in somatic cells. In addition to its role in somatic cell regulation of imprinted gene expression, DNA methylation at certain sites in imprinted genes is established differentially during spermatogenesis and oogenesis to serve another role as a key component of the gametic memory for imprinted genes (64, 80). In intron 7 of the human *SNURF-SNRPN* locus, several CpG sites have been found that are methylated in sperm and in somatic cells on the active paternal allele but are unmethylated in fetal oocytes (30). Similar gametic methylation differences occur within mouse *Snurf-Snrpn* but are also found in the 5' CpG-island (promoter and first exon) where the maternal allele is methylated specifically in the oocyte (74).

Six "phylogenetic footprints," DNA sequences of 7–10 bp, are conserved in the human and mouse *SNURF-SNRPN* 5' promoter (62). Some of these likely form transcription factor (TF) binding sites for regulation of somatic expression of the gene, but one or more may also have IC regulatory function as sites for binding certain TFs during imprint switching (see below). This region can also silence gene expression over short distances in transgenic *Drosophila* (53), which may be relevant for the imprint mechanism (53) but may alternatively reflect an element for reducing expression in nonneural tissues (62). Not surprisingly, within the promoter region and a site at the 5' end of intron 1, *SNURF-SNRPN* displays paternal-specific open chromatin, as revealed by nuclease hypersensitivity (Figure 5e, see color insert), that suggests binding of TFs (62, 73). The intron 1 region is close to G-rich locally repeated sequences in the human and mouse *SNURF-SNRPN* loci (27). Similar repeats occur in *MAGEL2* and the mouse ortholog (6, 49) but do not occur in all imprinted genes in the PWS region [(44); J.H. Chai, D.P. Locke, T. Ohta, J.M. Greally, & R.D. Nicholls, submitted] nor in a proportion of imprinted loci from other genomic regions (64), raising doubts about the relevance of the repeats. Nevertheless, these elements may form local structures required for recognition by DNA methylases or other proteins that form heterochromatic-like DNA (see 27). Surprisingly, flanking the paternal-specific nuclease hypersensitive sites at 5' *SNURF-SNRPN* is a broader region of maternal-specific nuclease hypersensitivity that spans 3' of intron 1 to exon 9 and two sites ~30 kb and ~130 kb upstream of exon 1, though these patterns were relatively weak in the lymphoblastoid cells tested (73). The maternal-specific site in intron 7 corresponds to a region of differential gametic DNA methylation (see above), whereas the two sites 5' of *SNURF-SNRPN* map to the upstream part of the IC (73). Performing similar experiments in the mouse using brain, the tissue of highest expression for the PWS and AS genes, as well as germ cells and the early embryo (see below) will be important.

In addition to the "localized" features of chromatin discussed to date, additional regional chromosomal features may play a role in imprinted gene regulation. The regions spanning 30 kb upstream and 20 kb downstream of the human and mouse *SNURF-SNRPN* exon 1 and intron 1 are, as shown by a biochemical assay, composed of an unusually high density of matrix attachment regions (MARs), with maternal-specific condensation and attachment to the nuclear matrix observed in vivo using a FISH halo method (Figure 5*f*, see color insert) (34). A similar density of MARs was also found at the "target" imprinted gene *MKRN3*, located ~1 Mb away (34). The high density of MARs is hypothesized to form heterochromatic-like DNA specifically on the maternal chromosome, such that this state could be spread during imprint establishment in germ cells or the early embryo by "position effect" (34). Another feature of imprinted chromosome regions, observed in lymphocytes, is homologous association (48), but this property is not required for imprint establishment and maintenance in the mouse PWS/AS deletion model (28). In contrast, imprinted chromosome regions generally undergo asynchronous DNA replication during S phase of the cell cycle, and importantly, the allelic difference in timing of DNA replication is erased and reset during oogenesis and spermatogenesis and maintained in the early embryo (76). However, the paternally derived allele is usually earlier replicating than the maternal allele (Figure 5*f*), even when maternally expressed genes are present, and the differential replication timing extends across the entire 15q11-q13 domain, including nonimprinted loci (Figure 3*a*) (45, 46). This suggests that replication timing may be an essential but not sufficient event for establishing the full imprint. Therefore, the establishment of allelic replication timing in germ cells, as well as allele-specific DNA methylation patterns, may be coordinately regulated during gametogenesis and postzygotic development.

In contrast to other imprinted loci in 15q11-q13, *UBE3A* imprinted expression is tissue-specific (see Imprinted PWS/AS Domain), and no methylation imprint occurs; the CpG-island at the 5' end of the gene is fully unmethylated in all somatic tissues and germ cells (51a). Paternal-only expression of a brain-specific antisense gene, *UBE3A-AS*, might prevent transcription of the paternal *UBE3A* allele and thus result in imprinted expression (66). However, the transcriptional level, spatial and temporal expression pattern, and the extent of overlap with *UBE3A* are undetermined in humans and mouse. Indeed, it is tempting to speculate that *UBE3A-AS* may result from read-through transcription from the *HBII-52/MBII-52* host gene (Figure 3*a*, *b*). Although an antisense transcript makes for an attractive model and is observed as a common event in imprinted domains (64), other possible mechanisms to silence paternal *UBE3A* include roles for chromatin structure, insulators, or transcriptional enhancer competition (80).

Regulation in *cis* Based on Inherited Imprinting Mutations

As described in Inheritance in PWS and AS (above), defects in the imprinting process occur in up to 5% of PWS and AS cases, and DNA analysis shows biparental inheritance but with uniparental DNA methylation and gene function within the 15q11-q13 imprinted domain (7, 20, 60–62). A substantial proportion (25% or more) of PWS and AS ID are familial, and in these cases, the affected probands and their father or mother, respectively, carry a microdeletion of the IC. In 10 PWS ID mutations with fully characterized deletion breakpoints, the shortest region of deletion overlap (SRO) is ~4 kb spanning the first exon and promoter of the *SNURF-SNRPN* locus (Figure 3*a*) (4, 62). Nevertheless, in each case, all of the PWS-region genes that are normally paternally expressed show a maternal methylation imprint and gene silencing. Likewise, a targeted deletion of 42 kb at the 5′ end of *Snurf-Snrpn* in the mouse [(Figure 3*b*), note deletion e] led to an equivalent ID (86). Similarly, in 10 AS ID patients, the AS-SRO has been narrowed to 880 bp at a site 35 kb upstream of *SNURF-SNRPN* (9, 20, 61). These AS patients show biparental expression of genes normally expressed from the paternal allele and presumably silencing the *UBE3A* gene on the maternal allele in the brain (69). These expression patterns, coupled with the patterns of inheritance in PWS and AS families, led to an initial hypothesis that the IC is a bipartite regulatory element that controls the erasure and reestablishment of parental imprints in the germline (20, 60–62). The PWS-SRO and AS-SRO elements of the IC would thus control the maternal to paternal imprint switch during spermatogenesis or the paternal to maternal switch during oogenesis, respectively.

Recent comparative studies of a rare PWS family in which the father is mosaic for an IC microdeletion on the paternal chromosome and similar studies on chimeric mice with an IC deletion provided the important observation that the deleted chromosome acquires a maternal methylation imprint with accompanying silencing of paternal gene expression across the imprinted domain (4). The deletion likely arose in the early blastomere in the mosaic father and was induced in ES cells in the mouse, which is prior to the global demethylation and the subsequent de novo methylation after implantation (4, 64). As methylation at *Ndn* was not affected in ES cells, these results suggest that the PWS-SRO element of the IC has a postzygotic function to ensure survival of the paternal imprint (4).

These data are also consistent with a model in which the IC only functions in somatic cells (54). Significantly, recent studies in human and mouse have shown that sperm DNA from males with a maternally inherited deletion of the IC (encompassing the PWS-SRO and *SNURF-SNRPN*) have a normal paternal methylation imprint throughout 15q11-q13/7C, and hence, the imprint was correctly reset on the mutant chromosome (22). Therefore, the PWS-SRO element of the IC does not appear to play a role in the germline methylation switch and may only function in maintenance of the paternal imprint during embryonic development (4, 22). However, the differentially methylated region in 3′ *Snrpn* (30, 74) has not been analyzed in the mouse IC deletion and is deleted in the PWS father, and this may be an important germline methylation target of the IC. Other differential epigenetic chromatin events that occur in or close to the PWS-SRO (see Imprinted Features of Expression and Chromatin Regulation), such as replication timing (3, 76), may also emerge as key markers of parental origin and elements of IC function in imprint switching in gametes and preimplantation somatic development.

The discussion above focuses on events at the PWS-SRO element of the IC on the paternally derived allele. An intriguing recent observation indicates that maternal stores of an oocyte-specific form of DNA methyltransferase (Dnmt1o) are required at the 8-cell stage of preimplantation development in the mouse, for proper imprint maintenance at *Snurf-Snrpn* and a proportion of other imprinted genes at other chromosomal loci (36a). Whereas the maternal methylation imprint at the *Snurf-Snrpn* 5′ end (equivalent to the PWS-SRO) is correctly set in *Dnmt1o*-deficient oocytes, embryos lacking Dnmt1o protein fail to maintain methylation on 50% of maternal alleles, leading to biparental expression of *Snurf-Snrpn* (36a). Combined, the preceding discussion highlights the complexity of *cis* and *trans* events on both parental alleles of one component (the PWS-SRO) of the IC during preimplantation development.

Despite these genetic advances in understanding IC function, the molecular basis of how imprints are reset and parental origin is identified remains elusive. As the IC has a bipartite structure, the role of the AS-SRO must also be considered further. In the human, there is a complex pattern of alternative splicing of upstream (U) exons to *SNURF-SNRPN* (Figure 3*a*) (20, 23). These 5′ transcripts are imprinted with paternal expression and initiate from two copies of a duplicated upstream promoter (20, 23). The AS-SRO, based on microdeletions in AS ID patients, includes exon U5 of the 5′ transcript (23), and a potential splice mutation of exon U2 was found in an AS ID patient (20). These data suggest one model in which the alternative 5′ *SNURF-SNRPN* transcripts could be a major component of the paternal to maternal IC switch mechanism (20). It will be necessary to examine expression of these transcripts in germ cells and to identify equivalent transcripts in the mouse to further assess this model. However, if the AS-SRO element and the transcripts are conserved in the mouse is still unknown. In this regard, a 76-kb clone spanning the human AS-SRO, PWS-SRO, and *SNURF-SNRPN* did not imprint in transgenic mice, whereas an equivalent mouse clone with 2 copies did imprint, leading to the suggestion that the human and mouse imprinting mechanism may have diverged (5).

Alternatively, the AS-SRO may encode a *trans*-acting factor binding site that is involved in the imprint switch process (20, 60, 61). Although there is no methylation imprint in the AS-SRO (72), the finding of maternal-specific open chromatin in the AS-SRO region is consistent with *trans*-factors (73). To further examine the *trans*-factor model for AS-SRO function, Shemer et al. (75) generated transgenes (Tg's) in mouse with 1 kb of the human AS-SRO sequence fused to 0.2 kb of the mouse *Snurf-Snrpn* minimal promoter (homologous to the PWS-SRO). Expression, DNA methylation, and replication asynchrony data indicated that these Tg's were appropriately imprinted on both maternal and paternal transmission, whereas

Tg's with 1.2–5 kb of the mouse *Snurf-Snrpn* minimal promoter were not imprinted (75). These results suggest that the AS-SRO binds a *trans*-factor during oogenesis and interacts in *cis* with the PWS-SRO to confer imprinting in the PWS/AS domain (75). Nevertheless, confirming these results in the endogenous state, as many nonimprinted sequences become imprinted in small Tg's as a consequence of either sequence context or the strain background (14), will be necessary. In this regard, control Tg's with the AS-SRO alone, the *Snurf-Snrpn* minimal promoter alone, or the AS-SRO and PWS-SRO separated by ∼35 kb as in the endogenous state may be useful. Identification of the *trans* factors binding the AS-SRO and PWS-SRO in somatic and germ cells will also be critical. Though further studies will be necessary to determine the roles of *trans* factors and transcription within the IC during the imprint switch mechanism, answers and perhaps more surprises may soon be available.

OTHER GENETIC DISEASE IN 15q11-q13/7C

Several additional phenotypes map within the human and mouse PWS/AS chromosome regions, and these genes may modify the PWS or AS phenotype. The first recognized was the pink-eyed dilution (*p*) locus in mouse chromosome 7C and corresponding oculocutaneous albinism type II (OCA2) locus in human (Figure 3*a, b*) (see 77). Although OCA2 is recessive in human and mouse, hypopigmentation is a common finding in PWS and AS patients with the common deletion (see Figure 1*a*). This semidominant, nonimprinted phenotype is associated with deletion of one *P* (OCA2) gene allele (77), but the molecular basis is still not understood.

The mouse *p* (oca2) locus has served as a visible marker for the generation and maintenance of a collection of radiation-induced chromosome deletions extending different distances from *p* (see 16, 19, 40, 43). Two recessive phenotypes mapped close to oca2 (Figure 3*b*) are a neonatally lethal cleft palate (cp1) and juvenile development and fertility (jdf2) (16, 40). The cp1 phenotype is 95% penetrant for both radiation-induced deletions and targeted knockout mutations in the *Gabrb3* gene, and the former could be rescued by *Gabrb3* transgenes (16, 17). Animals surviving the neonatal period have a severe neurological disorder characterized by tremor, seizures, jerky gait, abnormal EEG, and other AS-like behaviors, consistent with the idea that *GABRB3* loss in AS deletions could modify the AS phenotype (17, 57). Similarly, the *Herc2* gene is specifically mutated in jdf2 animals, characterized by progressive neuromuscular degeneration, shortened lifespan, and sperm and ovary defects (40, 51). Mouse, human, and *Drosophila HERC2* encode highly conserved, giant proteins (39, 40) and may have biochemical function in guanine nucleotide exchange and as an E3 ubiquitin ligase, likely in protein trafficking pathways (40, 51).

Mice heterozygous for several of the *p*-locus deletions have significantly increased body fat [*p*-locus-associated obesity, plo1; (Figure 3*b*)] when the deletion is of maternal as compared to paternal inheritance (19). Based on deletion mapping, the plo1 locus maps proximal to *Ube3a* in the region of a novel gene encoding a P-type ATPase, *Pfatp* (19), which may function as an aminophospholipid transporter (35). *Pfatp* is expressed in white adipose tissue and testis where it appears to be nonimprinted, but more complex imprinting regulation cannot be ruled out (19). Alternatively, body fat in progeny may be determined by maternal haploinsufficiency during pregnancy or lactation (19). As obesity is also associated with the AS deletion (Figure 1*d*) and patUPD (11) mouse models, as well as AS Classes II–IV in humans (see Inheritance in PWS and AS), understanding the potential contributions of the adjacent *Pfatp* and *Ube3a* loci to the genetic and physiological basis of obesity in plo1 and AS is important.

An imprinted susceptibility locus for autism may also be found within the PWS/AS region, as maternally inherited duplications of 15q11-q13 are associated with autism, whereas there is no phenotype with paternal inheritance of the same duplication (15a). Consequently, *UBE3A* and *PFATP*, should the latter be found to be imprinted and maternally expressed, are the most likely candidate genes. Nevertheless, because PWS UPD patients also have two active maternally derived alleles and do not usually show autism, a gene expression threshold effect or imbalance between parental alleles may be needed to explain the finding of autism in maternal duplications. Another disorder mapped to 15q11-q13 is dominant familial spastic paraplegia (SPG6), associated with axonal degeneration within the central nervous system (26). As there is no evidence that middle-aged individuals with PWS or AS deletions develop axonal degeneration, assuming that the SPG6 gene does map between BP1 and BP3 in 15q11-q13, it is more likely that the SPG6 mutation(s) are dominant negative rather than that the disease arises from haploinsufficiency. Finally, two methods of genome-wide linkage were used to identify a major tuberculosis-susceptibility locus in 15q11-q13 (2), and candidates for this role in infectious disease would be UBE3A (protein degradation), PFATP (ATPase), HERC2 (protein trafficking), SRA1 (membrane ruffling), and P (transporter expressed in immune cells) (M.S. Schaldach, J.L. Knepper, & R.D. Nicholls, unpublished data).

PWS/AS ARE GENOMIC DISEASES

PWS and AS can be classified among the large number of "genomic disorders" (38, 52) that arise due to dosage imbalance of one or more genes and are caused by chromosomal structural change commonly involving homologous recombination between chromosome-specific low-copy repeats, or "duplicons" (21). With a de novo frequency of 0.7–1/1000 births for genomic disorders as a group (38), the importance of this mechanism in human disease is remarkably clear. Most (>95%) PWS/AS deletion patients have clustered deletion breakpoints at BP1 or BP2 and BP3 (Figure 3*a*), suggesting that sequences near the breakpoints confer genomic instability (38). Large duplicons have been identified in these regions, using combinations of molecular and cytological methods (1, 15, 39).

FISH analyses with a phage probe corresponding to one part of the duplicon demonstrate a duplicated sequence at 15q11 and 15q13 for normal chromosomes, and demarcate each copy of 15q11-q13 in rearranged chromosomes (Figure 6*a*, see color insert). The intensity of signals for the 15-kb probe indicates that each "copy" is actually composed of multiple additional copies, a conclusion borne out by Southern analysis of YACs from BP1-BP2 and BP3 and of human chromosomes present in somatic cell hybrids that demonstrated at least 9–10 copies of 90%–99% identity in chromosome 15pter-q14 (Figure 6*b*, see color insert) (1, 39). Two copies of this duplicon also map to chromosome 16p11 (1, 15, 39). A large part of these duplicons derives from duplications of a large gene, *HERC2* (Figures 3*a*, 6*b*). *HERC2* spans 93 exons over 200–250 kb, with duplicons corresponding to the first 79 exons, although all copies have undergone internal rearrangements with respect to the ancestral sequence during the amplification and dispersal process in primates (38, 39). A model to explain the generation of PWS and AS common deletions invokes homologous misalignment and meiotic recombination between different *HERC2*-duplicons in proximal and distal 15q11-q13 (Figure 6*b*). *HERC2*-duplicons are transcribed in male and female germline tissues (1), and perhaps in germ cells, based on phenotypic studies of sperm from jdf2 mice (see 40), and thus, open chromatin may further stimulate recombination (1, 52). Supporting this model, a putative deletion breakpoint junction was identified using a *HERC2* probe in mRNA from a cell line from 1 of 5 PWS patients tested (1).

Christian et al. (15) also identified duplicons at the breakpoint regions of PWS/AS common deletions by STS and EST content mapping in YACs and by FISH studies using a PAC probe that identified signals in 15q11-q13, 16p11 and, with a weak signal, in 15q24 (see below). These workers also identified by FISH homologous duplicons in chimpanzee, gorilla, and an Old World monkey, indicating an origin for the genomic duplication events ∼20–25 mya. These results are consistent with estimates based on paralogous DNA sequences, which suggest that the ancestral duplicons arose from *HERC2* ∼14–20 mya (40). One end of the duplicons is clearly within the 3′ part of *HERC2* (40), with the *HERC2*-related content of different duplicons ranging from 36.6 kb in the most rearranged/deleted copy in 16p11.2 to an estimated >100 kb in some chromosome 15 duplicons (39), but the duplicon endpoint 5′ of *HERC2* is unknown. Christian et al. (15) also identified, at BP2 and BP3, additional homologous segments to the *MYLE* gene, an ATP-binding cassette protein, and a BEM-1/BUDS suppressor-like protein, which either are part of the *HERC2*-duplicons or form independent flanking duplicons. Indeed, independent duplicons flanking *HERC2*-duplicons at BP2 and BP3 include a truncated copy of the poly(A)-specific ribonuclease (*PARN*) gene from 16p13 (8), as well as the 13–22 kb LCR15 duplicons containing copies of the golgin-like protein (*GLP*) gene and *SH3P18* and that map in 15q26.1, 15q24, 15q13 adjacent to *HERC2* sequences, and to other chromosomes (63).

Recent analyses of the coverage and assembly of *HERC2*-duplicons within Gen-Bank suggest that the *HERC2*-duplicon sequences are underrepresented and that it has not been possible to correctly assemble closely related but different copies of these duplicons, and this has also been found for other, unrelated duplicons in the human genome [E.E. Eichler, personal communication; (36)]. Other 15q11-q13 rearrangements that include frequent inverted duplication (15) chromosomes with breakpoints at BP1, BP2, BP3, or a more distal BP4, as well as direct duplications, triplications, and inversions of 15q11-q13, may also arise from recombination involving duplicons found at the breakpoint regions (38, 83). Future studies will be required to determine the exact structural arrangement and nature of duplicons in 15q11-q13 as well as the mechanisms underlying chromosome rearrangements involving these sequences.

CONCLUDING COMMENTS: THE PAST AND THE FUTURE

From the work discussed here, it is clear that the genetic and epigenetic makeup of human chromosome 15q11-q13 and mouse 7C is complex. Large duplications of DNA sequences flanking 15q11-q13 significantly impact the risk for abnormal recombination, leading to disease (PWS or AS). Additional duplications of snoRNA genes and sequences within and upstream from the IC occur and may have unknown functional significance. Our understanding of the structure and genomic fluidity during the evolution of the PWS/AS domain may be increased when the gaps in 15q11-q13 DNA sequence are filled because these gaps may reflect additional duplicated segments resistant to cloning. Furthermore, the imprinted domain is evolutionarily young [(23, 32, 44, 59); J.H. Chai, D.P. Locke, T. Ohta, J.M. Greally, R.D. Nicholls, submitted] and is flanked on either side by ancient nonimprinted loci (*SRA1*, *GABR* cluster, *P*, *HERC2*). *UBE3A* may also classify among the latter, as its mechanism of imprinting may be a secondary consequence of silencing by an antisense gene expressed from the paternal allele. The evolutionary basis for imprinting in the PWS/AS domain may have come from phenotypic selection for a paternally derived factor involved in postnatal growth as a consequence of genetic conflict over maternal resources (59, 64, 80). The PWS neonatal phenotype is consistent with this model as is the molecular evidence that localizes the imprinting regulatory element (the IC) to a complex paternally expressed locus.

At least two paternally expressed genes may underlie the PWS neonatal phenotype (29, 55, 58, 82). Combining deficiencies for *Ndn*, the PWS-region snoRNAs, and other single PWS candidate gene mutations, as well as transgene rescue of PWS mouse models, may establish better PWS models and allow dissection of the relative roles of each gene in the neonatal and adult PWS phenotypes. However, whether an adult mouse model will show additional features of PWS, such as hyperphagia and obesity, remains to be established. Generation of single gene mutations in the mouse, and the testing of each mutant for behavioral phenotypes, will also play an important role in understanding the pathophysiological basis of PWS. Nevertheless, evolutionary maintenance is an even stronger guide to important in vivo function, and better ways to assess phenotypes might be needed before the contributions of each relevant gene are understood. Finally, understanding the evolution and functional role of the complex polycistronic IC-*SNURF-SNRPN*-snoRNA locus and the other snoRNA repeats in the PWS phenotype and/or the imprint mechanism will be important. Further analyses of the complex genomic structure and function of the PWS/AS domain and the molecular mechanisms by which maternal and paternal origin are specified and exerted over a chromosomal domain are certain to continue to bring novel concepts to the field of mammalian genetics.

ACKNOWLEDGMENTS

We thank Drs. Merlin Butler, Howard Cedar, Todd Gray, John Greally, Alexander Hüttenhofer, Shinji Saitoh, and Stuart Schwartz for images reproduced in the figures and Drs. Juergen Brosius, Karin Buiting, Madhu Dhar, Daniel Driscoll, Bernhard Horsthemke, Alexander Hüttenhofer, Dabney Johnson, and Rachel Wevrick for kindly sending preprints of manuscripts and/or discussion prior to publication. We apologize that because of space limitations it was sometimes necessary to limit references to more recent publications. Work in the authors' laboratory is supported by grants from the National Institutes of Health (HD31491, HD36079, and HD36436), the March of Dimes Birth Defects Foundation (6-FY99-390), and the Muscular Dystrophy Association.

Visit the Annual Reviews home page at www.AnnualReviews.org

LITERATURE CITED

1. Amos-Landgraf JM, Ji Y, Gottlieb W, Depinet T, Wandstrat AE, et al. 1999. Chromosome breakage in the Prader-Willi and Angelman syndromes involves recombination between large, transcribed repeats at proximal and distal breakpoints. *Am. J. Hum. Genet.* 65:370–86

1a. Banbury Conf. Genet. Backgr. Mice. 1997. Mutant mice and neuroscience: recommendations concerning genetic background. *Neuron* 19:755–59

2. Bellamy R, Beyers N, McAdam KP, Ruwende C, Gie R, et al. 2000. Genetic susceptibility to tuberculosis in Africans: a genome wide scan. *Proc. Natl. Acad. Sci. USA* 97:8005–9

3. Ben-Porath I, Cedar H. 2000. Imprinting: focusing on the center. *Curr. Opin. Genet. Dev.* 10:550–54

4. Bielinska B, Blaydes SM, Buiting K, Yang T, Krajewska-Walasek M, et al. 2000. *De novo* deletions of *SNRPN* exon 1 in early human and mouse embryos result in a paternal to maternal imprint switch. *Nat. Genet.* 25:74–78

5. Blaydes SM, Elmore M, Yang T, Brannan CI. 1999. Analysis of murine *Snrpn* and human SNRPN gene imprinting in transgenic mice. *Mamm. Genome* 10:549–55

6. Boccaccio I, Glatt-Deeley H, Watrin F, Roeckel N, Lalande M, Muscatelli F. 1999. The human *MAGEL2* gene and its mouse homologue are paternally expressed and mapped to the Prader-Willi region. *Hum. Mol. Genet.* 8:2497–505

7. Buiting K, Dittrich B, Groß S, Lich C, Färber C, et al. 1998. Sporadic imprinting

defects in Prader-Willi syndrome and Angelman syndrome: implications for imprint switch models, genetic counseling, and prenatal diagnosis. *Am. J. Hum. Genet.* 63:170–80

8. Buiting K, Korner C, Ulrich B, Wahle E, Horsthemke B. 1999. The human gene for the poly(A)-specific ribonuclease (PARN) maps to 16p13 and has a truncated copy in the Prader-Willi/Angelman syndrome region on 15q11-q13. *Cytogenet. Cell. Genet.* 87:125–31

9. Buiting K, Lich C, Cottrell S, Barnicoat A, Horsthemke B. 1999. A 5-kb imprinting center deletion in a family with Angelman syndrome reduces the shortest region of deletion overlap to 880 bp. *Hum. Genet.* 105:665–66

10. Butler MG. 1990. Prader-Willi syndrome: current understanding of cause and diagnosis. *Am. J. Med. Genet.* 35:319–32

11. Cattanach BM, Barr JA, Beechey CV, Martin J, Noebels J, Jones J. 1997. A candidate model for Angelman syndrome in the mouse. *Mamm. Genome* 8:472–78

12. Cattanach BM, Barr JA, Evans EP, Burtenshaw M, Beechey CV, et al. 1992. A candidate mouse model for Prader-Willi syndrome which shows an absence of Snrpn expression. *Nat. Genet.* 2:270–74

13. Cavaillé J, Buiting K, Kiefmann M, Lalande M, Brannan CI, et al. 2000. Identification of brain-specific and imprinted small nucleolar RNA genes exhibiting an unusual genomic organization. *Proc. Natl. Acad. Sci. USA* 97:14311–16

14. Chaillet JR. 1994. Genomic imprinting: lessons from mouse transgenes. *Mutat. Res.* 307:441–49

15. Christian SL, Fantes JA, Mewborn SK, Huang B, Ledbetter DH. 1999. Large genomic duplicons map to sites of instability in the Prader-Willi/Angelman syndrome chromosome region (15q11-q13). *Hum. Mol. Genet.* 8:1025–37

15a. Cook EH Jr, Lindgren V, Leventhal BL, Courchesne R, Lincoln A, et al. 1997. Autism or atypical autism in maternally

but not paternally derived proximal 15q duplication. *Am. J. Hum. Genet.* 60:928–34

16. Culiat CT, Stubbs LJ, Woychik RP, Russell LB, Johnson DK, Rinchik EM. 1995. Deficiency of the beta 3 subunit of the type A gamma-aminobutyric acid receptor causes cleft palate in mice. *Nat. Genet.* 11:344–46

17. DeLorey TM, Handforth A, Anagnostaras SG, Homanics GE, Minassian BA, et al. 1998. Mice lacking the beta 3 subunit of the GABAA receptor have the epilepsy phenotype and many of the behavioral characteristics of Angelman syndrome. *J. Neurosci.* 18:8505–14

18. de los Santos T, Schweizer J, Rees CA, Francke U. 2000. Small evolutionarily conserved RNA, resembling C/D box small nucleolar RNA, is transcribed from *PWCR1*, a novel imprinted gene in the Prader-Willi deletion region, which is highly expressed in brain. *Am. J. Hum. Genet.* 67:1067–82

19. Dhar M, Webb LS, Smith L, Hauser L, Johnson D, West DB. 2000. A novel ATPase on mouse chromosome 7 is a candidate gene for increased body fat. *Physiol. Genomics* 4:93–100

20. Dittrich B, Buiting K, Korn B, Rickard S, Buxton J, et al. 1996. Imprint switching on human chromosome 15 may involve alternative transcripts of the *SNRPN* gene. *Nat. Genet.* 14:163–70

21. Eichler EE. 1998. Masquerading repeats: paralogous pitfalls of the human genome. *Genome Res.* 8:758–62

22. El-Maarri O, Buiting K, Peery EG, Kroisel PM, Balaban B, et al. 2001. Methylation imprints on human chromosome 15 are established around or after fertilization. *Nat. Genet.* 27:341–44

23. Färber C, Dittrich B, Buiting K, Horsthemke B. 1999. The chromosome 15 imprinting center (IC) region has undergone multiple duplication events and contains an upstream exon of *SNRPN* that is deleted in all Angelman syndrome patients with

an IC deletion. *Hum. Mol. Genet.* 8:337–43

24. Färber C, Groß S, Neesen J, Buiting K, Horsthemke B. 2000. Identification of a testis-specific gene (*C15orf2*) in the Prader-Willi syndrome region of chromosome 15. *Genomics* 65:174–83

25. Filipowicz W. 2000. Imprinted expression of small nucleolar RNAs in brain: time for RNomics. *Proc. Natl. Acad. Sci. USA* 97:14035–37

26. Fink JK, Wu CT, Jones SM, Sharp GB, Lange BM, et al. 1995. Autosomal dominant familial spastic paraplegia: tight linkage to chromsome 15q. *Am. J. Hum. Genet.* 56:188–92

27. Gabriel JM, Gray TA, Stubbs L, Saitoh S, Ohta T, Nicholls RD. 1998. Structure and function correlations at the imprinted mouse *Snrpn* locus. *Mamm. Genome* 9:788–93

28. Gabriel JM, Merchant M, Ohta T, Ji Y, Caldwell RG, et al. 1999. A transgene insertion creating a heritable chromosome deletion mouse model of Prader-Willi and Angelman syndromes. *Proc. Natl. Acad. Sci. USA* 96:9258–63

29. Gérard M, Hernandez L, Wevrick R, Stewart CL. 1999. Disruption of the mouse *Necdin* gene results in early postnatal lethality. *Nat. Genet.* 23:199–202

30. Glenn CC, Saitoh S, Jong MTC, Filbrandt MM, Surti U, et al. 1996. Gene structure, DNA methylation and imprinted expression of the human *SNRPN* gene. *Am. J. Hum. Genet.* 58:335–46

31. Gray TA, Hernandez L, Carey AH, Schaldach MA, Smithwick MJ, et al. 2000. The ancient source of a distinct gene family encoding proteins featuring RING and C(3)H zinc finger motifs with abundant expression in developing brain and nervous system. *Genomics* 66:76–86

32. Gray TA, Saitoh S, Nicholls RD. 1999. An imprinted, mammalian bicistronic transcript encodes two independent proteins. *Proc. Natl. Acad. Sci. USA* 96:5616–21

33. Gray TA, Smithwick MJ, Schaldach MA, Martone DL, Graves JAM, et al. 1999. Concerted regulation and molecular evolution of the duplicated *SNRPB'/B* and *SNRPN* loci. *Nucleic Acids Res.* 27:4577–84

34. Greally JM, Gray TA, Gabriel JM, Song L, Zemel S, Nicholls RD. 1999. Conserved characteristics of heterochromatin-forming DNA at the 15q11-q13 imprinting center. *Proc. Natl. Acad. Sci USA* 96:14430–35

35. Halleck MS, Lawler JF Jr, Blackshaw S, Gao L, Nagarajan P, et al. 1999. Differential expression of putative transbilayer amphipath transporters. *Physiol. Genomics* 1:139–50

35a. Herzing LB, Kim SJ, Cook EH Jr, Ledbetter DH. 2001. The human aminophospholipid-transporting ATPase gene *ATP10C* maps adjacent to *UBE3A* and exhibits similar imprinted expression. *Am. J. Hum. Genet.* 68:1501–5

36. Horvath JE, Schwartz S, Eichler EE. 2000. The mosaic structure of human pericentromeric DNA: a strategy for characterizing complex regions of the human genome. *Genome Res.* 10:839–52

36a. Howell CY, Bestor TH, Ding F, Latham KE, Mertineit C, et al. 2001. Genomic imprinting disrupted by a maternal effect mutation in the dnmt1 gene. *Cell* 104:829–38

37. Huang L, Kinnucan E, Wang G, Beaudenon S, Howley PM, et al. 1999. Structure of an E6AP-UbcH7 complex: insights into ubiquitination by the E2-E3 enzyme cascade. *Science* 286:1321–26

38. Ji Y, Eichler EE, Schwartz S, Nicholls RD. 2000. Structure of chromosomal duplicons and their role in mediating human genomic disorders. *Genome Res.* 10:597–610

39. Ji Y, Rebert NA, Joslin JM, Higgins MJ, Schultz RA, Nicholls RD. 2000. Structure of the highly conserved *HERC2* gene and of multiple partially duplicated paralogs in human. *Genome Res.* 10:319–29

40. Ji Y, Walkowicz MJ, Buiting K, Johnson DK, Tarvin RE, et al. 1999. The ancestral gene for transcribed, low-copy repeats in the Prader-Willi/Angelman region encodes a large protein implicated in protein trafficking, which is deficient in mice with neuromuscular and spermiogenic abnormalities. *Hum. Mol. Genet.* 8:533–42

41. Jiang YH, Armstrong D, Albrecht U, Atkins CM, Noebels JL, et al. 1998. Mutation of the Angelman ubiquitin ligase in mice causes increased cytoplasmic p53 and deficits of contextual learning and long-term potentiation. *Neuron* 21:799–811

42. Jiang Y, Lev-Lehman E, Bressler J, Tsai TF, Beaudet AL. 1999. Genetics of Angelman syndrome. *Am. J. Hum. Genet.* 65:1–6

43. Johnson DK, Stubbs LJ, Culiat CT, Montgomery CS, Russell LB, Rinchik EM. 1995. Molecular analysis of 36 mutations at the mouse pink-eyed dilution (p) locus. *Genetics* 141:1563–71

44. Jong MT, Gray TA, Ji Y, Glenn CC, Saitoh S, et al. 1999. A novel imprinted gene, encoding a RING zinc-finger protein, and overlapping antisense transcript in the Prader-Willi syndrome critical region. *Hum. Mol. Genet.* 8:783–93

45. Kitsberg D, Selig S, Brandeis M, Simon I, Keshet I, et al. 1993. Allele-specific replication timing of imprinted gene regions. *Nature* 364:459–63

46. Knoll JHM, Cheng S, Lalande M. 1994. Allele specificity of DNA replication timing in the Angelman/Prader-Willi syndrome imprinted chromosomal region. *Nat. Genet.* 6:41–46

47. Krawczak M, Cooper DN. 1997. The human gene mutation database. *Trends Genet.* 13:121–22

48. LaSalle JM, Lalande M. 1996. Homologous association of oppositely imprinted chromosomal domains. *Science* 272:725–28

49. Lee S, Kozlov S, Hernandez L, Chamberlain SJ, Brannan CI, et al. 2000. Expression and imprinting of MAGEL2 suggest a role in Prader-Willi syndrome and the homologous murine imprinting phenotype. *Hum. Mol. Genet.* 9:1813–19

50. Lee S, Wevrick R. 2000. Identification of novel imprinted transcripts in the Prader-Willi syndrome and Angelman syndrome deletion region: further evidence for regional imprinting control. *Am. J. Hum. Genet.* 66:848–58

51. Lehman AL, Nakatsu Y, Ching A, Bronson RT, Oakey RJ, et al. 1998. A very large protein with diverse functional motifs is deficient in rjs (runty, jerky, sterile) mice. *Proc. Natl. Acad. Sci. USA* 95:9436–41

51a. Lossie AC, Whitney MM, Amidon D, Chen P, Theriaque D, et al. 2001. Distinct phenotypes distinguish the molecular classes of Angelman syndrome. *J. Med. Genet.* In press

52. Lupski JR. 1998. Genomic disorders: structural features of the genome can lead to DNA rearrangements and human disease traits. *Trends Genet.* 14:417–22

53. Lyko F, Buiting K, Horsthemke B, Paro R. 1998. Identification of a silencing element in the human 15q11-q13 imprinting center by using transgenic *Drosophila*. *Proc. Natl. Acad. Sci. USA* 95:1698–702

54. Mann MRW, Bartolomei MS. 2000. Maintaining imprinting. *Nat. Genet.* 25:4–5

54a. Meguro M, Kashiwagi A, Mitsuya K, Nakao M, Kondo I, et al. 2001. A novel maternally expressed gene, *ATP10C*, encodes a putative aminophospholipid translocase associated with Angelman syndrome. *Nat. Genet.* 28:19–20

55. Muscatelli F, Abrous DN, Massacrier A, Boccaccio I, Le Moal M, et al. 2000. Disruption of the mouse *Necdin* gene results in hypothalamic and behavioral alteration reminiscent of the human Prader-Willi syndrome. *Hum. Mol. Genet.* 9:3101–10

56. Nawaz Z, Lonard DM, Smith CL, Lev-Lehman E, Tsai SY, et al. 1999. The Angelman syndrome-associated protein, E6-AP, is a coactivator for the nuclear

hormone receptor superfamily. *Mol. Cell Biol.* 19:1182–89

57. Nicholls RD. 1998. Strange bedfellows? Protein degradation and neurological dysfunction. *Neuron* 21:647–49

58. Nicholls RD. 1999. Incriminating gene suspects, Prader-Willi style. *Nat. Genet.* 23:132–34

59. Nicholls RD, Ohta T, Gray TA. 1999. Genetic abnormalities in Prader-Willi syndrome and lessons from mouse models. *Acta Paediatr.* 88(Suppl.):99–104

60. Nicholls RD, Saitoh S, Horsthemke B. 1998. Imprinting in Prader-Willi and Angelman syndromes. *Trends Genet.* 14:194–200

61. Ohta T, Buiting K, Kokkonen H, McCandless S, Heeger S, et al. 1999. Molecular mechanism of Angelman syndrome in two large families involves an imprinting mutation. *Am. J. Hum. Genet.* 64:385–96

62. Ohta T, Gray TA, Rogan PK, Buiting K, Gabriel JM, et al. 1999. Imprinting-mutation mechanisms in Prader-Willi syndrome. *Am. J. Hum. Genet.* 64:397–413

63. Pujana MA, Nadal M, Gratacos M, Peral B, Csiszar K, et al. 2001. Additional complexity on human chromosome 15q: identification of a set of newly recognized duplicons (LCR15) on 15q11-q13, 15q24, and 15q26. *Genome Res.* 11:98–111

64. Reik W, Walter J. 2001. Genomic imprinting: parental influence on the genome. *Nat. Rev. Genet.* 2:21–32

65. Robinson WP, Christian SL, Kuchinka BD, Penaherrera MS, Das S, et al. 2000. Somatic segregation errors predominantly contribute to the gain or loss of a paternal chromosome leading to uniparental disomy for chromosome 15. *Clin. Genet.* 57:349–58

66. Rougeulle C, Cardoso C, Fontes M, Colleaux L, Lalande M. 1998. An imprinted antisense RNA overlaps *UBE3A* and a second maternally expressed transcript. *Nat. Genet.* 19:15–16

67. Russo D, Cogliati F, Viri M, Cavalleri F,

Selicorni A, et al. 2000. Novel mutations of ubiquitin protein ligase 3A gene in Italian patients with Angelman syndrome. *Hum. Mutat.* 15:387

68. Saitoh S, Buiting K, Cassidy SB, Conroy JM, Driscoll DJ, et al. 1997. Clinical spectrum and molecular diagnosis of Angelman and Prader-Willi syndrome patients with an imprinting mutation. *Am. J. Med. Genet.* 68:195–206

69. Saitoh S, Buiting K, Rogan PK, Buxton J, Driscoll DJ, et al. 1996. Minimal definition of the imprinting center and fixation of a chromosome 15q11-q13 epigenotype by imprinting mutations. *Proc. Natl. Acad. Sci. USA* 93:7811–15

70. Saitoh S, Kubota T, Ohta T, Jinno Y, Niikawa N, et al. 1992. Familial Angelman syndrome caused by imprinted submicroscopic deletion encompassing GABAA receptor beta 3-subunit gene. *Lancet* 339:366–67

71. Saitoh S, Wada T. 2000. Parent-of-origin specific histone acetylation and reactivation of a key imprinted gene locus in Prader-Willi syndrome. *Am. J. Hum. Genet.* 66:1958–62

72. Schumacher A, Buiting K, Zeschnigk M, Doerfler W, Horsthemke B. 1998. Methylation analysis of the PWS/AS region does not support an enhancer-competition model. *Nat. Genet.* 9:324–25

73. Schweizer J, Zynger D, Francke U. 1999. *In vivo* nuclease hypersensitivity studies reveal multiple sites of parental origin-dependent differential chromatin conformation in the 150 kb *SNRPN* transcription unit. *Hum. Mol. Genet.* 8:555–66

74. Shemer R, Birger Y, Riggs AD, Razin A. 1997. Structure of the imprinted mouse *Snrpn* gene and establishment of its parental-specific methylation pattern. *Proc. Natl. Acad. Sci. USA* 94:10267–72

75. Shemer R, Hershko AY, Perk J, Mostoslavsky R, Tsuberi B, et al. 2000. The imprinting box of the Prader-Willi/Angelman syndrome domain. *Nat. Genet.* 26:440–43

76. Simon I, Tenzen T, Reubinoff BE, Hillman D, McCarrey JR, Cedar H. 1999. Asynchronous replication of imprinted genes is established in the gametes and maintained during development. *Nature* 401:929–32

77. Spritz RA, Bailin T, Nicholls RD, Lee ST, Park SK, et al. 1997. Hypopigmentation in the Prader-Willi syndrome correlates with *P* gene deletion but not with haplotype of the hemizygous *P* allele. *Am. J. Med. Genet.* 71:57–62

78. Sutcliffe JS, Nakao M, Christian S, Orstavik KH, Tommerup N, et al. 1994. Deletions of a differentially methylated CpG island at the SNRPN gene define a putative imprinting control region. *Nat. Genet.* 8:52–58

79. Taniguchi N, Taniura H, Niinobe M, Takayama C, Tominaga-Yoshino K, et al. 2000. The postmitotic growth suppressor Necdin interacts with a calcium-binding protein (NEFA) in neuronal cytoplasm. *J. Biol. Chem.* 275:31674–81

80. Tilghman SM. 1999. The sins of the fathers and mothers: genomic imprinting in mammalian development. *Cell* 96:185–93

81. Tsai TF, Armstrong D, Beaudet AL. 1999. Necdin-deficient mice do not show lethality or the obesity and infertility of Prader-Willi syndrome. *Nat. Genet.* 22:15–16

82. Tsai TF, Jiang Y, Bressler J, Armstrong D, Beaudet AL. 1999. Paternal deletion from *Snrpn* to *Ube3a* in the mouse causes hypotonia, growth retardation and partial lethality and provides evidence for a gene contributing to Prader-Willi syndrome. *Hum. Mol. Genet.* 8:1357–64

83. Ungaro P, Christian SL, Fantes JA, Mutirangura A, Black S, et al. 2001. Molecular characterization of four cases of intrachromosomal triplication of chromosome 15q11-q14. *J. Med. Genet.* 38:26–34

84. Wevrick R, Francke U. 1997. An imprinted mouse transcript homologous to the human imprinted in Prader-Willi syndrome (*IPW*) gene. *Hum. Mol. Genet.* 6:325–32

85. Wirth J, Back E, Hüttenhofer A, Nothwang HG, Lich C, et al. 2001. A translocation breakpoint cluster disrupts the newly defined 3' end of the *SNURF-SNRPN* transcription unit on chromosome 15. *Hum. Mol. Genet.* 10:201–10

86. Yang T, Adamson TE, Resnick JL, Leff S, Wevrick R, et al. 1998. A mouse model for Prader-Willi syndrome imprinting-centre mutations. *Nat. Genet.* 19:25–31

87. Zeschnigk M, Schmitz B, Dittrich B, Buiting K, Horsthemke B, Doerfler W. 1997. Imprinted segments in the human genome: different DNA methylation patterns in the Prader-Willi/Angelman syndrome regions as determined by the genomic sequencing method. *Hum. Mol. Genet.* 6:387–95

NOTE ADDED IN PROOF

The *PFATP* gene is now known as *ATP10C* and has been shown to display predominantly maternal expression in human brain and lymphocytes (35a, 54a).

Annu. Rev. Genomics Hum. Genet. 2001. 2:177–211

GENE THERAPY: Promises and Problems

Alexander Pfeifer and Inder M. Verma

*The Salk Institute, La Jolla, California 92037; e-mail: verma@salk.edu,
apfeifer@ems.salk.edu*

Key Words   gene transfer, viral vectors, gene therapy trials

■ **Abstract**   Gene therapy can be broadly defined as the transfer of genetic material to cure a disease or at least to improve the clinical status of a patient. One of the basic concepts of gene therapy is to transform viruses into genetic shuttles, which will deliver the gene of interest into the target cells. Based on the nature of the viral genome, these gene therapy vectors can be divided into RNA and DNA viral vectors. The majority of RNA virus-based vectors have been derived from simple retroviruses like murine leukemia virus. A major shortcoming of these vectors is that they are not able to transduce nondividing cells. This problem may be overcome by the use of novel retroviral vectors derived from lentiviruses, such as human immunodeficiency virus (HIV). The most commonly used DNA virus vectors are based on adenoviruses and adeno-associated viruses. Although the available vector systems are able to deliver genes in vivo into cells, the ideal delivery vehicle has not been found. Thus, the present viral vectors should be used only with great caution in human beings and further progress in vector development is necessary.

INTRODUCTION

Biologists will remember Monday, June 19, 2000, as an historic day. Flanking Bill Clinton, the 42nd President of the United States of America, were Francis Collins of the National Institutes of Health (NIH), leader of the publicly funded Human Genome project, and Craig Venter, CEO of Celera Genomics of Rockville, Maryland, to announce the near-completion of the sequencing of the human genome. Imagine: The entire 3 billion nucleotides of our genome are decoded—an impossible task just a few years ago. The estimate of the number of genes ranges from a low of 35,000 to a high of more than 100,000.

What a bonanza for gene therapy. The science of gene therapy relies on the introduction of genes to cure a defect or slow the progression of the disease and thereby improve the quality of life. Therefore, we need genes. Suddenly, we have tens of thousands of them at hand. Though gene therapy holds great promise for the achievement of this task, the transfer of genetic material into higher organisms still remains an enormous technical challenge. Presently available gene delivery vehicles for somatic gene transfer can be divided into two categories: viral and

1527-8204/01/0728-0177$14.00

nonviral vectors. Viruses evolved to depend on their host cell to carry their genome. They are intracellular parasites that have developed efficient strategies to invade host cells and, in some cases, transport their genetic information into the nucleus either to become part of the host's genome or to constitute an autonomous genetic unit. The nonviral vectors, also known as synthetic gene delivery systems (45), represent the second category of delivery vehicles and rely on direct delivery of either naked DNA or a mixture of genes with cationic lipids (liposomes). In this review, we focus on viral vectors and highlight some examples of their use in clinical trials. A complete, constantly updated list of human gene therapy trials in the United States is available at the Office of Biotechnology Activities, NIH (http://www4.od.nih.gov/oba/rdna.htm).

General Concept of Viral Vectors

The first step in viral vector design is to identify the viral sequences that are required for the assembly of viral particles, the packaging of the viral genome into the particles, and the delivery of the transgene to the target cells. Next, dispensable genes are deleted from the viral genome to reduce patho- and immunogenicity. The residual viral genome and the gene of interest (also termed transgene) are integrated into the vector construct (Figure 1).

Viral vectors can be divided into two general categories: (*a*) integrating vectors, capable of providing life-long expression of the transgene, and (*b*) nonintegrating vectors. Examples for integrating vectors are retroviral and adeno-associated virus (AAV)–derived vectors. The major nonintegrating vector currently employed is based on adenoviruses, and the viral DNA is maintained as an episome in the infected cell. Each of these vectors has specific advantages and major limitations. What, then, would be an ideal vector? We believe that it should fulfill the following requirements (147):

1. Efficient and easy production: High-titer preparations of vector particles should be reproducibly available. The efficient transduction of cells within tissues is only possible if a sufficient number of infectious particles reaches the target cells. For the widespread use of viral vectors, facile production procedures have to be developed.

Figure 1   Basic principal of viral vector design. (*A*) Structure of a generic viral genome. (*B*) Strategy of gene therapy vectors. The viral genome is separated into the packaging construct, which contains the viral sequences encoding proteins required for packaging of the vector genome and its replication. The vector construct contains the transgene and *cis*-acting sequences (*hatched boxes*) that are essential for encapsidation of the vector genome and for viral transduction of the target cell. (*C*) The vector and packaging constructs are expressed in the packaging cells, which produce the recombinant viral particles.

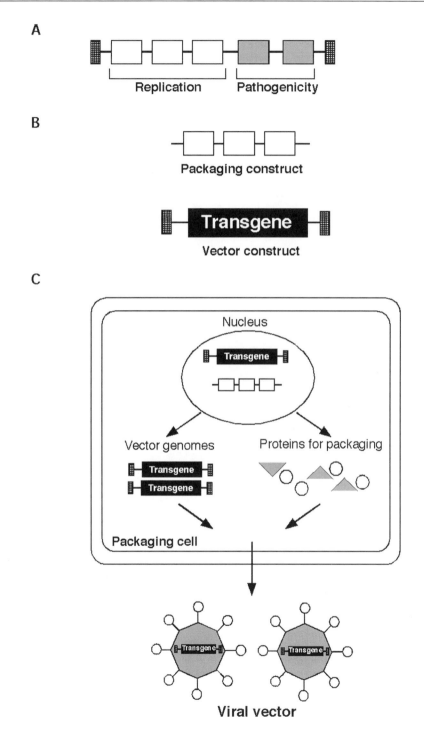

2. Safety aspects: The vector should neither be toxic to the target cells nor induce unwanted effects, including immunological reactions against the viral vector or its cargo. The latter carries not only the threat of eliminating the vector and/or the infected cells but also may lead to life-threatening complications, such as septic shock.

3. Sustained and regulated transgene expression: The gene delivered by the viral vector has to be expressed in a proper way. Permanent or even life-long expression of the therapeutic gene is desired only in a minority of diseases (e.g., treatment of hemophilia). Controlled expression of the transgene in a reversible manner would be highly desirable in many cases (e.g., gene therapy for insulin-dependent diabetes mellitus).

4. Targeting of the viral vectors: Preferential or exclusive transduction of specific cell types is very desirable.

5. Infection of dividing and nondividing cells: Because the majority of the cells in an adult human being are in a postmitotic, nondividing state, viral vectors should be able to efficiently transduce these cells.

6. Site-specific integration: Integration into the host genome at specific site(s) could enable us to repair genetic defects, such as mutations and deletions, by insertion of the correct sequences. Thus, replacing defective gene expression by introducing foreign genes and cDNAs would be unnecessary.

RNA VIRUS VECTORS

RNA viruses are a large and diverse group of viruses (150) that have either a single-stranded or a double-stranded RNA genome. They can infect a broad spectrum of cells, ranging from prokaryotes to many eukaryotic cells. Among the RNA-containing viruses, one group has attracted much attention as a gene delivery vehicle: the *Retroviridae* (30). Retroviruses comprise a diverse family of enveloped RNA viruses and can be divided into two categories according to the organization of their genome: simple and complex retroviruses (29). All retroviruses contain three major viral proteins: *gag*, *pol*, *env* [Figure 2, *right*; for review see (30, 160)]. *Gag* encodes the structural virion proteins that form the matrix, capsid, and the nucleoprotein complex. *Pol* codes for the essential viral enzymes reverse transcriptase and integrase. *Env* encodes the viral glycoproteins that are displayed on the surface of the virus. Moloney murine leukemia virus (MLV), a prototypic simple retrovirus, carries only a small set of genetic information, whereas the complex retroviruses like lentiviruses [e.g., human immunodeficiency virus (HIV)] contain additional regulatory and accessory genes. Initially, gene therapy vectors were developed from simple retroviruses. The lessons learned from simple retroviral vectors provided an invaluable basis for the development of vectors derived from complex retroviruses. Emerging vectors based on other RNA viruses, such as alphaviruses, are reviewed elsewhere (71).

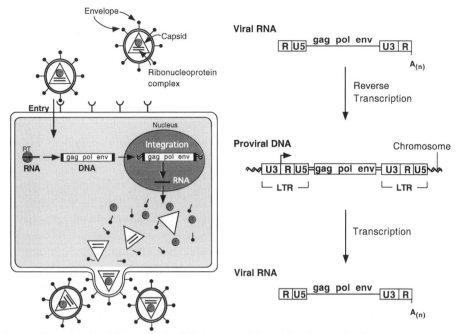

Figure 2   Retroviral lifecycle. (*Left*) Overview of the replication cycle of a prototypic retrovirus, MLV. (*Right*) Organization of the retroviral genome and its transition from RNA → DNA → RNA during viral replication. The prototypic retroviral genome contains the *gag*, *pol*, and *env* genes flanked by the R/U5 region at the 5′ end and the U3/R region at the 3′ end. Reverse transcription results in the proviral DNA that contains long terminal repeats (LTRs) at each end. The LTRs comprise U3, R, and U5 elements in the provirus. Transcription between (not including) the 5′ U3 and the 3′ U5 regions generates the identical organization of the terminal domains as in the parental virus (*top*). $A_{(n)}$, polyA tail.

Retroviral Life Cycle

Knowledge of the viral life cycle was crucial for the development of retroviral vectors. Following infection of the cell, the genomic RNA is reverse transcribed into linear double-stranded DNA by the virion reverse transcriptase (156). Reverse transcription involves two jumps of the transcriptase enzyme from the 5′ end to the 3′ end of the viral template, causing a duplication of the sequences located at the ends of the viral RNA. Thus, the viral DNA is significantly longer than the viral genome at both the 5′ and 3′ ends. The resulting tandem repeats in the viral DNA are termed long terminal repeats (LTRs) (Figure 2). Reverse transcription takes place in the cytoplasm and the viral DNA is translocated into the nucleus.

Simple and complex retroviruses enter the nucleus of the host cell by two different mechanisms: Nuclear entry of simple retroviruses can only occur when the nuclear membrane is disassembled and is, therefore, mitosis dependent. In contrast, lentiviruses can access the nucleus of nondividing cells by import through the nuclear pore [for review see (19)]. This has practical implications for the spectrum of target cells that can be transduced by viral vectors derived from simple or complex retroviruses (see below). After entry into the nucleus, the viral DNA integrates into the host genome to form a provirus. The formation of a provirus is a unique genetic strategy: The DNA intermediate stage mimics a cellular gene and uses the host-cell machinery for gene expression. Therefore, the provirus requires *cis*-acting elements that control the host transcriptional machinery. Most of these elements are situated within the proviral LTRs. The complex retroviruses additionally contain *trans*-acting factors that serve as activators of RNA transcription (e.g., HIV-1 Tat).

The LTRs are divided into the U3, R, and U5 regions (Figure 2, *right*). The R region is defined as the transcription start site in the 5′ LTR. In the 3′ LTR, the R region is the target of 3′-end processing (Figure 2, *right panel*). The U3 region that is found upstream of the transcription start site in the 5′ LTR contains the majority of *cis*-acting control elements. These elements regulate transcriptional initiation by the cellular RNA polymerase II. In addition, the sequence at the immediate 5′ end of the U3 region contains the so-called *att* site that is necessary for integration. This sequence motif is also found in the 3′ LTR: The 3′ end of the U5 region contains an inverted copy of the *att* site. In addition, the 3′ LTR contains the *cis*-acting control elements involved in posttranscriptional processing of the 3′ end of the viral RNA (e.g., polyadenylation). Therefore, the transgene present in retroviral vectors should not contain a polyA signal sequence because this would lead to the replacement of the 3′ U3/R region with a polyA tail during transcription of the vector RNA.

The regulation of RNA processing differs between simple and complex retroviruses and is an important aspect of gene therapy approaches that require unspliced transcripts for the expression of therapeutic genes. Retroviral RNAs are subject to the same processing events as cellular RNAs: cap addition at the 5′ end, cleavage and polyadenylation of the 3′ end, and splicing [for review see (132)]. Simple retroviruses regulate the cytoplasmic ratio of full-length versus spliced RNAs through *cis*-acting elements within the RNAs, whereas complex viruses encode proteins that regulate the transport and, presumably, the splicing of the viral RNA. For example, the Rev protein of HIV promotes the efficient transport of unspliced RNAs that contain Rev response elements from the nucleus to the cytoplasm (also see section on Lentiviral Vectors).

After translation of the viral messages, the resulting protein products and the progeny RNA are assembled into viral particles that are released from the cell by budding of the plasma membrane (Figure 2, *left*).

Retrovirus Vectors

The majority of the retroviral vectors [see also (33, 100, 175)] presently used in gene therapy models are derived from MLV, and they were among the first viral vectors to be used in human gene therapy trials (18).

To generate retroviral vectors, all of the protein-encoding sequences were removed from the virus and replaced by the transgene of interest (Figure 3). The essential *cis*-acting sequences, such as the packaging signal sequences (Ψ), which are required for encapsidation of the vector RNA, have to be included in the vector construct. The viral sequences necessary for reverse transcription of the vector RNA and integration of the proviral DNA, the LTRs, the transfer RNA-primer binding site, and the polypurine tract (PPT) [for a detailed description see (30, 100)] have to be present in the vector construct. Several modifications have been introduced into this basic retroviral vector system. For instance, inclusion of the 5' end of the *gag* domain leads to a 50–200-fold increase in vector titers, because of an increased efficacy of vector RNA encapsidation. MLV-based vectors carrying this extended packaging signal [nucleotides 215–1039 of *gag* (9)] are called Ψ^+ or gag$^+$ vectors (7, 9)

The 5' as well as 3' LTR regions of the vector constructs have been subject to a number of modifications. Replacement of the U3 region in the 5' LTR with the immediate early region of the human cytomegalovirus (CMV) enhancer-promoter (Figure 3) resulted in an almost 100-fold increase (42, 116) in viral titers. The CMV/LTR hybrid has a high transcriptional activity, especially when introduced in the appropriate cell lines (42), e.g., human embryonic kidney (HEK), 293 cells. This cell line expresses the adenoviral E1 gene products (51) that superactivate the CMV promoter (48). The effects of this U3 modification are restricted to the packaging cells because the U3 region of the 5' LTR of the provirus is derived from the U3 region of the 3' end of the vector RNA (Figure 2). This hallmark of the retroviral life cycle is the basis for the development of transcriptionally silenced vectors, so-called self-inactivating (SIN) vectors (176, 181) (Figure 3). The SIN vectors were developed to cope with the problem of insertional activation of cellular oncogenes through the promoter and enhancer elements of the proviral LTR. The strategy of transcriptional inactivation of the provirus is based on the fact that deletions of the promoter/enhancer sequences of the U3 region of the 3' LTR of the viral vector are carried over to the 5' LTR during reverse transcription; in other words, the vector inactivates itself. The major drawback of the loss of transcriptional regulatory elements is a substantial reduction (10–100-fold lower) of viral titers. On the other hand, a partial deletion of the viral transcription control regions, with retention of the TATA box, results in only a partial abolition of LTR-driven transcription (176). However, the use of SIN vectors may be necessary to avoid interference of the viral promoter and enhancer regions with internal promoters (see below) within the vector in the target cells (40). Taken together, the modifications of the vector LTRs significantly increased vector yields and improved biosafety of the retroviral vectors.

REGULATION OF TRANSGENE EXPRESSION   In the most simple vector design, the 5' LTR of the integrated provirus drives the expression of the transgene (Figure 3). However, inclusion of an internal, heterologous promoter that drives transcription of the transgene in the target cells (Figure 3D) can either achieve an increase in

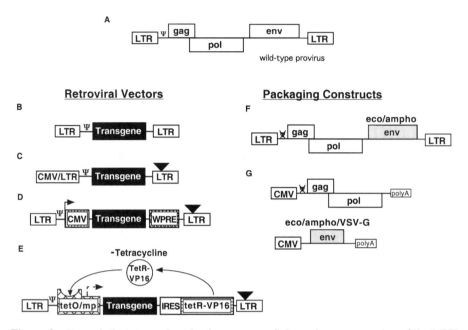

Figure 3   Retroviral vector and packaging systems. Schematic representation of the MLV provirus. (*B–E*) The different retroviral vectors. (*B*) In the first generation of retroviral vectors based on MLV, the transgene replaces most of the viral sequences. In addition, it contains the necessary *cis*-acting factors, such as the LTRs and the packaging signal (Ψ). (*C*) Modifications of the 5′ LTRs result in an increase in vector yield (e.g., inclusion of a CMV/LTR hybrid). SIN-mutations in the U3 region of the 3′ LTR (*black triangle*) improve the biosafety of the recombinant virus. (*D*) The latest generation of retroviral vectors incorporates internal promoters (e.g., CMV) that drive transgene expression in the target cells. In addition, a posttranscriptional element (WPRE) can be included 3′ of the transgene, which enhances expression three- to fivefold. (*E*) Tetracycline(tet)-regulated expression system. In the absence of tet, transgene expression is activated by the tet repressor (tetR)-VP16 fusion protein, which binds to the tet operator (tetO) fused to a minimal human CMV promoter (mp). Both the transgene as well as the tetR-VP16 coding regions are included in one vector construct. Bicistronic expression is achieved by incorporation of an internal ribosomal entry site (IRES). (*F, G*) Retroviral packaging constructs. (*F*) First-generation packaging constructs contain deletion mutations of the packaging signal (Ψ). Wild-type *env* (eco) can be replaced with amphotropic *env* (amph), resulting in a broadened host range. (*G*) Latest generation of split-genome packaging system: *gag/pol* and *env* are encoded on separate plasmids. The 5′ LTR is replaced with a strong promoter (e.g., CMV), and the 3′ LTR is replaced by a polyadenylation signal (polyA).

transgene expression and/or restriction of expression to a specifc cell type or tissue. In addition, this approach allows the incorporation of regulatable transcriptional elements that may be switched on and off via exogenous stimuli (Figure 3*E*).

The regulatable systems used in viral vectors either are based on naturally occuring inducible promoters that exhibit tissue specificity or consist of chimeric systems, which contain pro- and eukaryotic elements from different organisms [for review see (2)]. Among the chimeric regulatable systems, the tetracycline-(tet)-regulatable system (49) is one of the best characterized and most widely used systems. It is based on the inhibitory action of the tet repressor (tetR) of *Escherichia coli* on the tet operator sequence (*tet*O). The tet-regulatable system most widely used in mammalian cells carries two modifications: The tetR is fused to the carboxy terminus of VP16 (a herpes virus transactivator), and the *tet*O-repeats are fused to a minimal human CMV promoter (Figure 3*E*). In the presence of tet, the tetR-VP16 fusion protein cannot bind to and activate *tet*O (tet-off system), whereas in the absence of tet, the tetR-VP16 protein can bind to *tet*O, resulting in increased expression levels of the gene of interest. An elegant way to deliver both the tet-off system and the gene of interest to the target cell is to use a single retroviral vector that contains both elements (66). The regulated expression of both the gene of interest as well as the regulatory sytem can be achieved by using an internal ribosomal entry site (IRES), resulting in bicistronic expression (Figure 3*E*). A drawback of this approach is that the tetR-VP16 fusion protein is toxic to cells. However, this problem can be overcome by placing tetR-VP16 under the control of the *tet*O-containing promoter (145). Also, the reverse tet-regulated system (50) in which the addition of tet induces transactivation (tet-on) has been successfully used in the context of retroviral vectors (90).

Another family of chimeric-regulated systems is based on steroid hormones and their nuclear receptors. The organisms from which the hormones and their receptors have been isolated range from insects (ecdysone hormone of *Drosophila melanogaster* and *Bombyx mori*) to mammals (e.g., progesterone). The progesterone system is based on a mutated human progesterone receptor (164). The binding domain of this receptor is fused to the yeast GAL4 DNA binding domain and the VP16 domain, and it is activated by the antiprogesterone mifepristone (RU486), but not by the endogenous molecule present in mammals. In the presence of mifepristone, this chimeric regulator binds to the target gene, which contains the 17-mer GAL4 binding site, and activates transcription of the transgene (164). The use of the insect ecdysone-responsive system (117, 151) has the potential advantage that the ecdysone hormones are neither toxic nor known to affect mammalian physiology. In the *Drosophila* and the *Bombyx*-derived ecdysone-responsive system (DmEcR and BmEcR, respectively), the insect ecdysone receptor forms a heterodimer with the mammalian retinoid X receptor (RXR) (117). However, the DmEcR yields high levels of transactivation only if supraphysiological levels of RXR are present (117, 151). In contrast, the *Bombyx*-derived ecdysone-responsive system (BmEcR), presumably due to a higher affinity for RXR than the DmEcR, is capable of full transactivation with no added exogenous

RXR (151). Both ecdysone systems function in the context of retroviral vectors (151).

Recently, a rapamycin-regulated transcriptional system was described for retroviral vectors (128). Rapamycin mediates the formation of heterodimers between the immunophilin FK506-binding protein (FKBP) and the lipid kinase homolog FRAP (129). Fusion of the FKBP domains to a DNA-binding domain (called ZFHD1) and the rapamycin-binding domain of FRAP to a transcriptional activation domain (derived from the p65 subunit of human NF-κB) forms a functional transcription factor that can be activated through rapamycin-dependent dimerization (135). This dimerizer-responsive transcription factor cassette can be incorporated with the regulatable transgene into a single retroviral vector, which results in low basal expression levels and high dose-dependent induction of transgene expression. The induction ratios are in the range of three orders of magnitude and are comparable to the tet system.

Retroviral transgene expression can also be controlled at the level of translation by inclusion of *cis*-acting posttranscriptional regulatory elements (PREs). PREs are present in herpes simplex, hepatitis B virus, and the woodchuck hepatitis virus. The latter increases reporter gene expression at least fivefold if placed 3′ of the transgene in MLV-derived vectors (184).

PACKAGING OF RETROVIRAL VECTORS   To package the replication-defective vector into virions, the necessary viral proteins are provided in *trans* in the packaging cell (Figures 1 and 3). Retroviral packaging contructs are either transfected transiently into the packaging cells or a cell line [for a review over the presently available packaging cell lines, see (100)] is established that stably expresses the viral proteins. In either case, the packaging constructs are modified to reduce the chances of generating replication-competent virus (RCV) through recombination in the packaging cells. The *cis*-acting sequences required for packaging of the RNA (Ψ), the polypurine tract, and the 3′ LTR were deleted in the packaging constructs (93, 101). Furthermore, the 5′ end of the 5′ LTR that contains the *cis*-acting signal sequence required for integration [the *att* element; for details see (30)] was removed. Replacement of the 5′ LTR with the CMV promoter resulted in a CMV-driven packaging system that is compatible with the CMV/LTR hybrid vectors (see above) and results in high-titer virus preparations, especially if 293 cells are used (42, 116).

To further minimize the extent of homology between the packaging constructs and the retroviral vectors, which could lead to the production of helper virus after a single recombination event in the packaging cells, a split genome packaging strategy was developed. In this case, two packaging constructs, one containing *gag* and *pol* and the other carrying *env*, are used (34, 94) (Figure 3*G*). Splitting the packaging genome into multiple units not only increases the safety of retroviral vectors but also facilitates pseudotyping of retroviral vectors with the envelope of different viruses.

PSEUDOTYPING OF RETROVIRUS VECTORS   The tropsim of retroviruses is determined by the envelope glycoprotein, which binds to the receptor on the target cells. Ecotropic MLV can only infect murine cells that express the receptor for Env (a sodium-independent cationic amino acid transporter) (3). Fortunately, the host range of the vectors can be easily expanded by replacing the MLV *env* gene with envelope sequences from other retroviruses, e.g., amphotropic viruses that are able to infect mouse and nonmurine species (31, 102).

Pseudotyping of retroviral vectors is not restricted to the envelope glycoproteins of other retroviruses. The G protein of the vescular stomatitis virus (VSV-G), a member of the rhabdovirus family, can substitute for the viral Env protein (21). The two major advantages of incorporation of the VSV-G protein are (*a*) the extremely broad host range of VSV, which enters the host cell by membrane fusion via the interaction with phospholipid components of the cell membrane (95) and (*b*) the ability to concentrate VSV-G pseudotyped particles more than 1000-fold (titers > 10^9 IU/ml) by ultracentrifugation (21), which has important practical implications. The major disadvantage of VSV-G is that it is toxic to the packaging cells (21). Therefore, stable cell lines with inducible (e.g., tet-off system, see Regulation of Transgene Expression) VSV-G expression systems (26, 120) are required. On the other hand, transient transfection of the VSV-G expression plasmid, together with packaging constructs, circumvents this problem because harvesting is restricted to a relatively short period of several days following transfection.

Lentiviral Vectors

Lentiviruses are complex retroviruses, which have been named (*lente*, Latin for slow) according to the prototypic slowly progressing neurologic disease in sheep caused by the maedi/visna virus (70). An important genetic difference between simple retroviruses and lentiviruses are regulatory (*tat* and *rev*) and auxillary genes (*vpr, vif, vpu,* and *nef*) that have important functions during the viral life cycle and viral pathogenesis (for details see 30, 64, 70). An outstanding feature of lentiviruses is their ability to infect nondividing, terminally differentiated mammalian cells, including lymphocytes and macrophages. This feature of lentiviruses makes them a very attractive tool for gene delivery (115, 158).

HIV-BASED VECTORS   The first lentiviral vectors were derived from HIV-1 (114, 124, 125, 134), the most extensively studied lentivirus. The HIV vector and packaging system are constantly evolving and serve as templates for the other lentiviral vectors. Apart from HIV-1, lentivirus vectors have been derived from HIV-2 (126), feline immunodeficiency virus (FIV) (127), equine infectious anemia virus (119), simian immunodeficiency virus (SIV) (92), and maedi/visna virus (12). Most of the lentiviral vectors presently in use for gene therapy approaches are HIV-derived vectors; therefore, we focus on these vectors. Similar to simple retroviruses, the *cis-* and *trans*-acting factors of lentiviruses can be separated while preserving their functions. The lentiviral packaging systems provide in *trans* the viral proteins that are required for the assembly of viral particles in the packaging cells. The vector constructs contain the viral *cis* elements, packaging sequences (Ψ), the Rev response element (RRE), and the transgene (Figure 4).

HIV is able to infect nondividing and terminally differentiated cells without requiring the disassembly of the nuclear membrane (165). The HIV preintegration

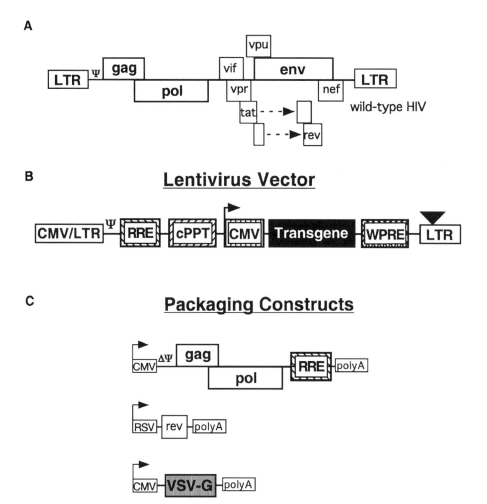

Figure 4   Lentiviral vectors. (*A*) Schematic representation of the wild-type HIV provirus. (*B*) The latest generation of SIN-lentiviral vector constructs incorporates a central polypurine tract (cPPT) to enhance nuclear translocation of the vector in the target cell. In addition, a WPRE is included. Black triangle, SIN mutation; RRE, Rev response element. (*C*) Third generation, tat-free, nonoverlapping split genome packaging system. One plasmid codes for *gag* and *pol*, whereas *rev* is expressed in *trans* from another plasmid (38).

nucleoprotein complex can enter the nucleus via nuclear localization signal-mediated, energy-dependent import [for references see (19)]. HIV-1 contains multiple, partially redundant nuclear localization signals. At least three karyophilic proteins have been identified in HIV-1: the *gag*-encoded matrix protein (MA) (20), integrase (IN) (46), and Vpr (61). Both MA and IN are provided by the packaging cells (Figure 4). However, the latest generations of lentiviral packaging constructs (38, 186) lack Vpr (see next section). Analysis of the effect of Vpr-deficiency on lentiviral gene transfer revealed that Vpr is dispensable for the transduction of neurons in vivo (72, 186). Quiescent human and murine hematopoietic stem cells (22, 96) and murine liver cells (123) are also transduced by Vpr-deficient vectors. These findings underline the notion that the signals required for nuclear entry can be provided by either MA or Vpr (61). Apart from karyophilic proteins, HIV-1, as well as SIV, contain genomic *cis*-acting sequences that enhance the nuclear DNA import (43, 92, 182). In addition to the 3' polypurine tract (PPT), located just 5' of the U3 region of the LTR [for details see (30, 164)], all lentiviruses contain a central copy of the PPT (cPPT) at which synthesis of the downstream plus strand is initiated (182). Inclusion of the upstream cPPT element enhances nuclear import of the HIV-1–derived vector genome (43, 182). Although the cPPT element, which is situated in the *pol* gene, is deleted in the *pol*-deficient lentiviral vectors (Figure 4), these vectors are able to tranduce a broad spectrum of terminally differentiated, nondividing cells, such as neurons (114), muscle (72), hepatocytes (72, 123), retinal photoreceptors (107, 155), and hematopoietic stem cells (22, 106). Thus, at least in these cell types, the cPPT element is not necessary for import into the interphase nucleus. However, inclusion of a cPPT-element in lentiviral vectors significantly improves their transduction efficiency in vivo. HIV- and SIV-based vectors containing a cPPT exhibit a two- to threefold enhancement in transduction efficacy (43, 92).

The same basic principel used to generate retroviral SIN vectors (see above) (Figure 3*C*) can be employed for the production of lentiviral SIN vectors (Figure 4). The SIN lentivectors have a reduced risk of insertional activation of cellular oncogenes. In addition, the chances of mobilizing integrated proviruses by co- or superinfection with a replication-competent virus is considerably reduced (105, 185).

The development of regulated lentiviral vectors has focused so far only on the tetracyciline-regulated system (75) (see Regulation of Transgene Expression). With this regulatable lentiviral vector, tet-dependent induction of transgene expression (i.e., GFP) was demonstrated in vitro and in vivo (rat brain). Another important improvement of the lentiviral vectors is the inclusion of *cis*-acting transcriptional regulatory elements, such as the WPRE, which enhances transgene expression in the target cells. Incorporation of WPRE in HIV-derived vectors increases reporter gene (GFP and luciferase) expression five- to eightfold after transduction of both dividing and arrested 293T cells (184). Similar to vectors based on MLV, the WPRE has to be present within the transgene transcript in sense orientation and is placed 3' of the transgene cDNA upstream of the 3' LTR (Figures 3*D*, 4*B*).

DEVELOPMENT OF LENTIVIRAL PACKAGING SYSTEMS   Earlier studies on HIV-1 demonstrated that structural components of lentiviruses can be provided in *trans* by packaging plasmids, and virus particles can be assembled by expressing viral proteins in packaging cells (130, 144). However, these initial packaging systems incorporated the endogenous HIV envelope. A crucial step for the development of HIV-based vectors was the incorporation of envelope proteins of other viruses, such as amphotropic retroviruses, in HIV-1 particles (148). Replacement of the HIV envelope glycoprotein with VSV-G (Figure 4) led to the design of the first useful lentiviral gene therapy vectors (113, 134). The major advantage provided by VSV-G pseudotyped lentiviruses—apart from broadening the host-range of the vectors—is that the viral particles can be easily concentrated (~1000-fold) by ultracentrifugation of the cell culture medium harvested from the packaging cells (113, 134).

Although the first generation of packaging systems incorporated frameshift mutations (113) or deletions (134) of the *env* coding domain together with truncations of accessory genes (e.g., *vpu*), more than 80% of the HIV genome was still intact. To minimize the amount of HIV coding regions present in the packaging system, a second version of the packaging system was introduced with extensive deletions of the viral genome. The accessory genes (*vpr*, *vpu*, *vif*, and *nef*) are not required for efficient production of viral particles (186). Thus, packaging systems lacking all the accessory genes that contain only the *gag*, *pol*, *tat*, and *rev* genes of HIV-1 were developed (72, 186). Vectors packaged with this system are able to deliver genes into dividing and nondividing cells in vitro and transduce adult neurons and muscle in vivo (72, 186). The third generation of packaging constructs lack the HIV-1 *tat* gene, and the residual HIV genome is divided into two expression constructs (38) (Figure 4C). Tat is a strong transcriptional activator of the HIV-1 LTR promoter and is essential for viral replication (32, 39). Because interaction of Tat with the LTR promoter region is essential for transcription of the vector plasmid, *tat* had to be included in the packaging plasmids (38). Replacement of the 5' LTR promoter/enhancer regions with strong constitutive promoters, such as CMV (38, 105) or RSV (38), creates vectors that are efficiently transcribed in the packaging cells in the absence of Tat. In addition, the remaining HIV genome was split into two separate plasmids (38): One plasmid carries *gag*, *pol*, and the HIV RRE. The expression of *gag* and *pol* are, therefore, dependent on the presence of Rev, which is encoded on a separate expression plasmid (Figure 4C). Packaging systems have also been developed for other lentiviruses, including FIV (127), SIV (92), HIV-2 (126), EIAV (119), and visna virus (12).

So far, most of the lentiviral vectors have been produced by transient transfection of packaging and vector plasmids. Using the highly transfectable 293T cells, one routinely obtains titers of 1×10^9–1×10^{10} IU/ml after transient transfection with the latest generations of packaging and vector constructs, followed by concentration of the virus particles by ultracentrifugation. However, standardization of the virus production is not easily achieved with transient transfections, and each preparation should be tested for possible contaminations with RCVs. Thus, stable producer cell lines may be of advantage, especially if HIV-derived vectors are used in clinical trials.

The development of a packaging cell line for lentiviral vectors was hampered by the fact that VSV-G and some lentiviral proteins (Vpr, Gag, and Tat) are toxic. Therefore, packaging cell lines were established that express the toxic components from inducible plasmids (74, 180) using tetracycline-regulated systems (see above). In addition, stable cell lines for the production of SIN vectors (172) and stable third-generation lentiviral packaging sytems have been generated (80).

DNA VIRUS VECTORS

The currently prominent DNA virus vectors are based on adeno-associated viruses (AAV) and adenoviruses. Adeno-associated viruses (AAVs) contain a single-stranded, relatively small ($\sim$4.7 kb) genome, whereas adenoviruses contain a double-stranded DNA genome . of $\sim$36 kb [for review see (59)]. The basic principals of vector design (Figure 1) also apply to DNA virus vectors. However, there are important practical differences, e.g., in contrast to retroviral vector particles that accumulate in the culture medium, adenovirus and AAV particles accumulate in the producer cells, which have to be lysed to liberate the viral particles.

Adenovirus-Based Vectors

Adenoviruses are medium-sized DNA viruses and have been isolated from avian and mammalian species (for review see 143). The human adenovirus (Ad) family consists of more than 50 serotypes that can infect and replicate in a wide range of organs, such as the respiratory tract, the eye, urinary bladder, gastrointestinal tract, and liver (for review see 67). Adenoviral infection causes an initial nonspecific host response with synthesis of cytokines (tumor necrosis factor, interleukin 1 and 6) followed by a specific response of cytotoxic T lymphocytes (CTLs) directed against virus-infected cells that display viral peptide antigens (143). In addition, there is activation of B cells and the necessary CD4$^+$ T cells, leading to a humoral response. Serologic surveys found antibodies against Ad serotypes 1, 2, and 5 (the latter is widely used as a gene therapy vector) in 40%–60% of children (67). The host's immune response against adenoviral proteins is the major hurdle to the efficient and safe use of adenoviral vectors. Even an inactivated recombinant adenoviral vector can elicit a potent CTL response (73). The safety issue is underlined by the massive immune response and death of a patient enrolled in a gene therapy trial using Ad vectors at the University of Pennsylvania, Philadelphia (see also 158a).

The adenoviral genome consists of a single, double-stranded linear DNA molecule ($\sim$36 kb) that codes for 11 virion proteins (Figure 5). Transcription of the adenoviral chromosome is divided into three major units (143) that are dependent

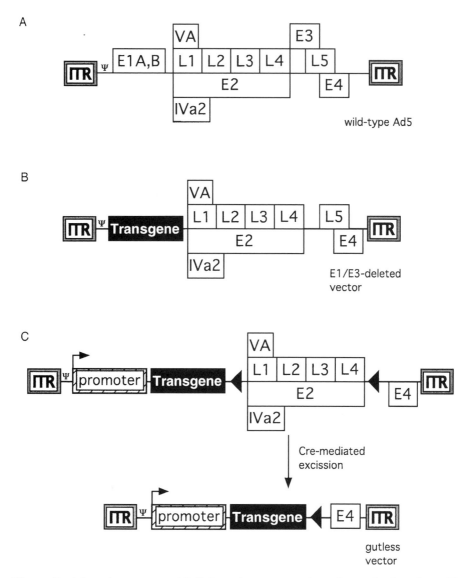

Figure 5 Adenovirus vectors. (*A*) Schematic representation of type 5 adenovirus. (*B*) E1/E3-deleted adenoviral vector. (*C*) Generation of helper-dependent vectors through Cre-mediated excision of most of the adenoviral genome (flanked by loxP recognition sites, *black triangles*).

on the time course of their expression during the viral replicative cycle: early (E1A, E1B, E2, E3, and E4), delayed (IX and Iva2), and late unit (major late). The latter is processed into five mRNAs (L1–L5) that share the same carboxy terminus. All these transcription units are transcribed by RNA polymerase II, whereas the VA gene(s) is transcribed by RNA polymerase III. The viral chromosome contains two origins for DNA replication in each terminal repeat, which are identical and are used for the two reading strands (rightward: E1A, E1B, IX, major late, VA, and E3; leftward: E4, E2, and IVa2) [for details see (143)].

The Ad genome is packaged in a nonenveloped icosahedral protein capsid that emanates fiber coat proteins. The carboxy-terminal knob domain of the fiber protein forms a high-affinity complex with a host cell surface receptor protein. For the majority of Ad serotypes, this receptor is the Coxsackie-adenovirus receptor (CAR) (10). Apart from attachment, efficient virus internalization requires the interaction of integrin αv with the viral penton base (167). Within minutes, the virus appears in the cytosol—after escaping the endosome—and is transported to the nucleus where the viral DNA associates with the nuclear matrix. The Ad chromosome does not integrate into the host genome.

DEVELOPMENT OF ADENOVIRAL VECTORS   Most adenoviral vectors (for review see also 55, 65, 82, 161, 168) are derived from serotype 5. However, Ad vectors have also been derived form Ad2, Ad7, and Ad4 [reviewed in (65)]. Replication-defective Ad vectors were designed by replacing crucial adenoviral coding regions. The E1 gene was targeted in the first generation of Ad vectors because the E1A gene plays a crucial role for viral replication and is the principal protein that activates the expression of other Ad transcription units [for review see (143)]. However, E1 is not required for Ad replication in 293 cells (51), which makes them convenient producer cells for E1-deleted vectors. The 293 cells express the left (according to the conventional Ad map; see also Figure 5) 11% of the Ad5 genome, including the E1 region. Up to 3.2 kb of the E1 region of the viral vector can be replaced by transgenes (15). However, even larger transgenes can be accommodated, as the packaging capacity of Ad5 virions is 105% of the wild-type genome (16). Thus, transgenes of up to 4.7–4.9 kb can be incorporated into E1-deleted adenoviral vectors. To further increase the capacity of Ad vectors, additional sequences were deleted from the Ad genome that are dispensable (11, 56). The E3 gene products are not essential for viral replication, although they play an important immune-modulatory and immune-suppressive role [for details see (65, 143)]. The largest tolerated E3-deletion is 3.1 kb (15), resulting in a total cloning capacity of 8.3 kb if combined with an E1-deletion in one mutant virus.

Because of the size of the Ad genome, the gene of interest is usually incorporated into the vector by homologous recombination in the packaging cells. The transgene is inserted into a shuttle vector that contains adenoviral genomic sequences flanked by the recombination target site (i.e., the E1 coding region). Through homologous recombination, the transgene is integrated into the Ad vector DNA, thereby replacing the E1 gene (Figure 5*B*). Although it is possible to produce up to 1×10^{11} viral particles per 10-cm cell culture dish with this approach, a major drawback is the contamination with wild-type virus that did not recombine with the shuttle vector. To minimize the wild-type Ad contamination, two major changes were introduced into the Ad genomic plasmids. Either the packaging signal can be removed from the Ad genome (15) or its size can be increased beyond the packagable maximum (98). As an alternative, the Ad vector can be produced in yeast (as a yeast artifical chromosome) (77) or in bacteria (e.g., by homologous recombination in *Escherichia coli*) (25). Infectious particles can than be generated by transfection of the purified vector DNA into permissive cells (e.g., 293 cells).

The second generation of Ad vectors are derived from E1-deleted vectors by further deleting the E2 and/or E4 regions. Because the E2 region encodes proteins that are essential for the replication of the viral chromosome, it has to be provided in *trans* in the packaging cells. However, owing to the toxicity of the E2 protein (81), either inducible systems (41) or cell lines that complement only for crucial E2 components (e.g., the E2-coded DNA polymerase) (5) have been developed. Furthermore, a number of deletions have been introduced into the E4 region, which encodes proteins required—among others—for viral DNA replication and late protein synthesis. Partial deletions of the E4 region, in particular, give rise to viable vectors (47, 162, 177). In order to propagate E1/E4-deleted vectors, E4-complementing cell lines have been developed (85, 163). Although deletions in the E4 region increase the cloning capacity of the Ad vector, some reports indicate that the E4 region may exert a positive effect on long-term expression (6, 8).

Given the fact that Ad proteins can induce a fulminant immune response of the host, the aim of Ad vector development was to reduce the genomic sequences present in the vectors to an absolute minimum. Theoretically, it should be possible to create mini- or "gutless" vectors that carry no viral sequences, apart from the inverted terminal repeats (ITRs) and the *cis*-acting packaging signal (60). However, initial attempts to create such a vector were not very successful because of the size constraints of Ad packaging. The optimum packaging length for the Ad genome ranges from 75% to 105% of the wild-type (~36 kb) genome (122). Although vectors smaller than 75% can be packaged, they are unstable and undergo DNA rearrangements (122).

An important step toward a third generation of Ad vectors was the development of high-capacity, helper-dependent vectors based on the Cre/*loxP*-system of site-specific DNA excision (Figure 5*C*) (58, 89, 121). The Cre-recombinase of bacteriophage P1 excises DNA sequences that are flanked by *loxP* recognition sites and has been successfully used for the targeted deletion of genomic sequences of up to 3–4 cM (54, 133). Using the Cre/*loxP*-system, researchers deleted 25 kb of adenoviral genome from an Ad vector containing *loxP* sites in 293 cells stably expressing the Cre recombinase (89). Although the resulting vector is devoid of most of the Ad genome, its size is only 9 kb (Figure 5*C*) and, therefore, prone to DNA rearrangements (89). One way to circumvent this problem is to include sufficient stuffer sequences that keep the vector size above 27 kb.

Another problem with these third-generation vectors is their dependence on helper viruses, which contaminate the vector preparations. The amount of helper virus contamination can be reduced to $<1\%$ by deleting the Ad packaging signal from the helper virus (83). Although further reductions (to $\sim 0.01\%$) of helper contaminations have been achieved using a Cre/*loxP* system (121), the titer of helper Ad still amounts to 10^8–10^{12} pfu (plaque-forming unit), owing to the high titers (10^{12}–10^{14} pfu) normally used in human trials. Thus, the helper-dependent Ad vectors still carry a high risk of adverse immunological or toxic side effects. Nevertheless, these vectors have a great potential for efficient gene delivery.

REPLICATION-COMPETENT ADENOVIRAL VECTORS   A completely different approach to dealing with the immunological problems and the toxicity of Ad is to use Ad vectors in a setting where these normally adverse effects may be a beneficial therapeutic feature. Although Ad can establish a long-term persistence within its host, the final consequence of viral replication is host cell shut-off and disruption of the cytoskeleton as well as lysis of the infected cell [see also (67, 143)].

Also known as viral oncolysis, the combination of the lytic action of replicating Ad together with the delivery of a therapeutic or toxic gene is a promising approach to cancer gene therapy (63). In addition, the immunogenic properties of Ad vectors can elicit an antitumor effect. The use of replicating, oncolytic viruses in humans requires targeting of these viruses to the tumor cells, and therefore, mutant adenoviruses that replicate preferentially in tumor cells have been isolated (4). An example for such a virus is an Ad mutant (called dl1520 or ONYX-015) that lacks the E1B encoded 55-kDa (E1B-55K) protein, which normally binds to and inactivates the tumor-suppressor p53 (17, 62). Because inactivation of p53 is required for efficient virus replication (178), ONYX-015 cannot replicate in normal, p53-positive cells. In contrast, ONYX-015 can productively infect and lyse a broad spectrum of human tumors with p53 abnormalities (e.g., p53 mutations and/or deletions) in vitro and in vivo (62). Recently, the first results of a Phase II clinical trial using intratumoral administration ONYX-015 in combination with standard intravenous chemotherapy in patients with recurrent squamous cell carcinoma of the head and neck were published (78). The combined therapy caused a $>50\%$ decrease in tumor volume in $>60\%$ of the patients, and none of the responding tumors progressed over 6 months, whereas all tumors treated with chemotherapy alone had progressed.

Adeno-Associated Virus–Derived Vectors

Adeno-associated viruses are nonpathogenic parvoviruses. They contain a single-stranded DNA genome of only 4.7 kb with ITRs (Figure 6). Heparan sulfate proteoglycan is the primary receptor for AAV type 2 (153), the staple of most AAV vector backbones. Binding of AAV to heparan sulfate proteoglycan mediates its attachment to the cell membrane, whereas integrin $\alpha v \beta 5$ and/or fibroblast growth factor receptor 1 (FGFR1) act as coreceptors for AAV infection (131, 152). The

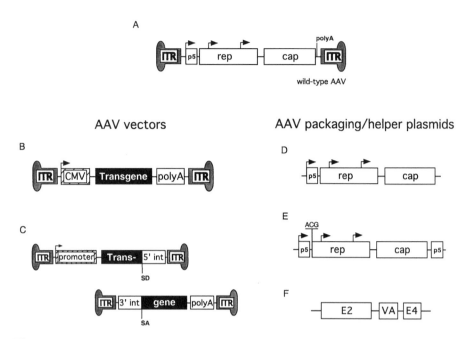

Figure 6   AAV-derived vectors. (*A*) Schematic representation of the wild-type AAV. (*B*) AAV-derived vector, containing the CMV-driven transgene and polyadenylation signal sequences (polyA) flanked by the inverted terminal repeats (ITRs). (*C*) Dual AAV vector. The split transgene is delivered by two separate AAV vectors that contain either the promoter or the polyA signal. In addition, splice donor (SD) and splice acceptor (SA) signals are included that allow the proper assembly of the transgene mRNA. (*D*) First-generation AAV packaging construct. (*E*) AAV packaging construct with modifications of the *rep* expression cassette. (*F*) Minimal helper construct derived from adenovirus.

latter are required for efficient binding of AAV to the cell surface and entry. For productive infection, AAV is dependent (therefore, these viruses are also termed *Dependoviruses*) on co-infection of the target cell with a helper virus (adeno- or herpesvirus) (for details see 13, 179). In addition, AAV is also able to establish latent infection in the absence of helper viruses. Under such nonpermissive conditions, AAV integrates into the host genome in a specifc site on human chromosome 19, a highly desirable feature of a gene therapy vector (159). However, efficient, site-specific integration requires the presence of the viral Rep protein (166), which is normally not included in the AAV vectors. Nevertheless, recombinant AAV integrates into the genome of mammalian cells in vitro (169, 174) as well as in vivo (87, 111). In addition, episomal circular forms of recombinant AAV genomes that persisted for up to 9 months have been found in muscle and brain tissue (36, 111).

DEVELOPMENT OF AAV-BASED VECTORS   Although six serotypes of AAV have been found in human and primates, the vast majority of AAV vectors are derived from

AAV-2 (for review see 14, 53, 108, 110, 141, 149). These vectors exhibit a broad host range and infect a wide variety of cells, including nondividing cells like hepatocytes (146), muscle cells (170), and neurons (97).

In contrast to other viral vectors, the initial AAV vector design (99, 139) was conserved for a decade without major changes (Figure 6). Because all the required *cis*-acting functions are located within the ITR and the immediately adjacent 45 nucleotides (140), the two viral open reading frames (rep and cap) can be completely replaced with the transgene of interest and its promoter (Figure 6). Vectors of up to 5.2 kb can be packaged; however, the optimum size of AAV vector genomes is between 4.1 and 4.9 kb (35). To increase the packaging capacity of the AAV vectors, a dual vector, *trans*-splicing system was developed (37, 112, 154, 173). The basic priniciple of this system (Figure 6) is to clone a split transgene into two separate vectors. If both vectors infect the target cell, they can form episomal circular multimers in the nucleus that concatamerize in a head-to-tail orientation (36). Because the two parts of the transgene would be separated by ITR sequences, splicing signals have to be included that will allow the removal of the interrupting ITR sequences during RNA maturation (Figure 6). Although this is a promising development, conflicting results have been published regarding the efficacy of the dual vector system ranging from 16% (154) to 70% (112).

Finally, inclusion of WPRE also increases transgene expression in the context of AAV-derived vectors (91). Although the WPRE enhances transgene expression three- to fivefold, it further reduces the size of the possible transgene by almost 20%.

AAV PACKAGING/HELPER SYSTEMS   According to the basic priniciple of viral vector design, packaging systems have been developed for the production of AAV vectors, which provide viral components in *trans*. In the first generation of AAV packaging systems, the packaging plasmid, which encodes the *rep* and *cap* genes, is cotransfected together with the vector constructs into adenovirus-infected cells (140). Under these conditions, the vector is excised from the plasmid backbone, amplified, and packaged into AAV virions. In addition, helper virus is produced by the producer/packaging cells (Figure 6). To reduce the amount of contaminating helper Ad, multiple CsCl gradients and/or inactivation procedures (e.g., heat-inactivation) have been used.

To overcome this problem, an Ad-free packaging system was developed (171). The identification of the essential helper functions of Ad [reviewed in (13, 110)] allowed the development of a packaging system that provides these functions— i.e., the E2A, E4, and VA RNA regions (Figure 6)—on an expression plasmid (171), whereas the E1 gene products are provided by 293 cells (51). This packaging system was further improved by modifying the expression of *rep* (171). Paradoxically, overexpression of the Rep68/78 proteins has a negative effect on the titer yield (88). To decrease Rep68/78 protein levels, the ATG translation start codon of the *rep* gene was mutated into an ACG codon and a second copy of the p5 promoter that inhibits its own transcriptional activity was inserted $3'$ of *cap* (Figure 6). These changes increase vector yields approximately 15-fold over the conventional packaging plasmid, resulting in viral titers of up to $1 \times 10^{10}/10$ cm plate (171).

In analogy to retroviral vectors, stable packaging cell lines have been established for AAV vectors. Again, due to the toxicity of viral proteins, inducible cell lines were developed in which the expression of Rep and Cap proteins is regulated by the tetracycline-inducible system (69). However, this packaging system requires the infection with a helper virus and extensive purification procedures to minimize the amount of helper virus contamination.

AAV PRODUCTION AND PURIFICATION   The AAV purification protocol, which normally involved a stepwise precipitation of AAV particles with ammonium sulfate, followed by CsCl density gradient centrifugation (170), has been recently improved by the use of iodixanol (183), which is an iso-osmotic, inert X-ray contrast reagent. Crude producer cell lysates can be purified at least 100-fold and 70%–80% of the starting starting material is recovered (183).

An important further development is the use of ligand affinity chromatography purification techniques. Based on the identification of the cellular receptor for AAV-2 (153), purification procedures were developed that use commercially available heparin columns (28, 183). Combination of iodixanol centrifugation and heparin chromatography procedure results in 50%–70% recovery of virus that has a purity of over 99%. In addition, this procedure results in higher particle-to-infectivity ratios (100:1 for iodixanol/heparin) as compared to the traditional Cs/Cl method (1000:1) (183). Instead of heparin affinity columns, immunoaffinity columns can also be used (52).

Other Vectors

The quest for the ideal gene therapy vehicle (159) sparked the development of vectors from a variety of other viruses. Vectors derived from herpes simplex viruses (HSVs) are among the most promising. Most of the HSV vectors have been derived from HSV-1, which is is a neurotropic DNA virus that can establish either a lytic, replicative cycle or latency. In humans, HSV-1 causes a number of diseases, ranging from herpes labialis to disseminated visceral infections in immunocompromised patients, hepatitis, and encephalitis after spread to the central nervous system. HSV-1 is of interest for gene therapy because of the size and complexity of its ~150-kb genome [for a review of the HSV life cycle see (136)]. The HSV genome can accommodate larger transgenes than other presently available vectors, with the exception of helper-dependent, "gut-less" Ad vectors. In addition, HSV vectors that contain multiple (e.g., five) transgenes that are independently regulated (84) have been described. On the other hand, the complexity of the HSV genome, which contains more than 80 genes (136), is a major hurdle to the construction of efficient and safe vectors. Three major HSV vector systems are presently available, amplicon vectors, replication-defective HSV, and attenuated HSV. Amplicon vectors [reviewed in (44)] are based on replication-defective HSV mutants and contain less than 1% of the wild-type genome. The production of amplicon vectors requires essential HSV functions, which must be provided in *trans*. This can be achieved either by infection of the producer cells with HSV helper viruses, which carries the risk of helper virus contaminations, or by the use of reconstituted HSV packaging systems [e.g., a bacterial artificial chromosome–encoded HSV genome (137)]. However, the latter only results in a low vector yield of only up to 10^7 pfu/ml. The second HSV vector system is based on replication-defective HSV, which carries deletions in the immediate-early (IE) genes. Although vectors have been described that do not express any IE proteins and are noncytotoxic, CMV-driven transgene expression in the context of these vectors is very low (138). Finally, attenuated replication-competent HSV vectors have been developed. High-titer ($>10^9$ pfu/ml) preparations of these vectors can be produced in the absence of helper viruses, but they exhibit residual cytophathology in vivo. This cytopathic effect of attenuated vectors, together with the inherent tropisms of HSV for neuronal cells, led to the successful use of these vectors in animal models for human brain tumors (24, 104). Although attenuated HSV vectors are a promising candidate for oncolytic gene therapy, the threat of neurologic complications and of host immune responses have to be unequivocally ruled out prior to their use in humans.

Among the emerging gene therapy vectors are those derived from baculoviruses (142) and alpha-viruses (68). [For review of these and other vectors we refer the reader to (71).]

CLINICAL TRIALS   Worldwide, over 400 clinical trials have been conducted or are underway, with enrollment of over 6000 patients. A substantial portion of these clinical trials (over 70%) are cancer related and often carried out on terminal patients. The most commonly used vectors are retroviral vectors based on MLV, which were the first viral vectors to be used in a gene therapy trial (18). The targets of that first clinical trial were the T-lymphocytes of two children suffering from severe combined immunodeficiency (SCID) (18). Unfortunately, a clear judgment of the success of this trial was not possible because the patients received supportive conventional therapy. This trial, however, did set the stage for other gene therapy trials.

Gene transfer into multipotent hematopoietic stem cells has received much attention owing to its relevance for a broad variety of human diseases, ranging from hematological disorders to cancer. In addition, it allows the use of ex vivo transduction protocols, thereby minimizing the exposure of the patient with viral particles. The use of retroviral vectors in this setting is hampered by the low frequency of gene delivery, as transduction by retrovirus vectors occurs only in cells that are replicating at the time of infection (103), and therefore, the transduction of slowly or nondividing stem cells and progenitors is inefficient. A number of growth factor combinations have been used to prestimulate hematopoietic stem/progenitor cells and have been shown to increase transduction efficacy (118). The drawback of this approach is that exposure of progenitors to growth factors over several days markedly impaires their ability for long-term engraftment (157). However, a brief (<24 h) exposure to specific cytokines and stromal support allows engraftment with a high number (10%–20%) of retrovirally modified cells.

Given the positive effect of bone marrow stroma on retroviral gene delivery to stem cells, attempts were made to mimic the bone marrow milieu by the addition of purified extracellular matrix molecules. Certain fibronectin fragments (e.g., fibronectin CH296) (109) proved to significantly increase the gene transfer efficiency as a result of colocalization of retroviral particles and target cells (57). This procedure resulted in a relatively high level (median 14%) of gene transfer in human CD34-positive cells as assayed by the number of transduced progenitor colonies (1). Although a median engraftment level of 12% transduced cells was observed in human bone marrow 1 month after transplantation, the number of transduced cells fell to 5% over the next 11 months (1). The combination of fibronectin and cytokine cocultivation is a promising approach to retroviral transduction of stem cells, which has resulted in an improved gene transfer to baboon marrow stem cells (79), and was used for the correction of the SCID phenotype in two patients (23).

Among the clinical trials presently conducted in the United States that involve AAV, one has received much attention. In this Phase I trial, patients suffering from hemophilia B were injected into the skeletal muscle with an AAV vector carrying the cDNA for factor IX (76). Preliminary data on three patients that received the starting dose of the dose-escalation study—i.e., 10^{11} I.U./kg—suggest that the transduced muscle cells expressed factor IX protein for at least 2–3 months. In addition, an 80% reduction of factor IX infusion was observed in one patient and was interpreted as evidence for a modest clinical response. However, the reduction in factor usage is a quite surprising finding, given the fact that the factor IX levels were not affected by the AAV-factor IX injection in the same patient. The fact that so far neither vector-related toxicity (at least at the initial dose) nor evidence for germline transmission of vector sequences was found in these patients is an important finding, but it must be confirmed at higher, therapeutic vector doses. Another important question involves the immune response of the patients against the vector. Studies addressing this issue (27, 76) show a preexisting humoral immunity against AAV in virtually all patients. In addition, administration of the AAV vector elicited a 10- to 1000-fold increase in the neutralizing antibody titer (76). This boost of the humoral immunity against the vector is of major concern, especially if the AAV vectors must be readministered.

POTENTIAL OF CLINICAL TRIALS USING LENTIVIRAL VECTORS   Two features of lentiviral vectors are of outstanding interest in regard to their use in human beings: (*a*) the ability to transduce nondividing cells in vivo and (*b*) the ability to efficiently deliver large (~8 kb) and complex transgenes to the target cells and tissues.

HIV-derived lentiviral vectors transduce a broad spectrum of nondividing cells in vivo, such as neurons (114), retinal cells (105, 155), muscle cells (72), and hepatocytes (72, 123). An important finding is that lentiviral vectors can transduce human CD34$^+$ hematopoietic stem cells without cytokine prestimulation (22, 43,

106). The transduced CD34$^+$ cells provided long-term repopulation and were capable of engrafting and differentiation into multiple hematopoietic lineages after transplantation into NOD/SCID mice (106). Because bone marrow stem cells can be transduced ex vivo, there is no requirement to use "live" virus in patients, making hematopoietic stem cells the most likely target for lentiviral gene therapy trials in humans.

Viral vectors based on simple retroviruses exhibit genetic instability if genomic fragments like introns are included in the vector, mainly due to inappropriate splicing and/or incorrect polyadenylation (86). In contrast, HIV-based vectors should allow the faithful expression of unspliced transgene RNAs owing to the presence of a RRE in the vector RNA. The Rev protein is delivered in sufficient amount to the target cells by the recombinant virus particles. An important example for the use of HIV-derived vectors to deliver a complex transgene is the successful delivery of the human β-globin gene, together with intronic sequences containing the locus-control region, into murine hematopoietic stem cells, which resulted in β-globin expression and the "cure" of β-thalassemic mice (96).

PERSPECTIVES

The young field of gene therapy promises major medical progress toward the cure of a broad spectrum of human diseases, ranging from immunological disorders to heart disease and cancer, and has, therefore, generated great hopes and great hypes. The idea to use the genetic information obtained from the sequencing of the human genome for the treatment of diseases is compelling. However, gene therapy will only be added to the daily-use therapeutical arsenal if scientists from many different disciplines participate and pull together as a team: Geneticists must identify target genes that contribute to specific diseases or that can influence the disease course. The task for the virologists is to develop efficient and safe vectors that are able to deliver the genes of interest to the target cells and assure the proper expression of the transferred genetic material. Cell biologists will establish ways to facilitate the gene transfer and identify stem cells that may be used to regenerate failing organs. Bioengineers will be needed to show the biologists how three-dimensional tissues and even whole organs may be generated in a test tube. Clinicians will carry out clinical trials with vectors optimized for the disease and the medical requirements of the patients.

Gene therapy has undergone extreme scrutiny in the recent past. It is our responsibility to assure the public that the patient's welfare and health is the major goal. Strict adherence to the guidelines is incumbent on all scientists and investigators involved in a clinical trial. The gene therapy community will need to meet the challenge of new regulations and guidelines introduced by the NIH and FDA to ensure both the quality of clinical trials and the protection of volunteers enrolled in the trials. We want to continue to participate and lead the nation in harnessing and providing the benefits of the unprecedented golden age of biomedical research.

ACKNOWLEDGMENTS

Alexander Pfeifer receives a Heisenberg Scholarship from the Deutsche Forsch-
ungsgemeinschaft (Pf301/6-1) and is supported by Fonds der Chemischen Indus-
trie. Inder Verma is an American Cancer Society Professor of Molecular Biology.
He is supported by grants from the NIH, the March of Dimes, the Wayne and
Gladys Valley Foundation, and the H.N. and Frances C. Berger Foundation. During preparation of this manuscript, we have heavily relied on previous review
articles (124, 147) published from our laboratory.

Visit the Annual Reviews home page at www.AnnualReviews.org

LITERATURE CITED

1. Abonour R, Williams DA, Einhorn L, Hall KM, Chen J, et al. 2000. Efficient retrovirus-mediated transfer of the multi-drug resistance 1 gene into autologous human long-term repopulating hemato-poietic stem cells. *Nat. Med.* 6:652–58

2. Agha-Mohammadi S, Lotze MT. 2000. Regulatable systems: applications in gene therapy and replicating viruses. *J. Clin. Invest.* 105:1177–83

3. Albritton LM, Tseng L, Scadden D, Cunningham JM. 1989. A putative murine ecotropic retrovirus receptor gene encodes a multiple membrane–spanning protein and confers susceptibility to virus infection. *Cell* 57:659–66

4. Alemany R, Balague C, Curiel DT. 2000. Replicative adenoviruses for cancer therapy. *Nat. Biotechnol.* 18:723–27

5. Amalfitano A, Hauser MA, Hu H, Serra D, Begy CR, Chamberlain JS. 1998. Production and characterization of improved adenovirus vectors with the E1, E2b, and E3 genes deleted. *J. Virol.* 72:926–33

6. Armentano D, Smith MP, Sookdeo CC, Zabner J, Perricone MA, et al. 1999. E4ORF3 requirement for achieving long-term transgene expression from the cytomegalovirus promoter in adenovirus vectors. *J. Virol.* 73:7031–34

7. Armentano D, Yu SF, Kantoff PW, von Ruden T, Anderson WF, Gilboa E. 1987. Effect of internal viral sequences on the utility of retroviral vectors. *J. Virol.* 61:1647–50

8. Armentano D, Zabner J, Sacks C, Sookdeo CC, Smith MP, et al. 1997. Effect of the E4 region on the persistence of transgene expression from adenovirus vectors. *J. Virol.* 71:2408–16

9. Bender MA, Palmer TD, Gelinas RE, Miller AD. 1987. Evidence that the packaging signal of Moloney murine leukemia virus extends into the gag region. *J. Virol.* 61:1639–46

10. Bergelson JM, Cunningham JA, Droguett G, Kurt-Jones EA, Krithivas A, et al. 1997. Isolation of a common receptor for Coxsackie B viruses and adenoviruses 2 and 5. *Science* 275:1320–23

11. Berkner KL, Sharp PA. 1983. Generation of adenovirus by transfection of plasmids. *Nucleic Acids Res.* 11:6003–20

12. Berkowitz RD, Ilves H, Plavec I, Veres G. 2001. Gene transfer systems derived from visna virus: analysis of virus production and infectivity. *Virology* 279:116–29

13. Berns KI. 1996. Parvoviridae: the viruses and their replication. See Ref. 41a, pp. 2173–97

14. Berns KI, Linden RM. 1995. The cryptic lifestyle of adeno-associated virus. *BioEssays* 17:237–45

15. Bett AJ, Haddara W, Prevec L, Graham FL. 1994. An efficient and flexible system for construction of adenovirus vectors

with insertions or deletions in early regions 1 and 3. *Proc. Natl. Acad. Sci. USA* 91:8802–6

16. Bett AJ, Prevec L, Graham FL. 1993. Packaging capacity and stability of human adenovirus type 5 vectors. *J. Virol.* 67:5911–21

17. Bischoff JR, Kirn DH, Williams A, Heise C, Horn S, et al. 1996. An adenovirus mutant that replicates selectively in p53-deficient human tumor cells. *Science* 274:373–76

18. Blaese RM, Culver KW, Miller AD, Carter CS, Fleisher T, et al. 1995. T lymphocyte–directed gene therapy for ADA-SCID: initial trial results after 4 years. *Science* 270:475–80

19. Brown PO. 1997. Integration. See Ref. 30a pp. 161–203

20. Bukrinsky MI, Haggerty S, Dempsey MP, Sharova N, Adzhubel A, et al. 1993. A nuclear localization signal within HIV-1 matrix protein that governs infection of nondividing cells. *Nature* 365:666–69

21. Burns JC, Friedmann T, Driever W, Burrascano M, Yee JK. 1993. Vesicular stomatitis virus G glycoprotein pseudotyped retroviral vectors: concentration to very high titer and efficient gene transfer into mammalian and nonmammalian cells. *Proc. Natl. Acad. Sci. USA* 90:8033–37

22. Case SS, Price MA, Jordan CT, Yu XJ, Wang L, et al. 1999. Stable transduction of quiescent CD34$^{(+)}$CD38$^{(-)}$ human hematopoietic cells by HIV-1–based lentiviral vectors. *Proc. Natl. Acad. Sci. USA* 96:2988–93

23. Cavazzana-Calvo M, Hacein-Bey S, de Saint Basile G, Gross F, Yvon E, et al. 2000. Gene therapy of human severe combined immunodeficiency (SCID)-X1 disease. *Science* 288:669–72

24. Chambers R, Gillespie GY, Soroceanu L, Andreansky S, Chatterjee S, et al. 1995. Comparison of genetically engineered herpes simplex viruses for the treatment of brain tumors in a scid mouse model

of human malignant glioma. *Proc. Natl. Acad. Sci. USA* 92:1411–15

25. Chartier C, Degryse E, Gantzer M, Dieterle A, Pavirani A, Mehtali M. 1996. Efficient generation of recombinant adenovirus vectors by homologous recombination in *Escherichia coli. J. Virol.* 70:4805–10

26. Chen ST, Iida A, Guo L, Friedmann T, Yee JK. 1996. Generation of packaging cell lines for pseudotyped retroviral vectors of the G protein of vesicular stomatitis virus by using a modified tetracycline inducible system. *Proc. Natl. Acad. Sci. USA* 93:10057–62

27. Chirmule N, Propert K, Magosin S, Qian Y, Qian R, Wilson J. 1999. Immune responses to adenovirus and adeno-associated virus in humans. *Gene Ther.* 6:1574–83

28. Clark KR, Liu X, McGrath JP, Johnson PR. 1999. Highly purified recombinant adeno-associated virus vectors are biologically active and free of detectable helper and wild-type viruses. *Hum. Gene Ther.* 10:1031–39

29. Coffin JM. 1992. Structure and classification of retroviruses. In *The Retroviridae*, ed. JA Levy, pp. 19–49. New York: Plenum

30. Coffin JM. 1996. Retroviridae: the viruses and their replication. See Ref. 41a, pp. 1767–848

30a. Coffin JM, Hughes SH, Varmus HE, eds. 1997. *Retroviruses.* New York: Cold Spring Harbor Lab.

31. Cone RD, Mulligan RC. 1984. High-efficiency gene transfer into mammalian cells: generation of helper-free recombinant retrovirus with broad mammalian host range. *Proc. Natl. Acad. Sci. USA* 81:6349–53

32. Cullen BR. 1998. HIV-1 auxiliary proteins: making connections in a dying cell. *Cell* 93:685–92

33. Daly G, Chernajovsky Y. 2000. Recent developments in retroviral-mediated gene transduction. *Mol. Ther.* 2:423–34

34. Danos O, Mulligan RC. 1988. Safe and efficient generation of recombinant retroviruses with amphotropic and ecotropic host ranges. *Proc. Natl. Acad. Sci. USA* 85:6460–64

35. Dong JY, Fan PD, Frizzell RA. 1996. Quantitative analysis of the packaging capacity of recombinant adeno-associated virus. *Hum. Gene Ther.* 7:2101–12

36. Duan D, Sharma P, Yang J, Yue Y, Dudus L, et al. 1998. Circular intermediates of recombinant adeno-associated virus have defined structural characteristics responsible for long-term episomal persistence in muscle tissue. *J. Virol.* 72:8568–77

37. Duan D, Yue Y, Yan Z, Engelhardt JF. 2000. A new dual-vector approach to enhance recombinant adeno-associated virus–mediated gene expression through intermolecular *cis* activation. *Nat. Med.* 6:595–98

38. Dull T, Zufferey R, Kelly M, Mandel RJ, Nguyen M, et al. 1998. A third-generation lentivirus vector with a conditional packaging system. *J. Virol.* 72:8463–71

39. Emerman M, Malim MH. 1998. HIV-1 regulatory/accessory genes: keys to unraveling viral and host cell biology. *Science* 280:1880–84

40. Emerman M, Temin HM. 1984. Genes with promoters in retrovirus vectors can be independently suppressed by an epigenetic mechanism. *Cell* 39:449–67

41. Engelhardt JF, Ye X, Doranz B, Wilson JM. 1994. Ablation of E2A in recombinant adenoviruses improves transgene persistence and decreases inflammatory response in mouse liver. *Proc. Natl. Acad. Sci. USA* 91:6196–200

41a. Fields BN, Knipe DM, Howley PM, eds. 1996. *Fields Virology.* Philadelphia: Lippincott-Raven

42. Finer MH, Dull TJ, Qin L, Farson D, Roberts MR. 1994. kat: a high-efficiency retroviral transduction system for primary human T lymphocytes. *Blood* 83:43–50

43. Follenzi A, Ailles LE, Bakovic S, Geuna M, Naldini L. 2000. Gene transfer by lentiviral vectors is limited by nuclear translocation and rescued by HIV-1 pol sequences. *Nat. Genet.* 25:217–22

44. Frenkel N, Singer O, Kwong AD. 1994. Minireview: the herpes simplex virus amplicon—a versatile defective virus vector. *Gene Ther.* 1:S40–46

44a. Friedmann T, ed. 1999. *The Development of Human Gene Therapy.* New York: Cold Spring Harbor Lab.

45. Friedmann T, Roblin R. 1972. Gene therapy for human genetic disease? *Science* 175:949–55

46. Gallay P, Swingler S, Song J, Bushman F, Trono D. 1995. HIV nuclear import is governed by the phosphotyrosine-mediated binding of matrix to the core domain of integrase. *Cell* 83:569–76

47. Gao GP, Yang Y, Wilson JM. 1996. Biology of adenovirus vectors with E1 and E4 deletions for liver-directed gene therapy. *J. Virol.* 70:8934–43

48. Gorman CM, Gies D, McCray G, Huang M. 1989. The human cytomegalovirus major immediate early promoter can be *trans*-activated by adenovirus early proteins. *Virology* 171:377–85

49. Gossen M, Bujard H. 1992. Tight control of gene expression in mammalian cells by tetracycline-responsive promoters. *Proc. Natl. Acad. Sci. USA* 89:5547–51

50. Gossen M, Freundlieb S, Bender G, Muller G, Hillen W, Bujard H. 1995. Transcriptional activation by tetracyclines in mammalian cells. *Science* 268:1766–69

51. Graham FL, Smiley J, Russell WC, Nairn R. 1977. Characteristics of a human cell line transformed by DNA from human adenovirus type 5. *J. Gen. Virol.* 36:59–74

52. Grimm D, Kern A, Rittner K, Kleinschmidt JA. 1998. Novel tools for production and purification of recombinant adeno-associated virus vectors. *Hum. Gene Ther.* 9:2745–60

53. Grimm D, Kleinschmidt JA. 1999. Progress in adeno-associated virus type 2 vector production: promises and prospects for

clinical use. *Hum. Gene Ther.* 10:2445–50

54. Gu H, Zou YR, Rajewsky K. 1993. Independent control of immunoglobulin switch recombination at individual switch regions evidenced through Cre-loxP–mediated gene targeting. *Cell* 73:1155–64

55. Hackett NR, Crystal RG. 2000. Adenovirus vectors for gene therapy. In *Gene Therapy*, ed. NS Templeton, DD Lasic, pp. 17–39. New York: Marcel Dekker

56. Haj-Ahmad Y, Graham FL. 1986. Development of a helper-independent human adenovirus vector and its use in the transfer of the herpes simplex virus thymidine kinase gene. *J. Virol.* 57:267–74

57. Hanenberg H, Xiao XL, Dilloo D, Hashino K, Kato I, Williams DA. 1996. Colocalization of retrovirus and target cells on specific fibronectin fragments increases genetic transduction of mammalian cells. *Nat. Med.* 2:876–82

58. Hardy S, Kitamura M, Harris-Stansil T, Dai Y, Phipps ML. 1997. Construction of adenovirus vectors through Cre-lox recombination. *J. Virol.* 71:1842–49

59. Harrison SC, Skehel JJ, Wiley DC. 1996. Virus structure. See Ref. 41a, pp. 59–99

60. Hearing P, Samulski RJ, Wishart WL, Shenk T. 1987. Identification of a repeated sequence element required for efficient encapsidation of the adenovirus type 5 chromosome. *J. Virol.* 61:2555–58

61. Heinzinger NK, Bukinsky MI, Haggerty SA, Ragland AM, Kewalramani V, et al. 1994. The Vpr protein of human immunodeficiency virus type 1 influences nuclear localization of viral nucleic acids in nondividing host cells. *Proc. Natl. Acad. Sci. USA* 91:7311–15

62. Heise C, Sampson-Johannes A, Williams A, McCormick F, Von Hoff DD, Kirn DH. 1997. ONYX-015, an E1B gene–attenuated adenovirus, causes tumor-specific cytolysis and antitumoral efficacy that can be augmented by standard chemotherapeutic agents. *Nat. Med.* 3:639–45

63. Hermiston T. 2000. Gene delivery from replication-selective viruses: arming guided missiles in the war against cancer. *J. Clin. Invest.* 105:1169–72

64. Hirsch MS, Curran J. 1996. Human immunodeficiency viruses. See Ref. 41a, pp. 1953–75

65. Hitt MM, Parks RJ, Graham FL. 1999. Structure and genetic organization of adenovirus vectors. See Ref. 44a, pp. 61–86

66. Hofmann A, Nolan GP, Blau HM. 1996. Rapid retroviral delivery of tetracycline-inducible genes in a single autoregulatory cassette. *Proc. Natl. Acad. Sci. USA* 93:5185–90

67. Horwitz MS. 1996. Adenoviruses. See Ref. 41a, pp. 2149–71

68. Huang HV. 1996. Sindbis virus vectors for expression in animal cells. *Curr. Opin. Biotechnol.* 7:531–35

69. Inoue N, Russell DW. 1998. Packaging cells based on inducible gene amplification for the production of adeno-associated virus vectors. *J. Virol.* 72:7024–31

70. Joag SV, Stephens EB, Narayan O. 1996. Lentiviruses. See Ref. 41a, pp. 1977–96

71. Jolly DJ. 1999. Emerging viral vectors. See Ref. 44a, pp. 209–40

72. Kafri T, Blomer U, Peterson DA, Gage FH, Verma IM. 1997. Sustained expression of genes delivered directly into liver and muscle by lentiviral vectors. *Nat. Genet.* 17:314–17

73. Kafri T, Morgan D, Krahl T, Sarvetnick N, Sherman L, Verma I. 1998. Cellular immune response to adenoviral vector infected cells does not require de novo viral gene expression: implications for gene therapy. *Proc. Natl. Acad. Sci. USA* 95:11377–82

74. Kafri T, van Praag H, Ouyang L, Gage FH, Verma IM. 1999. A packaging cell line for lentivirus vectors. *J. Virol.* 73:576–84

75. Kafri T, von Praag H, Gage FH, Verma IM. 2000. Lentiviral vectors—regulated gene expression. *Mol. Ther.* 1:516–21

76. Kay MA, Manno CS, Ragni MV, Larson PJ, Couto LB, et al. 2000. Evidence for gene transfer and expression of factor IX in haemophilia B patients treated with an AAV vector. *Nat. Genet.* 24:257–61

77. Ketner G, Spencer F, Tugendreich S, Connelly C, Hieter P. 1994. Efficient manipulation of the human adenovirus genome as an infectious yeast artificial chromosome clone. *Proc. Natl. Acad. Sci. USA* 91:6186–90

78. Khuri FR, Nemunaitis J, Ganly I, Arseneau J, Tannock IF, et al. 2000. A controlled trial of intratumoral ONYX-015, a selectively-replicating adenovirus, in combination with cisplatin and 5-fluorouracil in patients with recurrent head and neck cancer. *Nat. Med.* 6:879–85

79. Kiem HP, Andrews RG, Morris J, Peterson L, Heyward S, et al. 1998. Improved gene transfer into baboon marrow repopulating cells using recombinant human fibronectin fragment CH-296 in combination with interleukin-6, stem cell factor, FLT-3 ligand, and megakaryocyte growth and development factor. *Blood* 92:1878–86

80. Klages N, Zufferey R, Trono D. 2000. A stable system for the high-titer production of multiply attenuated lentiviral vectors. *Mol. Ther.* 2:170–76

81. Klessig DF, Brough DE, Cleghon V. 1984. Introduction, stable integration, and controlled expression of a chimeric adenovirus gene whose product is toxic to the recipient human cell. *Mol. Cell Biol.* 4:1354–62

82. Kochanek S. 1999. High-capacity adenoviral vectors for gene transfer and somatic gene therapy. *Hum. Gene Ther.* 10:2451–59

83. Kochanek S, Clemens PR, Mitani K, Chen HH, Chan S, Caskey CT. 1996. A new adenoviral vector: replacement of all viral coding sequences with 28 kb of DNA independently expressing both full-length dystrophin and beta-galactosidase.

Proc. Natl. Acad. Sci. USA 93:5731–36

84. Krisky DM, Marconi PC, Oligino TJ, Rouse RJ, Fink DJ, et al. 1998. Development of herpes simplex virus replication–defective multigene vectors for combination gene therapy applications. *Gene Ther.* 5:1517–30

85. Krougliak V, Graham FL. 1995. Development of cell lines capable of complementing E1, E4, and protein IX defective adenovirus type 5 mutants. *Hum. Gene Ther.* 6:1575–86

86. Leboulch P, Huang GM, Humphries RK, Oh YH, Eaves CJ, et al. 1994. Mutagenesis of retroviral vectors transducing human beta-globin gene and beta-globin locus control region derivatives results in stable transmission of an active transcriptional structure. *EMBO J.* 13:3065–76

87. Lee HC, Kim SJ, Kim KS, Shin HC, Yoon JW. 2000. Remission in models of type 1 diabetes by gene therapy using a single-chain insulin analogue. *Nature* 408:483–88

88. Li J, Samulski RJ, Xiao X. 1997. Role for highly regulated *rep* gene expression in adeno-associated virus vector production. *J. Virol.* 71:5236–43

89. Lieber A, He CY, Kirillova I, Kay MA. 1996. Recombinant adenoviruses with large deletions generated by Cre-mediated excision exhibit different biological properties compared with first-generation vectors in vitro and in vivo. *J. Virol.* 70:8944–60

90. Lindemann D, Patriquin E, Feng S, Mulligan RC. 1997. Versatile retrovirus vector systems for regulated gene expression in vitro and in vivo. *Mol. Med.* 3:466–76

91. Loeb JE, Cordier WS, Harris ME, Weitzman MD, Hope TJ. 1999. Enhanced expression of transgenes from adeno-associated virus vectors with the woodchuck hepatitis virus posttranscriptional regulatory element: implications for gene therapy. *Hum. Gene Ther.* 10:2295–305

92. Mangeot PE, Negre D, Dubois B, Winter AJ, Leissner P, et al. 2000. Development of minimal lentivirus vectors derived from simian immunodeficiency virus (SIVmac251) and their use for gene transfer into human dendritic cells. *J. Virol.* 74:8307–15

93. Mann R, Mulligan RC, Baltimore D. 1983. Construction of a retrovirus packaging mutant and its use to produce helper-free defective retrovirus. *Cell* 33:153–59

94. Markowitz D, Goff S, Bank A. 1988. A safe packaging line for gene transfer: separating viral genes on two different plasmids. *J. Virol.* 62:1120–24

95. Mastromarino P, Conti C, Goldoni P, Hautecoeur B, Orsi N. 1987. Characterization of membrane components of the erythrocyte involved in vesicular stomatitis virus attachment and fusion at acidic pH. *J. Gen. Virol.* 68:2359–69

96. May C, Rivella S, Callegari J, Heller G, Gaensler KM, et al. 2000. Therapeutic haemoglobin synthesis in beta-thalassaemic mice expressing lentivirus-encoded human beta-globin. *Nature* 406:82–86

97. McCown TJ, Xiao X, Li J, Breese GR, Samulski RJ. 1996. Differential and persistent expression patterns of CNS gene transfer by an adeno-associated virus (AAV) vector. *Brain Res.* 713:99–107

98. McGrory WJ, Bautista DS, Graham FL. 1988. A simple technique for the rescue of early region I mutations into infectious human adenovirus type 5. *Virology* 163:614–17

99. McLaughlin SK, Collis P, Hermonat PL, Muzyczka N. 1988. Adeno-associated virus general transduction vectors: analysis of proviral structures. *J. Virol.* 62:1963–73

100. Miller AD. 1997. Development and application of retroviral vectors. See Ref. 30a, pp. 437–73

101. Miller AD, Buttimore C. 1986. Redesign of retrovirus packaging cell lines to avoid recombination leading to helper virus production. *Mol. Cell Biol.* 6:2895–902

102. Miller AD, Law MF, Verma IM. 1985. Generation of helper-free amphotropic retroviruses that transduce a dominant-acting, methotrexate-resistant dihydrofolate reductase gene. *Mol. Cell Biol.* 5:431–37

103. Miller DG, Adam MA, Miller AD. 1990. Gene transfer by retrovirus vectors occurs only in cells that are actively replicating at the time of infection. *Mol. Cell Biol.* 10:4239–42. Erratum. 1992. *Mol. Cell Biol.* 12(1):433

104. Mineta T, Rabkin SD, Yazaki T, Hunter WD, Martuza RL. 1995. Attenuated multimutated herpes simplex virus 1 for the treatment of malignant gliomas. *Nat. Med.* 1:938–43

105. Miyoshi H, Blomer U, Takahashi M, Gage FH, Verma IM. 1998. Development of a self-inactivating lentivirus vector. *J. Virol.* 72:8150–57

106. Miyoshi H, Smith KA, Mosier DE, Verma IM, Torbett BE. 1999. Transduction of human CD34+ cells that mediate long-term engraftment of NOD/SCID mice by HIV vectors. *Science* 283:682–86

107. Miyoshi H, Takahashi M, Gage FH, Verma IM. 1997. Stable and efficient gene transfer into the retina using an HIV-based lentiviral vector. *Proc. Natl. Acad. Sci. USA* 94:10319–23

108. Monahan PE, Samulski RJ. 2000. AAV vectors: Is clinical success on the horizon? *Gene Ther.* 7:24–30

109. Moritz T, Dutt P, Xiao X, Carstanjen D, Vik T, et al. 1996. Fibronectin improves transduction of reconstituting hematopoietic stem cells by retroviral vectors: evidence of direct viral binding to chymotryptic carboxy-terminal fragments. *Blood* 88:855–62

110. Muzyczka N. 1992. Use of adeno-associated virus as a general transduction vector for mammalian cells. *Curr. Top. Microbiol. Immunol.* 158:97–129

111. Nakai H, Iwaki Y, Kay MA, Couto LB. 1999. Isolation of recombinant adeno-associated virus vector—cellular DNA junctions from mouse liver. *J. Virol.* 73:5438–47

112. Nakai H, Storm TA, Kay MA. 2000. Increasing the size of rAAV-mediated expression cassettes in vivo by intermolecular joining of two complementary vectors. *Nat. Biotechnol.* 18:527–32

113. Naldini L, Blomer U, Gage FH, Trono D, Verma IM. 1996. Efficient transfer, integration, and sustained long-term expression of the transgene in adult rat brains injected with a lentiviral vector. *Proc. Natl. Acad. Sci. USA* 93:11382–88

114. Naldini L, Blomer U, Gallay P, Ory D, Mulligan R, et al. 1996. In vivo gene delivery and stable transduction of nondividing cells by a lentiviral vector. *Science* 272:263–67

115. Naldini L, Verma IM. 1999. Lentiviral vectors. See Ref. 44a, pp. 47–60

116. Naviaux RK, Costanzi E, Haas M, Verma IM. 1996. The pCL vector system: rapid production of helper-free, high-titer, recombinant retroviruses. *J. Virol.* 70:5701–5

117. No D, Yao TP, Evans RM. 1996. Ecdysone-inducible gene expression in mammalian cells and transgenic mice. *Proc. Natl. Acad. Sci. USA* 93:3346–51

118. Nolta JA, Smogorzewska EM, Kohn DB. 1995. Analysis of optimal conditions for retroviral-mediated transduction of primitive human hematopoietic cells. *Blood* 86:101–10

119. Olsen JC. 1998. Gene transfer vectors derived from equine infectious anemia virus. *Gene Ther.* 5:1481–87

120. Ory DS, Neugeboren BA, Mulligan RC. 1996. A stable human-derived packaging cell line for production of high titer retrovirus/vesicular stomatitis virus G pseudotypes. *Proc. Natl. Acad. Sci. USA* 93: 11400–6

121. Parks RJ, Chen L, Anton M, Sankar U, Rudnicki MA, Graham FL. 1996. A hel-per-dependent adenovirus vector system: removal of helper virus by Cre-mediated excision of the viral packaging signal. *Proc. Natl. Acad. Sci. USA* 93:13565–70

122. Parks RJ, Graham FL. 1997. A helper-dependent system for adenovirus vector production helps define a lower limit for efficient DNA packaging. *J. Virol.* 71:3293–98

123. Pfeifer A, Kessler T, Yang M, Baranov E, Kootstra NA, et al. 2001. Transduction of liver cells by lentiviral vectors: analysis in living animals by fluorescence imaging. *Mol. Ther.* 3:319–22

124. Pfeifer A, Verma IM. 2001. Virus vectors and their applications. In *Fields Virology*, ed. PM Howley, DM Knipe, D Griffin, RA Lamb, A Martin, et al. Philadelphia: Lippincott-Raven. In press

125. Poeschla E, Corbeau P, Wong-Staal F. 1996. Development of HIV vectors for anti-HIV gene therapy. *Proc. Natl. Acad. Sci. USA* 93:11395–99

126. Poeschla E, Gilbert J, Li X, Huang S, Ho A, Wong-Staal F. 1998. Identification of a human immunodeficiency virus type 2 (HIV-2) encapsidation determinant and transduction of nondividing human cells by HIV-2–based lentivirus vectors. *J. Virol.* 72:6527–36

127. Poeschla EM, Wong-Staal F, Looney DJ. 1998. Efficient transduction of nondividing human cells by feline immunodeficiency virus lentiviral vectors. *Nat. Med.* 4:354–57

128. Pollock R, Issner R, Zoller K, Natesan S, Rivera VM, Clackson T. 2000. Delivery of a stringent dimerizer-regulated gene expression system in a single retroviral vector. *Proc. Natl. Acad. Sci. USA* 97:13221–26

129. Pollock R, Rivera VM. 1999. Regulation of gene expression with synthetic dimerizers. *Methods Enzymol.* 306:263–81

130. Poznansky M, Lever A, Bergeron L, Haseltine W, Sodroski J. 1991. Gene transfer into human lymphocytes by

a defective human immunodeficiency virus type 1 vector. *J. Virol.* 65:532–36

131. Qing K, Mah C, Hansen J, Zhou S, Dwarki V, Srivastava A. 1999. Human fibroblast growth factor receptor 1 is a coreceptor for infection by adeno-associated virus 2. *Nat. Med.* 5:71–77

132. Rabson AB, Wills JW. 1997. Synthesis and processing of viral RNA. See Ref. 30a, pp. 205–61

133. Ramirez-Solis R, Liu P, Bradley A. 1995. Chromosome engineering in mice. *Nature* 378:720–24

134. Reiser J, Harmison G, Kluepfel-Stahl S, Brady RO, Karlsson S, Schubert M. 1996. Transduction of nondividing cells using pseudotyped defective high-titer HIV type 1 particles. *Proc. Natl. Acad. Sci. USA* 93:15266–71

135. Rivera VM, Clackson T, Natesan S, Pollock R, Amara JF, et al. 1996. A humanized system for pharmacologic control of gene expression. *Nat. Med.* 2:1028–32

136. Roizman B, Sears AE. 1996. Adenoviruses. See Ref. 41a, pp. 2231–95

137. Saeki Y, Ichikawa T, Saeki A, Chiocca EA, Tobler K, et al. 1998. Herpes simplex virus type 1 DNA amplified as bacterial artificial chromosome in *Escherichia coli*: rescue of replication-competent virus progeny and packaging of amplicon vectors. *Hum. Gene Ther.* 9:2787–94

138. Samaniego LA, Neiderhiser L, DeLuca NA. 1998. Persistence and expression of the herpes simplex virus genome in the absence of immediate-early proteins. *J. Virol.* 72:3307–20

139. Samulski RJ, Chang LS, Shenk T. 1987. A recombinant plasmid from which an infectious adeno-associated virus genome can be excised in vitro and its use to study viral replication. *J. Virol.* 61:3096–101

140. Samulski RJ, Chang LS, Shenk T. 1989. Helper-free stocks of recombinant adeno-associated viruses: normal integration does not require viral gene expression. *J. Virol.* 63:3822–28

141. Samulski RJ, Sally M, Muzyczka N. 1999. Adeno-associated viral vectors. See Ref. 44a, pp. 131–72

142. Sarkis C, Serguera C, Petres S, Buchet D, Ridet JL, et al. 2000. Efficient transduction of neural cells in vitro and in vivo by a baculovirus-derived vector. *Proc. Natl. Acad. Sci. USA* 97:14638–43

143. Shenk T. 1996. Adenoviruses: the viruses and their replication. See Ref. 41a, pp. 2111–48

144. Shimada T, Fujii H, Mitsuya H, Nienhuis AW. 1991. Targeted and highly efficient gene transfer into CD^{4+} cells by a recombinant human immunodeficiency virus retroviral vector. *J. Clin. Invest.* 88:1043–47

145. Shockett P, Difilippantonio M, Hellman N, Schatz DG. 1995. A modified tetracycline-regulated system provides autoregulatory, inducible gene expression in cultured cells and transgenic mice. *Proc. Natl. Acad. Sci. USA* 92:6522–26

146. Snyder RO, Miao CH, Patijn GA, Spratt SK, Danos O, et al. 1997. Persistent and therapeutic concentrations of human factor IX in mice after hepatic gene transfer of recombinant AAV vectors. *Nat. Genet.* 16:270–76

147. Somia N, Verma IM. 2001. Gene therapy: trials and tribulations. *Nat. Rev. Genet.* 1:91–99

148. Spector DH, Wade E, Wright DA, Koval V, Clark C, et al. 1990. Human immunodeficiency virus pseudotypes with expanded cellular and species tropism. *J. Virol.* 64:2298–308

149. Srivastava A. 1994. Parvovirus-based vectors for human gene therapy. *Blood Cells* 20:531–36

150. Strauss EG, Strauss JH, Levine AJ. 1996. Virus evolution. See Ref. 41a, pp. 153–71

151. Suhr ST, Gil EB, Senut MC, Gage FH. 1998. High level transactivation by a modified *Bombyx* ecdysone receptor in mammalian cells without exogenous retinoid X receptor. *Proc. Natl. Acad. Sci. USA* 95:7999–8004

152. Summerford C, Bartlett JS, Samulski RJ. 1999. αV β5 integrin: a co-receptor for adeno-associated virus type 2 infection. *Nat. Med.* 5:78–82

153. Summerford C, Samulski RJ. 1998. Membrane-associated heparan sulfate proteoglycan is a receptor for adeno-associated virus type 2 virions. *J. Virol.* 72:1438–45

154. Sun L, Li J, Xiao X. 2000. Overcoming adeno-associated virus vector size limitation through viral DNA heterodimerization. *Nat. Med.* 6:599–602

155. Takahashi M, Miyoshi H, Verma IM, Gage FH. 1999. Rescue from photoreceptor degeneration in the rd mouse by human immunodeficiency virus vector-mediated gene transfer. *J. Virol.* 73:7812–16

156. Telesnitsky A, Goff SP. 1997. Reverse transcriptase and the generation of retroviral DNA. See Ref. 30a, pp. 121–60

157. Tisdale JF, Hanazono Y, Sellers SE, Agricola BA, Metzger ME, et al. 1998. Ex vivo expansion of genetically marked rhesus peripheral blood progenitor cells results in diminished long-term repopulating ability. *Blood* 92:1131–41

158. Trono D. 2000. Lentiviral vectors: turning a deadly foe into a therapeutic agent. *Gene Ther.* 7:20–23

158a. US Dept. Health. 1999. *Minutes of Symp. Meet., Recombinant DNA Advisory Comm., Dec. 8–10*. Washington, DC: Natl. Inst. Health. http://www4.od.nih.gov/oba/1299rac.pdf

159. Verma IM, Somia N. 1997. Gene therapy—promises, problems and prospects. *Nature* 389:239–42

160. Vogt PK. 1997. Historical introduction to the general properties of retroviruses. See Ref. 30a, pp. 1–25

161. Wang Q, Finer MH. 1996. Second-generation adenovirus vectors. *Nat. Med.* 2:714–16

162. Wang Q, Greenburg G, Bunch D, Farson D, Finer MH. 1997. Persistent transgene expression in mouse liver following in vivo gene transfer with a deltaE1/deltaE4 adenovirus vector. *Gene Ther.* 4:393–400

163. Wang Q, Jia XC, Finer MH. 1995. A packaging cell line for propagation of recombinant adenovirus vectors containing two lethal gene-region deletions. *Gene Ther.* 2:775–83

164. Wang Y, O'Malley BW Jr, Tsai SY, O'Malley BW. 1994. A regulatory system for use in gene transfer. *Proc. Natl. Acad. Sci. USA* 91:8180–84

165. Weinberg JB, Matthews TJ, Cullen BR, Malim MH. 1991. Productive human immunodeficiency virus type 1 (HIV-1) infection of nonproliferating human monocytes. *J. Exp. Med.* 174:1477–82

166. Weitzman MD, Kyostio SR, Kotin RM, Owens RA. 1994. Adeno-associated virus (AAV) Rep proteins mediate complex formation between AAV DNA and its integration site in human DNA. *Proc. Natl. Acad. Sci. USA* 91:5808–12

167. Wickham TJ, Mathias P, Cheresh DA, Nemerow GR. 1993. Integrins αV β3 and αV β5 promote adenovirus internalization but not virus attachment. *Cell* 73:309–19

168. Wivel NA, Gao G-P, Wilson JM. 1999. Adenovirus vectors. See Ref. 44a, pp. 87–111

169. Wu P, Phillips MI, Bui J, Terwilliger EF. 1998. Adeno-associated virus vector–mediated transgene integration into neurons and other nondividing cell targets. *J. Virol.* 72:5919–26

170. Xiao X, Li J, Samulski RJ. 1996. Efficient long-term gene transfer into muscle tissue of immunocompetent mice by adeno-associated virus vector. *J. Virol.* 70:8098–108

171. Xiao X, Li J, Samulski RJ. 1998. Production of high-titer recombinant adeno-associated virus vectors in the absence of helper adenovirus. *J. Virol.* 72:2224–32

172. Xu K, Ma H, McCown TJ, Verma IM, Kafri T. 2001. Generation of a stable cell line producing high-titer self-inactivating lentiviral vectors. *Mol. Ther.* 3:97–104

173. Yan Z, Zhang Y, Duan D, Engelhardt JF. 2000. From the cover: *trans*-splicing vectors expand the utility of adeno-associated virus for gene therapy. *Proc. Natl. Acad. Sci. USA* 97:6716–21

174. Yang CC, Xiao X, Zhu X, Ansardi DC, Epstein ND, et al. 1997. Cellular recombination pathways and viral terminal repeat hairpin structures are sufficient for adeno-associated virus integration in vivo and in vitro. *J. Virol.* 71:9231–47

175. Yee J-K. 1999. Retroviral vectors. See Ref. 44a, pp. 21–46

176. Yee JK, Moores JC, Jolly DJ, Wolff JA, Respess JG, Friedmann T. 1987. Gene expression from transcriptionally disabled retroviral vectors. *Proc. Natl. Acad. Sci. USA* 84:5197–201

177. Yeh P, Dedieu JF, Orsini C, Vigne E, Denefle P, Perricaudet M. 1996. Efficient dual transcomplementation of adenovirus E1 and E4 regions from a 293-derived cell line expressing a minimal E4 functional unit. *J Virol* 70:559–65

178. Yew PR, Berk AJ. 1992. Inhibition of p53 transactivation required for transformation by adenovirus early 1B protein. *Nature* 357:82–85

179. Young NS. 1996. Parvoviruses. See Ref. 41a, pp. 2199–220

180. Yu H, Rabson AB, Kaul M, Ron Y, Dougherty JP. 1996. Inducible human immunodeficiency virus type 1 packaging cell lines. *J. Virol.* 70:4530–37

181. Yu SF, von Ruden T, Kantoff PW, Garber C, Seiberg M, et al. 1986. Self-inactivating retroviral vectors designed for transfer of whole genes into mammalian cells. *Proc. Natl. Acad. Sci. USA* 83:3194–98

182. Zennou V, Petit C, Guetard D, Nerhbass U, Montagnier L, Charneau P. 2000. HIV-1 genome nuclear import is mediated by a central DNA flap. *Cell* 101:173–85

183. Zolotukhin S, Byrne BJ, Mason E, Zolotukhin I, Potter M, et al. 1999. Recombinant adeno-associated virus purification using novel methods improves infectious titer and yield. *Gene Ther.* 6:973–85

184. Zufferey R, Donello JE, Trono D, Hope TJ. 1999. Woodchuck hepatitis virus posttranscriptional regulatory element enhances expression of transgenes delivered by retroviral vectors. *J. Virol.* 73:2886–92

185. Zufferey R, Dull T, Mandel RJ, Bukovsky A, Quiroz D, et al. 1998. Self-inactivating lentivirus vector for safe and efficient in vivo gene delivery. *J. Virol.* 72:9873–80

186. Zufferey R, Nagy D, Mandel RJ, Naldini L, Trono D. 1997. Multiply attenuated lentiviral vector achieves efficient gene delivery in vivo. *Nat. Biotechnol.* 15:871–75

Annu. Rev. Genomics Hum. Genet. 2001. 2:213–33

HUMAN GENETICS ON THE WEB

Alan E. Guttmacher

*National Human Genome Research Institute, National Institutes of Health, Bethesda,
Maryland 20892-2152; e-mail: guttmach@mail.nih.gov*

Key Words   Internet, World Wide Web, computers, genomics, medical genetics

■ **Abstract**   Use of the World Wide Web ("the web") and our knowledge of human
genetics are both currently expanding rapidly. By allowing swift, universal, and free
access to data, the web has already played an important role in human genetics research.
It has also begun to change the way that information is shared in clinical genetics and,
to a lesser degree, affect how education in human genetics occurs. There are scores
of web sites helpful to those interested in either research or clinical aspects of human
genetics. The web and related communication technologies should continue to play
increasingly important roles in human genetics.

INTRODUCTION

The past decade has witnessed unprecedented growth in our understanding of genetics. It has also witnessed unprecedented growth in the power and pervasiveness of computers. Indeed, this second fact is partially responsible for the first; without the last decade's increase in computing power, many recent keystones of advances in genetics, including determining the draft sequence of the human genome, would have been impossible. Yet, it is not only through enabling such new fields of basic science research that computer-based activity is reshaping the face of genetics. Use of computer-based communication is also dramatically changing, if not revolutionizing, the way researchers, health professionals, and the public each access information about many aspects of genetics.

This significant impact of computer-based communication has largely been due to the expansion, improvement, and increased uptake of the Internet. Although some people herald telemedicine as a use of the Internet (and other communication modes) that will dramatically alter the practice of clinical medicine, despite its decades of existence, telemedicine has yet to become pervasive in healthcare. In genetics specifically, its use has been quite limited (5). Similarly, some other uses of the Internet, such as Internet telephony, have thus far been of relatively little importance in genetics. Instead, it has been through its two major components, e-mail and the World Wide Web, that the Internet has, so far, had a significant impact in genetics.

1527-8204/01/0728-0213$14.00

E-mail has accelerated communication among researchers, whether in the same laboratory or different hemispheres. Particularly as the size of research teams has grown, the ability e-mail affords for near instantaneous communication with any number of colleagues, regardless of distances, time zones, or variations in individual diurnal rhythms, has changed the way much research is conducted. However, e-mail has changed the conduct not only of research, but also of other aspects of biomedicine. In clinical medicine, including clinical genetics, e-mail has allowed providers to communicate more effectively, both with one another and with patients. Moreover, it has provided patients a new way to communicate with not only their own healthcare providers but also with healthcare providers they would otherwise never know. Importantly, e-mail has also encouraged patients to connect with each other. Especially for many rare genetic disorders, e-mail has allowed patients and their families to share knowledge and form important support alliances in ways more effective than could have been imagined in the pre-Internet era.

As much as e-mail has changed the way researchers, clinicians, and the public communicate about genetics, it is not the only part of the Internet that has wrought such change. The World Wide Web, or simply "the web," as it is often called, has changed (some would claim revolutionized) the way information of many sorts is accessed. Web-based databases are now an important source of genetics information for many researchers; the daily deposit of data into a publicly accessible database has been one of the key, and some would say most important, characteristics of the Human Genome Project. By making the data it develops available almost instantaneously to anyone in the world with a computer and a modem, the Human Genome Project has not only made further advances in genetics easier and more rapid, it has also set a high-profile standard for collegial sharing of research data that may have long lasting impact even in fields beyond genetics.

However, rapid worldwide free sharing of such information via the web is not only a key characteristic of the Human Genome Project, it is also a key tool in the project's success. Without such near instantaneous sharing of data, the Human Genome Project would, at best, have taken much longer to accomplish its ambitious sequencing goals; at worst, it would have failed at achieving these goals at all.

Like e-mail, the web has made a difference not only for researchers but also for clinicians and patients. Over half of Americans with Internet access, or more than 50,000,000 people, have used the web to access health or medical information, and most do so at least once a month (4). It is interesting that, although there are few reliable data, the experience of many people suggests that patients are ahead of providers in utilizing the web to access information about genetics. Clinicians increasingly report patients who come to them waving downloads from the web about their particular genetic concerns. For patients to report that their providers have shared similar web-based information with them appears to be less common. Nonetheless, it is clear that clinicians are using the web increasingly to access information about genetics and that there are a growing number of web sites that aim to supply useful genetics information to clinicians.

Another use of the web in relationship to genetics is informing not patients, per se, but the general public. As is true in many other areas of knowledge, the web has an increasing wealth of information about genetics aimed for various lay audiences, ranging from pages directed at teaching high school students about the "central dogma" of genetics to those seeking to inform concerned citizens about ethical, legal, and social aspects of the field.

One can often categorize web sites accurately by type of content. However, categorizing sites accurately by type of audience is not so easy, as sites often reach audiences who are different from those they intend or claim to attract. An important characteristic of the web is that, with the partial exception of the relatively few password-protected sites, it is neither the designers nor the "owners" but the users who determine a site's actual audience. A site targeted to health professionals may get more traffic from health consumers. A page that seeks to teach high schoolers about genetics may be visited not only by their younger siblings but also by their parents.

Much is made of the "digital divide," the observation that the Internet is not equally accessible to all members of society. Especially because access to health information and healthcare is not equitable in our society, a major challenge confronting use of "the net" as an important means of communication about health, including genetics, is achieving equitable access. It is important to realize that, as regards healthcare, this digital divide exists not only among the general public (1) but also among health care providers (2).

Although there are important and valid concerns about socioeconomic and other disparities in access to the Internet, at the same time, the Internet and its components are, in many ways, egalitarian. This egalitarianism is important in understanding the web's impact on communication. The web makes fewer distinctions about age, academic degrees, or professional status than do many other forms of communication, particularly as those other forms have been historically used in science and health care.

Moreover, standard print media are characterized by a number of factors, including their long-established status and the expense and length of the publication process, which tend to make them prioritize values, such as "authoritativeness" and clear identification of authors, especially in subject areas like human genetics. As a different and more recent communication medium, the web often prioritizes different and "newer" values, such as creativity and more equal weighting of all opinions.

Such aspects of web culture, as well as the increased ease of mounting an impressive-looking web site rather than publishing and distributing a textbook, for instance, can often make it more difficult to establish the source and/or quality of information on the web than is true for print publications. In using the web, one tends to move much more quickly, easily, and less consciously from one information source to another than is true in more established media, such as print.

Whether such characteristics are a benefit or a curse of the web may depend on both the specific situation and the evaluator, but it is certainly important to realize that, in many ways, web-based communication is not merely the printed page viewed on a computer screen.

Of course, the culture of the larger society, not only of the web, will help determine the eventual role of the web and related new technologies in communication. As with the advent of older communication media, a combination of the technology, its own culture, and larger societal changes will come together to determine the technology's eventual role in communication. At the turn of the twentieth century, the technological innovation of linotype machines, aspects of newspapers' own culture, such as comic strips, and the spread of literacy throughout society made newspapers such an important means of communication in America. At the turn of this century, the technological innovations of the web, its own developing culture, and larger societal forces will determine how effective a communication medium the web becomes in general and in human genetics specifically.

ASSESSING WEB SITES

It is often difficult to assess the quality and accuracy of information, no matter what the medium by which it is received. Many find this an even greater challenge when accessing information from the web and a particularly important one when the information has to do with health and related issues. However, answering a few basic questions can help evaluate the quality of information on web sites.

- Disclosure
  Does the site openly disclose its mission and sources of ownership and/or support?
- Confidentiality
  Does the site openly disclose its policy regarding confidentiality of any information it receives from users? Does that policy adequately safeguard users?
- Timely updating
  Does the site update its information frequently, and does it plainly label the date on which information was last revised?
- Expertise
  Does the site list its staff and consultants? Is authorship of information clear? Do the staff, consultants, or authors have appropriate expertise?
- Ethics codes
  Does the site subscribe to any recognized codes of conduct, such as the HON code of conduct (http://www.hon.ch/HONcode/), the Hi-Ethics Principles (http://www.hiethics.org/Principles/index.asp), or the eHealth Code of Ethics (http://www.ihealthcoalition.org/ethics/ehcode.html)?

THE WEB AND GENETICS RESEARCH

The web has already proven a valuable resource in human genetics research. As mentioned previously, perhaps the central tool of such research, the Human Genome Project, relies heavily on the web both to amass and to distribute its data. But, a number of other web-based resources are also of use to the genetics researcher. Indeed, as most biomedical research institutions maintain web sites, there are scores of sites that may be of help in pursuing various sorts of genetics-related research inquiries. Although it is beyond the scope of this article to consider all of the many web-based resources pertinent to human genetics research, several warrant particular attention.

In terms of the Human Genome Project and closely related activities, the single most valuable resource at present is probably Human Genome Central hosted by the National Center for Biotechnology Information (http://www.ncbi.nlm.nih.gov/genome/guide/central.html) and the European Bioinformatics Institute (http://www.ensembl.org/genome/central/). This site is so valuable because, as its name implies, it is a centralized site that links to many other useful ones. Actually, describing Human Genome Central as a pair of valuable resources may be more accurate because both sites on which it is located, although they are referred to by the same sobriquet and offer similar information, vary somewhat in both presentation and content. The European Bioinformatics Institute's version of Human Genome Central currently links to 11 other sites, as shown in Table 1. The National Center for Biotechnology Information's version of Human Genome Central also links to most of those 11 sites as well as a number of others, including 21 genome sequencing centers. It also offers "quick links" to 16 other genetics sites hosted by the National Center for Biotechnology Information.

In addition to those sites found through Human Genome Central, a number of other web-based resources have already made important contributions in genetics research and will continue to do so. Databases will play a prominent role among such resources. For instance, particularly with the growing importance of pharmacogenomics, the Pharmacogenetics Knowledge Base (PharmGKB) (http://www.pharmGKB.org/) can be a valuable research tool.

An increasing number of commercial ventures also offer web-based resources that are of interest to genetics researchers. Such sites offer some information that is free of charge and can be used without restriction. Other information at commercial sites is subject to either subscription costs or license requirements. Among research-oriented sites offered by the commercial sector are those of Celera Genomics (http://www.celera.com/) and DoubleTwist (http://www.doubletwist.com/).

THE WEB AND CLINICAL GENETICS

Both health care professionals and the general public increasingly use the web as an important source of information about clinical genetics, and they have a growing number of useful sites from which to choose. As noted above, a general

TABLE 1 Human Genome Central links (http://www.ensembl.org/genome/central/)

Resource	URL	Content
Ensembl	http://www.ensembl.org	Access to DNA and protein sequences with automatic baseline annotation
National Center for Biotechnology Information	http://www.ncbi.nlm.nih.gov/genome/guide/	Views of chromosomes, maps, and loci; links to other NCBI resources
Oak Ridge Genome Channel	http://compbio.ornl.gov/tools/channel/	Java viewers for human genome data
BLAST	http://www.ensembl.org/Data/blast.html	Searches of protein or DNA sequence against human draft data
The SNP consortium	http://snp.cshl.org/	A variety of ways to query for SNPs in the human genome
OMIM	http://www.ncbi.nlm.nih.gov/Omim/	Information about human genes and disease
Gene Map '99	http://www.ncbi.nlm.nih.gov/genemap99/	Data and viewers for radiation hybrid maps
Human genome maps	http://genome.wustl.edu/gsc/human/Mapping/	Links to clone and accession maps of the human genome
Human Genome Project Working Draft Sequence	http://genome.cse.ucsc.edu/	An assembly of the current draft of the human genome
Ethical, Legal, & Social Issues	http://www.ornl.gov/hgmis/elsi/elsi.html and http://www.nhgri.nih.gov/ELSI/	Information, articles, and links on a wide range of issues

characteristic of the web is that a site apparently targeted for one audience may be visited heavily, even predominantly, by another audience. This is an important aspect of web sites devoted to clinical genetics. Although some sites are clearly designed primarily for health professionals and others for the general public, users of either background often find useful information on sites targeted to the other group.

Health providers should realize that they will find much of use on sites designed predominantly for lay users. They also need to understand that their patients are increasingly reading and downloading information from sites that may have been designed primarily for health professionals. As when patients access printed medical texts or journals, this practice can lead to a more informed patient, but it can also lead to a more confused patient. Because of the growing frequency with which patients and their families explore resources on the web that were originally targeted for professionals, providers should ask patients whether they have read such information, offer to "interpret" it, and consider recommending it to patients.

Sites with medical genetics information that are targeted primarily at health professionals have varying emphases and will appeal to different types of users (see Table 2). Some may tend to appeal more to primary care provides, whereas others may appeal more to genetics specialists.

One authoritative site of use to many sorts of providers is GeneClinics (http://www.geneclinics.org/). This site features expert-authored, peer-reviewed clinical synopses of a number of disorders with genetic bases. At the time of this writing, it offers information on approximately 100 disorders, focusing primarily, but not exclusively, on Mendelian disorders.

Another authoritative site that focuses on Mendelian disorders, Online Mendelian Inheritance in Man (OMIM) (http://www3.ncbi.nlm.nih.gov/Omim/), may appeal more to the genetics specialist. Like its print counterpart, *Mendelian Inheritance in Man*, this site is an encyclopedic resource that discusses several thousand entities thought to have a Mendelian basis. Individual entries vary somewhat in terms of content but tend to focus more on the genetics aspects of disorders than on their wider clinical aspects. Like many, but not all, of the web sites mentioned in this article, one of OMIM's valuable attributes is that it allows its users to explore pertinent articles from the biomedical literature quickly and effectively. OMIM does so by providing, via PubMed (http://www.ncbi.nlm.nih.gov/PubMed/), direct links to articles it cites.

The Human Genome Epidemiology Network (HuGENet) (http://www.cdc.gov/genetics/hugenet/) is a resource based at the Centers for Disease Control and Prevention that provides epidemiologic information on the human genome. The site includes population-specific prevalence data on human gene variants, epidemiologic data on the association between genetic variation and diseases in different populations, quantitative population–based data on gene-environment interactions, information on the population impact of genetic tests and services, and information on disease prevention and health promotion aspects of human gene variants.

The Office of Rare Diseases at the National Institutes of Health maintains a site (http://rarediseases.info.nih.gov/ord/index.html) that offers information on

TABLE 2   Clinical and public health genetics

Resource	URL	Content
GeneClinics	http://www.geneclinics.org/	Authoritative synopses of disorders that have a significant genetic component
Online Mendelian Inheritance in Man	http://www3.ncbi.nlm.nih.gov/Omim	Authoritative information about thousands of Mendelian conditions
Human Genome Epidemiology Network	http://www.cdc.gov/genetics/hugenet/	Epidemiologic information on the human genome
Office of Rare Diseases, National Institutes of Health	http://raredisease.info.nih.gov/ord/index.html	Information on thousands of rare disorders
The Genetics Resource Center	http://www.pitt.edu/~edugene/resource/	Clinical and educational information related to genetic counseling
National Cancer Institute's CancerNet	http://cancernet.nci.nih.gov/genetics_prevention.html	Authoritative information about cancer genetics
INFOGENETICS	http://www.infogenetics.org/	Wide array of information for the clinician
National Newborn Screening & Genetics Resource Center	http://GENES-R-Us.uthsca.edu	Clinical genetics and newborn screening

TABLE 3   Some for-profit organizations offering clinical genetics information and/or services

Resource	URL
GeneSage	http://www.genesage.com/
Genetic Health	http://www.genetichealth.com/
Genetic Resources Medical Group, Inc.	http://www.dnaMD.com/
Genomic Health	http://www.genomichealth.com/

more than 6000 rare diseases, including current research, publications, completed research, ongoing studies, ethical trials, and patient support groups.

The Genetics Resource Center (http://www.pitt.edu/~edugene/resource/) is a site maintained by the University of Pittsburgh's graduate program in genetic counseling. It offers a wide array of information related to both clinical genetics and genetics education, focusing particularly, but not exclusively, on areas related to genetic counseling.

A number of sites focus on the genetic aspects of cancer. One authoritative site in this area is the genetics section of the National Cancer Institute's CancerNet (http://cancernet.nci.nih.gov/genetics_prevention.html).

Like genetics research, clinical genetics is also of interest to for-profit entities, several of which currently offer web sites (see Table 3) that provide a range of information and services, usually with the former, but not the latter, available at no cost.

The use of the web to sell goods or services related to human genetics raises concern in some quarters. However, marketplaces themselves are rarely forces for good or for evil; it is how they are utilized that determines their impact, moral and otherwise. Commercial entities can serve a valuable role in providing excellent web content that informs users in an unbiased way. Obviously, they can also pervert the web by providing false, misleading, or significantly incomplete information. Of course, the same is true for sites operated by nonprofit organizations. At least two forces may help assure that genetics content on the web, no matter what its source, is of high quality. The first is vigilance by users who apply criteria like those cited earlier in this article to assess web sites and guide their use of them (see Assessing Web Sites above). The second is identification and adoption by the genetics community of clear and high standards of conduct for web-based entities, followed by a willingness to support only those sites that meet such standards (see Genetics Resources on the Web below).

Whether a site is operated by a nonprofit or for-profit organization, it is important, of course, to be wary about issues of privacy and confidentiality when "surfing" the web. This is particularly true when one provides any personal information to a site. However, even if one provides no information, through "cookies" and related mechanisms, sites can track an individual user, often without that user's knowledge. In areas as personal as health and genetics, such issues warrant particular care.

TABLE 4   Genetic testing

Resource	URL	Content
GeneTests	http://www.genetests.org	Information for health professionals about hundreds of genetic tests
Secretary's Advisory Committee on Genetic Testing	http://www4.od.nih.gov/oba/sacgt.htm	Policy issues regarding genetic testing
A Question of Genes: Inherited Risks	http://www.Pbs.org/gene/	1997 PBS documentary about genetic testing
Understanding Gene Testing	http://www.accessexcellence.org/ AE/AEPC/NIH/index.html	Primer on genetic testing

Several web sites primarily provide information about genetic testing (see Table 4). GeneTests (http://www.genetests.org) is of particular clinical relevance, providing information about genetic tests currently offered, whether on a clinical or a research basis. Unlike its sister site, GeneClinics, GeneTests is available at present only to health professionals, requiring free registration before use. The Secretary's Advisory Committee on Genetic Testing (SACGT) is an advisory body to the Secretary of Health and Human Services that considers such questions as what oversight should various sorts of genetic testing require and what bodies should provide this oversight. Its site (http://www4.od.nih.gov/oba/sacgt.htm) provides information about these sorts of considerations. A Question of Genes: Inherited Risks (http://www.Pbs.org/gene/) is the repository site for a two-hour program, which aired on public television in 1997, that focused on the social, ethical, and emotional consequences of genetic testing. The National Institutes of Health developed both the text and web site for Understanding Gene Testing, (http://www.accessexcellence.org/AE/AEPC/NIH/index.html) to explain many aspects of genetic testing.

Disease-specific and more general lay support groups have long been of great help in clinical genetics, both to patients and their families and to providers. The Internet has allowed such groups to recruit new members and, through e-mail, message boards, listservs, etc., to offer new ways for members and others to communicate with each other. The web has allowed literally hundreds of such groups to provide information to members of the public and health professionals from around the world. Listing all such web sites is beyond the purview of this paper (see Table 5). However, just as Human Genome Central offers links to numerous genetics research uniform resource locators (URL), the sites of the Genetic Alliance (http://www.geneticalliance.org/) and of the National Organization for Rare Disorders (NORD) (http://www.rarediseases.org/) each provide links to numerous other patient support and advocacy organizations' sites.

TABLE 5   Support and advocacy groups' web sites

Resource	URL	Content
The Genetic Alliance	http://www.geneticalliance.org/	Wide array of genetic-related information
National Organization for Rare Disorders	http://www.raredisease.org/	Wide array of genetic-related information
Family Village	http://www.familyvillage.wisc.edu/index.htmlx	Disability-related resources
Rare Genetic Diseases in Children	http://mcrcr2.med.nyu.edu/murphp01/homenew.htm	Focuses on pediatric genetic conditions
The Genome Action Coalition	http://www.tgac.org	Federal genetics legislation
Coalition for Genetic Fairness	http://www.nationalpartnership.org/healthcare/genetic/coalition.htm	Federal legislation regarding genetics discrimination

Both of these sites also offer a wide array of information of use to individuals with genetic concerns, their families, and their providers. Family Village (http://www.familyvillage.wisc.edu/index.htmlx) is a valuable site that focuses on disability-related resources.

THE WEB AND GENETICS EDUCATION

The Internet has great potential for both informing and educating. However, although the Internet, its denizens, and perhaps, modern society in general sometimes obfuscate the distinction, distinguishing between informing and educating is important. Supplying information has a real value, and the Internet is becoming an increasingly significant means, indeed for some sectors of society the dominant means, of communicating information. Education, however, implies more than supplying information alone. True education includes a pedagogy and curricular structure that, when successful, imbues the student with an ability to synthesize and analyze separate pieces of information and to interpret such information in context with other information or knowledge. Above, we have considered web-based communication of genetics information; here, we specifically examine web-based education in genetics.

The web is a promising medium for educating a number of audiences that may have differing, if overlapping, educational needs regarding genetics. Among such audiences are scientists, practicing health professionals, students, and the general public. Of course, each of these groupings subsumes disparate individuals with different educational needs. For instance, among students, medical students may require materials different from those aimed at physical therapy students; high school students in an introductory biology course may have distinct needs from those of advanced undergraduates. Among practicing health professionals, as well, there are subgroups that may require or respond best to different material; a primary care nurse practitioner and a radiation oncologist may have different content in mind when they use the web to educate themselves about genetics.

A potential attraction of web-based educational material is that the medium lends itself to reaching such different audiences efficiently. Thus, a curriculum in genetics might include certain core content applicable to learners of many backgrounds and other modules targeted to more specific needs. Indeed, another advantage of the web as a means for education is that, at its most effective, it allows the individual user to define his or her individual needs. For instance, someone who finds that a web-based educational program primarily designed for users of another professional background best fits their educational needs can use that personally preferred curriculum, or any of its elements, as a supplement or even replacement for the curriculum originally targeted to them by the web page provider. On the web, one can transfer between classrooms at the touch of a keystroke. Particularly in a swiftly expanding area of knowledge, such as genetics, educational flexibility and tailoring to the individual student's desires is particularly useful.

Another potential advantage that the web brings to education in genetics is its ability to update content rapidly and constantly. In an area in which knowledge is quickly changing, such as genetics, timeliness is of particular importance.

Despite the great potential for the web as a means of education in genetics, actual attempts to use it in this manner have been limited thus far. However, considering the newness of the web, this may not be a condemnation of the web's eventual significance in genetics education. Indeed, one might argue that the fact that the web already offers a number of opportunities for genetics education is an indicator of the important role it will play in this arena within a few years.

GENETICS EDUCATION FOR HEALTH PROFESSIONALS

Because a computer-based system represents an efficient means for sharing and updating content among many people, it is no surprise that a large number of educational institutions use them to post course syllabi or even to administer examinations. Much of this activity, however, is Intranet-based, rather than Internet-based. That is, it is available over a closed system only to formally enrolled students. However, a small but growing number of syllabi, some in human genetics, are freely available on the web. On the Federation of American Societies for Experimental Biology's web site is a page, Genetics: Educational Information, (http://www.faseb.org/genetics/careers.htm), that links to a number of examples of human genetics syllabi, some of which include lecture notes.

Of course, it is not only as a convenient means for distributing course syllabi that the web plays a role in genetics education. The web offers a number of other educational resources pertinent to human genetics that vary in both form and content (see Table 6).

The Core Competencies in Genetics Essential for All Health-Care Professionals contained on the web site of the National Coalition for Health Professional Education in Genetics (NCHPEG) is a valuable resource for those interested in educating health professionals in genetics (see http://www.nchpeg.org/newsbox/corecompetencies000.html). This web page details areas of knowledge, skills, and attitudes regarding genetics that NCHPEG, an interdisciplinary umbrella organization that includes more than 100 member organizations from a wide array of health professions, deems important for today's health professionals. Although specific health disciplines may need to individualize this list, it serves as an excellent starting point for designing curricula in genetics, whether web-based or not. Worth noting is that NCHPEG apparently plans to add material to its site that would be of use to many health professionals interested in using the web for genetics education.

On its web site (http://www.modimes.org/Programs2/ProfEd/gyp.htm), the March of Dimes offers Genetics & Your Practice. This resource features an

TABLE 6   Genetics education for health professionals—web sites

Resource	URL	Content
Genetics: Educational Information	http://www.faseb.org/ genetics/careers.htm	Medical school courses in genetics, some with syllabi
National Coalition for Health Professional Education in Genetics	http://www.nchpeg.org/	Core competencies in genetics
Genetics and Your Practice	http://www.modimes.org/ Programs2/ProfED/gyp.htm	Electronic textbook
Clinical Genetics: A Self Study for Health Care Providers	http://www.vh.org/Providers/ Textbooks/ClinicalGenetics/ Contents.html	Electronic textbook
HuGEM II	http://www.gucdc.georgetown. edu/hugem/	Manual and other materials
Genetics Program for Nursing Faculty	http://www.gpnf.org	Links to genetics resources of particular interest to nurses
Foundation for Genetic Education and Counseling	http://www.fgec.org	Genetics and common diseases, especially psychiatric disorders
Information for Genetics Professionals	http://www.kumc.edu/ gec/geneinfo.html	Educational, clinical, and research resources

electronic textbook designed to meet the genetics education needs of a variety of health care providers.

Another electronic medical genetics textbook, Clinical Genetics: A Self Study for Health Care Providers, is offered by the Virtual Hospital® of University of Iowa Health Care on its site (http://www.vh.org/Providers/Textbooks/ClinicalGenetics/Contents.html).

The Human Genome Education Model Project II (HuGEM II) maintains a web site (http://www.gucdc.georgetown.edu/hugem/) that contains a variety of useful materials from the project, which seeks to address the genetics education need of a variety of health professionals, including dietitians, occupational therapists, psychologists, physical therapists, social workers, and speech-language pathologists.

The Genetics Program for Nursing Faculty at Children's Hospital Medical Center of Cincinnati has a web site (http://www.gpnf.org) that offers links to a number of resources that its faculty and students have found helpful.

The web site of The Foundation for Genetic Education and Counseling (http://www.fgec.org) focuses on educational materials for both the general public and health professionals about the genetics of common, complex diseases, especially psychiatric disorders.

Information for Genetic Professionals is a regularly updated site (http://www.kumc.edu/gec/geneinfo.html) at the University of Kansas Medical Center with a wide array of clinical, research, and educational resources targeted at genetic counselors, clinical geneticists, and medical geneticists, but of interest to other audiences as well.

GENETICS EDUCATION FOR THE GENERAL PUBLIC

Although the above sites are designed primarily for health professionals, several of them also offer content of value to the general public. However, a number of other sites offer material geared primarily for a lay audience (see Table 7).

The DNA Files is a site (http://www.dnafiles.org/) based on the 9-hour public radio series of the same name. The site offers audio excerpts from the series and related information.

A fairly large number of sites are targeted to students in elementary and, particularly, secondary education. Just as it offers an excellent genetics education site geared toward health professionals, the University of Kansas Medical Center also offers a genetics site for educators, the Genetics Education Center (http://www.kumc.edu/gec/), which contains an assortment of valuable information and links, including curricula, lesson plans, and classroom activities.

The Hispanic Educational Genome Project (http://caldera.calstatela.edu/hgp/Bienvenido.html) is unusual among these sites in that it provides a curriculum and related materials in both English and Spanish.

The New Genetics: A Resource for Students and Teachers (http://www4.umdnj.edu/camlbweb/teachgen.html) provides links to many human genetics education sites.

TABLE 7 Genetics education for the general public—web sites

Resource	URL	Content
Hispanic Educational Genome Project	http://caldera.calstatela.edu/ hgp/Bienvenido.html	English and Spanish versions of a high school curricualm
Genetic Science Learning Centre	http://www.genetics.utah.edu/ section5/sc5afrm.html	Basic genetics, genetic disorders, genetics in society, and several thematic units
Gene Almanac	http://vector.cshl.org/	A variety of resources, including an animated primer on genetics
The New Genetics: A Resource for Students and Teachers	http://www4.umdnj.edu/ camlbweb/teachgen.html	Links
The Human Genome Project: Exploring Our Molecular Selves	http://www.nhgri.nih.gov/ educationkit/	Video about Human Genome Project, timeline about genetics, talking glossary, classroom activities, 3-D animation of cell
MendelWeb	http://www.netspace.org/ MendelWeb/	Mendel's papers in English and German and related materials
Genetics Education Centre	http://www.kumc.edu/gec/	Material for educators
The DNA Files	http://www.dnafiles.org/	Nine-hour audio series and related information
Foundations of Classical Genetics	http://www.esp.org/ foundations/genetics/ classical	Complete versions of classic genetics works written between 350 A.D. and 1932

Gene Almanac (http://vector.cshl.org/) is the web site of Cold Spring Harbor Laboratory's DNA Learning Center. It offers a number of resources, including DNA from the Beginning, an animated primer on genetics.

The Human Genome Project: Exploring Our Molecular Selves (http://www. nhgri.nih.gov/educationkit/) is a site featuring a multimedia educational kit aimed primarily at high school students that was released by the National Human Genome Research Institute in February, 2001, to coincide with the publication of scientific papers about the draft sequence of the human genome. This site contains classroom activities, an award-winning video documentary interviewing many of the leaders of the Human Genome Project, an interactive timeline of genetics that includes many of the classic articles in human genetics, a 3-D animation of molecules and cells, a "talking" glossary of genetics terms, and access to hundreds of geneticists throughout the United States who are willing to participate as "mentors" to classes using the kit.

MendelWeb (http://www.netspace.org/MendelWeb/) features both German and English versions of Mendel's papers as well as links to Estonian and Italian translations. The site also contains a number of commentaries and other resources related to Mendel and his work.

Electronic Scholarly Publishing's Foundations of Classical Genetics (http://www.esp.org/foundations/genetics/classical/) presents several dozen seminal works in genetics that were originally published between 350 A.D. and 1932. It includes, for instance, complete versions of several of Darwin's books.

The Genetic Science Learning Center (http://www.genetics.utah.edu/section5/sc5afrm.html) has separate areas of its site designed for teachers, for students, and for families. Its content areas include basic genetics, genetic disorders, genetics in society, and several thematic units.

ETHICAL, LEGAL, AND SOCIAL ISSUES—WEB SITES

Along with their other content, some of the web sites already cited offer information about the ethical, legal, and social implications of human genetics; however, a number of other sites focus primarily, if not exclusively, on these issues (see Table 8).

GENETIC PROFESSIONAL SOCIETIES—WEB SITES

Like many other organizations, most of the major genetics societies and groups have established web sites (see Table 9). Some of these sites are fairly rudimentary and may be useful only to the groups' own members, whereas

TABLE 8   Ethical, legal, and social issues—web sites

Resource	URL
Bioethics Resources on the Web	http://www.nih.gov/sigs/bioethics/
The Communities of Color & Genetics Policy Project	http://www.sph.umich.edu/genome/webprojectsummary/index.htm
The Council for Responsible Genetics	http://www.gene-watch.org/org.html
The Gene Media Forum	http://www.genemedia.org/
Genetics & Ethics	http://www.ethics.ubc.ca/brynw/index.html
Genome Technology and Reproduction: Values and Public Policy	http://www.sph.umich.edu/genome/initialprojectsummary/index.html
HumGen	http://www.humgen.umontreal.ca/en/
National Information Resources on Ethics & Human Genetics	http://bioethics.georgetown.edu/nirehg.html
Your Genes Your Choice	http://ehrweb.aaas.org/ehr/books/index.html

TABLE 9   Genetics professional groups' web sites

Resource	URL
American Board of Genetics Counseling	http://www.faseb.org/genetics/abgc/abgcmenu.htm
American Board of Medical Genetics	http://www.faseb.org/genetics/abmg/abmgmenu.htm
American College of Medical Genetics	http://www.faseb.org/genetics/acmg/
American Society for Human Genetics	http://www.ashg.org/
Genetics Society of America	http://www.faseb.org/genetics/gsa/gsamenu.htm
International Society of Nurses in Genetics	http://nursing.creighton.edu/isong/
National Society of Genetic Counselors	http://www.nsgc.org/
Society for the Study of Inborn Errors of Metabolism	http://www.ssiem.org.uk/

others have more robust offerings that contain information of value to wider audiences.

GENETICS JOURNALS—WEB SITES

Many scientific journals maintain web sites. At least two genetics journals provide free access, via the web, to some of their material. *Nature Genetics* and its parent *Nature* maintain a joint site (http://www.nature.com/genomics/) that features original articles from both, as well as links to other sites. *Genetics in Medicine* has a site (http://www.wwilkins.com/GIM/) on which it posts its table of contents, abstracts of articles, and policy statements from its owner, the American College of Medical Genetics. Many other genetics journals offer electronic versions by subscription.

U.S. GOVERNMENT AGENCIES—WEB SITES

The principal U.S. government agencies with interest in human genetics maintain web sites that offer a wide variety of information and helpful links (see Table 10).

GENETICS NEWS—WEB SITES

Although not focused primarily on human genetics, many general news and business web sites carry frequent articles about human genetics. However, there are several web sites that provide regularly updated links that specifically track

TABLE 10   U.S. Government genetics agencies' web sites

Resource	URL
National Human Genome Research Institute	http://www.nhgri.nih.gov/
National Institutes of Health	http://www.nih.gov/
Department of Energy	http://www.ornl.gov/hgmis/medicine/assist.html
Center for Disease Control and Prevention Office of Genetics & Disease Prevention	http://www.cdc.gov/genetics/default.htm
Health Resources and Services Administration Genetics Services Branch	http://www.mchb.hrsa.gov/html/genetics.html

current news articles about human genetics (see Table 11). The sites sponsored by the American Medical Association (http://www.ama-assn.org/ama/pub/category/1799.html) and the Pharmaceutical Research & Manufacturers of America (http://genomics.phrma.org/today/index.html) supply links to such articles. The sites sponsored by Celera Genomics (http://www.celera.com/genomics/genomics.cfm) and GeneSage, Inc. (http://www.geneletter.org/index.epl) provide links to such articles but also include original articles.

GENETICS RESOURCES ON THE WEB (GROW)

The web offers the potential, already partially realized, for a multitude of voices to play important roles in telling the story of genetics. However, it is still too early to know whether these many voices will form a harmonious chorus or a

TABLE 11   Sites that track or report on what's new in genetics

Resource	URL	Content
The Gene Letter (GeneSage, Inc.)	http://www.geneletter.org/index.epl	Original articles and links
Genetics and Molecular Medicine Front Page (American Medical Association)	http://www.amaassn.org/ama/pub/category/1799.html	Links to current articles
Genome News Network (Celera Genomics)	http://www.celera.com/genomics/genomics.cfm	Original articles and links
Genomics Today (Pharmaceutical Research & Manufacturers of America)	http://genomics.phrma.org/today/index.html	Links to current articles

cacophonous caterwaul. The availability of numerous sources of web-based content about genetics geared toward various audiences offers the benefits that come from wide and easy availability of multiple information resources. On the other hand, the large number of such sites may confuse, rather than empower, users and may provide information that varies greatly, and not always obviously, in accuracy and quality.

In an effort to further coordination among the multiple genetics sites, their users, and their content writers, in August of 1999, the National Human Genome Research Institute of the National Institutes of Health, with financial collaboration from the Office of Rare Diseases at the National Institutes of Health, convened a meeting of over two dozen organizations with particular interest in use of the web to inform health professionals and the public about human genetics. One of the products of this initial meeting was the formation of a group, Genetics Resources on the Web (GROW), whose mission is "to optimize utilization of the web to provide health professionals and the public with high quality information about human genetics, especially those aspects of human genetics dealing with health." (6)

As of early 2001, GROW was still finalizing its membership criteria, programs, and standards. Approximately three dozen organizations have participated thus far in its activities. Although these organizations have in common a deep interest in the provision of human genetics information on the web, they vary widely in their nature. Participating organizations include a range of health professional groups, patient support groups, federal agencies interested in human genetics, and foundations and other nonprofits, as well as for-profit, entities.

GROW has had two more conferences, in February and November 2000, and plans to continue to meet once to twice each year. Among its priorities is creation of a search engine that would search simultaneously across all its member organizations' web sites. GROW hopes to make this search engine available publicly sometime in 2001 or 2002. Another priority for GROW is to help organizations identify and adopt quality practices for their web sites. GROW hopes that in doing so it can help assure that users will have a large number of human genetics related web sites that they can trust to offer accurate, timely, and unbiased information. Among other GROW projects is identifying unmet needs in terms of genetics information on the web and to encourage organizations to address such needs effectively.

THE FUTURE

In few, if any, areas of science is knowledge currently expanding as rapidly as it is in human genetics. At the same time, the web and related communication media are among the most rapidly evolving areas of technology. Because of the rapidly changing nature of each, it is difficult to foresee precisely what the intersection of the two, human genetics on the web, will look like, even in the relatively near future. However, one can discern certain opportunities and challenges ahead.

Increasingly sophisticated computer-based databases will continue to be vital tools in human genetics and genomics research. Indeed, bioinformatics should be one of the most important enablers of genetics research in the next few decades. If researchers at great distance are to create, access, and utilize such databases effectively, they will need to do so over the web or newer Internet-related communication networks. As databases get increasingly large and complex, their effective use will require that the web or its successors offer increasingly more rapid and robust transmission of data. That the technology of Internet communication will continue to improve is certain; however, whether this improvement will be sufficiently swift and far-reaching so as not to constrain genomics research and applications is less definite.

The web, abetted by its Internal sibling, e-mail, has arguably helped change the culture of clinical medicine. The web has empowered patients by making a wider range of information more easily available. Along with other recent societal trends that have also made patients more able and eager to take a leading role in their own healthcare, this effect of the web should continue to have a significant impact on many aspects of medicine in years to come. However, it may have particular importance in genomic medicine, in which the old healthcare paradigm of the doctor treating disease in a passive, uninformed patient gives way to an acitve, informed patient implementing a plan that they have helped create to prevent disease. Much of the power of genomic medicine in the coming decades will rest on the ability of patients, working in concert with their healthcare providers, to utilize knowledge of personal genetic predispositions to create and adopt presymptomatic lifestyles, dietary, screening and pharmaceutical interventions to maintian health (3). For patients to understand and act optimally on such presymptomatic knowledge, they will need to be optimally informed and educated about genetics. The web and related technologies should play a key role in providing such information and education.

CONCLUSION

We have only seen the beginning of the World Wide Web's impact on how we create, access, and share information. The web and related technologies are still evolving, and in ways that are hard to predict. What they do today will seem outdated tomorrow. Similarly, our knowledge of genetics continues to evolve. What it tells us today about how we remain healthy, how we become ill, and who we are will seem outdated tommorrow. The web and related technologies have already begun to play an important role in both human genetics research and clinical genetics. Even if it is not clear precisely how these roles will develop in the years to come, that they will become increasingly important is clear.

Visit the Annual Reviews home page at www.AnnualReviews.org

LITERATURE CITED

1. Brodie M, Flournoy RE, Altman DE, Blendon RJ, Benson JM, et al. 2000. Health information, the internet and the digital divide. *Health Aff.* 19:225–65
2. Burstin H. 2000. Traversing the digital divide. *Health Aff.* 19:245–49
3. Collins F. 1999. Shattuck lecture—medical and societal consequences of the human genome project. *N. Engl. J. Med.* 341:28–37
4. Fox S, Rainien L, Horrigan J, Spooner T, Burke M, et al. 2000. *The Online Health Care Revolution: How the Web Helps Americans Take Better Care of Themselves.* Washington, DC: Pew Internet & Am. Life Proj. 3 pp.
5. Gray J, Brain K, Iredale R, Alderman J, France E, et al. 2000. A pilot study of telegenetics. *J. Telemed. Telecare.* 6:245–47
6. Guttmacher AE, Collins FS. 2000. Genetics resources on the web (GROW). *Genet. Med.* 2:296–99

Annu. Rev. Genomics Hum. Genet. 2001. 2:235–58

METHODS FOR GENOTYPING SINGLE NUCLEOTIDE POLYMORPHISMS

Pui-Yan Kwok
Washington University School of Medicine, St. Louis, Missouri 63110;
e-mail: kwok@genetics.wustl.edu

Key Words   genetic markers, genetic mapping, large-scale genetic studies

■ **Abstract**   One of the fruits of the Human Genome Project is the discovery of millions of DNA sequence variants in the human genome. The majority of these variants are single nucleotide polymorphisms (SNPs). A dense set of SNP markers opens up the possibility of studying the genetic basis of complex diseases by population approaches. In all study designs, a large number of individuals must be genotyped with a large number of markers. In this review, the current status of SNP genotyping is discussed in terms of the mechanisms of allelic discrimination, the reaction formats, and the detection modalities. A number of genotyping methods currently in use are described to illustrate the approaches being taken. Although no single genotyping method is ideally suited for all applications, a number of good genotyping methods are available to meet the needs of many study designs. The challenges for SNP genotyping in the near future include increasing the speed of assay development, reducing the cost of the assays, and performing multiple assays in parallel. Judging from the accelerated pace of new method development, it is hopeful that an ideal SNP genotyping method will be developed soon.

INTRODUCTION

As of the end of 2000, over 1.5 million single nucleotide polymorphisms (SNPs) have been found in the human genome and deposited to public databases (20a, 37). The availability of an ultra-high density SNP map opens the possibility of studying by association genetic factors important in complex genetic traits in the human, taking advantage of the fact that genetic markers in close proximity to mutant genes may be in linkage disequilibrium (LD) to them (21, 42). Association studies can be done with a genome-wide approach (without assuming one region of the genome is more likely to harbor the associated genetic factor) or with a candidate gene approach (using some biological knowledge to prioritize the parts of the genome for the study) (11). Because the strength of LD diminishes rapidly with distance, hundreds to thousands of individuals must be genotyped, regardless of the approach used, with a large number of SNPs (25). Genotyping of SNPs will likely be a major part of every genetic

1527-8204/01/0728-0235$14.00

association study, and the appropriate genotyping method is critical to the success of the study.

The ideal genotyping method must possess the following attributes: (a) The assay must be easily and quickly developed from sequence information; (b) the cost of assay development must be low in terms of marker-specific reagents and time spent by expert personnel on optimization; (c) the reaction must be robust, such that even suboptimal DNA samples will yield reliable results; (d) the assay must be easily automated and must require minimal hands-on operation; (e) the data analysis must be simple, with automated, accurate genotype calling; (f) the reaction format must be flexible and scalable, capable of performing a few hundred to a million assays per day; and (g) once optimized, the total assay cost per genotype (including equipment, reagents, and personnel) must be low.

To date, no such ideal genotyping method exists. Further improvements in biochemistry, engineering, and analytical software must occur before SNP genotyping methods closer to the ideal can be developed. In this review, the three aspects of a genotyping assay—allelic discrimination, assay format, and detection methodology—are discussed. Where appropriate, genotyping methods currently in use are described in the context of these three areas.

ALLELIC DISCRIMINATION

SNPs can be detected in either a sequence-specific or sequence-nonspecific way. Sequence-nonspecific detection is based on the capture, cleavage, or mobility change during electrophoresis or liquid chromatography of mismatched heteroduplexes formed between allelic DNA molecules or single-stranded DNA molecules that assume slightly different conformations under nondenaturing conditions (28). Although sequence-nonspecific detection of polymorphisms is the mainstay in polymorphism/mutation discovery, it is not an acceptable approach to genotyping because one is never certain if the inferred genotyping is the true genotype.

Sequence-specific detection relies on four general mechanisms for allelic discrimination: allele-specific hybridization, allele-specific nucleotide incorporation, allele-specific oligonucleotide ligation, and allele-specific invasive cleavage (27). All four mechanisms are reliable, but each has its pros and cons.

Hybridization

With the hybridization approach, two allele-specific probes are designed to hybridize to the target sequence only when they match perfectly (Figure 1). Under optimized assay conditions, the one-base mismatch sufficiently destabilizes the hybridization to prevent the allelic probe from annealing to the target sequence. Because no enzymes are involved in allelic discrimination, hybridization is the simplest mechanism for genotyping. The challenge to ensure robust allelic discrimination lies in the design of the probe. With ever more sophisticated probe

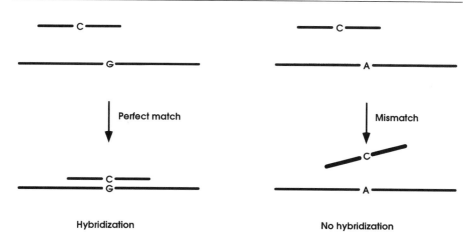

Figure 1   Allele-specific hybridization.

design algorithms and the use of hybridization enhancing moieties, such as DNA minor groove binders, allele-specific probes can be designed with high success rate.

When the allele-specific probes are immobilized on a solid support, labeled target DNA samples are captured, and the hybridization event is visualized by detecting the label after the unbound targets are washed away. Knowing the location of the probe sequences on the solid support allows one to infer the genotype of the target DNA sample. Allele-specific hybridization is also the basis of several elegant homogeneous genotyping assays. These assays differ in the way they report the hybridization event. In the 5′ nuclease assay, a probe annealed to target DNA that is amplified is cleaved during the polymerase chain reaction (PCR) (32). Thus, monitoring the cleavage event is a way to determine whether hybridization has occurred. With molecular beacon detection, hybridization to target DNA opens the stem-loop structure (24, 49). Determining the open-closed status of the stem-loop structure is, therefore, a way to figure out if hybridization has occurred. With "light-up" probes, the thiazole orange derivative linked to a peptide nucleic acid (PNA) oligomer fluoresces only when the PNA oligomer hybridizes specifically to complementary nucleic acids (46). Fluorescence is, therefore, evidence of hybridization.

Primer Extension

Primer extension is a very robust allelic discrimination mechanism. It is highly flexible and requires the smallest number of primers/probes. Probe design and optimization of the assay are usually very straightforward. There are numerous variations in the primer extension approach that are based on the ability of DNA polymerase to incorporate specific deoxyribonucleosides complementary to the sequence of the template DNA. However, they can be grouped into two categories.

First is a sequencing (allele-specific nucleotide incorporation) approach where the identity of the polymorphic base in the target DNA is determined. Second is an allele-specific PCR approach where the DNA polymerase is used to amplify the target DNA only if the PCR primers are perfectly complementary to the target DNA sequence.

In the sequencing approach, one can either determine the sequence of amplified target DNA directly by mass spectrometry (29) or perform primer extension reactions with amplified target DNA as a template and analyze the products to determine the identity of the base(s) incorporated at the polymorphic site (allele-specific nucleotide incorporation; see Figure 2). A number of ingenious ways have been devised for primer extension product analysis in homogeneous assays. Most of these approaches combine novel nucleic acid analogs and monitoring of interesting differences in physical properties between starting reagents and primer extension products.

In the allele-specific PCR approach, one relies on the DNA polymerase to extend a primer only when its 3′ end is perfectly complementary to the template (Figure 3). When this condition is met, a PCR product is produced. By determining whether a PCR product is produced or not, one can infer the allele found on the target DNA. Several innovative approaches have been utilized to detect the formation of specific PCR products in homogeneous assays. Some are based on melting curve analysis, and some are based on hybridization of target specific probes. A variation of this approach is the allele-specific primer extension. Here, the PCR product containing the polymorphic site serves as template, and the 3′ end of the primer extension probe consists of the allelic base. The primer is extended only if the 3′ base complements the allele present in the target DNA. Monitoring the primer extension event, therefore, allows one to infer the allele(s) found in the DNA sample.

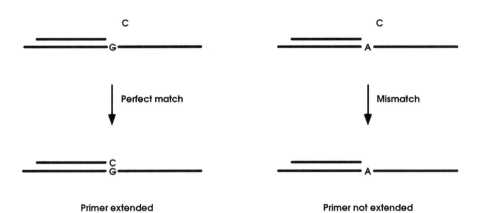

Figure 2   Allele-specific nucleotide incorporation.

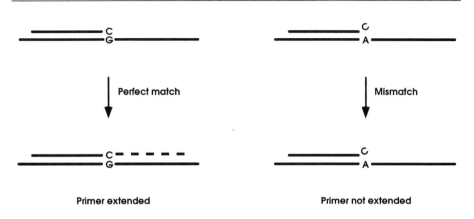

Figure 3   Allele-specific primer extension.

Ligation

DNA ligase is highly specific in repairing nicks in the DNA molecule. When two adjacent oligonucleotides are annealed to a DNA template, they are ligated together only if the oligonucleotides perfectly match the template at the junction (Figure 4). Allele-specific oligonucleotides can, therefore, interrogate the nature of the base at the polymorphic site. One can infer the allele(s) present in the target DNA by determining whether ligation has occurred. Although ligation has the highest level of specificity and is easiest to optimize among all allelic discrimination mechanisms, it is the slowest reaction and requires the largest number of modified probes. However, ligation as a mechanism has the potential of genotyping without prior target amplification by PCR. This can be accomplished either by the ligation

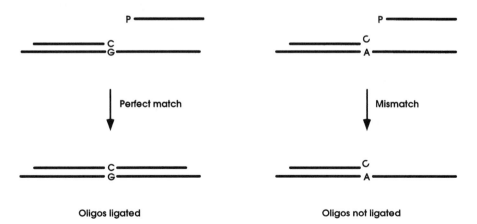

Figure 4   Allele-specific oligonucleotide ligation.

chain reaction (LCR) (3) or by the use of ligation (padlock) probes that are first circularized by DNA ligase followed by rolling circle signal amplification (2, 33).

Invasive Cleavage

Structure-specific enzymes cleave a complex formed by the hybridization of overlapping oligonucleotide probes. When probes are designed such that the polymorphic site is at the point of overlap, the correct overlapping structure is formed only with the allele-specific probe but not with the probe with a one-base mismatch. Elevated temperature and an excess of the allele-specific probe enable the cleavage of multiple probes for each target sequence present in an isothermal reaction. In an innovative application of this method, the cleaved allele-specific probes are used in a second reaction where a labeled secondary probe is cleaved (Figure 5). This signal amplification step helps boost the amount of labeled cleavage product produced to 10^6–10^7 per target sequence per hour, an amount sufficient for detection without the need for a target amplification process such as PCR (17).

The major advantages of this approach are the isothermal nature of the reaction and the potential for genotyping without PCR amplification. There are a number of technical issues that need further refinement. First, the amount of genomic DNA needed in the reaction is high. Second, the purity of the marker specific probes must be extremely high or nonspecific reactions become a nuisance. Third, probe design is somewhat tricky because the sequential reactions have to work under the

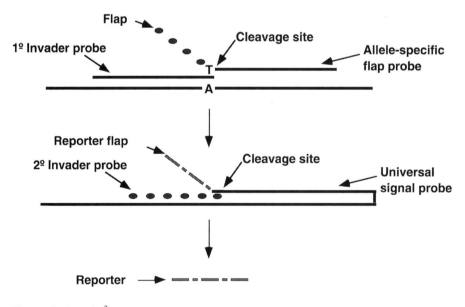

Figure 5   Invader^2 assay.

same conditions, but the sequence context of the SNPs is fixed. With improvements in probe design algorithms and further development, these technical concerns will likely be overcome.

REACTION FORMATS

Starting with genomic DNA, each genotyping method undergoes a series of bio-chemical steps and a product detection step. The reaction format mostly reflects the requirements of the detection modality. In general, biochemical reactions are more robust in solution, but capturing the reaction products on solid support allows for detection in parallel and increases the throughput substantially.

Homogeneous Reactions

A number of innovative genotyping methods are done in solution from beginning to end and are therefore designated as homogeneous reactions. Some of them require no further manipulations once the reaction is set up initially. Others call for a number of reagent addition steps, but no separation or purification steps are needed. Homogeneous assays are usually robust, highly flexible, and not labor intensive. The major drawback is the limited amount of multiplexing one can do with homogeneous assays.

Solid Phase Reactions

Solid supports used in genotyping can be a latex bead, a glass slide, a silicon chip, or just the walls of a microtiter well. In some cases, marker specific oligonucleotides are placed on the solid support, and the allelic discrimination reaction is done on the support (Figure 6); in other cases, generic oligonucleotides are placed on the solid support, and they are used to capture complementary sequence tags conjugated to marker specific probes (Figure 7). In the former strategy, the oligonucleotide arrays act as a collection of reactors where the target DNA molecules find their counterparts, and the allelic discrimination step for numerous markers proceeds in parallel. In the latter, the arrayed oligonucleotides are used to sort the products of the allelic discrimination reactions (also done in parallel) performed in homogeneous solution. In both cases, the identity of an oligonucleotide on a latex bead or at a particular location on the microarray (on a glass slide or silicon chip) is known, and the genotypes are inferred by determining which immobilized oligonucleotide is associated with a positive signal. The major advantage of performing genotyping reactions on solid supports is that many markers can be interrogated at the same time. Besides saving time and reagents, performing numerous reactions in parallel also decreases the probability of sample/result mix-ups. The drawback of performing genotyping reactions on solid support is that design of the arrays and optimization of the multiplex reactions requires substantial capital and time

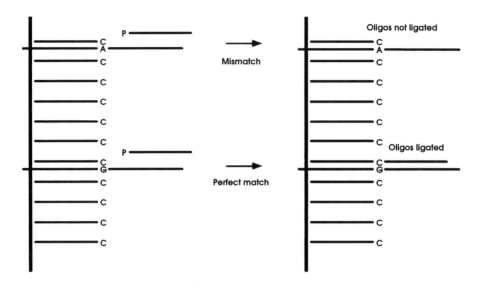

Figure 6   Ligation on solid support.

investment. With better algorithms for multiplex PCR design, this limitation may be alleviated in the near future.

DETECTION MECHANISMS

Detection of a positive allelic discrimination reaction is done by monitoring the light emitted by the products, measuring the mass of the products, or detecting a change in the electrical property when the products are formed. Numerous labels with various light-emitting properties have been synthesized and utilized in detection methods based on light detection or electrical detection. In general, only one label with ordinary properties is needed in genotyping methods where the products are separated or purified from the excess starting reagents. For homogeneous reactions, where no separation or purification is needed, the property of the label has to be changed when a product is formed. This usually requires interaction of the label with another component of the reaction when a product is formed. A number of elegant genotyping methods have been developed to take advantage of certain physical characteristics of the labels.

Monitoring light emission is the most widely used detection modality in genotyping, and there are many ways to do so. Luminescence, fluorescence, time-resolved fluorescence, fluorescence resonance energy transfer (FRET), and fluorescence polarization (FP) are useful properties of light utilized in a host of genotyping methods.

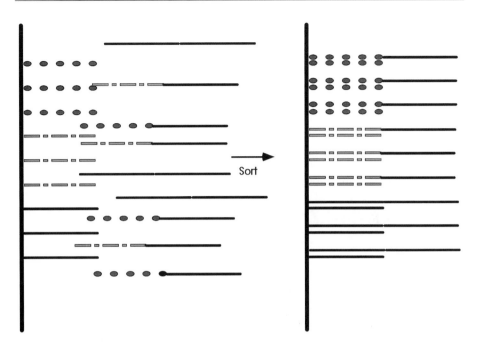

Figure 7 Multiplex homogeneous assay with assortment on solid support.

Luminescence Detection

Luminescence is emitted in an ATP-dependent luciferase reaction. When ATP production is coupled with a primer extension reaction, luminescence is observed every time a deoxyribonucleoside is added in the primer extension reaction. Because the background is extremely low, luminescence has a very good signal to noise ratio. However, the additional enzymatic steps and substrates required complicate the experimental procedure and increase the cost of the assay.

Fluorescence Detection

Fluorescence detection is straightforward and easy to implement. Besides using it to capture fluorescent labels on a solid support or separate the fluorescent product from the label by gel or capillary electrophoresis, fluorescence detection can be used to monitor the formation of double-stranded DNA with a DNA intercalating dye that only fluoresces in the presence of double-stranded DNA. Direct fluorescence detection is very versatile and can be done in multiplex to a certain extent. However, the need for product purification or separation when fluorescent labels are used and the interference by nonspecific double-stranded DNA species when intercalating dyes are used are some of the drawbacks of direct fluorescence detection.

Time-Resolved Fluorescence Detection

Time-resolved fluorescence as a detection approach is feasible when the emission half-life of the fluorescent dye is long. With this class of dyes (mostly compounds of rare earth elements, such as Lanthanides), the fluorescence reading is done sufficiently long after excitation, such that autofluorescence (which has a very short half-life) is not observed (18, 23). The background in time-resolved fluorescence detection is almost nonexistent; thus, this is a very sensitive detection modality. The drawback is that the lanthanides are inorganic compounds that cannot be used to label nucleic acids directly. An organic chelator conjugated to the probe must be used to bind the lanthanides in the reaction.

Fluorescence Resonance Energy Transfer

Fluorescence resonance energy transfer is a popular detection method in homogeneous genotyping assays. FRET occurs when two conditions are met. First, the emission spectrum of the fluorescent donor dye must overlap with the excitation wavelength of the acceptor dye. Second, the two dyes must be in close proximity to each other because energy transfer drops off quickly with distance. The proximity requirement is what makes FRET a good detection method for a number of allelic discrimination mechanisms. Basically, any reaction that brings together or separates two dyes can use FRET as its detection method. FRET detection has, therefore, been used in primer extension and ligation reactions where the two labels are brought into close proximity to each other. It has also been used in the 5' nuclease reaction, the molecular beacon reaction, and the invasive cleavage reactions where the neighboring donor/acceptor pair is separated by cleavage or disruption of the stem-loop structure that holds them together (17, 24, 32, 49). The major drawback of this method is the cost of the labeled probes required in all the genotyping approaches with FRET detection. In the cleavage approaches, the probes are doubly labeled, further increasing the cost of probe synthesis.

Fluorescence Polarization

Fluorescence polarization (FP) has been used in clinical diagnosis and numerous binding assays for years, but its use as a detection method for SNP genotyping has a very short history. This is because instruments sensitive enough for detecting small amounts of dyes were not available until recently. When a dye is excited by plane polarized light, the emitted fluorescence is also polarized. The degree of polarization is determined by the temperature, the viscosity of the solvent, and the molecular volume of the fluorescent molecule. All these factors affect molecular motion, and in general, the faster a molecule tumbles and rotates in solution, the less polarized is its fluorescence. Because molecular volume is proportional to molecular weight, fluorescence polarization is, therefore, a good method to detect changes in molecular weight. In principle, any genotyping method in which the product of the allelic discrimination reaction is substantially larger or smaller than the starting fluorescent molecule can use FP as a detection method. Indeed, FP is used as the detection method in the primer extension reaction where small fluorescent dye terminators are incorporated into a larger probe (8). Furthermore, FP is a good detection method in the 5' nuclease reaction where small fluorescent molecules are formed when large fluorescent probes are cleaved in the reaction (30). Our group also found that FP can be used as a detection method for the invasive cleavage reaction where the large fluorescent signal probe is cleaved, producing a small fluorescent tag (T.M. Hsu, S. Law, S. Duan, B. Neri, & P-Y. Kwok, unpublished results). The advantages of the FP detection method include the much smaller amount of fluorescent dyes needed compared to FRET or direct fluorescence detection methods, the cheaper probes used, and the potential for utilizing the full visible spectrum in multiplex reactions. The drawback is mainly that any nonspecific products will increase the noise in the signal.

Mass Spectrometry

Unlike all other detection methods that infer the identity of the products generated in the allelic discrimination reaction by monitoring the fate of some label, mass spectrometry (MS) measures the molecular weight of the products formed and is, therefore, the most direct method of detection. Because MS determines the fundamental property of the DNA molecule, no labels are needed. High resolution MS can easily distinguish between DNA molecules that differ by only one base (4, 5, 31, 44). A further advantage of MS is that it takes only milliseconds to analyze each sample, so even though MS analyzes each sample serially, the throughput is still very high. Furthermore, by appropriately designing the probes, moderate multiplexing is possible (44). The main disadvantage of the MS detection method is the exquisite purity required of the analyte for it to work. With further refinement of the product purification process, it may be possible to overcome this drawback.

Electrical Detection

A promising detection method is one that monitors a change in the electrical properties of the products of the allelic discrimination reaction. Currently, this is done on solid support where oligonucleotides are deposited on electrodes (12, 51). The electrical property of the probe is altered when the DNA complementary to the probe is annealed to it. This is exaggerated if a ferromagnetic label is used. Electrical detection combines semiconductor technology with biochemistry and eliminates the need for light detection or extensive product processing. This area is still in its infancy, and there are still a number of biochemical and engineering obstacles to overcome before the throughput of genotyping methods based on this detection mechanism is high enough and the cost low enough for its wide acceptance.

EXAMPLES OF GENOTYPING METHODS

A number of SNP genotyping methods are discussed in some detail to highlight how the allelic discrimination mechanisms, reaction formats, and detection modalities can be combined in various ways to produce the many SNP genotyping approaches in use today. Because of space limitations, the list of genotyping methods is not exhaustive. Interested readers are encouraged to examine the primary references for in-depth description of the methods.

Microarray Genotyping: Hybridization on Solid Support with Fluorescence Detection

The first large-scale genotyping method was developed jointly by the Whitehead Institute and Affymetrix, Inc. (50). The GeneChip HuSNP Mapping Array contains 1494 SNPs that can be genotyped in one experiment (35). The major breakthrough of this approach is the degree of multiplexing achieved by designing PCR assays that amplify very small products and a second round of PCR with a common set of primers. In the current version of the assay, the entire set of 1494 SNPs is amplified in just 24 multiplex reactions (average of 62 SNPs in one multiplex reaction). In a rather long protocol, the PCR products are pooled, concentrated, hybridized to the DNA microarray, stained, and visualized. Because each SNP is interrogated by a set of "tiling" oligonucleotides, the genotypes called are quite accurate. Other advantages of this approach include the low requirement for starting genomic DNA (120 ng total for all 1494 markers), the large number of SNPs that are typed in one experiment, and the minimal manual steps. The major drawback of this approach is that because the design and manufacture of the microarray are quite expensive, the set of markers selected cannot be changed quickly or arbitrarily. Furthermore, it is a common experience that about 20% of the SNPs on the HuSNP chip do not yield confident results (J. Fan, U. Surti, T. Taillon-Miller, G. Kennedy, T. Ryder, & P-Y. Kwok, unpublished results). This level of failure rate is too high for many applications.

Another assay based on hybridization on solid support is the DASH (dynamic allele-specific hybridization) assay (19). Instead of monitoring the hybridization at a constant temperature, this approach looks for the melting temperature differences between the allele specific probe when it is annealed to the matched and mismatched targets by monitoring the hybridization over a range of temperatures. Using a DNA intercalating dye, such as Sybr Green I, that fluoresces in the presence of double-stranded DNA, fluorescence is observed only when hybridization occurs and double-stranded DNA species is formed. With recent improvements in probe design, DASH assays can now be designed for nearly 100% of SNPs that can be uniquely amplified by PCR (41). The specificity and robustness of the assay come with a price. Because double-stranded DNA species will cause the dye to fluoresce, the PCR products have to be rendered single-stranded. This is accomplished by utilizing a biotinylated PCR primer and capturing the biotinylated

PCR product on solid support, followed by denaturing and washing away the unlabeled strand of PCR product. Furthermore, the single reporter used means that two reactions must be run in parallel for each SNP.

Molecular Beacon Genotyping: Homogeneous Hybridization with FRET Detection

Molecular beacons are stem-loop structures that hold a fluorescent reporter in close association with a universal quencher such that fluorescence is only observed when the stem-loop structure opens up (24). With proper design, a DNA target that is perfectly complementary to the sequence of the loop portion of the molecular beacon hybridizes to the molecular beacon and forces open the stem, leading to the emergence of fluorescence. The molecular beacon with a one base mismatch will not hybridize to the target strongly enough to disrupt the stem-loop structure, and no fluorescence is observed (49). Because the presence or absence of fluorescence reflects the open or close status of the stem-loop structure, no purification or separation steps are needed. In fact, once the assay is set up, no more manual manipulations are needed. As long as one can monitor the fluorescence, one can infer the genotype of the DNA target. This "closed-tube" system has real advantages because cross contamination is minimized and automation is easily achievable. An added advantage is that when real-time fluorescence monitoring is possible, the assay can be used to quantify the amount of DNA present in an unknown sample. With many fluorescent dyes available in the visible spectrum, multiplex analysis is possible to some extent. The one drawback of this approach is the cost of the two doubly labeled molecular beacons needed for each SNP marker. Until design algorithms are perfected, a fraction of molecular beacons will not work without optimization.

5' Nuclease Assay: Homogeneous Hybridization with FRET or FP Detection

Taq DNA polymerase possesses a 5' nuclease activity that displaces and cleaves oligonucleotides hybridized to a DNA segment undergoing replication. Based on this observation, the TaqMan assay was developed with a doubly labeled probe consisting of a reporter fluorescent dye and a minor groove binder (MGB)/universal quencher complex (32). During the extension phase of PCR, the TaqMan probes hybridize only to the perfectly matching DNA target and not to those with a one-base mismatch. Cleavage of the hybridized probe separates the quencher from the reporter, and fluorescence is observed. One can, therefore, infer the genotype of a test sample by monitoring the fluorescence of the reaction mixture. It was shown recently that fluorescence polarization is a good detection method for this assay (30). Because the starting probe has a much higher molecular weight than the cleavage products, the fluorescence polarization changes drastically in a positive reaction. Just like the molecular beacon assay, the TaqMan assay is a closed-tube system and can be used for quantification of unknown DNA samples. This assay has been in use and thoroughly tested over several years now. With a number of improvements in assay design, such as the incorporation of the minor groove binder that enhances the discriminating power between the TaqMan probes and more reliable primer design algorithms, the assay is easier to optimize. The cost of the labeled probes is the main obstacle to the widespread adoption of this method by the average laboratory.

Allele-Specific PCR: Homogeneous Primer Extension with Fluorescence or FRET Detection

Allele-specific PCR has been used for quite some time to genotype SNPs. It is a relatively simple technique, and when coupled with gel analysis, the genotypes can be called easily. Because the products are of the same size, two parallel reactions must be performed for each marker when gel electrophoresis is used as the detection method. Because there is no way to control for false-negative results, allele-specific PCR is not used in large-scale projects. Everything changed when three groups devised novel ways to detect PCR products in homogeneous solution.

Germer & Higuchi took advantage of several recent advances to achieve single-tube genotyping by allele-specific PCR (15). First, a DNA intercalating dye was used to detect the presence of double-stranded DNA. Second, real-time fluorescence detection was used to determine the melting curve of a PCR product. Third, a GC-rich sequence was added to one of the allele-specific PCR primers to increase the melting temperature of one of the PCR products. Fourth, the stoffel fragment of Taq polymerase with two important attributes for allele-specific PCR was used in this method. Under assay conditions, the stoffel fragment of Taq polymerase only extends the primer where the 3′ end matches the target sequence and only yields amplicons <100 bps. Taken together, these four advances allow for highly specific PCR amplification, with the resulting products easily distinguishable by melting curve analysis. Because both alleles can be assayed in the same reaction, this is another closed-tube method where one only has to set up the reaction and the instrument will take care of the rest. The advantage of this method is the low cost of the unlabeled primers and the simplicity of the assay. The only drawback is that not all SNPs can be assayed by this method because the amplicons must be small. Under the best circumstances, kinetic PCR cannot be designed for about 20% of SNPs.

Todd et al. developed a method based on the ability of a DNA enzyme that can cleave an RNA-containing reporter probe (47). Specifically, the antisense sequence of a 10–23 DNAzyme is added to one of the PCR primers for the assay such that the active DNAzyme is formed only if PCR amplification occurs. A DNA/RNA chimeric reporter substrate containing fluorescent and quencher dye molecules on opposite sides of the cleavage site is added to the reaction mixture and is cleaved as the DNAzyme forms during PCR amplification. The accumulation of PCR products is monitored in real time via changes in the fluorescence that is released by the separation of fluoro/quencher dye molecules as the newly formed DNAzyme cleaves the reporter substrate. The DzyNA-PCR DNA detection is novel and attractive because the only specialty reagent, the energy transfer DNA/RNA hybrid reporter substrate of the DNAzyme, can be used in any assay. The only target specific reagents are the two PCR primers, with one modified with the antisense sequence of the DNAzyme. Because this assay can be monitored in real time, DNA quantification is possible. At this point, the limitations of this assay are that only one reporter substrate is used and SNP genotyping has to be done in parallel reactions.

Myakishev et al. recently described a similar method in which the two allele-specific PCR primers are tailed with sequences that introduce priming sites for universal energy-transfer-labeled primers (36). The energy-transfer-labeled primer contains a reporter dye that is quenched by a universal quencher in its natural stem-loop structure. When allele-specific PCR products are formed, the priming site is formed and the energy-transfer-labeled primer is extended. The extension product serves as a template for the next round of PCR, and the stem-loop structure is opened up as PCR proceeds, thereby releasing the reporter dye from the quencher and fluorescence is observed. Like kinetic PCR, only one reaction is required for each SNP and it is done in a closed-tube format. With one set of universal energy-transfer-labeled primers that can be used for any assay, this approach is quite cost-effective.

All three allele-specific PCR methods have nice features and are relatively inexpensive to develop. However, allele-specific PCR cannot be designed for every SNP because of local sequence constraints and because of the difficulty of performing the reaction in multiplex.

Allele-Specific Primer Extension: Primer Extension on Solid Support with Fluorescence Detection

Instead of performing allele-specific PCR, one can utilize allele-specific primers to detect the presence or absence of a SNP within a PCR product. Because the PCR step is separated from the allele detection step, this method is more versatile and has two levels of specificity. It is, therefore, very robust, and assays can be designed for almost all SNPs. In a recently described rendition of this method, some clever modifications were made to simplify the reaction procedure (40). By attaching a T7 RNA polymerase promoter sequence to one of the PCR primers, researchers generated RNA templates from the PCR products. The RNA templates then serve as the target in an allele-specific primer extension reaction mediated by a reverse transcriptase. Dye-labeled rNTPs are used in the reaction, and they are incorporated when the immobilized allele-specific primer's 3′ base matches the allele found on the RNA template. Because of the large number of RNA templates produced by the T7 RNA polymerase and the multiple dye-labeled rNTPs incorporated, very small amounts of PCR products are needed. This approach solved some of the problems that plagued solid phase primer extension reaction in the past, namely, the need for PCR product purification and generation of single-stranded DNA template for robust reaction. The good attributes of this approach include simple reaction procedure, small reaction volume, and low requirements for genomic DNA templates. The reaction mixture is complex, however, with multiple enzymes and specialty rNTPs.

Arrayed Primer Extension: Primer Extension on Solid Support with Fluorescence Detection

Unlike allele-specific PCR, which assays for the presence or absence of a PCR product, the "generic" primer extension approach assays for the specific nucleotide that is incorporated onto the primer at the polymorphic site. In the arrayed primer extension (APEX) approach (13, 39, 45, 48), oligonucleotides that sequence correspond to the those neighboring the polymorphic sites of their corresponding SNPs and are immobilized via their 5' end on glass surface. These probes are extended by one base in the presence of PCR products containing the SNP sequences. With each of the four dideoxyterminators labeled with different fluorescent dyes, the identity of the incorporated base can be inferred easily. Using fluorescence imaging techniques, the genotypes can be determined simply by noting the colors found on the various spots on the array. The APEX assay is quite robust and can be multiplexed. Furthermore, a universal master mix containing the four dye-labeled terminators and DNA polymerase is used for all SNPs, making it a very simple reaction to set up. The challenge is that thermal cycling is generally not easily achieved in solid phase reactions, so single-stranded templates are needed for robust primer extension. This requires a larger amount of PCR products as target and a strand separation step that increases the cost of the reaction. In addition, placing SNP-specific probes on the solid support decreases the flexibility of the approach.

Homogeneous Primer Extension Assays with FRET or FP Detection

These approaches take advantage of the fact that dye-labeled terminators are incorporated covalently onto an oligonucleotide as the reaction proceeds. If a donor dye is found on the probe, excitation of the donor dye will cause the acceptor dye attached to the incorporated terminator to fluoresce (7). Observation of FRET is, therefore, an indication that primer extension has occurred. With two different acceptor dye-labeled terminators, the genotype of a sample can be determined in one reaction (10). When a thermostable DNA polymerase is used and the fluorescence monitored in real-time, the assay is very sensitive and robust. However, the dye-labeled probe is relatively costly.

The primer extension reaction greatly increases the molecular weight of the dye-labeled terminator when it is incorporated onto the oligonucleotide probe. This change is reflected in the observed FP value. When the reaction is driven to completion, the FP value is maximally increased and the genotypes can be determined easily (8). Because the probe is unlabeled and is used only to add molecular weight to the primer extension product, the start-up cost of the assay is among the cheapest of all genotyping assays in use to date. With highly sensitive instruments readily available for FP detection, the throughput of this assay can be very high.

The major drawback of the homogeneous primer extension assays is the need to degrade the excess PCR primers and dNTPs after the PCR step. This is necessary because the primers and dNTPs will interfere with the primer extension reaction. In the current reaction protocol, the excess primers and dNTPs are degraded enzymatically with exonuclease I and shrimp alkaline phosphatase. After a short incubation, the enzymes are heat inactivated before the primer extension reaction mix is added, and the reaction is allowed to proceed. Although the possibility for multiplex reaction is quite limited, the versatility and simplicity of the assay, with minimal requirements for optimization, make the assay an attractive choice for many applications.

Primer Extension with Detection by Mass Spectrometry

Mass spectrometry (MS), a method well suited for detection of small DNA molecules, has been used as a detection method for a number of primer extension genotyping assays. It is used for both the generic primer extension reaction and the allele-specific primer extension reactions.

To date, the most successful applications of MS detection in the primer extension reaction are found in the biotechnology industry. For example, Ross et al. utilize matrix-assisted laser desorption/ionization time-of-flight mass spectrometry (MALDI-TOF MS) to detect primer extension products in multiplex (44). Because the mass resolution is high, the few mass units that differentiate the primer extension products of the two alleles can be distinguished handily. Careful design of the primers used in the primer extension reaction ascertains that all the extended and unextended primers are in well-resolved mass windows. As a result, researchers can genotype 12 SNP markers in one multiplex PCR/primer extension/MS detection sequence. Buetow et al. (5) describe a variation of the generic primer extension method in which only one dideoxyterminator is used in the reaction such that the primer extension products for the two alleles differ by at least one base (over 300 mass units). Unlike methods that depend on some form of fluorescence detection, no labeling is necessary in MS detection. The intrinsic mass differences between the primer extension products are assayed. Although MS is highly accurate and a moderate degree of multiplex is possible, a number of obstacles need to be overcome. First and foremost, the MS instrument can only handle one sample at a time. Even with multiplex PCR, each MS instrument will probably not be able to genotype more than 10,000 marker assays per day. Second, MS detection requires purified samples that are free from ions and other impurities. This increases both the cost and time required for sample processing. However, many talented groups are working to improve the approach and these obstacles may be overcome in the near future.

Pyrosequencing: Homogeneous or Solid Phase Primer Extension with Luminescence Detection

Pyrosequencing is a new DNA sequencing method based on detecting the formation of pyrophosphate, the by-product of DNA polymerization (43). When DNA polymerase takes a deoxynucleoside triphosphate and incorporates it onto the extending primer, pyrophosphate is formed. By a number of wisely designed enzymatic steps, the pyrophosphate is converted to ATP that fuels a luciferase reaction (1, 38). Therefore, light is observed when a nucleotide is added to the growing chain of DNA. Because stepwise addition of nucleoside triphosphates is needed in this procedure, utilizing this method to give long sequencing reads is difficult. It is a robust method for the primer extension detection, especially if there are a number of closely spaced SNPs (as in HLA typing). The major advantage of this method is the fact that, if needed, multiple bases in the vicinity of the polymorphic site can be determined and so the placement of sequencing primer is more flexible. As in other primer extension methods, excess PCR primers and dNTPs must be removed prior to the pyrosequencing reaction. Furthermore, seven enzymes and two specialized reagents (APS and luciferin) are required in the homogeneous assay (four enzymes are needed in the solid phase reaction format). These requirements make it almost impossible to keep the cost of genotyping with this method lower than other approaches.

Multiplex Primer Extension Sorted on Genetic Arrays: Homogeneous Reaction with Separation/Capture on Solid Support and Fluorescence Detection

To increase the SNP genotyping throughput, a number of groups have devised ways to perform the allelic discriminating primer extension reaction in multiplex in solution and separate the products by capturing them on solid support. This is done by utilizing chimeric primer in the primer extension reaction with 3' complementarity to the specific SNP loci and 5' complementarity to specific capture probes on solid support (6, 14). The solid support can be color-coded microspheres or a silicon chip. The dye-labeled terminators provide the identity of the base incorporated, whereas the specific capture probe sequence provides the identity of the SNP being assayed. Flow cytometry is used for sorting the microspheres, and CCD imaging is used for microarray analysis. With this approach, one takes advantage of the more robust homogeneous reaction format in the allelic discrimination step and the capture probes on solid support to allow for multiplex reactions. Because the capture probes are generic, they can be designed in such a way that all of them will anneal optimally to their complementary sequences at the same temperature. Furthermore, the collection of capture probes on microspheres and on the silicon chip or glass slide can be made in high volume, therefore lowering the cost of the reagent. The only drawback of these approaches is that multiplex PCR and multiplex primer extension reaction are still not easy to optimize.

Ligation with Rolling Circle Amplification: Solid Phase Reaction with Fluorescence Detection

Formation of a circular DNA molecule by ligation provides a means to perform SNP genotyping without PCR amplification. Two groups have shown that the circular DNA ligation product serves as a template for a rolling circle amplification (RCA) step that yields a product thousands of times the size of the original circle. If fluorescent nucleotides are used as the building blocks of the RCA reaction, the signal achieved is bright enough to allow one to detect single molecules. This approach has been used to determine the allele found on a single chromosome in FISH (fluorescent in situ hybridization) analysis and to genotype SNPs on solid support (2, 33). The ability to obtain the genotype of SNPs directly from genomic DNA is a major advantage. This is perhaps the only method one can use to determine long-range haplotypes by FISH analysis. The high cost of the probes and the fact that only one level of specificity is involved make this an approach for special applications but not for general use.

Homogeneous Ligation with Fret Detection

When an allele-specific oligonucleotide labeled with an acceptor dye is ligated to an oligonucleotide bearing a donor dye in the presence of the complementary target DNA, FRET is observed. Because PCR and ligation are different reactions, the assay can be done in a closed-tube format by thermally isolating the two reactions (9). The PCR primers are designed to be long and anneal at a higher temperature, whereas the ligation probes are designed to be short and therefore anneal only at a lower temperature. If the PCR reaction is allowed to proceed at high temperature, the ligation probes will not anneal and the 5′ probe will not be extended (and, thus, will not be taken out of the ligation reaction). After the PCR is largely completed, the thermal cycling conditions are changed, and ligation is allowed to run at a lower temperature. FRET is monitored in real time, and the genotypes can be determined quite easily by measuring the rate of emergence of fluorescence for the two dyes found on the allele-specific ligation probes. Because the ligation reaction is very specific, this reaction is probably the easiest of all closed-tube reactions to optimize. However, all 3′ ligation probes are labeled with dyes and the start-up reagent cost is high.

Multiplex Ligation Reaction Sorted on Genetic Arrays: Homogeneous Reaction with Separation/Capture on Solid Support and Fluorescence Detection

Like the microsphere/microarray based primer extension described above, the ligation assay can also be done in multiplex and captured/separated on solid support. Once again, chimeric probes are used where the 5′ half of the sequence complements the capture probe and the 3′ half of the sequence is allele-specific for a particular SNP. A reporter dye is used to label the common ligation probe such that the dye is captured along with the allele-specific ligation probe only if ligated to the probe in the presence of a DNA target with the correct allele (20). In this approach, only one of the three ligation probes is labeled with a dye. The two allele-specific ligation probes are labeled only with inexpensive nucleotide sequences. Furthermore, because ligation is very specific, multiplex ligation requires minimal optimization. The major disadvantage of this method is the need for three ligation probes, compared to just one for the primer reaction.

Invader Assay: Homogeneous Invasive Cleavage with FRET, FP, or Mass Spectrometry Detection

The invader assay is based on the ability of a thermostable flap endonuclease to cleave a structure formed by the hybridization of two overlapping oligonucleotide probes to a target nucleic acid strand (22). By designing the flap probe with the allelic base at the overlapping site, the correct structure is formed only when the probes perfectly complement the DNA target. Upon cleavage, the flap released from the allele-specific probe (an arbitrary sequence unrelated to the SNP) serves as the "invader" probe in the secondary invader reaction. In the secondary reaction, a universal reporter probe is cleaved only when it forms the proper overlapping structure in the presence of the flap from the primary reaction. Taken together, the amplification is squared, and the assay can work from genomic DNA without the need for target amplification (17, 26, 34). The flap design of the universal reporter probe is based on the detection method employed. For example, if mass spectrometry is the detection method to be used, flaps with varying numbers of nucleotides serve as the reporter signal (16). If FRET is the detection method, a reporter dye is placed on the flap while the quencher is placed on the annealed portion of the universal probe (17). For FP detection, the dye reporter is placed at the end of the flap (T.M. Hsu, S. Law, S. Duan, B. Neri, & P-Y. Kwok, unpublished results).

The invader assay is an elegant SNP genotyping method and holds promise for directly assaying genomic DNA without PCR. However, the signal amplification approach suffers from the fact that only one level of specificity is utilized so that a significant fraction of SNPs in genomes with many repetitive sequences cannot be assayed by this method. Furthermore, although the primary probes are unlabeled, they must be exquisitely pure to work well in the system, thus increasing the cost of the starting reagents.

DISCUSSION AND CONCLUSIONS

Clever use of enzymatic and detection methods has produced a number of robust SNP genotyping methods. Despite recent advances in the field, none of the assays is ideally suited for all applications. Which genotyping assay to adopt, therefore, depends on the needs of the projects being pursued. In general, there are three scenarios to be considered. First, in a clinical diagnostic setting or when working with model organisms where a "canonical" set of markers are being tested on a stream of samples over an extended period of time, one can afford to invest in assay optimization. Here, the closed-tube assays that minimize contamination and sample mix-up are likely to dominate. Second, in a research study where new markers are constantly being identified and genotyped in hundreds of samples, assay development must be simple, and the initial cost of assay development must be low. In this instance, primer extension reactions are the most logical choice. Third, in cases where thousands of markers must be used to type thousands of samples, multiplex assays are needed at very low operating cost. Here, some form of array-based multiplex genotyping method or a high-density reaction vessel capable of handling thousands of homogeneous assays will be needed. Unlike the first two scenarios, the large-scale genotyping studies are currently not feasible.

Reflecting back on the attributes of the ideal SNP genotyping assay listed in the Introduction, it is fair to say that, with optimization, every genotyping method can be made to work robustly and produce accurate results. Furthermore, a number of companies have put together systems for automated SNP genotyping. However, three areas need special attention before an ideal genotyping assay, especially one that can handle large-scale genotpying projects, can be achieved. First, assay development has to be fast and inexpensive. Second, the cost of the assay (from instrumentation to reagents) must be affordable. Third, reactions must be done in a massively parallel fashion.

Although DNA is a simple polymer made up of four bases, the specific sequence of DNA affects its physical property and the way enzymes interact with it. Because we lack complete understanding of the relationship between DNA sequence and physical properties or enzymatic interactions, assay design is not always straightforward, and some optimization is needed for a significant number of SNPs. Furthermore, the redundancy of certain sequence motifs in some genomes (notably the human genome) imposes another level of complexity in assay design. A major challenge for SNP genotyping, in general, is the speed of assay development. If one day is needed to develop 100 assays, 3 years will be required before assays for a set of 100,000 markers (a projected number of markers needed for genome-wide studies) can be assembled for such a project. With increased sophistication in bioinformatics and a better understanding of enzymatic behavior, assay development will be faster and will have a higher rate of success. Given the fact that millions of SNPs have been found just in the human genome alone, this is a significant challenge to all working in the field.

Many of the new assays rely on detection methods that require highly specialized instrumentation. These include high-resolution mass spectrometers, plate readers capable of real-time detection of fluorescence, plate readers that can detect fluorescence polarization at high speed, integrated systems that can perform both the reaction and detection steps, and flow sorting instruments equipped with multiple lasers. These sophisticated instruments are very capable but come at a price too high for most laboratories. Even well-funded institutions cannot afford to acquire more than two or three different systems. Moreover, many assays require expensive probes or other reagents that keep the genotyping cost high. This double hurdle of high instrumentation and operating costs must be overcome before large-scale SNP genotyping becomes feasible. The hope is that with time, the cost of SNP genotyping will drop sufficiently so that even large studies can be done cost effectively.

The most difficult challenge may be the need to assay thousands of markers simultaneously. At present, multiplex PCR is developed by trial and error, and a 100-plex reaction is almost impossible to achieve routinely. Even with better understanding of the forces behind a successful multiplex reaction and improved software, one still has to synthesize two PCR primers for each marker. Now that the DNA sequences of whole genomes are available for a number of organisms, perhaps a new way of whole genome amplification with concomitant simplification can be achieved such that the target amplification step is done easily and cost effectively. A serious collaboration between molecular geneticists and those skilled in bioinformatics will be needed to make this happen. Judging from the speed with which new genotyping assays are being invented, there is great hope that the ideal assay will be developed in the near future.

Visit the Annual Reviews home page at www.AnnualReviews.org

LITERATURE CITED

1. Ahmadian A, Gharizadeh B, Gustafsson AC, Sterky F, Nyren P, et al. 2000. Single nucleotide polymorphism analysis by pyrosequencing. *Anal. Biochem.* 280:103–10

2. Baner J, Nilsson M, Mendel-Hartvig M, Landegren U. 1998. Signal amplification of padlock probes by rolling circle replication. *Nucleic Acids Res.* 26:5073–78

3. Barany F. 1991. Genetic disease detection and DNA amplification using cloned thermostable ligase. *Proc. Natl. Acad. Sci. USA* 88:189–93

4. Berlin K, Gut IG. 1999. Analysis of negatively "charge tagged" DNA by matrix-assisted laser desorption/ionization time-of-flight mass spectrometry. *Rapid Commun. Mass Spectrom.* 13:1739–43

5. Buetow KH, Edmonson M, MacDonald R, Clifford R, Yip P, et al. 2001. High-throughput development and characterization of a genome-wide collection of gene-based single nucleotide polymorphism markers by chip-based matrix-assisted laser desorption/ionization time-of-flight mass spectrometry. *Proc. Natl. Acad. Sci. USA.* 98:581–84

6. Chen J, Iannone MA, Li MS, Taylor JD, Rivers P, et al. 2000. A microsphere-based assay for multiplexed single nucleotide polymorphism analysis using single base chain extension. *Genome Res.* 10:549–57

7. Chen X, Kwok P-Y. 1997. Template-directed dye-terminator incorporation (TDI) assay: a homogeneous DNA diagnostic method based on fluorescence energy transfer. *Nucleic Acids Res.* 25:347–53

8. Chen X, Levine L, Kwok P-Y. 1999. Fluorescence polarization in homogeneous nucleic acid analysis. *Genome Res.* 9:492–98

9. Chen X, Livak K, Kwok P-Y. 1998. A homogeneous, ligase-mediated DNA diagnostic test. *Genome Res.* 8:549–56

10. Chen X, Zehnbauer B, Gnirke A, Kwok P-Y. 1997. Fluorescence energy transfer detection as a homogeneous DNA

diagnostic method. *Proc. Natl. Acad. Sci. USA* 94:10756–61

11. Collins FS, Guyer MS, Charkravarti A. 1997. Variations on a theme: cataloging human DNA sequence variation. *Science* 278:1580–81

12. Cornell BA, Braach-Maksvytis VL, King LG, Osman PD, Raguse B, et al. 1997. A biosensor that uses ion-channel switches. *Nature* 387:580–83

13. Dubiley S, Kirillov E, Mirzabekov A. 1999. Polymorphism analysis and gene detection by minisequencing on an array of gel-immobilized primers. *Nucleic Acids Res.* 27:e19

14. Fan JB, Chen X, Halushka MK, Berno A, Huang X, et al. 2000. Parallel genotyping of human SNPs using generic high-density oligonucleotide tag arrays. *Genome Res.* 10:853–60

15. Germer S, Higuchi R. 1999. Single-tube genotyping without oligonucleotide probes. *Genome Res.* 9:72–78

16. Griffin TJ, Hall JG, Prudent JR, Smith LM. 1999. Direct genetic analysis by matrix-assisted laser desorption/ionization mass spectrometry. *Proc. Natl. Acad. Sci. USA* 96:6301–6

17. Hall JG, Eis PS, Law SM, Reynaldo LP, Prudent JR, et al. 2000. From the cover: sensitive detection of DNA polymorphisms by the serial invasive signal amplification reaction. *Proc. Natl. Acad. Sci. USA* 97:8272–77

18. Hansen TS, Petersen NE, Iitia A, Blaabjerg O, Hyltoft-Petersen P, Horder M. 1995. Robust nonradioactive oligonucleotide ligation assay to detect a common point mutation in the CYP2D6 gene causing abnormal drug metabolism. *Clin. Chem.* 41:413–18

19. Howell WM, Jobs M, Gyllensten U, Brookes AJ. 1999. Dynamic allele-specific hybridization. A new method for scoring single nucleotide polymorphisms. *Nat. Biotechnol.* 17:87–88

20. Iannone MA, Taylor JD, Chen J, Li MS, Rivers P, et al. 2000. Multiplexed single nucleotide polymorphism genotyping by oligonucleotide ligation and flow cytometry. *Cytometry* 39:131–40

20a. Int. SNP Map Work. Group. 2001. A map of human genome sequence variation containing 1.4 million SNPs. *Nature* 409:928–33

21. Johnson GC, Todd JA. 2000. Strategies in complex disease mapping. *Curr. Opin. Genet. Dev.* 10:330–34

22. Kaiser MW, Lyamicheva N, Ma W, Miller C, Neri B, et al. 1999. A comparison of eubacterial and archaeal structure-specific 5′ exonucleases. *J. Biol. Chem.* 274:21387–94

23. Kirschstein S, Winter S, Turner D, Lober G. 1999. Detection of the ΔF508 mutation in the CFTR gene by means of time-resolved fluorescence methods. *Bioelectrochem. Bioenerg.* 48:415–21

24. Kostrikis LG, Tyagi S, Mhlanga MM, Ho DD, Kramer FR. 1998. Spectral genotyping of human alleles. *Science* 279:1228–29

25. Kruglyak L. 1999. Prospects for whole-genome linkage disequilibrium mapping of common disease genes. *Nat. Genet.* 22:139–44

26. Kwiatkowski RW, Lyamichev V, de Arruda M, Neri B. 1999. Clinical, genetic, and pharmacogenetic applications of the Invader assay. *Mol. Diagn.* 4:353–64

27. Kwok P-Y. 2000. High-throughput genotyping assay approaches. *Pharmacogenomics* 1:95–100

28. Kwok P-Y, Chen X. 1998. Detection of single nucleotide polymorphisms. In *Genetic Engineering, Principles and Methods*, ed. JK Setlow. New York: Plenum. 20:125–34

29. Laken SJ, Jackson PE, Kinzler KW, Vogelstein B, Strickland PT, et al. 1998. Genotyping by mass spectrometric analysis of short DNA fragments. *Nat. Biotechnol.* 16:1352–56

30. Latif S, Bauer-Sardiña I, Ranade K, Livak K, Kwok P-Y. 2001. Fluorescence polarization in homogeneous nucleic acid

analysis II: 5' nuclease assay. *Genome Res.* 11:436–40

31. Li J, Butler JM, Tan Y, Lin H, Royer S, et al. 1999. Single nucleotide polymorphism determination using primer extension and time-of-flight mass spectrometry. *Electrophoresis* 20:1258–65

32. Livak KJ. 1999. Allelic discrimination using fluorogenic probes and the 5' nuclease assay. *Genet. Anal.* 14:143–49

33. Lizardi PM, Huang X, Zhu Z, Bray-Ward P, Thomas DC, Ward DC. 1998. Mutation detection and single-molecule counting using isothermal rolling-circle amplification. *Nat. Genet.* 19:225–32

34. Lyamichev VI, Kaiser MW, Lyamicheva NE, Vologodskii AV, Hall JG, et al. 2000. Experimental and theoretical analysis of the invasive signal amplification reaction. *Biochemistry* 39:9523–32

35. Mei R, Galipeau PC, Prass C, Berno A, Ghandour G, et al. 2000. Genome-wide detection of allelic imbalance using human SNPs and high-density DNA arrays. *Genome Res.* 10:1126–37

36. Myakishev MV, Khripin Y, Hu S, Hamer DH. 2001. High-throughput SNP genotyping by allele-specific PCR with universal energy-transfer–labeled primers. *Genome Res.* 11:163–69

37. Natl. Center Biotechnol. Inf. *Single Nucleotide Polymorphism Database.* http://www.ncbi.nlm.nih.gov/SNP/

38. Nordstrom T, Nourizad K, Ronaghi M, Nyren P. 2000. Method enabling pyrosequencing on double-stranded DNA. *Anal. Biochem.* 282:186–93

39. Pastinen T, Kurg A, Metspalu A, Peltonen L, Syvanen AC. 1997. Minisequencing: a specific tool for DNA analysis and diagnostics on oligonucleotide arrays. *Genome Res.* 7:606–14

40. Pastinen T, Raitio M, Lindroos K, Tainola P, Peltonen L, Syvanen AC. 2000. A system for specific, high-throughput genotyping by allele-specific primer extension on microarrays. *Genome Res.* 10:1031–42

41. Prince JA, Feuk L, Howell WM, Jobs M, Emahazion T, et al. 2001. Robust and accurate single nucleotide polymorphism genotyping by dynamic allele-specific hybridization (DASH): design criteria and assay validation. *Genome Res.* 11:152–62

42. Risch NJ. 2000. Searching for genetic determinants in the new millennium. *Nature* 405:847–56

43. Ronaghi M. 2001. Pyrosequencing sheds light on DNA sequencing. *Genome Res.* 11:3–11

44. Ross P, Hall L, Smirnov I, Haff L. 1998. High level multiplex genotyping by MALDI-TOF mass spectrometry. *Nat. Biotechnol.* 16:1347–51

45. Shumaker JM, Metspalu A, Caskey CT. 1996. Mutation detection by solid phase primer extension. *Hum. Mutat.* 7:346–54

46. Svanvik N, Stahlberg A, Sehlstedt U, Sjoback R, Kubista M. 2000. Detection of PCR products in real time using light-up probes. *Anal. Biochem.* 287:179–82

47. Todd AV, Fuery CJ, Impey HL, Applegate TL, Haughton MA. 2000. DzyNA-PCR: use of DNAzymes to detect and quantify nucleic acid sequences in a real-time fluorescent format. *Clin. Chem.* 46:625–30

48. Tonisson N, Kurg A, Kaasik K, Lohmussaar E, Metspalu A. 2000. Unravelling genetic data by arrayed primer extension. *Clin. Chem. Lab Med.* 38:165–70

49. Tyagi S, Bratu DP, Kramer FR. 1998. Multicolor molecular beacons for allele discrimination. *Nat. Biotechnol.* 16:49–53

50. Wang DG, Fan JB, Siao CJ, Berno A, Young P, et al. 1998. Large-scale identification, mapping, and genotyping of single-nucleotide polymorphisms in the human genome. *Science* 280:1077–82

51. Wang J, Cai X, Rivas G, Shiraishi H, Dontha N. 1997. Nucleic-acid immobilization, recognition and detection at chronopotentiometric DNA chips. *Biosens. Bioelectron.* 12:587–99

Annu. Rev. Genomics Hum. Genet. 2001. 2:259–69

THE IMPACT OF MICROBIAL GENOMICS ON ANTIMICROBIAL DRUG DEVELOPMENT

Christoph M. Tang and E. Richard Moxon

University Department of Paediatrics, John Radcliffe Hosptial, Oxford OX3 9DU, United Kingdom; e-mail: christoph.tang@paediatrics.oxford.ac.uk

Key Words   pathogen, bacteria, antibiotic, resistance, genomes

■ **Abstract**   There is an urgent need to develop novel classes of antibiotics to counter the threat of the spread of multiply resistant bacterial pathogens. The availability of the complete genome sequence of many pathogenic microbes provides information on every potential drug target and is an invaluable resource in the search for novel compounds. Here, we review the approaches being taken to exploit the genome databases through a combination of bioinformatics, transcriptional analysis, and a further understanding of the molecular basis of the disease process. The emphasis is changing from compound screening to target hunting, as the latter offers flexible ways to design and optimize the next generation of broad-spectrum antibiotics.

INTRODUCTION

Approximately 30 years ago, the prevailing opinion was that the fight against infectious diseases, at least in developed countries, had been won. Rates of bacterial disease had been falling since the turn of the century, through a combination of improved public health and vaccination, and a variety of antimicrobial agents was still available to treat most important infections. Furthermore, synthetic analogues of antibiotics were being developed that were effective against wide ranges of microbes. Several pharmaceutical companies, therefore, decided to abandon antimicrobial drug development as there appeared to be little need for new compounds. Thus, until recently, no new classes of antibiotics had been licensed for clinical use against bacterial infections (15).

Under these circumstances, antimicrobial resistance inevitably became a major problem. Microorganisms have a remarkable ability to adapt, given their fast generation times, relatively high mutation rates (16), and sheer numbers; the number of bacteria carried by any individual far exceeds the number of their own cells. Selective pressure has been exerted not only by the unwarranted and injudicious use of antimicrobials in the clinical setting, but also by the widespread practice of feeding low-dose antibiotics as growth promoters to livestock. Microorganisms have now evolved that are able to resist the action of most antibiotics (17).

Pathogenic strains of *Staphylococcus aureus* and *Enterococcus* spp. have emerged that are resistant to all conventional antimicrobials (13, 23), and these organisms, along with other multiresistant bacteria, such as *Mycobacterium tuberculosis* and *Salmonella typhi*, now constitute a major public health threat (28, 39). The genetic information that confers the capacity of bacteria to survive in the face of selective pressure imposed by antimicrobial agents is frequently encoded on elements, such as plasmids, transposons, and phages, which can spread through bacterial populations by horizontal gene transfer (19). This has facilitated the rapid dissemination of resistance traits within species and across genera. Furthermore, resistance mechanisms that involve alterations of the target site of antibiotics have emerged, conferring resistance against whole classes of antibiotics rather than just single agents.

A further increasing threat is the ever-widening range of pathogens causing human disease. Advances in medical technology have provided these organisms, which had been previously regarded as harmless commensals, niches in which to flourish. Implanted prosthetic material, increasing populations of immunosuppressed patients, and the practice of prescribing prophylactic antibiotics have led to microbes, such as *Candida albicans* and *Pseudomonas aeruginosa*, becoming significant pathogens. These microbes pose most challenges in intensive care units, which have proven to be fertile breeding grounds for newly recognized, highly resistant organisms (7).

However, not all the news is bad. We now live in an age of genomic medicine in which the sequences of the human genome (33, 35) and many microbes are known. Following the publication, in 1995, of the first complete bacterial genome sequence, that of *Haemophilius influenzae* (11), a dramatic rise in the number of publicly available genomes has occurred. The success of microbial genome sequencing has been based on developing high-throughput sequencing machines and the use of shotgun cloning, without physical mapping, to generate the majority of the sequence information (36). Here, we outline ways in which the sequence information is being mined in the quest for new agents to thwart the seemingly inexorable rise in antimicrobial resistance.

THE ADVENT OF MICROBIAL GENOMICS

The complete genome sequences for over 20 microbes are now known, with many others nearing completion or in commercial databases (for further information, see http://wwwfp.mcs.anl.gov/~gaasterland/genomes.html; http://www.tigr.org/tdb/mdb/mdbcomplete.html; http://www.sanger. ac.uk/Projects/Microbes/). The list of available, sequenced genomes includes many important human pathogens, though there is a strong emphasis on those organisms that affect individuals in developed countries. For example, no genome sequence is available for pathogenic strains of *Escherichia coli* or *Shigella* spp., which contribute to diarrhea, malnutrition, and death in children throughout the tropics. The potential impact of the genome sequencing projects on the discovery of new antimicrobial agents is immediately

TABLE 1 Proportion of genes of unknown function in sequenced genomes

Organism	Genome size (Mb)	% of genes of unknown function
Escherichia coli	4.64	38
Haemophilus influenzae	1.83	37
Helicobacter pylori	1.7	31
Neisseria meningitidis	2.12	37
Staphylococcus aureus	2.8	>50

obvious. Contained within the sequence is information on every conceivable drug target against a microorganism, although selecting targets from sequence information is not a trivial matter. First, genome annotations are based on finding significant homology at the nucleotide or protein level to other sequences, and not necessarily to functionally defined proteins. Thus, once inaccuracies in annotations appear in the databases, they are replicated and perpetuated (through further rounds of homology searches and annotation), and they become extremely difficult to eradicate. Another striking finding from all of the microbial genome sequences is the limitation of current knowledge regarding gene function. For all sequenced bacteria, the function of between 30% and 40% of all genes is completely unknown (see Table 1); therefore, selecting effective targets from a significant proportion of genes is difficult. Despite years of intensive research on microbes, genetic analyses have failed to shed light on the role of the genes of unknown function (17). That many of these genes may be essential for the survival of the cell, and therefore, mutation in them is lethal, provides one possible explanation. Alternatively, genes may have roles for bacteria within nature that are not observed in the laboratory. Many genes of unknown function are clustered together and appear to form operons, suggesting that they might perform related functions in the cell. For example, the genome of enterohemorrhagic *E. coli* contains large blocks of genes of unknown function that may have been acquired by single genetic events (25). This may make analysis of their function simpler, as elucidation of the function of one of these genes may provide clues to the role of other genes within the same operon.

Bioinformatic approaches have also been used in an attempt to shed light on the genes of unknown function. A tremendous investment in this area, though mainly within the commercial sector, has been made. Therefore, many of the approaches and results are not widely available. However, there are examples in which bioinformatics has been applied to study the bacterial genomes. For instance, it may be possible to infer something about the function of a given gene by understanding the other proteins with which the gene product interacts within the cell. Potential interactions can be predicted by looking for "interactive domains," amino acid motifs that mediate the aggregation of single, individual proteins to form multisubunit enzymes. Identification of a gene product's partners within a cell may give a clue about its biochemical function. One way of finding interacting proteins is by interrogating complete genomes for gene fusion events (9, 21). This relies on the supposition that selection will lead to the close proximity of genes that interact and are coregulated. A composite protein in one organism may, therefore, have arisen through the fusion of two component proteins. In organisms in which the two components are encoded by distinct genes, the proteins are still likely to interact. A good example is the aromatic amino acid biosynthetic pathway. In yeast, this pathway is encoded by a single gene that encodes a multifunctional enzyme (6), whereas in bacteria, the five component genes are scattered around the genome (3). Analysis of the *E. coli, H. influenzae*, and *Methanococcus jannaschii* genome revealed 64 gene fusion events, with 11 in which both partners are hypothetical proteins (9). The function of another seven hypothetical proteins could be inferred by the identity of their interacting partner. Additionally, many experimental approaches, such as the yeast and the bacterial two-hybrid systems, can be used to detect interacting proteins within cells. Indeed, a concerted effort is underway to detect all interacting proteins in the budding yeast, *Saccharomyces cerevisiae* (18, 22).

Analysis of genome-wide gene expression profiles may give an indication of function. Studies in *S. cerevisiae* have shown that genes that perform related tasks within the cell tend to be coregulated at the transcriptional level. DNA microarray data collated the transcriptional response of *S. cerevisiae* to cell division, sporulation, and shock responses (8). The expression profiles tended to organize the genes into functional categories that, for the genes of known function, correlated with their annotations. Some of the genes of unknown function also fell into these expression clusters, providing a provisional assignment of functional category, which will allow for the testing of corresponding mutants in a directed fashion.

However, both bioinformatics and expression profiles only provide hypotheses about gene function that must be validated by experimentation. Systematic analyses of gene function are really needed to define the cellular location, profile of expression, interactions, biochemical function, and essentiality of gene products. Furthermore, when considering pathogenic microbes, examining the role of gene products in the disease process is also necessary (29). The relatively small and simple microbial genomes may provide valuable experimental models that will help to prepare the way for the exploitation of the genomes of more complex organisms. Only a few examples of efforts to systematically study gene function on a genome-wide scale exist. *S. cerevisiae* was the first eukaryote to be sequenced, and there is a collaborative project aimed at constructing a library of deletion mutants in which each of the 6200 genes has been knocked-out. A similar effort is being made with the Gram-positive bacterium, *Bacillus subtilis*. These projects share two features. First, the functional studies were initiated by a publically funded consortia prior to the completion of the genome sequence; and second, both of these microbes are nonpathogenic. The lack of similar approaches for pathogenic disease-causing bacteria is striking and disappointing, as the results should yield information that is fundamental to understanding the biology of infectious diseases. Even so, studies on model organisms, such as *S. cerevisiae* and *B. subtilis*, should yield a wealth of information that could be used to infer gene function in other microbes and inform experiments on clinically important microbes.

ANTIMICROBIAL TARGETS THROUGH UNDERSTANDING PATHOGENESIS

Identification of gene products that are involved in the disease process provides one way of finding new drug targets. A theoretical advantage is that resistance may be less likely to occur when using pathogen-specific agents. Such drugs would have little or no impact on the resident microbial flora, which often acts as a repository for the genetic elements conferring resistance (15). There are several methodologies that allow screens for genes to be performed in animal models rather than in less meaningful in vitro systems. In vivo expression technology and differential fluorescence induction are both used to pinpoint genes that are specifically switched on by an invading microbe while in the host (20, 34). The disadvantage of this approach is that the relationship between increased expression and essentiality in a given environmental condition is uncertain. Indeed, recent work compared the profile of *S. cerevisiae* gene expression in two media with the growth rates of the specific mutants in the two environments (38). Little correlation between the pattern of gene expression and the phenotype of the mutant was noted.

In contrast, signature tagged mutagenesis (STM) directly identifies the requirement for a gene product during pathogenesis (12). This is achieved by constructing insertional mutants with DNA sequence tags that allow the mutants to be differentiated from each other. This makes it possible to analyze large numbers of mutants in parallel using animal models of infection. STM has now been applied to understand the biological basis of Gram-negative and Gram-positive bacterial infections and fungal diseases. However, STM has the drawback in relation to finding new antimicrobials because it will not identify those genes that are needed for survival under all conditions, so-called essential genes, as mutations in these genes will be lethal to the cell.

Recent work with *Neisseria meningitidis*, the leading cause of bacterial meningitis, led to the isolation of 72 genes that are required during systemic infection of infant rats (30). A number of the genes identified are involved in the biosynthesis and/or modification of the targets of the usual drugs used to treat this infection. Four genes were identified that maintain the integrity of cell wall peptidoglycan (the target of the β-lactam antibiotics), while several genes in the *p*-aminobenzoic acid pathway (the target of the sulphonamides) were found. It is likely that the remainder of the genes includes others that encode further suitable targets for novel compounds. Also, some of the genes are highly conserved across bacterial genera, raising the possibility of finding compounds with broad spectrum activity. STM of *M. tuberculosis* identified attenuating insertions in a 50-kb virulence gene cluster that contains 13 open reading frames (ORFs) (4, 5). Seven of the ORFs may be involved in the biosynthesis of important components of the mycobacterial cell envelope, phthiocerol, phenolphthiocerol, and mycoside B. These lipid-based molecules have been found in eight mycobacterial species, seven of which are pathogenic. Therefore, the molecules appear to be largely restricted to pathogenic bacteria, and mutation in them results in specific defects for bacterial survival within the lung, the primary focus of tuberculosis. Pathways leading to the formation of these lipid-based molecules may, therefore, form the basis of antimicrobials that selectively inhibit growth of pathogenic strains in a key host environment.

An obvious proviso with disease-based approaches is that the target must be required at a stage when or after the disease becomes clinically manifest and treatment needs to be initiated. For instance, there is little point in blocking mucosal invasion by a microbe once systemic spread has occurred. Traditionally, the aim of pharmaceutical companies has been to find targets in which mutation is lethal. Preoccupation with essential genes, however, may miss valuable targets. For example, the sulphonmamide class of antibiotics, which were among the first discovered, inhibit a step in the p-aminobenzoic acid pathway; mutations in genes encoding the corresponding enzymes are not lethal for growth in the laboratory but do severely attenuate the capacity of microbes to cause disease in animal models. Therefore, pinpointing further drug targets through studies on microbial pathogenesis should be feasible, although this approach has not yet yielded a licensed drug.

GENE EXPRESSION PROFILES AND DRUG TARGETS

It may be possible to use alterations in gene expression to hunt for new antimicrobial targets. The availability of microarrays that contain DNA fragments corresponding to every ORF from a microbe allows the comprehensive analysis of the profile of gene expression. Thus, the transcriptional response of bacteria to certain stimuli, such as exposure to antimicrobial compounds, can be monitored. This provides an expression "signature" for a given inhibitor, which might identify the affected pathway and the specific target when the site of action of the compound is unknown. Identification of the target and pathway affected by known antimicrobials should provide the basis for the rational modification of existing compounds and the selection of new targets. For example, the transcriptional response of *Mycobacterium tuberculosis* after exposure to isoniazid (INH) was followed (37). INH is a component of standard antituberculosis therapy and is the agent against which resistance is most commonly found. The drug acts by inhibiting the synthesis of mycolic acid, a key component of the lipid envelope surrounding the cell. INH is also bactericidal for the organism. Examination of gene expression established a provisional pathway for mycolic acid biosynthesis and highlighted a number of potential new targets to treat this important infectious disease. These data now need to be compared with the results of inactivating genes encoding the putative targets. Similar approaches could be used to examine common final pathways that lead to the death of the bacterium after a variety of insults and to define the basis for antimicrobial resistance, allowing the design of compounds that retain bactericidal activity.

TARGET-BASED APPROACHES AND SELECTION

The available genome sequences have greatly enhanced "target-based" initiatives to compound finding (27). Traditionally, pharmaceutical companies had large collections of compounds, composed mostly of naturally occurring molecules, and screened them for antimicrobial activity against whole, live organisms. Such screens have a number of disadvantages. They are inherently insensitive and often led to the isolation of toxic compounds. These screens will only identify targets that are lethal for the bacteria under growth within the laboratory. Furthermore, as the precise nature of the target inhibited by any new compound is not defined, rational modification of the molecule to enhance its activity and moderate toxicity is difficult. Indeed, these screens have failed to isolate new classes of antimicrobials in recent years, merely leading to the identification of analogues of previously known drugs. This approach has been replaced by attempts to define effective drug targets, based on simple criteria, and then design antagonists to disrupt the activity of the target.

The microbial genomes have provided a wealth of potential pathways and targets against which antimicrobials can be developed. Targets can be selected using simple predefined criteria, which may include essentiality, their influence on pathogenesis, conservation among other pathogenic microbes, and the absence of related sequences in the human and other mammalian genomes. The latter provides one example of the power of combining sequence information from the host and the pathogen. Comparison between bacterial genomes can also be used to derive "minimal gene sets" for free-living organisms (24). When the *H. influenzae* and *Mycoplasma genitalium* genomes were aligned, approximately 250 common genes were identified. This group of genes should contain all genes that are essential for viability and are, therefore, candidates for new drug targets. Generation of large-scale libraries of transposon insertional mutants and sequencing of the insertion sites have also been used to identify essential genes. This can be accomplished by genome-wide transposon mutagenesis (14) or by focusing on specific regions using in vitro mutagenesis of PCR products, followed by allelic replacement (1). The genome-wide approach was used to find 265–350 potentially essential genes in *Mycoplasma*, the bacterium with the smallest genome sequenced so far, but the completeness of the results is entirely dependent on the efficiency of the insertional mutagen.

With the genome sequence of a range of pathogens in hand, a selection of syndrome-specific pathogens can be performed to identify those genes shared by microbes commonly causing pneumonia, for instance. This could provide compounds that are effective in treating defined clinical conditions. For instance, the genome sequence for the major causes of community-acquired pneumonia (*Streptococcus pneumoniae, Mycoplasma pneumoniae, Legionella pneumophila,* and *Haemophilus influenzae*) can be examined for conserved genes (31). This subset of ORFs can then be used as the starting point for the selection of drug targets that could be used to develop antipneumonia therapy. More ambitiously, treatments could be devised that treat a single pathogen. However, given the lack of specificity of the clinical presentation of most infectious diseases and the potentially small market for this type of agent, it is difficult to envisage this approach being taken by any large pharmaceutical company in the foreseeable future.

The negatively charged outer membrane of bacteria presents a significant obstacle to antimicrobial compounds reaching their active sites. A variety of bioinformatic programs are now available to predict and identify surface-located molecules within the genome, and these have been used for the preliminary selection of vaccine candidates. In *N. meningitidis*, informatics was used to produce a list of over 500 protein candidates from a genome of ~2100 genes (26). This approach could be adapted in the search for antimicrobial targets, although results could also be achieved experimentally by epitope-tagging essential genes and using the tag to demonstrate surface exposure of the protein. Although selecting drug targets that are located at the cell surface may be preferable, many current antimicrobials inhibit molecules and processes that occur within the cell. For example, the macrolide and aminoglycoside classes of antibiotic both exert their bacteriostatic effects by interfering with protein synthesis by inhibiting the subunits of the bacterial ribosome.

The target-based approach is entirely dependent on having a robust method for assaying the biochemical activity of a specific target. This is necessary for testing compounds that block its activity. Combinatorial chemistry provides a way to synthesize huge numbers of structurally diverse molecules. Thus, it is imperative that the activity can be assayed cheaply in a high-throughput format. This will facilitate both the initial identification of inhibitory molecules and the optimization of any lead compound.

Work on *S. pneumoniae*, an organism that is becoming increasingly resistant to first line agents, provides a good example of how multiple disciplines, including genomics, can be effectively brought to bear on developing antimicrobials. Bacteria detect and respond to varied host environments during different stages of the disease process. The ability to regulate appropriately the expression of key virulence factors is an important attribute of pathogenic bacteria. This is mediated in a number of important microbes by two-component regulators. Typically, two-component systems consist of a surface membrane–located sensor molecule that detects changes in the environment surrounding a bacterium by phosphorylating a response regulator, often a transcription factor, that then influences gene expression (10). Two-component systems, therefore, form a phosphorelay mechanism for converting environmental change into a phenotypic response by the bacterium and are widely distributed among pathogenic microbes. A systematic search of four *S. pnemoniae* sequence databases (only one of which is publically available) was performed for the conserved histidine and aspartate residues that are essential for the phosphorelay of exogenous stimuli (32). This search identified 14 pairs of genes predicted to encode two-component systems, 10 of which were novel. Each system was genetically inactivated and one, designated 492, was essential, as deletion of the response regulator was not possible. Seven of the remaining two-component systems are involved in pathogenicity, as the corresponding mutants were attenuated in a mouse respiratory tract model. These findings, as well as the realizations that signal transduction in mammalian cells is mediated through different mechanisms and that two-component systems are present in a range of bacteria, provide the rationale for using two-component systems as drug targets. An assay was developed to detect the phosphotransfer from ATP to KinA, a response regulator from *Bacillus subtilus* (2). A compound, RWJ-49815, was found that abolishes incorporation of the phosphate. RWJ-49815 is a hydrophobic tyramine that also has antibacterial activity, including a bacteriostatic effect on methicillin-resistant *S. aureus*. Further derivatives of RWJ-49815 were synthesized, though none was found to be as active as the original compound. The results did provide some indication of the key drug-target interactions, indicating how further modification could be successful.

SUMMARY

The next few years are likely to see an explosion in the number of drug targets and inhibitory compounds developed to treat bacterial infections, although the lag time to new antibiotics ready for use in the clinic will be considerable. Even though the availability of microbial genomes provides an enormous resource, the problems posed to human health by infectious diseases and the emergence of highly resistant pathogens will never go away. There is likely to be only one winner in the race between microbes and our efforts to find new pharmaceutical compounds. Effectively combating the threat of infectious diseases will not only involve disciplines, such as genomics, structural biology, genetics, and bioinformatics, but will also rely on establishing epidemiological networks of surveillance (to detect emerging pathogens) as well as improved public health measures (including vaccination) to protect individuals from an ever-increasing array of microbes.

ACKNOWLEDGMENTS

We are grateful to David Stroud, Sharmila Bakshi, Nick West, and Yao-hui Sun for their helpful comments and to Vicky Magee for preparing the manuscript.

Visit the Annual Reviews home page at www.AnnualReviews.org

LITERATURE CITED

1. Akerley BJ, Rubin EJ, Camilli A, Lampe DJ, Robertson HM, Mekalanos JJ. 1998. Systematic identification of essential genes by in vitro mariner mutagenesis. *Proc. Natl. Acad. Sci. USA* 95:8927–32
2. Barrett JF, Goldschmidt RM, Lawrence LE, Foleno B, Chen R, et al. 1998. Antibacterial agents that inhibit two-component signal transduction systems. *Proc. Natl. Acad. Sci. USA* 95:5317–22
3. Blattner FR, Plunkett G III, Bloch CA, Perna NT, Burland V, et al. 1997. The

complete genome sequence of *Escherichia coli* K-12. *Science* 277:1453–74

4. Camacho LR, Ensergueix D, Perez E, Gicquel B, Guilhot C. 1999. Identification of a virulence gene cluster of *Mycobacterium tuberculosis* by signature-tagged transposon mutagenesis. *Mol. Microbiol.* 34:257–67

5. Cox JS, Chen B, McNeil M, Jacobs WR Jr. 1999. Complex lipid determines tissue-specific replication of *Mycobacterium tuberculosis* in mice. *Nature* 402:79–83

6. Duncan K, Edwards RM, Coggins JR. 1988. *The Saccharomyces cerevisiae* ARO1 gene. An example of the coordinate regulation of five enzymes on a single biosynthetic pathway. *FEBS Lett.* 241:83–88

7. Edgeworth JD, Treacher DF, Eykyn SJ. 1999. A 25-year study of nosocomial bacteremia in an adult intensive care unit. *Crit. Care Med.* 27:1421–28

8. Eisen MB, Spellman PT, Brown PO, Botstein D. 1998. Cluster analysis and display of genome-wide expression patterns. *Proc. Natl. Acad. Sci. USA* 95:14863–68

9. Enright AJ, Iliopoulos I, Kyrpides NC, Ouzounis CA. 1999. Protein interaction maps for complete genomes based on gene fusion events. *Nature* 402:86–90

10. Finlay BB, Falkow S. 1989. Common themes in microbial pathogenicity. *Microbiol. Rev.* 53:210–30

11. Fleischmann RD, Adams MD, White O, Clayton RA, Kirkness EF, et al. 1995. Whole-genome random sequencing and assembly of *Haemophilus influenzae* Rd. *Science* 269:496–512

12. Hensel M, Shea JE, Gleeson C, Jones MD, Dalton E, Holden DW. 1995. Simultaneous identification of bacterial virulence genes by negative selection. *Science* 269:400–3

13. Hiramatsu K, Aritaka N, Hanaki H, Kawasaki S, Hosoda Y, et al. 1997. Dissemination in Japanese hospitals of strains of *Staphylococcus aureus* heterogeneously resistant to vancomycin. *Lancet* 350:1670–73

14. Hutchison CA, Peterson SN, Gill SR, Cline RT, White O, et al. 1999. Global transposon mutagenesis and a minimal *Mycoplasma genome. Science* 286:2165–69

15. Knowles DJ. 1997. New strategies for antibacterial drug design. *Trends Microbiol.* 5:379–83

16. LeClerc JE, Li B, Payne WL, Cebula TA. 1996. High mutation frequencies among *Escherichia coli* and Salmonella pathogens. *Science* 274:1208–11

17. Levy SB. 1998. The challenge of antibiotic resistance. *Sci. Am.* 278:46–53

18. Luban J, Goff SP. 1995. The yeast two-hybrid system for studying protein-protein interactions. *Curr. Opin. Biotechnol.* 6:59–64

19. Lupski JR. 987. Molecular mechanisms for transposition of drug-resistance genes and other movable genetic elements. *Rev. Infect. Dis.* 9:357–68

20. Mahan MJ, Slauch JM, Mekalanos JJ. 1993. Selection of bacterial virulence genes that are specifically induced in host tissues. *Science* 259:686–88

21. Marcotte EM, Pellegrini M, Thompson MJ, Yeates TO, Eisenberg D. 1999. A combined algorithm for genome-wide prediction of protein function. *Nature* 402:83–86

22. Mendelsohn AR, Brent R. 1999. Protein interaction methods—toward an endgame. *Science* 284:1948–50

23. Michel M, Gutmann L. 1997. Methicillin-resistant *Staphylococcus aureus* and vancomycin-resistant enterococci: therapeutic realities and possibilities. *Lancet* 349:1901–6

24. Mushegian AR, Koonin EV. 1996. A minimal gene set for cellular life derived by comparison of complete bacterial genomes. *Proc. Natl. Acad. Sci. USA* 93:10268–73

25. Perna NT, Plunkett G, Burland V, Mau B, Glasner JD, et al. 2001 Genome sequence of enterohaemorrhagic *Escherichia coli* O157:H7. *Nature* 409:529–33

26. Pizza M, Scarlato V, Masignani V, Giuliani MM, Arico B, et al. 2000. Identification of vaccine candidates against serogroup B meningococcus by whole-genome sequencing. *Science* 287:1816–20

27. Rosamond J, Allsop A. 2000. Harnessing the power of the genome in the search for new antibiotics. *Science* 287:1973–76

28. Rowe B, Ward LR, Threlfall EJ. 1997. Multidrug-resistant *Salmonella typhi*: a worldwide epidemic. *Clin. Infect. Dis.* 24(Suppl. 1):S106–9

29. Smith H. 1998. What happens to bacterial pathogens in vivo? *Trends Microbiol.* 6:239–43

30. Sun YH, Bakshi S, Chalmers R, Tang CM. 2000. Functional genomics of neisseria meningitidis pathogenesis. *Nat. Med.* 6:1269–73

31. Tang CM, Holden D. 1999. Pathogen virulence genes—implications for vaccines and drug therapy. *Br. Med. Bull.* 55:387–400

32. Throup JP, Koretke KK, Bryant AP, Ingraham KA, Chalker AF, et al. 2000. A genomic analysis of two-component signal transduction in *Streptococcus pneumoniae*. *Mol. Microbiol.* 35:566–76

33. The Genome International Sequencing Consortium. 2001. Initial sequencing and analysis of the human genome. *Nature* 409:860–921

34. Valdivia RH, Falkow S. 1997. Fluorescence-based isolation of bacterial genes expressed within host cells. *Science* 277:2007–11

35. Venter JC, Adams MD, Myers EW, Li PW, Mural RJ, et al. 2001. The sequence of the human genome. *Science* 291:1304–51

36. Venter JC, Adams MD, Sutton GG, Kerlavage AR, Smith HO, Hunkapiller M. 1998. Shotgun sequencing of the human genome. *Science* 280:1540–42

37. Wilson M, DeRisi J, Kristensen HH, Imboden P, Rane S, et al. 1999. Exploring drug-induced alterations in gene expression in *Mycobacterium tuberculosis* by microarray hybridization. *Proc. Natl. Acad. Sci. USA* 96:12833–38

38. Winzeler EA, Shoemaker DD, Astromoff A, Liang H, Anderson K, et al. 1999. Functional characterization of the *S. cerevisiae* genome by gene deletion and parallel analysis. *Science* 285:901–6

39. World Health Organization. 1997. *Antituberculosis Drug Resistance in the World: The WHO/IUATLD Global Project on Antituberculosis Drug Resistance Surveillance.* Geneva: WHO

Annu. Rev. Genomics Hum. Genet. 2001. 2:271–97

USHER SYNDROME: From Genetics to Pathogenesis

Christine Petit

*Unité de Génétique des Déficits Sensoriels, CNRS URA 1968 Institut Pasteur, 75724
Paris, Cedex 15, France; e-mail: cpetit@pasteur.fr*

Key Words syndromic deafness, retinitis pigmentosa, sensory cell disease,
photoreceptor cells, hair cells, lateral links, cell-cell junctions, unconventional
myosin, PDZ domain–containing protein, cadherin-like protein, planar polarity

■ **Abstract** Usher syndrome (USH) is defined by the association of sensorineural
deafness and visual impairment due to retinitis pigmentosa. The syndrome has three
distinct clinical subtypes, referred to as USH1, USH2, and USH3. Each subtype is
genetically heterogeneous, and 12 loci have been detected so far. Four genes have
been identified, namely, *USH1B*, *USH1C*, *USH1D*, and *USH2A*. *USH1B*, *USH1C*, and
USH1D encode an unconventional myosin (myosin VIIA), a PDZ domain–containing
protein (harmonin), and a cadherin-like protein (cadherin-23), respectively. Mutations
of these genes cause primary defects of the sensory cells in the inner ear, and probably
also in the retina. In the inner ear, the *USH1* genes, I propose, are involved in the same
signaling pathway, which may control development and/or maintenance of the hair
bundles of sensory cells via an adhesion force (*a*) at the junctions between these cells
and supporting cells and (*b*) at the level of the lateral links that interconnect the stereocilia. In contrast, the molecular pathogenesis of USH2A, which is owing to a defect
of a novel extracellular matrix protein, is likely to be different from that of USH1.

INTRODUCTION

Over 400 syndromes with deafness as one of the symptoms have been described
in the literature (46). They include a vast ensemble of hereditary monogenic diseases in which hearing loss is associated with miscellaneous disorders of the
musculoskeletal, cardiovascular, urogenital, nervous, endocrine, digestive, or in-
tegumentary systems. About 40 of these syndromes include an ocular disorder
(45). Among them is Usher syndrome (USH), defined by a bilateral sensorineural
deafness that originates in the cochlea (the auditory sensory organ) and a loss of
vision due to retinitis pigmentosa. This syndrome is the most frequent cause of
deafness accompanied by blindness. It accounts for more than 50% of individuals who are both deaf and blind (142), about 18% of retinitis pigmentosa cases,
and 3%–6% of congenital deafness cases (20). Its prevalence is between 1/16,000
and 1/50,000, based on studies of Scandinavian (48, 51, 84, 104, 120), Columbian
(132), British (57), and American populations (20); for individuals between the
ages of 30 and 49, the prevalence approaches 1/10,000 (57).

1527-8204/01/0728-0271$14.00

As early as 1858, the occurrence of a pigmentary retinopathy with deafness was observed in three siblings (out of a total of five) by the German ophthalmologist Alfred von Graefe and reported by his cousin Albrecht von Graefe (144). This discovery was made possible by the invention of the ophthalmoscope by German optical physiologists during the mid-nineteenth century (21). In 1861, Liebreich reported a survey of deaf inhabitants from Berlin in whom he noted the frequent presence of retinal pigmentation, in particular in Jewish consanguineous families (83). The existence of this dual sensory defect had been documented by several other reports, and its hereditability was well established when, in 1914, Charles Usher, a British ophthalmologist, looked at the incidence of deafness in 69 cases of retinitis pigmentosa (140). Following the description of the clinical heterogeneity of the syndrome, which was already outlined in the initial reports (12, 51), a classification into three subtypes, Usher type I (USH1), type II (USH2), and type III (USH3), was proposed in 1977 by Davenport & Omenn (31). During the 1990s, each of these clinical forms was shown to be genetically heterogeneous.

The study of USH may offer a unique opportunity to decipher the molecular bases of some of the developmental and functional similarities existing between the retinal and the cochlear sensory cells, i.e., the photoreceptor cells and the hair cells. Indeed, at least in some genetic forms of the disease, these cells have been shown to be the primary targets of the gene defect. What are the main structural features common to the photoreceptor cells (Figure 1, see color insert) and the auditory hair cell (Figure 2, see color insert)? First, both are ciliated cells: Photoreceptors possess a connecting cilium situated in a segment that links the internal segment to the external segment; embryonic auditory hair cells have an apical cilium, the kinocilium, which disappears soon after birth in mammals. Second, both cell types possess microvillar structures. In vertebrates, photoreceptors contain lamellar structures, forming discs where the phototransduction occurs; in invertebrates, discs are replaced by rhabdomeres, which are tightly packed genuine microvilli. The apical part of auditory hair cells carries up to 300 rigid microvilli, improperly named stereocilia, where the mechanotransduction takes place. Third, the synapses of both photoreceptors and auditory sensory cells have special characteristics: They are called ribbon synapses in reference to their plate-like presynaptic bodies to which synaptic vesicles are tethered. From a functional point of view, these synapses are characterized by a massive, rapid, and sustained discharge of neurotransmitters (147). Finally, the molecular analysis of Usher syndrome may also lead to the discovery of novel biological processes shared by cochlear and retinal sensory cells.

THREE CLINICAL SUBTYPES

The clinical classification of Usher syndrome, established by Davenport & Omenn is still in use (31). USH1 is the most severe form. It is characterized by severe to profound congenital sensorineural deafness, constant vestibular dysfunction (balance deficiency), and prepubertal onset retinitis pigmentosa. Vestibular dysfunction manifests clinically as a delay in motor development. Affected children cannot sit without support at 6 months of age and often do not walk before 18 months (100, 127). An ataxia of vestibular origin is occasionally observed. The retinopathy appears as a loss of night vision and a restriction of the visual field during childhood, and eventually, as a visual acuity loss that rapidly progresses to blindness. Anomalies of light-evoked electrical response of the retina can be detected as soon as 2 to 3 years of age by electroretinography, thus permitting early diagnosis of the disease. Later, fundus anomalies develop, which consist of pigment deposition likened to bone spicules around mid-periphery of the retina, thereafter extending inward and outward, and a narrowing of blood vessels. USH2 is mainly distinguished from USH1 by a less severe deafness, i.e., a mild hearing loss for low frequency sounds and a severe hearing loss for high frequency sounds, with the absence of the vestibular dysfunction. The loss of night vision develops around puberty (the age of onset partly overlaps that of USH1). The progression of the visual impairment seems to present a higher variability than in USH1 (145). Patients usually have fundoscopic anomalies beyond childhood (72). A third type, USH3, has been described as distinct from USH2 because of the progressiveness of the hearing loss and the occasional presence of vestibular dysfunction (47, 67, 68). In each of these forms, retinitis pigmentosa can be accompanied by cataract (9).

Initially, the respective proportions of USH1 and USH2 cases were reported to be 90% and 10% (51). This evaluation, however, was biased, as the examined patients were severely to profoundly deaf. In more recent analyses of patients from the United States and Northern Europe, USH1 and USH2 were reported between 33% and 44% and between 56% and 67%, respectively (38, 57, 108, 120). In Colombia however, 70% of the USH cases have been diagnosed as USH1 (132), a likely overestimated value as it is based on the screening of individuals attending schools for the deaf (whereas the degree of hearing loss of USH2 allows for a normal schooling). USH3 had first been evaluated at a very low percentage (31). Subsequently, the existence of this form was considered uncertain (127). The finding of USH3 in Finland, where it is the most common form of Usher syndrome (40% of cases) (107), due to a founder effect in this population, definitively established the reality of this form. Recently, USH3 was reported to account for 20% of the USH cases in the city of Birmingham (United Kingdom) (57).

Although the histopathological data on the syndrome are sparse and often related to undefined USH subtypes, examination of the temporal bone has been documented in some cases (131, 146). Whatever the supposed USH subtype, a severe degeneration of the organ of Corti (the auditory sensory epithelium) was observed, at least in the basal turn of the cochlea. The cochlear ganglion (devoted to afferent innervation) was atrophied. In USH1, the structure of the vestibular end organs varied from normal to a severe degeneration of certain neuroepithelial areas. No vestibular anomaly was detected in USH2 and USH3. In several patients affected by an undefined USH subtype, ciliar anomalies have been reported of both the connecting cilium of photoreceptor cells (11, 18, 60) and the cilium of olfactory receptor neurons (7). Slow sperm motility and an abnormal structure of the sperm cilium have also been described and proposed to account for patients' decreased fertility (60). Finally, bronchiectasis and reduced nasal mucociliary clearance, which were observed in some USH1 patients, might also be related to a ciliar dysfunction (18). Recently, sperm anomalies have been sought in USH2A patients, but no anomaly of the cilium could be detected (141); it is thus tempting to ascribe the reported ciliar anomalies to the USH1 form.

LOCI IDENTIFICATION*

The first USH locus to be mapped was USH2A, for Usher type II, in 1990. It was assigned to the distal part of chromosome 1q through the study of affected families from the United States and Europe (74). Next, the USH1A locus was found by analysis of a population from Poitou-Charentes, in the west of France (66); this locus has only been detected in this population so far. The first locus for USH3 was reported in 1995, in Finnish families (122). At present, evidence for 12 distinct genetic loci for Usher syndrome has been obtained (see Table 1), namely, 6 loci for USH1, USH1A at 14q32 (MIM 276900) (66), USH1B at 11q13.5 (MIM 276903) (73, 128), USH1C at 11p15.1 (MIM 276904) (128), USH1D at 10q (MIM 601067) (149), USH1E at 21q21 (MIM 602097) (24), USH1F (MIM 602083) on chromosome 10 (150); 4 loci for USH2, USH2A at 1q41 (MIM 276901) (74, 82), USH2B at 3p23-24.2 (MIM 276905) (56), USH2C at 5q14.3-21.3 (MIM 605472) (109), and a fourth locus not yet defined (156); 2 loci for USH3, USH3A (MIM 276902) at chromosome 3q21-25 (122) and 1 locus on the long arm of chromosome 20 (M. Mustapha, D. Weil, S. Chardenoux, R. Slim, & C. Petit, unpublished data). Of these 12 genetic forms, which all are transmitted on the autosomal recessive mode, USH1B and USH2A are the most prevalent. The former accounts for 30% to 60% of the USH1 cases (2, 8, 15, 63), the latter, for 74% to 82% of the USH2 cases (110, 156).

THE *USH* GENES

USH1B

The gene responsible for Usher 1B (*USH1B*) was the first USH gene to be isolated. This cloning was greatly facilitated by the study of the *shaker-1* (*sh-1*) mouse mutants (32, 118). Indeed, the *sh-1* phenotype had been assigned to the murine chromosomal region homologous to that defined for USH1B (22). *Sh-1*

---

*Disease loci and genes are referenced according to the Online Mendelian Inheritance in Man (OMIM) website nomenclature, (http://www.ncbi.nlm.nih.gov/Omim/searchomim.html).

TABLE 1   The genetic forms of Usher syndrome

Usher form	Chromosomal localization	Gene	Protein	DFNA/DFNB form	Mouse model
USH1A	14q32 (66)	—	—	—	—
USH1B	11q13.5 (73, 128)	*MYO7A* (152)	Myosin VIIA	DFNB2 (87, 153) DFNA11 (88)	*shaker-1* (42)
USH1C	11p15.1 (128)	*USH1C* (16, 143)	Harmonin	DFNB18?	—
USH1D	10q (149)	*CDH23* (17, 19)	Cadherin23	DFNB12 (19)	*waltzer* (33)
USH1E	21q21 (24)	—	—	—	—
USH1F	10 (150)	PCDH15 (4a)	Protocadherin15	DFNB23?	*Ames waltzer* (5)
USH2A	1q41 (74, 82)	*USH2A* (38)	Usherin	—	—
USH2B	3p23-24.2 (56)	—	—	DFNB6?	—
USH2C	5q14.3-21.3 (109)	—	—	—	—
USH3A	3q21-25 (122)	—	—	—	—

Princeps references are given.

mice exhibit head-tossing, circling behavior, and hyperactivity due to vestibular dysfunction. A rapidly progressive hearing loss accompanied by degeneration of the organ of Corti (70) is also exhibited. Although a lack of retinal degeneration had been reported in these mutants, the *sh-1* phenotype was hypothesized to involve the *USH1B* orthologue. A positional cloning strategy led to the identification of the murine *Myo7a* gene, predicted to encode the unconventional myosin VIIA, in the *sh-1* candidate interval. Because experimental evidence already supported a role of unconventional myosins in the auditory mechanotransduction process (43, 58), the myosin VIIA gene was considered a promising candidate for both *sh-1* and USH1B. Indeed, mutations in exons encoding the motor head of myosin VIIA were detected concurrently in *sh-1* mice (42) and in USH1B patients (152).

MYO7A consists of 48 coding exons (17 for the head, 3 for the neck, and 28 for the tail of the molecule) and encodes several alternatively spliced forms, mainly differing in their motor head. It extends over 100 kb (27, 154). Since the original report in 1995, a total of 81 different mutations have been detected in USH1B patients. They are distributed all along the gene, although somewhat clustered in the part encoding the motor head (2, 15, 29, 30, 37, 63, 81, 91, 152, 155, 157). About half of the mutations are missense or in-frame insertions/deletions. There is no indication for a hot spot of mutations. Of the 10 *sh-1* alleles, 7 have been analyzed; the mutations have roughly the same distribution and characteristics as the human mutations (42, 54, 96). Interestingly, among the category of zebrafish *circler* mutants that are characterized by defective hair cells of the auditory-vestibular and lateral line systems (i.e., two likely homologous sensory systems) (102), the five alleles of the *mariner* class carry *myo7a* mutations (36).

Unconventional myosins form a large family that is presently divided into 16 classes (14a). These motor proteins move along the actin filaments by using the energy generated by the hydrolysis of ATP. Their highly conserved N-terminal motor head possesses the actin- and ATP-binding sites. It is followed by a neck region and a tail that is highly divergent from one myosin to another. The tail sequence determines the functional specificity of each myosin, as it contains various putative protein-protein interacting domains that bind to cargo molecules, regulatory factors, and components of the transduction pathways (14a). Unconventional myosins have been implicated in the formation and the movements of cytoplasmic expansions, in the movements of vesicles, the transport of mRNA, and in signal transduction (14a, 92, 97, 135). The human myosin VIIA (Figure 3), 2215 amino acids in length (254 kDa), consists of a motor head domain of 729 amino acids, containing the ATP- and actin-binding motifs; a neck region of 126 amino acids, composed of five IQ (isoleucine-glutamine) motifs that are expected to bind calmodulin; and a long tail of 1360 amino acids (154). The actin-based motility of the head was demonstrated recently (139). The tail begins with a short coiled-coil domain (78 amino acids) that mediates the formation of homodimers (153). The coiled-coil domain is followed by two large repeats of about 460 amino acids. Each repeat contains a MyTH4 (myosin tail homology 4) domain, of unknown function, and a domain homologous to the membrane-binding domain of the proteins of the FERM (4.1, ezrin, radixin, moesin) superfamily (28). The tandem association of

Myosin VIIA

Figure 3   Schematic representation of myosin VIIA. This unconventional myosin consists of a motor head that contains the ATP- and actin-binding sites, a neck region composed of five isoleucine-glutamine (IQ) motifs expected to bind calmodulin, and a long tail of 1360 amino acids. The tail begins with a short coiled-coil domain, which is implicated in the formation of homodimers. The coiled-coil domain is followed by two large repeats, each containing a MyTH4 (myosin tail homology 4) and a FERM (4.1, ezrin, radixin, moesin) domain. The two repeats are separated by a poorly conserved SH3 (src homology-3) domain.

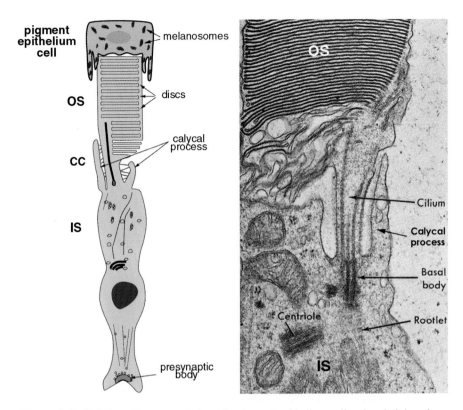

Figure 1 (*Left*) Schematic representation of a pigment epithelium cell and underlying photoreceptor cell. The photoreceptor cell (here a rod cell) is composed of an outer segment (OS), made up of a large number of discs stacked one above the other (containing the photosensitive molecule rhodopsin), a connecting cilium (CC) (see electron micrograph on the right), and an inner segment (IS) that contains the various cell organelles. The tip of the outer segment (OS) is continually being exfoliated and phagocytized by the overlying pigment epithelium cell. (*Right*) Electron micrograph of a rod cell connecting cilium [from Raviola (113)]. The outer and inner segments of the photoreceptor cell are connected by a modified cilium, which contains nine longitudinally oriented doublet microtubules that arise from a basal body in the end of the inner segment. In human, but not in mouse, the connecting cilium is surrounded by a calycal process.

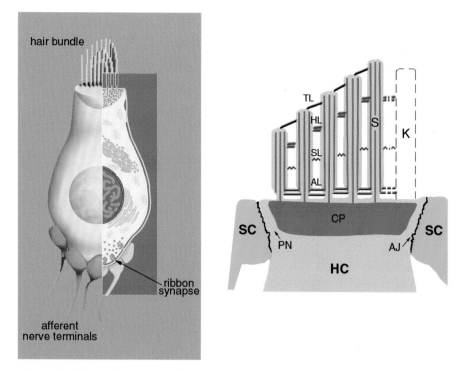

Figure 2 Schematic representation of an auditory inner hair cell (*left*) and detail of the cell apical region (*right*). Note the highly organized hair bundle, made up of several rows of stereocilia (S) at the apical pole of the cell. A stereocilium (K) is located at the vertex of the bundle during cell differentiation; it is no longer present in mature auditory hair cells. The stereocilia are interconnected by several types of lateral links, namely, the ankle links (AL), shaft links (SL), and horizontal links (HL). In addition, a tip link (TL) connects the apex of each stereocilium to the side of the taller adjacent one; tip links should gate the mechanotransduction channels. Three specific structures of the actin cytoskeleton are shown in red, namely, (*a*) the filaments of the stereocilia; (*b*) the cuticular plate (CP), a dense meshwork of horizontal filaments running parallel to the apical cell surface; and (*c*) the cortical network, beneath the plasma membrane. The tight junctions and adherens junctions (AJ) between the hair cell (HC) and the adjacent supporting cells (SC) are also shown. PN, periculticular necklace.

the MyTH4 and FERM domains, which is also present in myosin V, myosin XV, and a plant kinesin (14a, 27, 115, 148), argues in favor of a functional significance for this pairing. These two repeats are separated by a poorly conserved SH3 (src homology 3) domain (27, 96), a type of domain known to interact with proline-rich regions (101).

USH1C

In 1966, Kloepfer reported an unusually high frequency of USH patients in a small population of southwestern Louisiana who were descendents of French-speaking emigrants exiled from Acadia (Canada) in the eighteenth century (75). In 1994, this form was individualized as USH1C by linkage analysis and, for a long time, has only been observed in patients related to this community (10, 69, 128). The gene underlying USH1C was recently isolated using two different strategies, namely, a candidate gene approach (based on a subtracted cDNA library derived from microdissected sensory areas of the vestibular end organs) (143) and a positional cloning strategy (based on the analysis of a chromosomal deletion underlying a contiguous gene syndrome) (16). *USH1C* encodes a PDZ domain–containing protein, termed harmonin; it consists of 28 coding exons and spans about 51 kb (16, 143). So far, six mutations have been reported (16, 143, 163), none of which is a missense mutation. Which mutation(s) cause the disease in the Acadian affected individuals is still not clear. Indeed, two distinct causative mutations have been reported, namely, an expanded intronic VNTR proposed to affect transcriptional and/or posttranscriptional processes (143) and an exonic substitution that is consistent with the creation of a new splice site and leads to a small in-frame deletion and an unstable transcript (16). One mutation, 238-239insC, was detected in several European individuals (143) who share a common haplotype, thus arguing in favor of a founder effect (163). Mutation screening in a German population provided a minimal estimation of 12.5% of *USH1C* mutations among USH1 cases (163).

USH1C encodes a variety of alternatively spliced transcripts that result from the differential use of eight exons. In the murine inner ear, these transcripts predict at least eight harmonin isoforms, ranging from 420 to 910 amino acids in length (143). They can be distributed into three subclasses according to the number of domains they contain, namely, two or three PDZ domains, one or two coiled-coil domains, and the presence/absence of a PST (proline, serine, threonine) domain (Figure 4). The coiled-coil domain(s) might be implicated in the dimerization of the protein. Proline-rich domains serve as binding sites for SH3, WW (trp-trp), EVH1 [*Drosophila* enabled/VASP (vasodilator-stimulated phosphoprotein) homology 1] domains (129), and the actin-regulatory protein profilin (65, 98). PDZ domains are interacting domains, 80 to 90 amino acids in length, originally detected in P̲SD/SAP90 (postsynaptic density protein), D̲lg-A (*Drosophila* tumor suppressor), and Z̲O-1 (tight junction protein). The PDZ domain–containing proteins form a vast class of molecules, with one representative in bacteria (111). They are central organizers of high-order supramolecular complexes located at

Harmonin isoforms

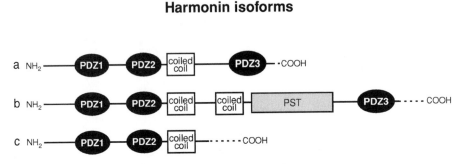

Figure 4   Schematic representation of the three predicted classes of harmonin isoforms. Class a isoforms contain three PDZ domains and one coiled-coil domain between PDZ2 and PDZ3. Class b isoforms contain an additional coiled-coil domain and a proline, serine, threonine (PST)-rich region. Class c isoforms contain only the first two PDZ domains and the first coiled-coil domain. Each class has two or three predicted isoforms, which differ in their internal sequences (class a) or in their C-terminal ends (class b and class c) [adapted from Verpy et al (143)].

specific emplacements of the plasma membrane. They bind not only to transmembrane proteins, in particular, ionic channels or transporters (50, 71, 126), but also to actin or actin-binding proteins. Hence, they cluster and coordinate the activity of various plasma membrane proteins and bridge them to the cortical cytoskeleton (39, 41, 125).

USH1D

The USH1D locus was defined by the study of a Pakistani family (149). The gene was recently isolated independently by two groups. One discovery was the result of a collaborative study aimed at identifying the gene defect underlying the phenotype of the *waltzer* (v) deaf mouse mutants (17). Indeed, based on mapping data (23, 162), these mutants had been proposed as the mouse model for USH1D, even though no retinal anomaly has been observed in these mice (5). The other group found the USH1D gene by a direct positional cloning strategy (19). This gene, *CDH23* (MIM 605516), comprises 69 coding exons and spans more than 300 kb (17). It encodes a cadherin-related protein.

Eight distinct mutations have been reported, namely, two nonsense mutations, one single-codon deletion, one missense mutation, three splice-site mutations that are predicted to preserve the reading frame, and one truncating splice-site mutation (17, 19). In the three v mutants analyzed, a nonsense and two splice-site mutations were found (33). USH1D seems to be the second most common genetic form of USH1 (8). Interestingly, individuals homozygous for the *CDH23* missense mutation present with a mild form of retinitis pigmentosa (almost undetectable in one patient) (17). A mild retinopathy was also observed in a family with a splice-site mutation preserving the reading frame (17).

Cadherin-23

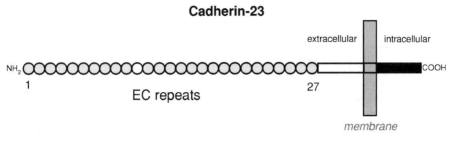

Figure 5   Schematic representation of cadherin-23. The protein ectodomain is composed of 27 extracellular cadherin (EC) repeats. The predicted cytoplasmic region (268 residues) has no homology with known proteins.

The predicted protein (Figure 5), termed cadherin-23, is composed of 3354 amino acids (17). Its ectodomain is composed of 27 extracellular cadherin (EC) repeats (17), most of which possess the characteristic Ca^{2+}-binding motifs of cadherins. These motifs are implicated in the calcium-dependent interaction between cadherins, which mediates cell-cell adhesion. The single transmembrane domain is followed by a 268 amino acid cytoplasmic region (17, 19). Cadherin-23 is more closely related to the subgroup of cadherins defined by the Drosophila fat, dachsous, and the human MEGF1 proteins, which are characterized by a large number of EC repeats (103, 161). However, its intracytoplasmic part shows no homology with known proteins. In particular, no clear β-catenin–binding motif can be detected.

USH2A

The gene underlying USH2A has been isolated by a positional cloning strategy; it comprises 21 exons, with the first exon being entirely noncoding (38). *USH2A* encodes a predicted 1551 amino acid (171 kDa) protein. This protein (Figure 6) begins with a signal peptide and has a modular structure. It is composed of a N-terminal 285 amino acid unclassified domain, followed by a laminin domain VI module (about 200 residues) (156), a tandem array of 10 laminin-type EGF-like (LE) modules, and four fibronectin-like type III (F3) repeats. Of particular interest are the other types of proteins that contain domain VI motifs, namely, laminin chains $\alpha 1$, $\alpha 2$, $\alpha 5$, $\beta 1$, $\beta 2$, $\beta 3$, $\gamma 1$, $\gamma 3$, and netrins. Laminins are the major noncollagenous components of basement membranes, and they are also present in some mesenchymal compartments. Netrins are small diffusable proteins involved in axon guidance. Therefore, the USH2A protein may be a novel extracellular matrix protein (38, 156).

A total of 20 pathogenic mutations have been reported in USH2A patients, namely, 7 nonsense, 7 frameshift, 1 splice site, and 5 missense (of which 4 are in the laminin type VI domain) mutations (3, 38, 156). Interestingly, the 2299delG (previously termed 2314delG) mutation appears particularly frequently, as it has been detected in 25% of a population of Spanish USH2 families (14) and in about

Usherin

Figure 6   Schematic representation of the *USH2A* encoded protein (usherin). The protein precursor has a signal peptide (SP), which is followed by a 285 amino acid unclassified domain, a laminin domain VI-like (L6) module, 10 laminin type EGF-like (LE) modules, and 4 fibronectin-like type III (F3) repeats. The C-terminal region (50 amino acids) has no detectable homology.

20% of USH2A families from Europe and the United States (156). Finally, the discovery of 4 different missense mutations in the laminin type VI domain of the USH2A protein marks this region for a potentially critical role in the cochlea and retina.

TOWARD PATHOGENESIS

Some clues for understanding the pathogenesis of a genetically heterogeneous disease can be obtained from the pathohistological anomalies observed in animal models, the expression patterns of the underlying genes, the cellular and subcellular localization of the encoded proteins, and the elucidation of the network of molecular interactions in which these proteins are involved. Because so far the genes causative of Usher syndrome have mainly been cloned by groups working in the field of hereditary deafness, the knowledge acquired about the pathogenesis of the disease relates more to the cochlear than to the retinal dysfunction. Presently, more information is available for USH1B than for USH1C and USH1D. Preliminary results for USH2A are also presented.

Usher 1B

COCHLEAR PHENOTYPE: A PRIMARY SENSORY CELL DEFECT   The hair bundle of inner ear sensory cells is a highly organized structure that displays a polarity axis in the plane of the epithelium. This planar polarity also manifests by the coordinated orientation of the hair bundles of the different hair cells. Within a bundle, the stereocilia are arranged in several rows of gradually increasing size, resembling a staircase (Figure 7A). The hair bundles have a U, V, or W shape. The kinocilium is located eccentrically at the vertex of the bundle; its position defines the vectorial orientation of the bundle. The hair bundle is also functionally polarized; its deflection in the direction of the kinocilium opens the mechanotransduction channels via the stretching of apical links, called the tip links (see Figure 2, see color insert), which should gate these channels (59). In the most severely affected

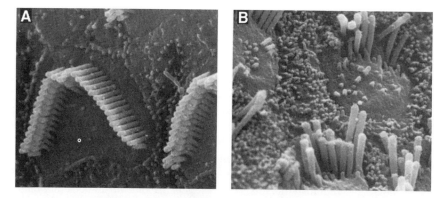

Figure 7   Scanning electron micrographs of the apical surface of auditory hair cells in a wild-type mouse (*A*) and a *shaker-1* mouse (*B*), showing the disorganization of the stereocilia bundle in the mutant lacking a functional myosin VIIA [from Self et al (124)].

sh-1 mouse mutants, the kinocilium is misplaced and the stereocilia display an irregular arrangement (124) (Figure 7*B*). This disorganization of the hair bundle progressively increases after birth; stereocilia form small clumps and splay out (124). Therefore, *sh-1* mutants can be regarded as planar polarity mutants (34). Into this category fall several Drosophila mutants affected in the polarity of hairs, sensory bristles, or ommatidia, which have permitted a genetic dissection of the underlying signaling pathway (99, 116). Erratic positions of the stereocilia are also observed in *mariner* zebrafish mutants (36). Moreover, although hair cells from *sh-1* mice are able to transduce (117, 124), whereas hair cells from *mariner* fishes seem unable to do so (102), this apparent discrepancy may in fact reflect differences in the experimental conditions of this testing because mechanotransduction in *sh-1* mutants requires much greater displacements of the hair bundle than in wild-type animals (76). Therefore, it is reasonable to assume that the disorganization of the hair bundle plays a major role in the hearing loss of USH1B patients. Finally, the hair cells of both *sh-1* mice and *mariner* zebrafish have lost the capacity to accumulate aminoglycosides. Whereas this has been initially interpreted as a defect in an apical endocytic pathway (117, 123), recent results suggest that aminoglycosides may enter the sensory cells via the mechanotransduction channel (G. Richardson, personal communication).

Myosin VIIA is synthesized by numerous cell types; all are epithelial cells that, in most cases, possess microvilli and/or cilia (121). Immunoelectron microscopy has shown that, in these cells, myosin VIIA is present all along the microvilli and the cilia (158). In the mouse inner ear, myosin VIIA is first observed at embryonic day 10 (E10) in the otic vesicle (121). Later, in the cochlea, the protein is restricted to the inner and outer hair cells in all the mammalian species tested (35, 53, 154). It is present in the whole cell body and along the kinocilium (158) and stereocilia

(35, 53, 154). However, the strongest immunoreactivity is associated with the pericuticular necklace; this is a narrow apical region situated between the actin apical belt (associated with the adherens junctions) and the cuticular plate (a dense apical network of horizontal filaments in which the actin filament rootlets of the stereocilia are anchored) (see Figure 2). This region is characterized by the presence of a large pool of coated vesicles (64). In the frog, an intense immunoreactivity has been detected at the base of the stereocilia (52), at the precise emplacements of the ankle links that, in this species, form the major subset of the lateral links that interconnect the stereocilia of the bundle [(44, 133); also see Figure 2].

Recent insight into the function of myosin VIIA at the cellular level provided clues to the understanding of some of the phenotypic anomalies observed in the mouse and zebrafish mutants. *Dictyostelium discoideum* amoeba null mutants for a myosin that may be the orthologue of the mammalian myosin VIIA display impaired phagocytosis (134) and migration, which is related to a defective adhesion to particles and to substrate, respectively (138). Moreover, two ligands of the C-terminal FERM domain of the mammalian myosin VIIA have been characterized. One is the type Iα regulatory subunit of the cAMP-dependent protein kinase (77), which suggests that the activity of myosin VIIA and/or associated molecules could be modulated by cAMP-dependent phosphorylation. The other ligand, vezatin, is a novel ubiquitous transmembrane protein of the adherens junctions (78). Vezatin interacts also with the cadherin-catenin junctional complex, which is involved in cell-cell adhesion and is linked to actin. At the junctions between hair cells and supporting cells (see Figure 2), myosin VIIA, anchored by vezatin, is expected to create a tension force and thus to strengthen cell-cell adhesion (1). The stabilization of these junctions could in turn stabilize the cuticular plate, hence the stereocilia, even though the cuticular plate seems to be attached to the adherens junctions only on the side of the shortest stereocilia (L. Tilney, personal communication). Interestingly, vezatin is also present at the base of the stereocilia, where it is tightly associated with the ankle links (78) (also see Figure 2). By its dynamic activity, the vezatin–myosin VIIA complex at the base of the stereocilia is expected to actively contribute to the cohesion of the hair bundle. Therefore, the loss of a functional vezatin–myosin VIIA complex at both the adherens junctions and the base of the stereocilia is likely to account for the splaying out of the stereocilia observed in myosin VIIA defective animals. Moreover, the presence of myosin VIIA all along the stereocilia in mammals suggests that myosin VIIA also interacts with the other lateral links of the stereocilia (see Figure 2). A possible role at the level of the tip link should also be considered, the loss of which may contribute to the abnormal transduction observed in the myosin VIIA defective mouse and zebrafish mutants (76, 102). Finally, as suggested by the abnormal position of the kinocilium in *sh-1* mice, myosin VIIA at cell-cell junctions might also be implicated in the, as yet undescribed, reorganization and asymetric distribution of the microtubules and microfilaments that should accompany the movement of the kinocilium toward its final peripheral position during the establishment of the planar polarity (55, 136).

RETINAL PHENOTYPE   In the mouse retina, myosin VIIA is first detected at E12 in the outer layer of the optic cup, which will give rise to the pigment epithelium cells. Thereafter, the protein is present in both the pigment epithelium cells and the photoreceptor cells in all species tested [although for unclear reasons, some antibodies failed to detect it in the rodent photoreceptors (35)]. Myosin VIIA is localized in the apical processes of the pigment epithelium cells, some of which are involved in the phagocytosis of the outer disks that are exfoliated from the photoreceptors (35, 53, 158) (see Figure 1, see color insert). In the photoreceptors, the protein is present at the periphery of the connecting cilium, i.e., the slender stalk that joins the inner and outer segments (see Figure 1). Finally, myosin VIIA has also been detected in the synaptic region of chick and macaque photoreceptor cells (35).

No sign of retinal degeneration at any age has ever been observed in the *sh-1* mouse mutants. Recently however, electroretinographic anomalies (a reduction of the amplitude of the wave response to light) have been reported in some *sh-1* mouse mutants carrying a deleterious allele. The pigment epithelium cells have only been examined in a *sh-1* mutant with a mildly deleterious allele. Phagosomes were found to be normal, but the melanosomes did not extend in the apical processes as they did in wild-type mice (85) (Figure 1). In the *sh-1* mice, no morphological anomaly of the photoreceptors has been reported to date (in particular, no misposition of the connecting cilium evoking a polarity defect). However, a significant decrease of the opsin flow at the level of the connecting cilium has been observed (86, 160). Myosin VIIA and actin, which are present along the connecting cilium (159), thus, may participate in the ciliary transport of opsins (86, 160).

The visual loss in retinitis pigmentosa corresponds to the degeneration of photoreceptor cells. The molecular studies have revealed that, in most cases, the primary defects causing these cells to die are either photoreceptor cell or pigment epithelium cell defects (112). Because both cell types express myosin VIIA, the cellular origin of the retinal disease in USH1B cannot be unambiguously established. The defect in melanosome transport in the pigment epithelium cells is not expected to be causative of the USH1B retinal degenerescence (85). However, with respect to the outer disk phagocytosis function of the pigment epithelium cells, the abnormal phagocytosis that has been observed in amoeba null mutants for the putative orthologue of myosin VIIA is intriguing (134).

Usher 1C

Ush1C is transcribed in all the murine tissues tested, with a higher expression in the eye, inner ear, kidney, and small intestine. Immunohistolabeling experiments have shown the presence of harmonin in all the hair cells of the inner ear (143); the protein is detected in the whole cell body and in the stereocilia. In the retina, harmonin is present in the photoreceptor layer (16). Therefore, the cochlear and retinal dysfunctions in USH1C may result from primary sensory cell defects. The various types of cells express different transcripts, with a greater variety in the inner ear than in the eye; moreover, some transcripts predicted to encode b isoforms (see Figure 4) have thus far only been detected in the inner ear. Very little information is available today on the pathogenesis of USH1C because no animal model exists. However, we consider the possibility of an interaction between myosin VIIA and harmonin as an attractive hypothesis, based on the following considerations: (*a*) These proteins, when defective, underlie the same phenotype; (*b*) they colocalize in the same cells, in particular in epithelial cells harboring microvilli; (*c*) PDZ domain–containing proteins are often associated with the sites of cell-cell contacts (80, 94, 99, 116); and (*d*) some PDZ domain–containing proteins interact with proteins of the FERM family (93, 114). The diverse harmonin isoforms predict their involvement in distinct molecular complexes. Major functional information will certainly be obtained from the identification of the transmembrane proteins to which harmonins bind.

Usher 1D

Cdh23 is transcribed in a large variety of murine tissues (19, 33). The gene is highly expressed in the retina. In the inner ear, *Cdh23* expression is restricted to the hair cells (33). In the *waltzer* mutants, the kinocilium is mispositioned, the hair bundle becomes progressively disorganized, and the stereocilia clump together (33), i.e., a series of anomalies resembling those of myosin VIIA defective mutants. Hence, *waltzer* can also be classified as a planar polarity mutant. Interestingly, the Drosophila fat and dachsous cadherins, which belong to the same cadherin subclass as cadherin-23 (161), are involved in epithelial planar polarity (4). Although the cellular distribution of cadherin-23 remains unknown, the usual localization of cadherins at membrane-membrane interacting sites in conjunction with the *waltzer* phenotype argue in favor of a role of cadherin-23 at the junctions with the supporting cells and/or at the level of lateral links. Interestingly, the cadherins present at the junctions between hair cells and supporting cells in the adult mouse have not been characterized yet (79). Moreover, it is noteworthy that in the presence of a calcium chelator, some of the lateral links are proteolyzed (44) and the tip link is disrupted (95). Because cadherins also undergo proteolysis in the absence of calcium (61, 106), cadherins might be components of these links. Because the myosin VIIA ligand vezatin interacts with the cadherin-catenin complex, an exciting possibility is that myosin VIIA and cadherin-23 belong to the same macromolecular complex required to transmit adhesion forces to the cytoskeleton. Arguing in favor of this interaction, double heterozygote *v/+ sh-1/+* mice become deaf at 3 to 6 months of age, whereas *v/+* and *sh-1/+* single heterozygote animals have normal hearing (32).

As discussed above, in the absence of a retinal degeneration in USH1B and USH1D mouse models and because of the lack of information on the cellular and/or subcellular localization of vezatin, harmonin, and cadherin-23 in the retina, the pathogenesis of the USH1 retinitis pigmentosa remains highly speculative. However, the absence of retinitis pigmentosa in both the *sh-1* and the *waltzer* deaf mouse mutants argues in favor of the involvement of the *USH1* gene products in a common process in the retina too. Similar to the adhesion defect in the auditory hair cells proposed in USH1, the myosin VIIA signaling pathway may act at some of the membrane-membrane adhesion sites of the photoreceptor cells. These include the synaptic junctions, the junctions with the Müller glial cells, and the contact sites of the calycal process (emerging from the inner segment) (44) with the connecting cilium and the basal part of the outer segment (see Figure 1). A possible role at this latter emplacement should be considered, especially because (*a*) mouse photoreceptor cells lack calycal processes, thus potentially explaining the retinal phenotype discrepancy between humans and mouse, and (*b*) the fibrous links that connect the calycal process to the connecting cilium and outer segment share some molecular components with the lateral links of the hair cells (44).

Usher 2A

No animal model of USH2A is yet available. In situ hybridization studies did not detect the *USH2A* transcript in either the photoreceptor cells or the auditory sensory cells; but rather, it was found in pigment epithelial cells and a variety of epithelial cells of the cochlea, including endothelial cells of the stria vascularis. Immunohistolabeling experiments have revealed usherin as a component of basement membranes in the retina and cochlea (15a). These preliminary data indicate that the cochlear and retinal sensory cells are unlikely to be the primary targets of the USH2A genetic defect.

USH GENES AND ISOLATED DEAFNESS OR RETINITIS PIGMENTOSA

In 1994, the second reported locus responsible for an autosomal recessive form of isolated deafness, DFNB2 (MIM 600060), was mapped to a 6 cM interval overlapping the candidate region for USH1B, through the study of a large consanguineous family living in the central part of Tunisia (49). Thereafter, several other USH loci and DFNB loci were assigned to overlapping genomic regions (see Table 1), namely, USH1C (69) and DFNB18 (MIM 602902) (62), USH1D (149) and DFNB12 (MIM 601386) (25), USH1F (150) and DFNB23 (http://www.uia.ac.be/dnalab/hhh/), USH2B (56) and DFNB6 (MIM 600971) (40), USH3 (122) and DFNB15 (MIM 601869) (26). Moreover, a dominant form of isolated deafness, DFNA11, was also mapped to the same chromosomal region as USH1B and DFNB2. Although this suggested that *USH* genes could also be causative genes for isolated deafness, the candidate intervals initially defined for most of the loci were very large (i.e., exceeded 10 cM). However, it is now established that *MYO7A*, which underlies USH1B, also underlies the isolated forms of deafness DFNB2 (87, 153) and DFNA11 (88). Moreover, strong indication supporting the involvement of *CDH23* and *USH1C* in DFNB12 (19) and DFNB18

(X-Z. Liu, personal communication), respectively, exists. In addition, *USH2A* underlies isolated retinitis pigmentosa (119). Finally, the idea of a continuum between the syndromic and nonsyndromic situations tends to emerge from several recent studies reporting variations of the phenotype within a given family (13, 89, 90, 119).

USH1B and Isolated Deafness

>In addition to the aforementioned Tunisian family (49), two small DFNB2 families from China have been identified (87). In the three DFNB2 families, some patients complained of balance problems and/or had abnormal vestibular tests. The hearing loss was severe to profound and congenital in the Chinese families, whereas it was progressive in the Tunisian family (the age of onset ranged from birth to 16 years). Notably, in the Tunisian family, one adult homozygous for the causative mutation had normal hearing. Moreover, a few of the 25 members of this family who carry the biallelic mutation subsequently developed signs of retinal degeneration, whereas others of the same age did not (13). In the Tunisian family, a G>A substitution in the last nucleotide of exon 16 of *MYO7A* was detected (153). This mutation is predicted both to substitute isoleucine for methionine 599 in the motor head of myosin VIIA and to affect the splicing of the mutated exon, resulting in the skipping of this exon in a large proportion of the mature transcripts (151). The expected product of the abnormally spliced transcript is a defective myosin VIIA chain with a 46 amino acid in-frame deletion in the motor head, which might dimerize with the normal chain and thus exert an intra/inter allelic dominant negative effect. One Chinese family was homozygous for the amino acid substitution Arg244Pro located in the motor head of myosin VIIA (87). Affected members of the second Chinese family were compound heterozygotes, who carried a 1 bp insertion in exon 28, leading to the truncation of the tail, on one allele and an acceptor splice site mutation of intron 3 on the other allele (87). Whether the type of the mutation accounts for the expression of either a USH or a DFNB phenotype is difficult to determine. Indeed, on the one hand, in none of the three DFNB2 families do the detected mutations unequivocally predict the complete absence of the protein in the patients. On the other hand, the large phenotypic heterogeneity observed in the Tunisian family, ranging from the lack of any sensory defect to complete Usher syndrome (13), strongly argues in favor of the role of genetic and/or environmental modulating factors. It is noteworthy that the dual sensory dysfunction that was observed in some of the Tunisian patients resembles that of USH3. Along the same line, in one family affected by an atypical Usher syndrome with a progressive hearing loss (i.e., most closely related to USH3), mutations in *MYO7A* have been detected. These patients were compound heterozygotes who carried two missense mutations, namely, a Leu651Pro substitution in the motor head and an Arg1602Gln substitution at the C-terminal end of the first FERM domain (90). As the latter mutation has also been observed in an authentic USH1 family (155), the milder phenotype in this USH3-like family has been proposed to be the result of a genetic background effect (even

though the precise impact of the motor head Leu651Pro mutation was unknown) (90).

In the single DFNA11 family reported to date, patients presented with a moderate bilateral hearing loss on all sound frequencies, which appeared in the first decade and then progressed. Vestibular anomalies were detected in some of the individuals. In this family, a 9 bp in-frame deletion located in exon 23, which encodes the coiled-coil domain, was detected (88). The mutated protein may form an inactive dimer with the wild-type myosin VIIA chain (153) and thus have a dominant negative effect. The persistence of some functional wild-type dimers would account for the moderate hearing loss and the absence of retinitis pigmentosa.

In summary, mutations in *MYO7A* can result not only in USH1 but also in USH3-like phenotypes, as well as in dominant or recessive isolated deafness with any degree of severity, congenital or progressive, with or without vestibular dysfunction. So far, no *MYO7A* mutation that leads to isolated retinitis pigmentosa has been discovered, meaning that, for unknown reasons (e.g., a myosin activity with a different critical threshold concentration, a different protein turnover, a cell restricted redundancy, etc.), mutations in myosin VIIA have a more deleterious effect in the inner ear than in the retina. At present, in some cases, the involvement of the genetic background and/or environmental factors in the clinical features can be inferred from distinct phenotypes either present within the same family or associated with the same mutation in different families. In contrast, in the absence of a functional testing of the activity of the various mutated forms of myosin VIIA, the individual effect of a given mutation on the phenotype remains speculative.

USH1D and Isolated Deafness

Whereas only deletions and nonsense or frameshift mutations in *CDH23* have been found in authentic USH1D patients (see above), missense mutations have been detected in several families affected by the isolated form of deafness DFNB12, which is characterized by severe to profound hearing loss (19). Although the entire gene has not yet been explored in these patients, we assume that these mutations are pathogenic. Accordingly, we suggest that the mutation type may play a crucial role in the phenotypic expression of *CDH23* defects. In addition, it is worth noting that the gene encoding protocadherin-15, which is defective in the deaf mouse mutant *Ames waltzer* (5) [which also displays a disorganization of the hair bundle (6)], has been proposed to underlie USH1F and the isolated deafness DFNB23 (5) (see Note Added in Proof, p. 297).

USH2A and Isolated Retinitis Pigmentosa

The *USH2A* 2299delG mutation is present at the homozygous state not only in typical USH2 patients, but also in USH3-like patients who present late onset progressive deafness that is occasionally associated with vestibular dysfunction (89).

The genetic background is not the only source of these differences because they were also observed in monozygotic twins (89). In addition, this mutation has also been detected at the heterozygous state in 4% of patients affected by a recessive form of isolated retinitis pigmentosa who only in the second instance reported mild subjective hearing impairment (119). Finally, a Cys759Phe missense mutation in the fifth LE motif of usherin has been observed in about 4.5% of the cases of recessive retinitis pigmentosa without an auditory defect in North America (119). These patients were either homozygotes for the mutation or compound heterozygotes for one of the aforementioned microdeletions detected in USH2 patients (38). Therefore, mutations in *USH2A* may account for a substantial proportion (i.e., more than 4.5%) of isolated retinitis pigmentosa (119).

CONCLUDING REMARKS

From the recent progress made in characterizing the molecular bases of Usher syndrome, some new ideas are currently emerging about the pathogenesis of this clinically and genetically heterogeneous disorder. Three of them deserve particular mention:

- The results obtained so far on both USH1 B, C, D genetic forms and USH2A indicate that distinct pathogenic processes are likely to account for the different clinical subtypes of the syndrome. In contrast, the various genetic forms of USH1 could result from molecular defects of the same signaling pathway.

- The misposition of the kinocilium during the development of the hair bundle of the auditory sensory cell, which is observed in the mouse models of USH1B and USH1D as well as the anomaly of the connecting cilium of the photoreceptor cell, which has been reported in some USH patients, suggest the existence of a planar polarity defect in USH1. Hence, clarifying USH1 pathogenesis might contribute to the elucidation of the mechanisms involved in the generation of planar polarity in mammals.

- At least three *USH* genes also underlie isolated auditory or visual sensory defects. Moreover, the idea of phenotypic continuums between the syndromic and nonsyndromic situations tends to emerge from several recent reports and to substitute for the classical nosologic frontiers. The analysis of additional mutations in *USH* genes, combined with the study of mutated gene products and careful clinical descriptions, should permit a more accurate estimation of possible direct phenotype-genotype correlation. Perhaps more importantly, these studies should also provide the basis for wide-scale segregation analyses aimed at identifying "modifier genes" in the rest of the genome. From this knowledge, we should gain additional insight into the pathogenic processes involved in Usher syndrome and thereby be guided in the search for targeted therapies.

ACKNOWLEDGMENTS

I am especially grateful to Jean-Pierre Hardelin for his many suggestions to improve the manuscript and to Jacqueline Levilliers for her warm and constant help in the preparation of the manuscript. I thank Aziz El-Amraoui and Sébastien Chardenoux for drawing the figures and Dominic Cosgrove for sharing unpublished results on *USH2A* expression. I also wish to thank Fondation R. & G. Strittmatter (France) and A. & M. Suchert and Forschung contra Blindheit-Initiative Usher Syndrome (Germany) for supporting our research on Usher syndrome.

Visit the Annual Reviews home page at www.AnnualReviews.org

LITERATURE CITED

1. Adams CL, Nelson WJ. 1998. Cytomechanics of cadherin-mediated cell-cell adhesion. *Curr. Opin. Cell Biol.* 10:572–77

2. Adato A, Weil D, Kalinski H, Pel-Or Y, Ayadi H, et al. 1997. Mutation profile of all 49 exons of the human myosin VIIA gene, and haplotype analysis, in Usher 1B families from diverse origins. *Am. J. Hum. Genet.* 61:813–21

3. Adato A, Weston MD, Berry A, Kimberling WJ, Bonné-Tamir A. 2000. Three novel mutations and twelve polymorphisms identified in the USH2A gene in Israeli USH2 families. *Hum. Mutat.* 15:388

4. Adler PN, Charlton J, Liu J. 1998. Mutations in the cadherin superfamily member gene *dachsous* cause a tissue polarity phenotype by altering *frizzled* signaling. *Development* 125:959–68

4a. Ahmed ZM, Riazaddin S, Bernstein SL, Ahmed Z, Khan S, et al. 2001. Mutations of the protocadherin gene *PCDH15* cause Usher syndrome type 1F. *Am. J. Hum. Genet.* 69: In press

5. Alagramam KN, Murcia CL, Kwon HY, Pawlowski KS, Wright CG, et al. 2001. The mouse Ames waltzer hearing-loss mutant is caused by mutation of *Pcdh15*, a novel protocadherin gene. *Nat. Genet.* 27:99–102

6. Alagramam KN, Zahorsky-Reeves J, Wright CG, Pawlowski KS, Erway LC, et al. 2000. Neuroepithelial defects of the inner ear in a new allele of the mouse mutation Ames waltzer. *Hear. Res.* 148:181–91

7. Arden GB, Fox B. 1979. Increased incidence of abnormal nasal cilia in patients with retinitis pigmentosa. *Nature* 279:534–36

8. Astuto LM, Weston MD, Carney CA, Hoover DM, Cremers CWRJ, et al. 2000. Genetic heterogeneity of Usher syndrome: analysis of 151 families with Usher type I. *Am. J. Hum. Genet.* 67: 1569–74

9. Auffarth GU, Tetz MR, Krastel H, Blankenagel A, Volcker HE. 1997. Complicated cataracts in various forms of retinitis pigmentosa. Type and incidence. *Ophthalmologe* 94:642–46

10. Ayyagari R, Nestorowicz A, Li Y, Chandrasekharappa S, Chinault C, et al. 1996. Construction of a YAC contig encompassing the Usher syndrome type 1C and familial hyperinsulinism loci on chromosome 11p14-15.1. *Genome Res.* 6:504–14

11. Barrong SD, Chaitin MH, Fliester PS, Possin DE, Jacobson SG, et al. 1982. Ultrastructure of connecting cilia in different forms of retinitis pigmentosa. *Arch. Ophthalmol.* 110:706–10

12. Bell J. 1922. Retinitis pigmentosa and allied diseases. In *Retinitis Pigmentosa and*

Allied Diseases, ed. K Pearson, pp. 1–29. London: Cambridge Press

13. Ben Zina Z, Masmoudi S, Ayadi H, Chaker F, Ghorbel AM, et al. 2001. From DFNB2 to Usher syndrome: variable expressivity of the same disease. *Am. J. Med. Genet.* 101:181–83

14. Beneyto MM, Cuevas JM, Millan JM, Espinos C, Mateu E, et al. 2000. Prevalence of 2314delG mutation in Spanish patients with Usher syndrome type II (USH2). *Ophthalmic Genet.* 21:123–28

14a. Berg JS, Powell BC, Cheney RE. 2000. A millennial myosin census. *Mol. Biol. Cell* 12:780–94

15. Bharadwaj AK, Kasztejna JP, Huq S, Berson EL, Dryja TP. 2000. Evaluation of the myosin VIIA gene and visual function in patients with Usher syndrome type I. *Exp. Eye Res.* 71:173–81

15a. Bhattacharya G, Miller C, Kimberling W, Jablonski MM, Cosgrove D. 2001. Localization and expression of usherin: a novel basement membrane protein defective in people with Usher syndrome IIa. *Hear. Res.* In press

16. Bitner-Glindzicz M, Lindley KJ, Rutland P, Blaydon D, Smith VV, et al. 2000. A recessive contiguous gene deletion causing infantile hyperinsulinism, enteropathy and deafness identifies the Usher type 1C gene. *Nat. Genet.* 26:56–60

17. Bolz H, von Brederlow B, Ramirez A, Bryda EC, Kutsche K, et al. 2001. Mutations of *CDH23*, encoding a new member of the cadherin gene family, causes Usher syndrome type 1D. *Nat. Genet.* 27:108–12

18. Bonneau D, Raymond F, Kremer C, Klossek J-M, Kaplan J, et al. 1993. Usher syndrome type I associated with bronchiectasis and immotile nasal cilia in two brothers. *J. Med. Genet.* 30:253–54

19. Bork JM, Peters LM, Riazuddin S, Bernstein SL, Ahmed ZM, et al. 2001. Usher syndrome 1D and nonsyndromic autosomal recessive deafness DFNB12 are caused by allelic mutations of the novel cadherin-like gene *CDH23*. *Am. J. Hum. Genet.* 68:26–37

20. Boughman JA, Vernon M, Shaver KA. 1983. Usher syndrome: definition and estimate of prevalence from two high-risk populations. *J. Chronic Dis.* 36:595–603

21. Brix PW, Decher G, von Feilitzsch FCO, Grashof F, Harms F, et al. 1856–1866. *Allgemeine Encyclopädie der Physik.* Leipzig: Leopold Voss

22. Brown KA, Sutcliffe MJ, Steel KP, Brown SDM. 1992. Close linkage of the olfactory marker protein gene to the mouse deafness mutation *shaker-1*. *Genomics* 13:189–93

23. Bryda EC, Ling H, Flaherty L. 1997. A high-resolution genetic map around waltzer on mouse chromosome 10 and identification of a new allele of waltzer. *Mamm. Genome* 8:1–4

24. Chaïb H, Kaplan J, Gerber S, Vincent C, Ayadi H, et al. 1997. A newly identified locus for Usher syndrome type I, *USH1E*, maps to chromosome 21q21. *Hum. Mol. Genet.* 6:27–31

25. Chaïb H, Place C, Salem N, Dodé C, Chardenoux S, et al. 1996. Mapping of DFNB12, a gene for a non-syndromal autosomal recessive deafness, to chromosome 10q21–22. *Hum. Mol. Genet.* 5:1061–64

26. Chen AH, Wayne S, Bell A, Ramesh A, Srisailapathy CRS, et al. 1997. New gene for autosomal recessive non-syndromic hearing loss maps to either chromosome 3q or 19p. *Am. J. Med. Genet.* 71:467–71

27. Chen Z-Y, Hasson T, Kelley PM, Schwender BJ, Schwartz MF, et al. 1996. Molecular cloning and domain structure of human myosin-VIIa, the gene product defective in Usher syndrome 1B. *Genomics* 36:440–48

28. Chishti AH, Kim AC, Marfatia SM, Lutchman M, Hanspal M, et al. 1998. The FERM domain: a unique module involved in the linkage of cytoplasmic

proteins to the membrane. *Trends Biochem. Sci.* 23:281–82

29. Cuevas JM, Espinos C, Millan JM, Sanchez F, Trujillo MJ, et al. 1998. Detection of a novel Cys628STOP mutation of the myosin VIIA gene in Usher syndrome type Ib. *Mol. Cell. Probes* 12:417–20

30. Cuevas JM, Espinos C, Millan JM, Sanchez F, Trujillo MJ, et al. 1999. Identification of three novel mutations in the MYO7A gene. *Hum. Mutat.* 14:181

31. Davenport SLH, Omenn GS. 1977. *The Heterogeneity of Usher Syndrome.* Int. Conf. Birth Defects, Vth, Montreal

32. Deol MS. 1956. The anatomy and development of the mutants pirouette, shaker-1 and waltzer in the mouse. *Proc. R. Soc. London Ser. B* 145:206–13

33. Di Palma F, Holme RH, Bryda EC, Belyantseva IA, Pellegrino R, et al. 2001. Mutations in *Cdh23*, encoding a new type of cadherin, cause stereocilia disorganization in waltzer, the mouse model for Usher syndrome type 1D. *Nat. Genet.* 27:103–7

34. Eaton S. 1997. Planar polarization of *Drosophila* and vertebrate epithelia. *Curr. Opin. Cell Biol.* 9:860–66

35. El-Amraoui A, Sahly I, Picaud S, Sahel J, Abitbol M, et al. 1996. Human Usher IB/mouse *shaker-1*; the retinal phenotype discrepancy explained by the presence/absence of myosin VIIA in the photoreceptor cells. *Hum. Mol. Genet.* 5:1171–78

36. Ernest S, Rauch G-J, Haffter P, Geisler R, Petit C, et al. 2000. *Mariner* is defective in *myosin VIIA*: a zebrafish model for human hereditary deafness. *Hum. Mol. Genet.* 9:2189–96

37. Espinos C, Millan JM, Sanchez F, Beneyto M, Najera C. 1998. Ala397Asp mutation of *myosin VIIA* gene segregating in a Spanish family with type-Ib Usher syndrome. *Hum. Genet.* 102:691–94

38. Eudy JD, Weston MD, Yao S, Hoover DM, Rehm HL, et al. 1998. Mutation of a gene encoding a protein with extracellular matrix motifs in Usher syndrome type IIa. *Science* 280:1753–57

39. Fanning AS, Anderson JM. 1999. PDZ domains: fundamental building blocks in the organization of protein complexes at the plasma membrane. *J. Clin. Invest.* 103:767–72

40. Fukushima K, Ramesh A, Srisailapathy CRS, Ni L, Wayne S, et al. 1995. An autosomal recessive non-syndromic form of sensorineural hearing loss maps to 3p-DFNB6. *Genome Res.* 5:305–8

41. Garner CC, Nash J, Huganir RL. 2000. PDZ domains in synapse assembly and signalling. *Trends Cell Biol.* 10:274–80

42. Gibson F, Walsh J, Mburu P, Varela A, Brown KA, et al. 1995. A type VII myosin encoded by the mouse deafness gene *Shaker-1. Nature* 374:62–64

43. Gillespie PG, Wagner MC, Hudspeth AJ. 1993. Identification of a 120 kd hair-bundle myosin located near stereociliary tips. *Neuron* 11:581–94

44. Goodyear R, Richardson G. 1999. The ankle-link antigen: an epitope sensitive to calcium chelation associated with the hair-cell surface and the calycal processes of photoreceptors. *J. Neurosci.* 19:3761–72

45. Gorlin RJ. 1995. Genetic hearing loss associated with eye disorders. See Ref. 46 pp. 105–40

46. Gorlin RJ, Toriello HV, Cohen MM, eds. 1995. *Hereditary Hearing Loss and Its Syndromes.* New York/Oxford: Oxford Univ. Press

47. Gorlin RJ, Tilsner TJ, Feinstein S, Duvall AJD. 1979. Usher's syndrome type III. *Arch. Otolaryngol.* 105:353–54

48. Grondahl J. 1987. Estimation of prognosis and prevalence of retinitis pigmentosa and Usher syndrome in Norway. *Clin. Genet.* 31:255–64

49. Guilford P, Ayadi H, Blanchard S, Chaïb H, Le Paslier D, et al. 1994. A human gene responsible for neurosensory, non-syndromic recessive deafness is a

candidate homologue of the mouse *sh-1* gene. *Hum. Mol. Genet.* 3:989–93

50. Hall RA, Premont RT, Chow CW, Blitzer JT, Pitcher JA, et al. 1998. The beta2-adrenergic receptor interacts with the Na⁺/H⁺-exchanger regulatory factor to control Na^+/H^+ exchange. *Nature* 392: 626–30

51. Hallgren B. 1959. Retinitis pigmentosa combined with congenital deafness, with vestibulo-cerebellar ataxia and mental abnormality in a proportion of cases. A clinical and genetico-statistical study. In *Acta Psychiatry and Neurologica Scandinavia*, pp. 5–101. Copenhagen: Munksgaard

52. Hasson T, Gillespie PG, Garcia JA, MacDonald RB, Zhao Y, et al. 1997. Unconventional myosins in inner-ear sensory epithelia. *J. Cell Biol.* 137:1287–307

53. Hasson T, Heintzelman MB, Santos-Sacchi J, Corey DP, Mooseker MS. 1995. Expression in cochlea and retina of myosin VIIa, the gene product defective in Usher syndrome type 1B. *Proc. Natl. Acad. Sci. USA* 92:9815–19

54. Hasson T, Walsh J, Cable J, Mooseker MS, Brown SD, et al. 1997. Effects of shaker-1 mutations on myosin-VIIa protein and mRNA expression. *Cell Motil. Cytoskel.* 37:127–38

55. Hirokawa N, Tilney LG. 1982. Interactions between actin filaments and between actin filaments and membranes in quick-frozen and deeply etched hair cells of the chick ear. *J. Cell Biol.* 95:249–61

56. Hmani M, Ghorbel A, Boulila-Elgaïed A, Ben Zina Z, Kammoun W, et al. 1999. A novel locus for Usher syndrome type II, USH2B, maps to chromosome 3 at p 23–24.2. *Eur. J. Hum. Genet.* 7:363–67

57. Hope CI, Bundey S, Proops D, Fielder AR. 1997. Usher syndrome in the city of Birmingham—prevalence and clinical classification. *Br. J. Ophthalmol.* 81:46–53

58. Howard J, Hudspeth AJ. 1987. Mechanical relaxation of the hair bundle mediates adaptation in mechanoelectrical transduction by the bullfrog's saccular hair cell. *Proc. Natl. Acad. Sci. USA* 84:3064–68

59. Hudspeth AJ. 1989. How the ear's works work. *Nature* 341:397–404

60. Hunter DG, Fishman GA, Mehta RS, Kretzer FL. 1986. Abnormal sperm and photoreceptor axonemes in Usher's syndrome. *Arch. Ophthalmol.* 104:385–89

61. Hyafil F, Babinet C, Jacob F. 1981. Cell-cell interactions in early embryogenesis: a molecular approach to the role of calcium. *Cell* 26:447–54

62. Jain PK, Lalwani AK, Li XC, Singleton TL, Smith TN, et al. 1998. A gene for recessive nonsyndromic sensorineural deafness (*DFNB18*) maps to the chromosomal region 11p14-p15.1 containing the Usher syndrome type 1C gene. *Genomics* 50:290–92

63. Janecke AR, Meins M, Sadeghi M, Grundmann K, Apfelstedt-Sylla E, et al. 1999. Twelve novel myosin VIIA mutations in 34 patients with Usher syndrome type I: confirmation of genetic heterogeneity. *Hum. Mutat.* 13:133–40

64. Kachar B, Battaglia A, Fex J. 1997. Compartmentalized vesicular traffic around the hair cell cuticular plate. *Hear. Res.* 107:102–12

65. Kaiser DA, Pollard TD. 1996. Characterization of actin and poly-L-proline binding sites of Acanthamoeba profilin with monoclonal antibodies and by mutagenesis. *J. Mol. Biol.* 256:89–107

66. Kaplan J, Gerber S, Bonneau D, Rozet JM, Delrieu O, et al. 1992. A gene for Usher syndrome type I (USH1A) maps to chromosome 14q. *Genomics* 14:979–87

67. Karjalainen S, Pakarinen L, Terasvirta M, Kaariainen H, Vartiainen E. 1989. Progressive hearing loss in Usher's syndrome. *Ann. Otol. Rhinol. Laryngol.* 98: 863–66

68. Karp A, Santore F. 1983. Retinitis pigmentosa and progressive hearing loss. *J. Speech Hear. Disord.* 48:308–14

69. Keats BJ, Nouri N, Pelias MZ, Deininger PL, Litt M. 1994. Tightly linked flanking microsatellite markers for the Usher syndrome type I locus on the short arm of chromosome 11. *Am. J. Hum. Genet.* 54:681–86

70. Kikuchi K, Hilding DA. 1965. The defective organ of Corti in shaker-1 mice. *Acta Otolaryngol.* 60:287–303

71. Kim E, Niethammer M, Rothschild A, Jan YN, Sheng M. 1995. Clustering of Shaker-type K^+ channels by interaction with a family of membrane-associated guanylate kinases. *Nature* 378:85–88

72. Kimberling WJ, Möller C. 1995. Clinical and molecular genetics of Usher syndrome. *J. Am. Acad. Audiol.* 6:63–72

73. Kimberling WJ, Möller CG, Davenport S, Priluck IA, Beighton PH, et al. 1992. Linkage of Usher syndrome type I gene (USH1B) to the long arm of chromosome 11. *Genomics* 14:988–94

74. Kimberling WJ, Weston MD, Möller C, Davenport SL, Shugart YY, et al. 1990. Localization of Usher syndrome type II to chromosome 1q. *Genomics* 7:245–49

75. Kloepfer HW, Laguaite JK. 1966. The hereditary syndrome of congenital deafness and retinitis pigmentosa. (Usher's syndrome). *Laryngoscope* 76:850–62

76. Kros CJ, Marcotti W, Richardson GP, Brown SDM, Steel KP. 1999. *Electrophysiology of Outer Hair Cells from Myosin VIIA Mutant Mice.* Assoc. Res. Otolaryngol. Meet., Feb. 15, 1999

77. Küssel-Andermann P, El-Amraoui A, Safieddine S, Hardelin J-P, Nouaille S, et al. 2000. Unconventional myosin VIIA is a novel A-kinase anchoring protein. *J. Biol. Chem.* 275:29654–59

78. Küssel-Andermann P, El-Amraoui A, Safieddine S, Nouaille S, Perfettini I, et al. 2000. Vezatin, a novel transmembrane protein, bridges myosin VIIA to the cadherin-catenins complex. *EMBO J.* 19:6020–29

79. Legan PK, Richardson GP. 1997. Extracellular matrix and cell adhesion molecules in the developing inner ear. *Semin. Cell Dev. Biol.* 8:217–24

80. Legouis R, Gansmuller A, Sookhareea S, Bosher JM, Baillie DL, et al. 2000. LET-413 is a basolateral protein required for the assembly of adherens junctions in *Caenorhabditis elegans. Nat. Cell Biol.* 2:415–22

81. Lévy G, Levi-Acobas F, Blanchard S, Gerber S, Larget-Piet D, et al. 1997. Myosin VIIA gene: heterogeneity of the mutations responsible for Usher syndrome type IB. *Hum. Mol. Genet.* 6:111–16

82. Lewis RA, Otterud B, Stauffer D, Lalouel JM, Leppert M. 1990. Mapping recessive ophthalmic diseases: linkage of the locus for Usher syndrome type II to a DNA marker on chromosome 1q. *Genomics* 7:250–56

83. Liebreich R. 1861. Abkunft und Ehen unter Blutsverwandten als Grund von Retinitis pigmentosa. *Dtsch. Klin.* 13:53–55

84. Lindenov H. 1945. *The Etiology of Deaf-Mutism with Special Reference to Heredity.* Copenhagen: Munksgaard

85. Liu X, Ondek B, Williams DS. 1998. Mutant myosin VIIa causes defective melanosome distribution in the RPE of shaker-1 mice. *Nat. Genet.* 19:117–18

86. Liu X, Udovichenko IP, Brown SD, Steel KP, Williams DS. 1999. Myosin VIIa participates in opsin transport through the photoreceptor cilium. *J. Neurosci.* 19:6267–74

87. Liu X-Z, Walsh J, Mburu P, Kendrick-Jones J, Cope MJTV, et al. 1997. Mutations in the myosin VIIA gene cause non-syndromic recessive deafness. *Nat. Genet.* 16:188–90

88. Liu X-Z, Walsh J, Tamagawa Y, Kitamura K, Nishizawa M, et al. 1997. Autosomal dominant non-syndromic deafness caused by a mutation in the myosin VIIA gene. *Nat. Genet.* 17:268–69

89. Liu XZ, Hope C, Liang CY, Zou JM, Xu LR, et al. 1999. A mutation

(2314delG) in the Usher syndrome type IIA gene: high prevalence and phenotypic variation. *Am. J. Hum. Genet.* 64:1221–25

90. Liu XZ, Hope C, Walsh J, Newton V, Ke XM, et al. 1998. Mutations in the myosin VIIA gene cause a wide phenotypic spectrum, including atypical Usher syndrome. *Am. J. Hum. Genet.* 63:909–12

91. Liu XZ, Newton VE, Steel KP, Brown SD. 1997. Identification of a new mutation of the myosin VII head region in Usher syndrome type 1. *Hum. Mutat.* 10:168–70

92. Long RM, Singer RH, Meng X, Gonzalez I, Nasmyth K, et al. 1997. Mating type switching in yeast controlled by asymmetric localization of *ASH1* mRNA. *Science* 277:383–87

93. Lue RA, Brandin E, Chan EP, Branton D. 1996. Two independent domains of hDlg are sufficient for subcellular targeting: the PDZ1-2 conformational unit and an alternatively spliced domain. *J. Cell Biol.* 135:1125–37

94. Mandai K, Nakanishi H, Satoh A, Obaishi H, Wada M, et al. 1997. Afadin: a novel actin filament-binding protein with one PDZ domain localized at cadherin-based cell-to-cell adherens junction. *J. Cell Biol.* 139:517–28

95. Marquis RE, Hudspeth AJ. 1997. Effects of extracellular Ca^{2+} concentration on hair-bundle stiffness and gating-spring integrity in hair cells. *Proc. Natl. Acad. Sci. USA* 94:11923–28

96. Mburu P, Liu XZ, Walsh J, Saw D, Cope MJTV, et al. 1997. Mutation analysis of the mouse myosin VIIA deafness gene. *Genes Funct.* 1:191–203

97. Mermall V, Post PL, Mooseker MS. 1998. Unconventional myosins in cell movement, membrane traffic, and signal transduction. *Science* 279:527–33

98. Metzler WJ, Bell AJ, Ernst E, Lavoie TB, Mueller L. 1994. Identification of the poly-L-proline-binding site on human profilin. *J. Biol. Chem.* 269:4620–25

99. Mlodzik M. 2000. Spiny legs and prickled bodies: new insights and complexities in planar polarity establishment. *Bioessays* 22:311–15

100. Möller CG, Kimberling WJ, Davenport SL, Priluck I, White V, et al. 1989. Usher syndrome: an otoneurologic study. *Laryngoscope* 99:73–79

101. Nguyen JT, Porter M, Amoui M, Miller WT, Zuckermann RN, et al. 2000. Improving SH3 domain ligand selectivity using a non-natural scaffold. *Chem. Biol.* 7:463–73

102. Nicolson T, Rusch A, Friedrich RW, Granato M, Ruppersberg JP, et al. 1998. Genetic analysis of vertebrate sensory hair cell mechanosensation: the zebrafish circler mutants. *Neuron* 20:271–83

103. Nollet F, Kools P, van Roy F. 2000. Phylogenetic analysis of the cadherin superfamily allows identification of six major subfamilies besides several solitary members. *J. Mol. Biol.* 299:551–72

104. Nuutila A. 1970. Dystrophia retinae pigmentosa-dysacusis syndrome (DRD): a study of the Usher or Hallgren syndrome. *J. Genet. Hum.* 18:57–88

105. Deleted in proof

106. Ozawa M, Engel J, Kemler R. 1990. Single amino acid substitutions in one Ca^{2+} binding site of uvomorulin abolish the adhesive function. *Cell* 63:1033–38

107. Pakarinen L, Karjalainen S, Simola KO, Laippala P, Kaitalo H. 1995. Usher's syndrome type 3 in Finland. *Laryngoscope* 105:613–17

108. Piazza L, Fishman GA, Farber M, Derlacki D, Anderson RJ. 1986. Visual acuity loss in patients with Usher's syndrome. *Arch. Ophthalmol.* 104:1336–39

109. Pieke-Dahl S, Moller CG, Kelley PM, Astuto LM, Cremers CW, et al. 2000. Genetic heterogeneity of Usher syndrome type II: localisation to chromosome 5q. *J. Med. Genet.* 37:256–62

110. Pieke-Dahl S, Weston MD, Kimberling WJ. 1997. Genetic heterogeneity of USH. *Am. J. Hum. Genet.* 61(Suppl.):A291

111. Ponting CP. 1997. Evidence for PDZ domains in bacteria, yeast, and plants. *Protein Sci.* 6:464–68
112. Rattner A, Sun H, Nathans J. 1999. Molecular genetics of human retinal disease. *Annu. Rev. Genet.* 33:89–131
113. Raviola E. 1994. The eye. In *A Textbook of Histology*, ed. DW Fawcett, pp. 872–918. New York: Chapman Hall
114. Reczek D, Berryman M, Bretscher A. 1997. Identification of EBP50: a PDZ-containing phosphoprotein that associates with members of the ezrin-radixin-moesin family. *J. Cell Biol.* 139:169–79
115. Reddy AS, Safadi F, Narasimhulu SB, Golovkin M, Hu X. 1996. A novel plant calmodulin-binding protein with a kinesin heavy chain motor domain. *J. Biol. Chem.* 271:7052–60
116. Reifegerste R, Moses K. 1999. Genetics of epithelial polarity and pattern in the *Drosophila* retina. *Bioessays* 21:275–85
117. Richardson GP, Forge A, Kros CJ, Fleming J, Brown SD, et al. 1997. Myosin VIIA is required for aminoglycoside accumulation in cochlear hair cells. *J. Neurosci.* 17:9506–19
118. Rinchik EM, Carpenter DA, Selby PB. 1990. A strategy for fine-structure functional analysis of a 6- to 11-centimorgan region of mouse chromosome 7 by high-efficiency mutagenesis. *Proc. Natl. Acad. Sci. USA* 87:896–900
119. Rivolta C, Sweklo EA, Berson EL, Dryja TP. 2000. Missense mutation in the USH2A gene: association with recessive retinitis pigmentosa without hearing loss. *Am. J. Hum. Genet.* 66:1975–78
120. Rosenberg T, Haim M, Hauch AM, Parving A. 1997. The prevalence of Usher syndrome and other retinal dystrophy-hearing impairment associations. *Clin. Genet.* 51:314–21
121. Sahly I, El-Amraoui A, Abitbol M, Petit C, Dufier J-L. 1997. Expression of myosin VIIA during mouse embryogenesis. *Anat. Embryol.* 196:159–70
122. Sankila E-M, Pakarinen L, Kääriäinen H, Aittomäki K, Karjalainen S, et al. 1995. Assignment of an Usher syndrome type III (USH3) gene to chromosome 3q. *Hum. Mol. Genet.* 4:93–98
123. Seiler C, Nicolson T. 1999. Defective calmodulin-dependent rapid apical endocytosis in zebrafish sensory hair cell mutants. *J. Neurobiol.* 41:424–34
124. Self T, Mahony M, Fleming J, Walsh J, Brown SDM, et al. 1998. Shaker-1 mutations reveal roles for myosin VIIA in both development and function of cochlear hair cells. *Development* 125:557–66
125. Sheng M, Pak DT. 2000. Ligand-gated ion channel interactions with cytoskeletal and signaling proteins. *Annu. Rev. Physiol.* 62:755–78
126. Shieh BH, Zhu MY. 1996. Regulation of the TRP Ca^{2+} channel by INAD in Drosophila photoreceptors. *Neuron* 16:991–98
127. Smith RJH, Berlin CI, Hejtmancik JF, Keats BJB, Kimberling WJ, et al. 1994. Clinical diagnosis of the Usher syndromes. Usher syndrome consortium. *Am. J. Med. Genet.* 50:32–38
128. Smith RJH, Lee EC, Kimberling WJ, Daiger SP, Pelias MZ, et al. 1992. Localization of two genes for Usher syndrome type I to chromosome 11. *Genomics* 14:995–1002
129. Sudol M. 1998. From Src homology domains to other signaling modules: proposal of the 'protein recognition code.' *Oncogene* 17:1469–74
130. Deleted in proof
131. Takasaki K, Balaban CD, Sando I. 2000. Histopathologic findings of the inner ears with Alport, Usher and Waardenburg syndromes. *Adv. Oto. Rhino. Laryngol.* 56:218–32
132. Tamayo ML, Bernal JE, Tamayo GE, Frias JL, Alvira G, et al. 1991. Usher syndrome: results of a screening program in Colombia. *Clin. Genet.* 40:304–11
133. Tilney LG, Tilney MS, Cotanche DA. 1988. Actin filaments, stereocilia, and hair

cells of the bird cochlea. V. How the staircase pattern of stereociliary lengths is generated. *J. Cell Biol.* 106:355–65

134. Titus MA. 1999. A class VII unconventional myosin is required for phagocytosis. *Curr. Biol.* 9:1297–303

135. Titus MA, Gilbert SP. 1999. The diversity of molecular motors: an overview. *Cell. Mol. Life Sci.* 56:181–83

136. Troutt LL, van Heumen WR, Pickles JO. 1994. The changing microtubule arrangements in developing hair cells of the chick cochlea. *Hear. Res.* 81:100–8

137. Deleted in proof

138. Tuxworth RI, Weber I, Wessels D, Addicks GC, Soll DR, et al. 2001. A role for myosin VII in dynamic cell adhesion. *Curr. Biol.* 11:318–29

139. Udovichenko IP, Vansant G, Williams DS. 2000. Actin-based motility of baculovirus-expressed and native retinal myosin VIIa. *Mol. Biol. Cell* 11:375 (Abstr. 1944)

140. Usher CH. 1914. On the inheritance of retinitis pigmentosa, with notes of cases. *R. Lond. Ophthalmol. Hosp. Rep.* 19:130–236

141. van Aarem A, Wagenaar M, Tonnaer E, Pieke Dahl S, Bisseling J, et al. 1999. Semen analysis in the Usher syndrome type 2A. *J. Oto-Rhino-Laryngol. Relat. Spec.* 61:126–30

142. Vernon M. 1969. Usher's syndrome—deafness and progressive blindness. Clinical cases, prevention, theory and literature survey. *J. Chronic Dis.* 22:133–51

143. Verpy E, Leibovici M, Zwaenepoel I, Liu X-Z, Gal A, et al. 2000. A defect in harmonin, a PDZ domain–containing protein expressed in the inner ear sensory hair cells, underlies Usher syndrome type 1C. *Nat. Genet.* 26:51–55

144. von Graefe A. 1858. Vereinzelte Beobachtungen und Bemerkungen. Exceptionelle Verhalten des Gesichtsfeldes bei Pigmentenartung des Netzhaut. *Albrecht Graefes Arch. Klin. Ophthalmol.* 4:250–53

145. Wagenaar M. 2000. *The Usher syndrome, a clinical and genetical correlation.* PhD thesis. Univ. Nijmegen, Netherlands

146. Wagenaar M, Schuknecht H, Nadol J Jr, Benraad-Van Rens M, Pieke-Dahl S, et al. 2000. Histopathologic features of the temporal bone in Usher syndrome type I. *Arch. Otolaryngol. Head Neck Surg.* 126:1018–23

147. Wagner HJ. 1997. Presynaptic bodies ("ribbons"): from ultrastructural observations to molecular perspectives. *Cell Tissue Res.* 287:435–46

148. Wang A, Liang Y, Fridell RA, Probst FJ, Wilcox ER, et al. 1998. Association of unconventional myosin *MYO15* mutations with human nonsyndromic deafness *DFNB3*. *Science* 280:1447–51

149. Wayne S, Der Kaloustian VM, Schloss M, Polomeno R, Scott DA, et al. 1996. Localization of the Usher syndrome type 1D gene (Ush1D) to chromosome 10. *Hum. Mol. Genet.* 5:1689–92

150. Wayne S, Lowry RB, McLeod DR, Knaus R, Farr C, et al. 1997. Localization of the Usher syndrome type 1F (Ush1F) to chromosome 10. *Am. J. Hum. Genet.* 61:A300

151. Weil D, Bernard M, Combates N, Wirtz MK, Hollister DW, et al. 1988. Identification of a mutation that causes exon skipping during collagen pre-mRNA splicing in an Ehlers-Danlos syndrome variant. *J. Biol. Chem.* 263:8561–64

152. Weil D, Blanchard S, Kaplan J, Guilford P, Gibson F, et al. 1995. Defective myosin VIIA gene responsible for Usher syndrome type 1B. *Nature* 374:60–61

153. Weil D, Küssel P, Blanchard S, Lévy G, Levi-Acobas F, et al. 1997. The autosomal recessive isolated deafness, DFNB2, and the Usher 1B syndrome are allelic defects of the myosin-VIIA gene. *Nat. Genet.* 16:191–93

154. Weil D, Lévy G, Sahly I, Levi-Acobas F, Blanchard S, et al. 1996. Human myosin VIIA responsible for the Usher 1B syndrome: a predicted membrane-associated

motor protein expressed in developing sensory epithelia. *Proc. Natl. Acad. Sci. USA* 93:3232–37

155. Weston MD, Carney CA, Rivedal SA, Kimberling WJ. 1998. Spectrum of myosin VIIA mutations causing Usher syndrome type Ib. *Assoc. Res. Otolaryngol.* 21:46

156. Weston MD, Eudy JD, Fujita S, Yao S, Usami S, et al. 2000. Genomic structure and identification of novel mutations in usherin, the gene responsible for Usher syndrome type IIa. *Am. J. Hum. Genet.* 66:1199–210

157. Weston MD, Kelley PM, Overbeck LD, Wagenaar M, Orten DJ, et al. 1996. Myosin VIIA mutation screening in 189 Usher syndrome type 1 patients. *Am. J. Hum. Genet.* 59:1074–83

158. Wolfrum U, Liu X, Schmitt A, Udovichenko IP, Williams DS. 1998. Myosin VIIa as a common component of cilia and microvilli. *Cell Motil. Cytoskelet.* 40:261–71

159. Wolfrum U, Schmitt A. 1999. Evidence for myosin VIIa-driven transport of rhodopsin in the plasma membrane of the photoreceptor-connecting cilium. In *Retinal Degenerative Diseases and Experimental Therapy*, ed. JG Hollyfield, pp. 3–14. New York: Kluwer

160. Wolfrum U, Schmitt A. 2000. Rhodopsin transport in the membrane of the connecting cilium of mammalian photoreceptor cells. *Cell Motil. Cytoskelet.* 46:95–107

161. Yagi T, Takeichi M. 2000. Cadherin superfamily genes: functions, genomic organization, and neurologic diversity. *Genes Dev.* 14:1169–80

162. Yonezawa S, Yoshiki A, Hanai A, Matsuzaki T, Matsushima J, et al. 1999. Chromosomal localization of a gene responsible for vestibulocochlear defects of BUS/Idr mice: identification as an allele of waltzer. *Hear. Res.* 134:116–22

163. Zwaenepoel I, Verpy E, Blanchard S, Meins M, Apfelstedt-Sylla E, et al. 2001. Identification of three novel mutations in the USH1C gene and detection of thirty-one polymorphisms used for haplotype analysis. *Hum. Mutat.* 17:34–41

NOTE ADDED IN PROOF

The gene encoding protocadherin15 has just been shown to underlie USH1F (4a).

Annu. Rev. Genomics Hum. Genet. 2001. 2:299–341

INBORN ERRORS OF STEROL BIOSYNTHESIS

Richard I. Kelley

Kennedy Krieger Institute, Baltimore Maryland 21205;
e-mail: kelle_ri@jhuvms.hcf.jhu.edu

Gail E. Herman

Children's Research Institute, Columbus, Ohio 43205;
e-mail: HermanG@pediatrics.ohio-state.edu

Key Words   cholesterol biosynthesis, Smith-Lemli-Opitz syndrome,
mevalonic aciduria, desmosterolosis, chondrodysplasia punctata

■ **Abstract**   The known disorders of cholesterol biosynthesis have expanded rapidly
since the discovery that Smith-Lemli-Opitz syndrome is caused by a deficiency of
7-dehydrocholesterol. Each of the six now recognized sterol disorders—mevalonic
aciduria, Smith-Lemli-Opitz syndrome, desmosterolosis, Conradi-Hünermann syndrome, CHILD syndrome, and Greenberg dysplasia—has added to our knowledge
of the relationship between cholesterol metabolism and embryogenesis. One of the
most important lessons learned from the study of these disorders is that abnormal
cholesterol metabolism impairs the function of the hedgehog class of embryonic
signaling proteins, which help execute the vertebrate body plan during the earliest
weeks of gestation. The study of the enzymes and genes in these several syndromes
has also expanded and better delineated an important class of enzymes and proteins
with diverse structural functions and metabolic actions that include sterol biosynthesis, nuclear transcriptional signaling, regulation of meiosis, and even behavioral
modulation.

INTRODUCTION

For its first hundred years, following the early work of Garrod at the turn of the
twentieth century, the study of inborn errors of metabolism emphasized disorders
of small, water soluble metabolites, such as phenylalanine in phenylketonuria,
or of macromolecule catabolism, such as mucopolysaccharide and sphingolipid
storage in the lysosomal storage diseases. Of the several hundred inborn errors
of metabolism discovered and characterized in the last hundred years, relatively
few involved abnormal de novo synthesis of an essential small metabolite. Furthermore, most inborn errors of metabolism have featured the postnatal evolution
of metabolic deficiencies or toxicities in children who are phenotypically normal at birth. However, the surprising discovery, in 1993, that Smith-Lemli-Opitz

syndrome—one of the best-known autosomal recessive malformation-mental retardation syndromes—is caused by a primary defect of cholesterol biosynthesis not only raised important questions about embryological links between abnormal sterol biosynthesis and congenital malformations but also focused attention on the special implications of a disorder involving an essential fetal metabolite that cannot be supplied in adequate amounts by the mother during gestation. Since 1993, the wider recognition that congenital malformations can result from an inborn error of cholesterol metabolism has led to the discovery of prenatal malformation syndromes caused by defects in most of the other steps in postsqualene cholesterol biosynthesis. The delineation of these new inborn errors of cholesterol biosynthesis has had obvious clinical importance, especially for diagnosis and prenatal detection of the individual syndromes. Moreover, the recognition of novel associations between aberrant cholesterol metabolism and diverse clinical problems—cerebral dysgenesis, cyclic inflammatory disease, ichthyosis, skeletal dysplasia, and the pharmacology of specific behavioral abnormalities—has opened many new avenues of biochemical and genetic investigation.

In this chapter, we present a brief overview of the essentials of human cholesterol biosynthesis and then follow with reviews of the phenotype, biochemistry, and molecular genetics of mevalonic aciduria, hyper-IgD syndrome, Smith-Lemli-Opitz syndrome (SLOS), desmosterolosis, Conradi-Hünermann syndrome, CHILD (congenital hemidysplasia with ichthyosiform erythroderma and limb defects) syndrome, and Hydrops-ectopic calcification (Greenberg) dysplasia. The larger biochemical and genetic implications of these newly discovered sterol disorders and the likely future directions of biochemical and clinical genetic research in this important biochemical pathway are also discussed.

NORMAL STEROL METABOLISM

Cholesterol Biosynthesis

Cholesterol is an essential metabolite and structural lipid in higher organisms and is synthesized by all nucleated mammalian cells. Although complex, the biosynthesis of cholesterol is only one element of the larger isoprenoid biosynthetic system, which incorporates the de novo synthesis of compounds as diverse as dolichol, ubiquinone, isopentenyladenine, and farnesyl pyrophosphate (Figure 1). Moreover, cholesterol is at once an end-product of isoprenoid metabolism and the starting substrate for the synthesis of yet another highly diverse group of metabolically active compounds, including all steroid hormones, bile acids, and signaling compounds, such as oxysterols. Cholesterol, via covalent linkage to an amino acid residue, also confers function to the hedgehog class of cell signaling proteins (discussed below in more detail).

The synthesis of sterols follows a complex series of reactions that begins with the condensation of acetyl-CoA and acetoacetyl-CoA to form 3-hydroxy-3-methylglutaryl(HMG)-CoA. The sequence reaches the halfway point with the

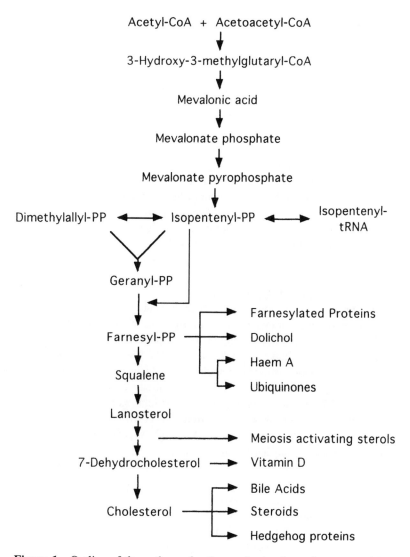

Acetyl-CoA + Acetoacetyl-CoA

↓

3-Hydroxy-3-methylglutaryl-CoA

↓

Mevalonic acid

↓

Mevalonate phosphate

↓

Mevalonate pyrophosphate

↓

Dimethylallyl-PP ⟷ Isopentenyl-PP ⟷ Isopentenyl-tRNA

Geranyl-PP

Farnesyl-PP → Farnesylated Proteins
Farnesyl-PP → Dolichol

Squalene → Haem A
Squalene → Ubiquinones

Lanosterol

→ Meiosis activating sterols

7-Dehydrocholesterol → Vitamin D

Cholesterol → Bile Acids
Cholesterol → Steroids
→ Hedgehog proteins

Figure 1   Outline of the pathway for the synthesis of sterols, nonsterol isoprenoids, and their derivatives.

formation of the 30-carbon precursor of all other sterols, lanosterol (4,4,14-trimethylcholesta-8(9),24-dien-3β-ol), and ending with the conversion of 7-dehydrocholesterol (7DHC) to cholesterol by the enzyme 7-dehydrocholesterol reductase (DHCR7) (Figure 2). HMG-CoA reductase, which catalyzes the conversion of HMG-CoA to mevalonic acid, is a major rate-determining step of cholesterol biosynthesis and is subject to both transcriptional and posttranslational regulation, largely in response to the level of cholesterol in the endoplasmic reticulum (ER) (19). The coordinated regulation of the levels of HMG-CoA reductase, acetoacetyl-CoA synthase, and the plasma membrane LDL receptor assures the maintenance of a stable cellular pool of cholesterol from both intracellular and extracellular sources (19). Cholesterol exists in several metabolically distinct pools, most importantly in the plasma membrane, lysosomal system, and ER.

Cholesterol (cholest-5-en-3β-ol), a 27-carbon, monounsaturated sterol, is synthesized from its 30-carbon sterol precursor, lanosterol, by a series of dehydrogenations, reductions, and demethylations (Figure 2). Although textbooks often indicate that desmosterol (cholesta-5,24-dien-3β-ol) is the penultimate sterol in this series of reactions, the finding, in 1993, (84) of markedly increased levels of 7DHC and almost no 7-dehydrodesmosterol in patients with SLOS indicated that reduction of the 24-25 double bond of lanosterol occurs early in the biosynthetic sequence (93). However, the relative abundance of desmosterol in neuronal tissues, the testes, and breast milk (16, 28, 116) suggests that desmosterol may have a special physiological role in these tissues.

Although all steps of cholesterol synthesis were once thought to take place in the ER, studies by Krisans and colleagues established that the second enzyme of the pathway, mevalonate kinase, is localized to the peroxisome and contains an N-terminal peroxisomal targeting amino acid sequence (PTS2) (171). Interestingly, many other enzymes of the presqualene cholesterol biosynthetic pathway also have peroxisomal targeting peptide sequences (PTS1 or PTS2) and appear to localize predominantly to the peroxisome (1). These include phosphomevalonate kinase, phosphomevalonate decarboxylase, isopentenyl diphosphate isomerase, and farnesyl pyrophosphate synthase. Sterol-carrier protein 2, a possible cofactor for DHCR7 and carrier protein for intracellular sterol transport, also appears to be targeted to and processed by peroxisomes (96). In contrast, squalene synthase resides exclusively in the ER (170). Although Appelkvist et al. published evidence that peroxisomes may harbor all the enzymes necessary for the conversion of lanosterol to cholesterol (6), the role of these putative peroxisomal activities in overall cellular cholesterol biosynthesis and homeostasis remains unknown. Nevertheless, the observations that patients with Zellweger syndrome, who have defective peroxisomal assembly, have markedly depressed serum cholesterol levels (97) and that Zellweger cells in vitro have depressed rates of cholesterol synthesis (73) strongly suggest that peroxisomes have an important role in cholesterol biosynthesis (73).

Another important, if poorly understood, aspect of cholesterol biosynthesis is the complex intracellular trafficking of cholesterol. As shown by Lange et al. (110), zymosterol (cholesta-8(9),24-dien-3β-ol), an obligatory intermediate in the synthesis of cholesterol, appears to move from the ER to the plasma membrane and then back to the ER-enriched microsomal fraction of the cell, where final conversion to cholesterol occurs. Other important pathways of intracellular cholesterol movement include transport from the plasma membrane to lysosomes and from lysosomes to the ER. A genetic disruption of the latter transport system is the cause of the abnormal lipid storage in type C Niemann-Pick disease (36, 117). Delineating intracellular trafficking of cholesterol may be important to understanding the

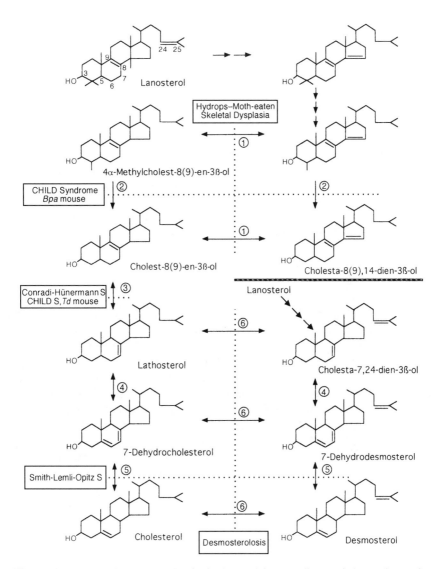

Figure 2 Enzymatic steps and principal sterol intermediates of the pathway for cholesterol biosynthesis. Clinical syndromes proven, or believed to be, caused by defects of sterol biosynthesis are indicated in the boxes. The enzymes denoted by the circled numbers are: (*1*) 3β-hydroxysteroid-Δ^{14}-reductase, (*2*) 4α-methylsterol-4-demethylase complex, (*3*) 3β-hydroxysteroid-Δ^{8},Δ^{7}-isomerase, (*4*) 3β-hydroxysteroid-Δ^{5}-desaturase (lathosterol dehydrogenase), (*5*) 3β-hydroxysteroid-Δ^{7}-reductase (7-dehydrocholesterol reductase, DHCR7), and (*6*) 3β-hydroxysteroid-Δ^{24}-reductase (desmosterol reductase).

cellular pathology of sterol disorders, in particular SLOS, because the low total sterol levels in SLOS suggest feedback inhibition of de novo sterol synthesis at one or more levels (78).

Prenatal and Postnatal Sterol Metabolism

Important to understanding the fetal pathology of disorders of cholesterol biosynthesis is evidence indicating that, in contrast to many other small metabolites, very little maternal cholesterol is transported to the fetus across the placenta (10, 22). Clinically, this is best illustrated by the finding that some SLOS newborns have plasma cholesterol levels as low as 1 mg/dl, barely 2% of the normal newborn cholesterol level (33). However, the situation may be quite different during the embryonic period, when, at least in mice, a substantial amount of LDL cholesterol reaches the developing neuroepithelium directly from the mother via a specialized, nonplacental LDL transport system mediated by megalin, a multifunctional transport protein in the embryonic neuroepithelium (46, 191). Thus, whereas delivery of cholesterol to the fetus after full development of the placenta may be minimal, critical aspects of embryonic tissue differentiation may be sensitive to maternal blood cholesterol and LDL levels.

Most cholesterol synthesized, both before and after birth, serves as a structural lipid of cell membranes. However, cholesterol also enters pathways for bile acid and steroid hormone synthesis during fetal as well as postnatal life. Cholesterol before birth is converted into a variety of fetal steroids for sexual differentiation and into estriol by combined action of the fetal adrenals and the placenta. Unconjugated estriol transferred to the mother from the fetus is believed to have a role in the maintenance of the pregnancy, but the true physiologic actions of this steroid remain unclear. Another recently discovered fate of cholesterol is its covalent linkage to hedgehog proteins, a family of embryonic signaling proteins that may be important targets of the abnormal sterol metabolism of SLOS (25, 101, 147) and possibly other sterol disorders. Less well studied than other fates of fetal cholesterol is its conversion to bile acids. However, the report of severe cholestasis and cirrhosis in one SLOS newborn (136) and the severe prenatal liver damage common to some primary defects of bile acid biosynthesis (160) suggest that abnormal species of bile acids may also have clinical consequences in defects of cholesterol biosynthesis.

MEVALONIC KINASE DEFICIENCY (MIM 251170, 260920)

History of Mevalonic Aciduria

Mevalonic aciduria (MIM 251170), caused by deficiency of mevalonate kinase (ATP:(R)-mevalonate 5-phosphotransferase; MVK), was the first reported disorder of cholesterol biosynthesis. Although elevated urinary mevalonic acid was described in 1985 by Berger et al. (11) in a one-year-old child with progressive ataxia, the first identification of mevalonate kinase deficiency (MKD), in a second child with severe mevalonic aciduria, was reported a year later by Hoffmann et al. (74). The second patient had severe failure-to-thrive, profound psychomotor retardation, ataxia, a dysmorphic appearance, rhizomelic shortness, cataracts, hepatosplenomegaly, anemia, and recurrent crises of fever, arthralgia, edema, rash, adenopathy, and hepatomegaly. Whereas that patient also had profound physical and neurological disease, several subsequently identified patients have had milder, apparently static neurological problems, such as developmental delay, hypotonia, myopathy, or ataxia (53, 75). Although recurrent inflammatory spells occurring every 3 to 6 weeks throughout life constitute one of the most distinctive signs of MKD, not until 1999 was the less impairing disorder with similar episodic inflammatory spells, hyperimmunoglobulinemia D syndrome (HIDS; MIM 260920), shown by two different groups to be caused by missense mutations in the MVK gene (41, 80). However, in contrast to the typical 5000- to 50,000-fold elevations of urinary mevalonate in classical MKD, patients with HIDS have only 10- to 100-fold increased levels of MVA, which, during nonfebrile periods, often are detectable only by isotope-dilution gas chromatography–mass spectrometry (GC/MS). In retrospect, a number of classical MKD patients also have had increased serum IgD levels as well as chronically increased levels of the inflammatory cytokine, leukotriene E4 (125).

Clinical Characteristics of Mevalonic Kinase Deficiency

Among the more interesting biological aspects of classical mevalonic aciduria is the abnormal morphogenesis, which is evident in the dysmorphic facies and skeletal dysplasia of most affected patients (75, 122). The dysmorphic features and other structural abnormalities somewhat resemble Zellweger syndrome, which, as a disorder of peroxisomal biogenesis, might be expected to have mevalonate kinase deficiency. Although urinary mevalonic acid levels are normal in Zellweger syndrome and Zellweger syndrome fibroblasts have normal total activities of mevalonate kinase and phosphomevalonate kinase, Zellweger liver has a marked deficiency of mevalonate kinase (187). Mevalonate kinase activity also is deficient in rhizomelic chondrodysplasia punctata (RCDP), a disorder of peroxisomal biogenesis wherein a subset of peroxisomal enzymes, including MVK, is not transported into peroxisomes (18, 186). Apparently, in both Zellweger syndrome and RCDP, mevalonate kinase is synthesized, but not delivered, to the peroxisome. This may be why cholesterol synthesis is impaired in Zellweger syndrome, even though there appears to be sufficient total cellular mevalonate kinase activity to prevent the accumulation of free mevalonate.

Recurrent or cyclic fever is a well-known, if uncommon, pediatric problem that often eludes etiologic diagnosis. Although many patients with MVK deficiency have recurrent but not truly cyclic fevers, others have very regular three- to six-week cycles of fever, lymphadenopathy, leukocytosis, arthralgia, bone pain, abdominal pain, and general debilitation (75). Some patients with MVA or HIDS have been thought to have familial Mediterranean fever or PFAPA [periodic fever, aphthous ulcers, and adenopathy syndrome (119)]. The recent recognition that MVK deficiency is the cause of the cyclic fevers of HIDS in otherwise normal individuals emphasizes that periodic fever and mevalonic aciduria are diagnostically the most important markers for MVK deficiency across the range of clinical severities (41, 80).

Pathophysiology and Treatment

Although MVK deficiency is a primary defect of cholesterol biosynthesis, plasma levels of cholesterol typically are normal or only mildly depressed (75). Considering the 10,000-fold or greater elevation of urinary mevalonic acid in classical mevalonic aciduria, HMG-CoA reductase appears to be able to upregulate mevalonate synthesis to a level sufficient to maintain adequate or nearly adequate flux through the pathway. However, cholesterol synthesis by an alternate pathway not involving mevalonate may be possible. Such a mevalonate-independent pathway for isoprenoid biosynthesis utilizing glyceraldehyde-3-phosphate and pyruvate occurs in the plant kingdom (114, 152), but no similar pathway has yet been found in higher animals.

As in any enzymatic disorder, the adverse effects of MVK deficiency may be mediated by precursor toxicity, product deficiency, or both. Unfortunately, treatment efforts directed at both metabolic consequences of MVK deficiency have met with little or no success (75). After the failure of end-product replacement therapy (cholesterol, coenzyme Q) to ameliorate the problems of mevalonic aciduria (75), an inhibitor of HMG-CoA reductase, lovastatin, was given to one child to lower the possibly toxic levels of mevalonic acid. Although lovastatin transiently lowered the level of mevalonic acid, further upregulation of HMG-CoA reductase activity seems to have occurred, followed by a life-threatening disease crisis. Pharmacological doses of corticosteroids can diminish the severity of febrile crises, but weaning MVA patients from treatment with high dose corticosteroids can be quite difficult. The finding of a direct correlation between urinary leukotriene E4 and MVA levels and the recent marketing of leukotriene receptor antagonists suggest a more direct approach to treatment of inflammation in MVA.

Genetics and Enzymology

MVA is an autosomal recessive disorder with an incidence of less than 1 in 100,000 births. Although there is no apparent ethnic predisposition for classical, severe MVA, most case reports of HIDS come from the Netherlands. The gene encoding MVK was first cloned from yeast in 1987 as *RAR1*, defined by an essential role in yeast chromosome replication (94). In 1990, the gene was independently cloned as *MVK* from yeast (ERG12) (140) and from rat (174). Two years later, a human *MVK* was cloned from a cDNA library by Schafer et al. (159). Human MVK is a homodimeric enzyme (Table 1) with a calculated monomeric weight of 42, 242 D (159). Cell localization studies by Krisans et al. (14, 171) indicate that MVK is a

TABLE 1 Molecular and enzymatic data for inborn errors of sterol biosynthesis

Disease	Mevalonic aciduria hyper IgD syndrome	Smith-Lemli-Opitz syndrome	Conradi-Hünermann, CHILD syndrome	Hydrops-ectopic calcification skeletal dysplasia	Desmosterolosis	CHILD syndrome
MIM numbers	251170 260920	268670, 270400 602858	300275 302960	215140	602938	300275 308050
Enzyme	Mevalonate kinase	7-Dehydrocholesterol reductase	Sterol-Δ7,Δ8-isomerase	Sterol Δ14-reductase	Desmosterol reductase	3β-OH-steroid dehydrogenase
EC number	2.7.1.36	1.3.1.21	5.3.3.5	1.3.1.—	1.3.—	1.1.—
Subcellular localization	Peroxisomes ?cytosol	ER	ER	ER	ER	ER
Greatest tissue abundance	Ubiquitous	Adrenal, liver, testis, brain	Liver, adrenal, gonads, uterus; low in brain & muscle		Ubiquitous	Liver, kidney, adrenal, ovary, testis
Cofactor	ATP	NADPH	NADPH	NADPH	FAD	NAD
Protein structure	Homodimer	?Homodimer	Homodimer	Homodimer (rat)		Multienzyme complex
Transmembrane domains (predicted)	0	9	4		1	1
Gene	*MVK*	*DHCR7*	*EBP*	*DHSR14*	*DHCR24/DWF1*	*NSDHL*
Yeast ortholog	ERG12	none	ERG2	ERG24		ERG26
Chromosomal localization	12q24	11q13	Xp11.22-23		1p31.1-p33	Xq28
Genomic DNA-kb	22	14	7		46.4	40
mRNA-bp	1785	2597	1073		4149	1563
Exons	11	9	5		8	8
Amino acids	396	475	230		516	373
Monomeric MW	42,424	54,516	26,335	38,000 (rat)	60,100	41,873

peroxisomal enzyme, although small amounts may normally exist in the cytosol. *MVK* has homology with ESTs from most phyla of both the plant and animal kingdoms, including organisms without the ability to make sterols (81). Genes encoding MVK in higher and lower organisms show homology in four conserved polypeptide domains (150). The second of these domains has homology with sequences in a large family of kinases for substrates as diverse as nonmevalonate isoprenoids, polypeptidyl serines, and galactose and is presumed to have a role in ATP-binding. The fourth conserved domain near the carboxy terminus of MVK likely determines mevalonate binding.

Mutation Studies

Before the *MVK* cDNA was cloned, enzymatic studies had shown a moderate correlation between the severity of clinical disease and the reduction in the MVK activity in cultured fibroblasts or peripheral lymphocytes. This phenotypic-enzymatic correlation became even more apparent in the study of HIDS. Whereas classical, severely affected MVA patients have no, or almost no, measurable MVK activity (<0.5% of control), patients with the HIDS phenotype have appreciable amounts, usually between 1% and 6% of normal in control lymphocytes (41, 80). Houten et al. (79) found that almost all patients with the Dutch-type familial periodic fever (HIDS) phenotype are heterozygous for a 1129G > A mutation (V377I). This mutation accounted for 52% of HIDS MVK alleles in their study and is presumed to be the allele responsible for most of the residual MVK activity in HIDS. To date, all MVK mutations, except one deletion mutation of exon 2, have been missense mutations distributed widely over the exons of MVK (Figure 3). The only alleles with a substantial frequency other than V377I are I268T, an allele present in both

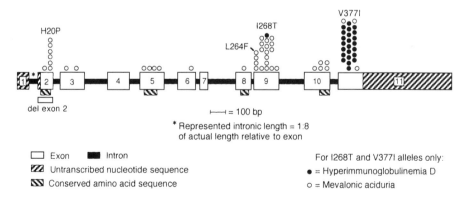

Figure 3   Distribution of *MVK* mutations in mevalonic aciduria and hyperimmunoglobulinemia D syndrome (HIDS). Approximately half of all missense mutations have been associated with HIDS. For mutations V377I and I268T only, the distrubtion of diagnoses between HIDS (closed circles) and mevalonic aciduria (open circles) is indicated.

classical MVA and HIDS patients, and H20P, which was found in one MVA patient (H20P/A334T) and five H20P/V377I HIDS compound heterozygotes. As a group, the H20P/V377I compounds had a significantly lower mean MVK activity in peripheral lymphocytes than V377I/I268T compound heterozygotes (1.8% vs. 5.0% of control). Because enzymatic assays of missense mutations in fibroblasts often overrepresent the in vivo residual activity, it is likely that I268T has intrinsically higher residual activity than H20P. That the V377I mutation is associated with high residual MVK activity also was evident in the identification of three patients heterozygous for V377I and a stop mutation (W62X, Y149X, and Y148X), two of whom were classified as HIDS and one as mild MVA (53). The inclusion of only a single homozygote for V377I among 22 HIDS patients for whom V377I represented 55% of MVK alleles suggests that homozygosity for V377I is associated with a normal phenotype in most individuals. Thus, V377I may constitute a much higher fraction of MVK alleles than indicated by its fraction of alleles among all MVA and HIDS patients. Moreover, that all four known null mutants were present in combination with only V377I suggests that truly absent MVK activity may be lethal in utero. Although one could argue that an enhanced inflammatory reaction afforded by the common V377I allele might offer a heterozygote advantage, a founder effect is also possible, especially in view of recent data concerning the apparently small number of founding Y chromosomes in Europe (153).

SMITH-LEMLI-OPITZ SYNDROME (MIM 270400; 268670)

Historical Overview

Smith-Lemli-Opitz syndrome (SLOS) is a classical, autosomal recessive multiple malformation syndrome first described in 1964 (168) and in more detail in 1969 as the RSH syndrome (139). The original SLOS patients had a distinctive phenotype consisting of a characteristic facial appearance, microcephaly, hypospadias, global developmental delay, and severe feeding problems (168). The description of new cases of SLOS in subsequent years added midline cleft palate, cataracts, postaxial polydactyly, and heart defects to the list of common abnormalities in this disorder. In 1986 and 1987, several papers described a new lethal syndrome, type II SLOS, that, in addition to the external anomalies of SLOS, had severe internal malformations, including pulmonary hypoplasia, complex congenital heart disease, renal hypoplasia or agenesis, and Hirschsprung disease (35, 40). Some 46,XY males had severe hypogenitalism or even female-appearing external genitalia.

Despite a relatively high population incidence of SLOS of about 1 in 40,000 (98), the genetic cause of the syndrome remained unsuspected until 1993 when Irons et al. (84) reported that patients with SLOS had a more than 100-fold increase in the plasma level of 7-dehydrocholesterol (cholesta-5,7-dien-3β-ol; 7DHC), the immediate precursor of cholesterol in the Kandutsch-Russell biosynthetic pathway (93). The same biochemical abnormality has since been found not only in most patients with an accepted clinical diagnosis of SLOS but also in patients with previously unknown or incorrect diagnoses now recognized as variant forms of SLOS. The apparent cause of the distinctive sterol abnormality, a deficiency of microsomal 7-dehydrocholesterol reductase (3β-hydroxysteroid-Δ^7-reductase; DHCR7), was supported by enzymatic assays in tissues and cultured cells (164). In 1998, Moebius et al. (130) cloned the human DHCR7 gene and localized it to chromosome 11q12-13. Shortly thereafter, the same authors reported that mutations of DHCR7 cause SLOS (49), a finding confirmed by two other groups (188, 190).

Clinical Characteristics

The clinical characteristics and phenotypic spectrum of SLOS have been described in detail in a number of clinical reviews (33, 88, 98, 99, 139, 158). The SLOS face, which combines microcephaly, bitemporal narrowing, ptosis, a short nasal root, anteverted nares, and micrognathia, is distinctive and easily recognized. Hypertelorism, epicanthal folds, cataracts, strabismus, midline cleft palate, and, more rarely, midline cleft lip (40, 101) are other important craniofacial findings in some patients. Characteristic skeletal abnormalities include postaxial polydactyly and syndactyly of the second and third toes (33, 158), limb shortness, and, more rarely, epiphyseal stippling (72, 127). Hypogenitalism in SLOS males, ranging from cryptorchidism to apparent complete sex reversal (12, 37, 91, 139), is one of the more important diagnostic characteristics of SLOS. In addition to the characteristic external anomalies of SLOS, there are important visceral and other internal malformations as well. In a cohort of 95 biochemically confirmed cases of SLOS with heart disease, Lin et al. (115) found atrioventricular canal (25%), primum atrial septal defect (20%), patent ductus arteriosus (18%), and membranous ventricular septal defect (10%) to be the most common defects. Both adrenal hyperplasia (126, 139) and hypoplasia (40) have been reported in SLOS. Pulmonary hypoplasia (24, 35, 104) and intestinal aganglionosis (35, 103, 196) are common anomalies among more severely affected patients, whereas pyloric stenosis is a prominent clinical problem for all degrees of severity of SLOS (139, 158). In addition to microcephaly, present in more than 90% of patients, common central nervous system (CNS) malformations include hypoplasia or aplasia of the corpus callosum, hypoplasia of the frontal lobes, and cerebellar hypoplasia, especially of the vermis (24, 47, 124, 158). Congenital sensorineural hearing deficits may affect as many as 10% of patients (158), and some form of the holoprosencephaly sequence—from a small midline notch of the upper lip to unilobar holoprosencephaly—occurs in about 5% of patients (101).

Natural History and Clinical Management

The neonatal period and infancy in SLOS usually are dominated by feeding problems, such as weak or abnormal suck, swallowing difficulties, vomiting, and lack of interest in feeding. As a result, more than 50% of patients require nasogastric tube feedings or prolonged or permanent gastrostomy feedings. However, failure-to-thrive is often misdiagnosed in children with SLOS, whose slow growth can usually be explained by genetic, not nutritional, factors. Infants with SLOS typically are small for gestational age and most continue to grow substantially below the third centile despite adequate caloric intake. Moreover, because of the significant muscle hypoplasia at birth, weight for height almost always is below the third centile. During both infancy and childhood, children with SLOS appear to have an increased number of infections, and sudden overwhelming pneumonia is not rare. Because stress-related augmentation of adrenal steroid synthesis is partly dependent on circulating LDL cholesterol (21, 141), SLOS children may have inadequate adrenal function during the stress of an infection or surgery (3).

During infancy, severe hypotonia is almost universal in SLOS and stems from both CNS abnormalities and muscle hypoplasia. However, muscle mass and tone usually improve with age. Children with SLOS characteristically have global psychomotor retardation that correlates with biochemical severity (33, 158). The average SLOS child is quite sociable, has substantially better receptive than expressive language, and may be surprisingly mechanically adept for the degree of mental retardation. Gross motor development is more severely delayed than fine motor development, but most children eventually learn to walk. Approximately 10% of the children fall into the mildly retarded range (IQ 50 to 70), with a rare patient testing in the low normal or even normal range [(120, 158); R.I. Kelley, unpublished]. Behavioral abnormalities that fall within the spectrum of autistic disorder include hand flapping, abnormal obsessions, rigidity and insistence on routine, and poor visual contact (178).

Although there are no recent estimates for life expectancy in SLOS, Johnson (88) found that 27% of patients died before age 2. Because ascertainment of SLOS, even in the era of biochemical diagnosis, remains incomplete, the true percentage of cases with early or later death remains uncertain. However, life expectancy in SLOS appears to be determined largely by the severity of the internal malformations and the quality of general supportive care, not by an intrinsic toxic or degenerative process.

Although clinical management of SLOS in the past was largely symptomatic, standard care now includes treatment of the cholesterol deficiency with supplemental dietary cholesterol. The estimated daily synthetic need for cholesterol for infants is about 30 mg/kg, whereas for adults the amount decreases to approximately 10 mg/kg (31, 89). Because infants can absorb from the diet almost their entire daily cholesterol requirement (31), a child with SLOS who is given supplementary cholesterol theoretically may be able to downregulate endogenous sterol synthesis substantially and, thereby, limit the de novo synthesis of 7DHC. However, because brain cholesterol in mammals appears to be synthesized entirely in situ (133), the biochemical improvement now seen in the plasma of treated SLOS patients probably has little direct effect on brain function, except perhaps, as influenced by peripheral hormonal or other biochemical changes. Even if cholesterol could reach the brain, cognitive improvement might be limited because the microcephaly and mental retardation of SLOS probably reflect more the abnormalities of embryonic and fetal cerebral development than any ongoing effect of 7DHC or the cholesterol deficiency.

Most treatment protocols for SLOS provide between 50 and 200 mg/kg/day cholesterol, either in natural form (eggs, cream, liver, meats, and meat-based formulas) or as purified food-grade cholesterol, sometimes with supplements of bile acids (cholic acid, chenodeoxycholic acid, or ursodeoxycholic acid) (42, 83, 184). In addition to improved growth and gastrointestinal function in many SLOS patients treated with cholesterol, often there is marked improvement in behavior. The behavioral improvement, which can occur within days of treatment, is the most important clinical benefit of dietary cholesterol therapy. Treatment of several SLOS patients with simvastatin, an inhibitor of HMG-CoA reductase, has now been reported (87). Although the initial goal of treatment with simvastatin was to lower 7DHC levels to a greater degree than cholesterol (based on the speculation that 7DHC is toxic), there was paradoxical increase in the level of cholesterol at the same time that the level of 7DHC decreased (87). Most likely, treatment with an HMG-CoA reductase inhibitor, which leads to increased activity of certain enzymes with sterol response elements, caused sufficient upregulation of residual DHCR7 activity to enhance conversion of 7DHC to cholesterol. In contrast, biochemical and clinical worsening of disease has been seen in severely affected SLOS children—who have very little residual DHCR7 activity—treated with HMG-CoA reductase inhibitors (R.I. Kelley, unpublished data).

Biochemistry of Smith-Lemli-Opitz Syndrome

In almost all plasma samples from SLOS patients, the most abundant precursor sterol is 7DHC. A second diene sterol, 8-dehydrocholesterol (cholesta-5,8-dien-3β-ol; 8DHC), which most likely derives from isomerization of 7DHC via sterol-Δ^8,Δ^7-isomerase, is normally present at about 75% of the amount of 7DHC (179). As shown in Figure 4, some SLOS patients have had cholesterol levels as low as 1 mg/dl, whereas others, between 10% and 15%, have normal cholesterol levels, even when the levels of 7DHC are substantially increased. Despite the apparently negligible transfer of cholesterol from mother to fetus (21, 22), SLOS patients with two null alleles have cholesterol levels 5 to 20 mg/dl at birth, and still higher levels in later months and years, even on cholesterol-free diets (33, 180). Thus, there may be another genetic source of DHCR7 activity, alternate splicing for certain *DHCR7* alleles, or a pathway for cholesterol synthesis not requiring DHCR7. A possible alternate source of cholesterol synthesis is the peroxisome, which, as described above, has an essential role in the early steps of cholesterol biosynthesis (13, 108, 171) and which, as shown by Appelkvist et al. (6), may have the capacity to synthesize cholesterol from lanosterol.

An interesting group of SLOS patients are those with a typical SLOS phenotype but normal plasma cholesterol levels and only mildly increased or even normal levels of 7DHC in plasma [(33); R.I. Kelley, unpublished observations]. However, if fibroblasts or lymphoblasts from these patients are grown in lipid-depleted culture medium, the level of 7DHC often rises to the same level found in cells from classical SLOS patients (2), indicating that rapidly growing cells in tissue culture

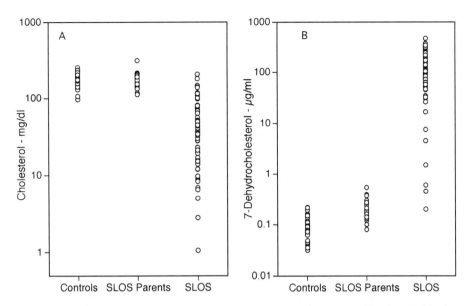

Figure 4 Distribution of plasma levels of (*A*) cholesterol and (*B*) 7-dehydro-cholesterol (7DHC) in patients with Smith-Lemli-Opitz syndrome (SLOS) and their parents. Only a very rare patient with SLOS has normal plasma levels of both cholesterol and 7DHC.

may reflect better the sterol metabolism of rapidly dividing and differentiating embryonic cells.

There have been very few formal investigations of steroid hormone metabolism in SLOS. One of the earliest (23) found abnormally high levels of DHEA sulfate and low levels of testosterone in two newborns with SLOS but abnormally low levels of DHEA sulfate in an older infant and child (30 months and 5 years). More recently, Shackleton et al. (161) found 7-dehydro homologs of most of the common urinary steroid metabolites in children with SLOS. What role, if any, these abnormal 7-dehydro steroid species play in the genital malformations of SLOS in unknown. However, the production of these abnormal steroids by the fetus is sufficient to allow prenatal diagnosis of SLOS by quantification of 7-dehydro steroids in maternal urine at midgestation (162, 163). The same abnormal fetal steroid metabolism is reflected in depressed maternal unconjugated serum estriol levels at midgestation (107, 126).

Biochemical Teratology

The discovery of DHCR7 deficiency as the cause of SLOS has drawn attention to the role of cholesterol metabolism in morphogenesis. Although children with the only previously described defect of cholesterol biosynthesis, mevalonic aciduria, can have dysmorphic facies and rhizomelic shortness, that a genetic disorder of cholesterol biosynthesis would be manifest principally as a severe multiple malformation syndrome affecting every organ system was unanticipated. However, Roux and colleagues had reported since the 1960s that, when given to pregnant rats or mice, inhibitors of enzymes of postsqualene cholesterol biosynthesis cause holoprosencephaly, microcephaly, pituitary agenesis, limb defects, and genital anomalies (154, 156). Such an apparent relationship between abnormal sterol metabolism and disordered embryologic development has been further supported by newer animal models using enzyme inhibitors and targeted genetic interruption of various steps in embryonic cholesterol metabolism (38, 112). Moreover, as noted earlier, holoprosencephaly, one of the major consequences of disrupted embryonic cholesterol metabolism, occurs in about 5% of patients with SLOS, often in patients not previously known to have SLOS (33, 101).

The mechanism by which inhibition of sterol biosynthesis disrupts normal embryologic development remained obscure until Chiang et al. (25) showed that targeted disruption of the *Sonic hedgehog* gene in mice causes not only holoprosencephaly, but also distal limb defects and other skeletal anomalies not unlike those of SLOS (25). The same group also showed that covalent addition of cholesterol to the N-terminal portion of Sonic hedgehog is an essential step in the formation of the actively signaling Shh-N hedgehog fragment (146, 147). Shortly thereafter, Roessler et al. (151) showed that haploinsufficiency for *Sonic hedgehog* in humans is one cause of sporadic as well as familial autosomal dominant holoprosencephaly. Moreover, targeted disruption of the gene for megalin (gp330), an important component of a system for delivery of maternal LDL cholesterol to the embryonic neuroepithelium, causes holoprosencephaly in homozygous deficient mice (191).

The discoveries linking holoprosencephaly, mutations in *Sonic hedgehog*, and abnormal cholesterol metabolism suggested that interference by 7DHC with cholesterol modification of Shh-N may be the cause of malformations in SLOS. However, Cooper et al. (29) published evidence that the apparent defect in Sonic hedgehog signaling instead may reside in the effect of the abnormal cellular sterol milieu on the response of the target tissue to Shh-N (29). Interestingly, the receptor for Shh, Patched, contains a sterol response element common to several proteins whose synthesis or function is regulated by the subcellular sterol environment (123). Patched, therefore, may be the element of the hedgehog signaling cascade primarily disturbed by the abnormal tissue sterol environment in SLOS tissues during embryogenesis. Because Sonic hedgehog is only one of a family of signaling proteins with a Patched receptor, similar impairment of the Desert hedgehog signaling cascade in the genital anlage (15) and of Indian hedgehog function in cartilage (85) may play a role in the abnormal morphogenesis of these structures in SLOS. Other less specific cellular disturbances, such as abnormal cell membranes and cell-cell interactions, may also contribute to the abnormal morphogenesis of SLOS and other defects of cholesterol biosynthesis (38).

Although a specific role of Sonic hedgehog in the embryological abnormalities of SLOS remains to be proven, there is now strong evidence that the malformations

of SLOS derives, at least in part, from impaired hedgehog function and the resulting downstream effects on the expression of homeobox genes. However, most
genetically characterized multiple anomaly syndromes that cause disturbances in
the body plan have been caused not by abnormalities of intermediary metabolism
but by mutations of homeobox genes or related transcriptional factors. Even the
multiple and severe metabolic abnormalities of Zellweger syndrome have little
effect on the fundamental embryonic body plan. Thus, SLOS may be an exception
among metabolic malformation syndromes because at least one action of its abnormal sterol biochemistry appears to disrupt embryonic signaling pathways and,
thereby, mimic the effects of mutations in homeobox genes.

Enzymology of 7-dehydrocholesterol Reductase

DHCR7, the enzyme that converts 7DHC to cholesterol, is a microsomal membrane-
bound enzyme with a mass of 55 kDa (Table 1). DHCR7 has been purified to
near homogeneity and some of its enzymatic characteristics have been described
(77, 164, 165). Important with regard to possible treatment strategies for SLOS
is that DHCR7 contains a sterol regulatory element among its polypeptide domains and may undergo phosphorylation/dephosphorylation regulation (8, 166).
The gene for DHCR7 was localized to 11q12-3, cloned, and sequenced in 1997 by
Moebius and colleagues (130), who also showed that the human DHCR7 enzyme
has strong homology with DHCR7s of both unicellular and other multicellular
eukaryotes as well as homology with 3β-hydroxysteroid-Δ^{14}reductases. Indeed,
there is considerable amino acid sequence homology of DHCR7 with at least other
five mammalian proteins, three of which have demonstrable sterol reductase activity: sterol-Δ^8-isomerase, lamin B receptor (LBR), and sterol 14 reductase (Table 1).
A comparison of the sequences of these enzyme proteins shows the greatest conservation of amino acid sequence in DHCR7 in exons 5 through 7, which, therefore,
may be the segments that invest DHCR7 with enzymatic activity (130). Although
four different sterol double bonds are acted on by the five homologous DHCR7-
like proteins with known enzymatic activity, the bonds cluster in one region of the
cholesterol molecule and the reaction sequence in all includes the formation of an
unstable carbocation intermediate. The fifth protein with homology to the sterol
reductases is LBR, a protein of the nuclear membrane. LBR has substantial homology with DHCR7 but even greater homology with TM7SF2, a protein whose
gene has an intron/exon structure almost identical to the nine $3'$ exons of LBR.
LBR differs from both DHCR7 and TM7SF2 in having an additional three exons
$5'$ to the DHCR7-homologous segments that appear to encode a nuclear envelope
targeting signal (169). Interestingly, LBR, which normally is tightly associated
with the inner surface of the nuclear envelope, redistributes to the ER when the
nuclear envelope disintegrates during mitosis (137). The complementary tissue
distributions of the sigma-1 receptor and TM7SF2 (Table 1) suggest that these
proteins may serve similar roles in lipid transport. That these proteins have strong
homology in the putative sterol binding carboxy termini of sterol Δ^{14}-reductase

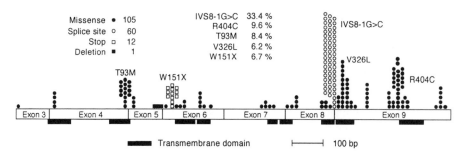

Figure 5 Structure of *DHCR7* and distribution of its mutations in Smith-Lemli-Opitz syndrome.

and sterol-Δ^8-isomerase further suggests that sterols may be preferred ligands for TM7SF2 and SIGMA1 (76).

DHCR7 Mutation Studies

Approximately 40% of known *DHCR7* mutations affect exon 9 and collectively constitute almost 75% of DHCR7 alleles in SLOS patients (Figure 5). These include four of the most common mutations, V326L, R352W, R404C, and IVS8-1G>C, which alone account for almost 50% of DHCR7 alleles. Of more than 70 known *DHCR7* mutations, almost two thirds occur in putative transmembrane regions. The most prevalent *DHCR7* mutation, IVS8-1G>C, activates a splice site 5' to the mutation site in intron 8, leading to a 134 bp insertion in the mRNA, a premature stop codon, and a protein product shortened by 154 amino acids (7, 49). Two other common mutations are T93M in exon 4 and W151X in exon 6. Heterologous expression of c-myc cDNA constructs of the four most common missense mutations (T93M, V326L, R404C, R352W) showed that all expressed normal amounts of mRNA but reduced amounts of immunoreactive protein (<10%). The premature stop mutation, W151X, creates a truncated protein with no enzymatic activity (7, 49). Unlike the relatively common W151X mutation, all other stop or frameshift mutations have so far been unique.

Epidemiology of DHCR7 Mutations

DHCR7 alleles are notable not only for the high prevalence of a small number of severe mutations, but also for a varying geographical distribution and for discrepancies between predicted and observed frequencies of certain genotypes. All larger studies of SLOS mutations in populations in or originating from Europe find the IVS8-1G>C splice site mutation to be the most common, usually accounting for about one third of alleles. In anonymous DNA samples from different populations, Yu et al. (194) found heterozygosity for IVS8-1G>C in 1 of 90 American Caucasians, but none in similar numbers of natives of Finland, Sierra Leone, China (Han), and Japan. In a group of 1503 anonymous newborn screening blood spots collected in the U.S. state of Oregon, Battaile found 16 carriers for IVS8-1G>C, giving an unusually high estimated carrier frequency for this single allele of 1% in a population that is between 80% and 90% of European origin. In a more detailed ethnic analysis of their previously reported DHCR7 mutation data (192), Witsch-Baumgartner et al. found evidence for an east to west European cline for IVS8-1G>C, with the highest frequency in the British Isles, and opposite west to east clines for W151X and V326L. Based on haplotype analyses, these authors speculated that the two common null mutations, IVS8-1G>C and W151X, are the most ancient among the common SLOS mutations in Europe. Interestingly, R404C and T93M, a mutation most prevalent in England and Italy, involve CpG islands and appear to have arisen on three and four different haplotypes, respectively.

Although no single DHCR7 mutation study has been large enough for a rigorous analysis of observed vs. expected frequency of *DHCR7* allele combinations, a metaanalysis of the larger studies—all of which reflect 90% or greater European ancestry—show some illuminating discrepancies. For example, among 179 SLOS patients from four major centers [(188, 192, 193); R.I. Kelley, unpublished data] there were 9 IVS8-1G>C homozygotes identified, compared to the 18 predicted by the allele fraction of 0.32 for IVS8-1G>C among all 358 alleles. In contrast, for combined heterozygosity of IVS8-1G>C with each of the four most common missense mutations (T93M, V326L, R352W, R404C), there was a twofold or greater number of patients than predicted, whereas the same relative increase was not found for IVS8-1G>C heterozygosity with the most common stop mutation, W151X (Table 2). Overall, there were 39 patients who were compound heterozygotes for IVS8-1G>C and one of the common missense alleles, compared to the 14 predicted. A similar relationship for the less common null allele, W151X, was found, with eight compound heterozygotes vs. three predicted.

Although some skewing of genotypic frequencies can be expected when combining different population samplings, the finding that all four common missense alleles show the same degree of overrepresentation in compounds with both null alleles argues that these null alleles constitute a much higher fraction of SLOS alleles in the reference populations than is evident from genotyping of identified SLOS patients. Because homozygosity for IVS8-1G>C usually leads

TABLE 2   Predicted vs. observed number of patients with specific *DHCR7* genotypes among 179 patients with Smith-Lemli-Opitz syndrome

| | | Mutation 2 | | | | | |
| | | Missense | | | Null | | |
Mutation 1		T93M	V326L	R352W	R404C	W151X	IVS8-1 G>C
IVS8-1	Predicted	5.6	3.3	2.2	3.3	2.9	18.2
G>C	Observed	18	8	6	7	2	9
W151X	Predicted	0.9	1.0	0.35	0.5	0.5	—
	Observed	4	2	2	0	1	—

to a prenatal or perinatal lethal form of SLOS, the lower than expected incidence of IVS8-1G>C homozygosity could reflect substantial fetal losses of null allele homozygotes or frequently missed patients because the diagnosis of SLOS is considered less often in the severely malformed fetus or SLOS newborn who dies in the immediate perinatal period without examination by a geneticist. Furthermore, because the opportunity for DNA sampling increases with length of survival, IVS8-1G>C and W151X homozygotes and IVS8-1G>C/W151X mixed heterozygotes may be substantially underrepresented in the SLOS DHCR7 mutation studies. If DHCR7 exists as a homodimer, another, if less likely, possibility to consider is a dominant positive phenomenon, whereby one mutant protein, such as the IVS8-1G>C–encoded truncated protein, in the dimer stabilizes a more labile DHCR7 protein encoded by a *DHCR7* missense mutation. However, a dominant positive phenomenon would be unlikely to explain the increase in the observed over the predicted number of patients found for each of the four common missense mutations.

The discrepancies in the reported frequencies of specific SLOS genotypes recommend a reexamination of some of the estimates of the birth incidence of SLOS. Combining their finding of a 1% heterozygote frequency for IVS8-1G>C with the knowledge that IVS8-1G>C accounts for one third of SLOS mutations in most SLOS patient studies, Battaile et al. estimated an overall *DHCR7* mutation heterozygote frequency of 1 in 30. This gives a predicted incidence of SLOS of between 1 in 1590 and 1 in 13,500 births, which is much higher than the estimated U.S. incidence of SLOS of 1 in 40,000 births (98). However, in view of the almost threefold excess of the observed over the predicted number of null/missense mixed heterozygotes, it is possible that IVS8-1G>C and W151X together account for as many as 80% of DHCR7 mutant alleles in the U.S. population and that, consequently, the carrier frequency of *DHCR7* mutations collectively may be less than 1.3%. Whereas a 1.3% carrier frequency would predict an incidence of 1 in 24,000 births, perhaps one third of SLOS births are lethal in utero, i.e., when null alleles are involved. Moreover, as reported by Langius et al. (111) some IVS8-1G>C/missense mixed heterozygotes are developmentally normal, minimally dysmorphic (e.g., only 2/3 toe syndactyly) school-aged children, despite abnormal sterol biochemistry in plasma. Also likely is that homozygotes and mixed heterozygotes for certain missense mutations are biochemically and clinically normal, as noted above for the common V377I *MVK* mutation. Thus, although missed patients with either very mild or very severe forms of DHCR7 deficiency may remain, it is likely that the actual birth incidence of clinically significant DHCR7 deficiency in the United States is close to the reported clinical experience of 1 in 40,000 births (98).

Phenotype-Genotype Correlation

In many inborn errors of metabolism, some correlation exists between the level of the primary abnormal metabolite and clinical severity. However, as first shown in 1997 by Cunniff et al. (33), the level of 7DHC in plasma poorly correlates with the SLOS physical severity score (33). Instead, the strongest correlation in SLOS exists between the physical severity and the dehydrosterol (7DHC + 8DHC) level expressed as a fraction of total sterols, a value that more accurately expresses the diminished cholesterol availability for both metabolic and structural fates (98). In their detailed study of phenotype-genotype correlation (192), Witsch-Baumgartner et al. divided the 40 mutations they found in 84 patients into four classes based on the structural effect or location of the mutation: missense mutations in (*a*) predicted transmembrane spanning segments and their border areas (TM), (*b*) the fourth cytoplasmic loop, (4L), (*c*) the C-terminus (C-T), and (*d*) null mutations (0), largely IVS8-1G>C and W151X. Overall, the predicted severity of the mutant allele class significantly correlated with both clinical and biochemical severity, with TM/TM compounds having the mildest biochemical and clinical phenotypes and 4L/4L and 0/0 compounds having the most severe. As would be expected, there was wide variability in the biochemical severity of 0/TM compounds. However, although 4L/4L and 0/0 were associated with the highest fractional levels of dehydrosterols, 0/4L compounds (almost all of which were IVS8-1G>C/4L compounds) had dehydrosterol fractions and clinical severities that ranged from the highest to lower values more characteristic of mild TM/TM compounds. Although the number of patients in each group was small, the unusual variability of the 0/4L compounds again raises the possibility of a dominant positive effect for some compound heterozygotes.

In addition to substantial phenotypic variability within each mutation class, much variability for single allelic combinations also exists, indicating significant contributions of other genetic or environmental factors to the clinical and biochemical phenotype of SLOS. For example, among 6 IVS8-1G>C/T93M compounds, the clinical severity score (0 = normal, 100 = most severe) ranged from 17 (mild) to 39 (moderate), whereas the pretreatment dehydrosterol fraction [(7DHC + 8DHC)/(7DHC + 8DHC + cholesterol)], ranged from 0.18 to 0.59 (mean 0.31), with no strong correlation between the clinical and biochemical measures of severity (R.I. Kelley, unpublished data). For 5 R404C homozygotes, the dehydrosterol fraction was higher and less variable—0.59 to 0.79, mean 0.67—but clinical variability was even greater, from 19 to 69. The variable dehydrosterol fraction within a single genotype and the substantial amount of cholesterol synthesized by IVS8-1G>C homozygotes suggest the existence of other DHCR7 activities or an alternate pathway of cholesterol biosynthesis, as discussed above.

DESMOSTEROLOSIS (MIM 602938)

In 1998, Fitzpatrick et al. (50) first described a third apparent defect of cholesterol biosynthesis in a 34-week gestation, 46,XX dysmorphic female infant who died 1 hour after birth from respiratory insufficiency. Anomalies included macrocephaly, thick alveolar ridges, gingival nodules, cleft palate, total anomalous pulmonary venous drainage, clitoromegaly, short limbs, and generalized osteosclerosis resembling Raine skeletal dysplasia syndrome (149). Although the infant had macrocephaly, not microcephaly, the similarity of some of the anomalies to those of SLOS suggested a possible disorder of sterol biosynthesis as the cause of the infant's condition. Indeed, instead of 7DHC, the infant had markedly increased tissue levels of desmosterol (cholesta-5,24-dien-3β-ol), an immediate precursor of cholesterol following an alternate synthetic pathway wherein the 24-double bond is reduced last, rather than early, as in the Kandutsch-Russell pathway (Figure 2). Although fibroblasts were not available for further study, both parents had mildly increased plasma levels of desmosterol, suggesting an autosomal recessive deficiency of desmosterol reductase (3β-hydroxysteroid-Δ^{24}-reductase; DHCR24). A second infant with increased plasma and cellular levels of desmosterol was reported by Andersson et al. (4). This three-year-old male had downslanting palpebral fissures, micrognathia, submucous cleft palate, clubfoot, a persistent patent ductus arteriosus, profound microcephaly, and complete agenesis of the corpus callosum. Although the level of desmosterol in plasma was only mildly increased to 60 μg/ml (nl < 1.1 μg/ml), or just 5% of the cholesterol level, the sterol fraction of desmosterol rose to 42% in lymphoblasts cultured in cholesterol-depleted medium, similar to the tissue fraction of desmosterol in the tissues of the patient of Fitzpatrick et al. (50). The parents of the second patient had two- to threefold increased levels of desmosterol in plasma (4), and desmosterol also was mildly increased in the mother's cultured lymphoblasts (R.I. Kelley, unpublished data).

Based on the description in 2000 of the human ortholog, *DWF1*, of *DIMIN-UTO/DWARF1*, a gene encoding a sterol-Δ^{24}-reductase–like activity in plants, Waterman identified a 169 kb human genomic clone from chromosome 1 that contained the human ortholog of *DIMINUTO/DWARF1* (189). The apparent human *DHCR24* (Table 1) mapped to 1p33.1 and spanned 46 kb of genomic DNA. Further analysis indicated a gene with eight exons that was expressed in all tissues and that, when expressed in naturally DHCR24-deficient yeast, conferred the ability to convert desmosterol to cholesterol. Testing of DNA from the two reported desmosterolosis patients and their parents revealed *DHCR24* mutations in all four alleles. Expression of the mutant alleles in yeast showed almost absent DHCR24 activity for the first patient and substantially more but still depressed activity in the second patient (19.9%). In first patient, the maternal allele carried a 1412C>T (Y471S) mutation, whereas the paternal allele had two mutations, 818A>C (N294T) and 918G>C (K306N). The DHCR24 activities of the identified mutations expressed in yeast paralleled the observed biochemical and phenotypic severity in the patients (189).

The embryologic abnormalities of the first desmosterolosis patient that parallel those of SLOS—cleft palate, ambiguous genitalia (although 46,XX), thick alveolar ridges with gingival nodules, and short limbs—support the conclusions of Cooper et al. (29) that the apparently impaired signaling of Sonic hedgehog in SLOS is not a specific effect of 7DHC or 8DHC but more likely reflects the intracellular cholesterol deficiency. Moreover, the studies of Roux and colleagues many years earlier had shown that Triparanol, a potent inhibitor of DHCR24, and AY-9944, an inhibitor of DHCR7, have similar malforming effects on mouse embryos

(154, 155, 157). However, other defects found in the first desmosterolosis patient, such as the marked osteosclerosis and severe limb shortness, suggest that there may be specific teratologic effects of desmosterol or, perhaps, other sterol metabolites with a 24-unsaturated bond that accumulate behind a deficiency of DHCR24. Interestingly, two other cholesterol precursors with 24-unsaturated bonds, 4,4′-dimethylcholesta-8,14,24-trien-3β-ol and 4,4′-dimethylcholesta-8,24-dien-3β-ol, have potent signaling activities in postfertilization meiosis (58, 86) and could have increased levels in desmosterolosis.

X-LINKED DOMINANT CHONDRODYSPLASIA PUNCTATA AND CHILD SYNDROME (MIM 302960, 300205, 308050)

Overview of the Chondrodysplasia Punctatas

The chondrodysplasia punctatas (CDPs) are a heterogeneous group of genetic disorders characterized by abnormal foci of calcification in the cartilaginous skeleton, termed chondrodysplasia punctata or epiphyseal stippling because of the predominant location of the lesions in the epiphyses. The severe autosomal recessive form of CDP (RCDP; MIM 215100) is associated with symmetric proximal (rhizomelic) limb shortness, ichthyosis, cataracts, growth and mental retardation, and peroxisomal abnormalities. Most patients with RCDP have mutations in the PEX7 gene, which encodes a receptor that directs proteins with a type 2 peroxisomal targeting signal (PTS2) to the peroxisomal matrix (18, 134, 148). Milder autosomal dominant, X-linked recessive (CDPX) and X-linked dominant (CDPX2) forms of CDP are also known (176). X-linked recessive CDP (MIM 302950) is often associated with terminal Xp deletions or X:Y translocations involving Xp22.32 (9). Recently, mutations in arylsulfatase E (ARSE) have been found in some CDPX patients with normal karyotypes (51). Epiphyseal stippling has also been noted in patients with a variety of other genetic and acquired disorders, including trisomy 21, congenital hypothyroidism, Zellweger syndrome, and prenatal maternal exposure to vitamin K antagonists, such as coumadin (176, 185). The occasional finding of epiphyseal stippling in Smith-Lemli-Opitz syndrome (90) and the report of a severe skeletal dysplasia in one of the two known patients with desmosterolosis (50) suggested a link between sterol metabolism and skeletal disease and led to the discovery that classic CDPX2 (102) and several other skeletal disorders (39, 56, 100, 118) are associated with a deficiency of an enzyme catalyzing one of the later steps of cholesterol biosynthesis.

Clinical Features and Genetics of X-linked Dominant CDP

CDPX2—also called Conradi-Hünermann or Happle syndrome—is a rare X-linked dominant disorder with presumed male lethality (60, 67). Affected females typically present with skeletal and skin abnormalities at birth. There is a flaky, usually erythematous, eruption (ichthyosiform erythroderma), which often resolves completely or substantially in the first months of life, leaving linear or whorled patches of atrophic and/or pigmented skin. In addition to hyperkeratosis, acanthosis, and patchy parakeratosis in the epidermis, there are neutrophilic or lymphocytic infiltrates in both the epidermis and dermis as well as characteristic follicular plugging (30, 36, 43, 60). Electron microscopy (EM) of the epidermis shows abnormal accumulations of lipid-laden vacuoles (43, 105). The hair is coarse and lusterless, and there may be patches of cicatricial alopecia. The nails occasionally are flattened, split, and hypoplastic, whereas the teeth usually are normal. As in many X-linked genodermatoses, the skin abnormalities of CDPX2 females characteristically follow the lines of Blaschko, which reflect the functional mosaicism caused by random X inactivation (181).

Skeletal abnormalities in CDPX2 include infantile epiphyseal stippling and asymmetric rhizomelic shortness of the limbs. Scoliosis, either congenital or later onset, and kyphosis are common. Clubfoot, postaxial polydactyly, joint contractures, and vertebral anomalies also are occasional findings. The stippling in CDPX2 often involves the vertebral and tracheal cartilages and is more widespread than in most other forms of CDP. Light microscopy reveals that sections of differentiating cartilage demonstrate patchy areas of chondrocyte loss, which by EM appear to be areas of apoptosis of prechondrocytes (W. Wilcox, personal communication). Craniofacial defects include frontal bossing, a flat nasal bridge, and midface hypoplasia. Cataracts, typically congenital and asymmetric or sectorial, are present in approximately two thirds of cases (61) and may be accompanied by microphthalmia, microcornea, or optic nerve hypoplasia (173, 175). Other features reported in some patients include congenital heart disease, developmental renal anomalies, including hydronephrosis, and sensorineural as well as conductive hearing losses. CNS malformations, although rare, have been reported, and most patients have normal intelligence [(60); G. Herman, unpublished data]. Polydactyly is relatively specific for the X-linked dominant form of CDP, and the frequency of ichthyosis and cataracts is reported to be higher in CDPX2 than in RCDP.

CDPX2 affects females almost exclusively. Most cases are sporadic, presumably the result of new mutations. Four males with CDPX2, one of whom had a 47,XXY karyotype, have been reported (63, 172, 195). Presumed X linkage with male lethality has been proposed as the mechanism to explain the inheritance pattern of CDPX2 (60), and somatic mosaicism or half-sister chromatid exchange has been invoked to account for the clinical findings in CDPX2 males with apparently normal chromosome complements (63).

Clinical diagnosis of CDPX2 in infancy is based on finding epiphyseal stippling in a female with other clinical features of the disorder and is confirmed by finding a characteristic pattern of cholesterol precursors (vide infra). Prenatal diagnosis can be performed by mutation analysis of chorionic tissue or cultured amniocytes. Although not yet reported, it is likely that, as in SLOS, prenatal diagnosis by measurement of cholesterol precursors in amniotic fluid will be possible.

Treatment of CDPX2 is symptomatic. Although the ichthyotic skin lesions usually resolve within weeks to months, they may persist to varying degrees in some patients and can be treated with a variety of emollients. Dry skin is not uncommon and alopecic patches persist throughout life. Early ophthalmologic examination to detect the presence of cataracts is essential to ensure normal visual development. Baseline head, renal, and cardiac sonographic examinations and hearing screening are also recommended. Involvement of the tracheal and laryngeal cartilages can lead to airway obstruction and difficulty with intubation. The skeletal asymmetries and, in particular, scoliosis may require surgical intervention. Also important to recognize is that congenital vertebral anomalies and underdevelopment of the foramen magnum can lead to congenital or postnatal neurological complications owing to vertebral body subluxations or spinal cord compression (34, 54).

Biochemistry and Mutation Analysis of CDPX2

In 1999, Kelley et al. reported markedly increased levels of 8DHC and 8(9)-cholestenol in five females with chondrodysplasia punctata (102). Subsequently, similar biochemical abnormalities have been detected in numerous additional females with CDPX2 (68, 82). This abnormal sterol pattern suggested a defect in 3β-hydroxysteroid-Δ^8,Δ^7-sterol isomerase, a microsomal enzyme that converts 8(9)-cholestenol to lathosterol in the terminal steps of normal cholesterol biosynthesis (Figure 2). The gene for human sterol-Δ^8,Δ^7-isomerase, also called emopamil binding protein (EBP), is X linked and maps to Xp11.2 (Table 1).

Nineteen different mutations in the human EBP gene in a total of 25 unrelated CDPX2 females have been published to date (17, 39, 68, 82) [Patients 2 and 6 in References (18) & (40), respectively, are the same.]. Biochemical studies were performed on patients in two of the reports (17, 82), and all of the females in whom mutations were found demonstrated the typical sterol pattern described above. The EBP mutations identified include 11 nonsense, 7 missense, 1 single amino acid deletion, 4 frameshift, and 2 splicing mutations. All of the missense mutations alter conserved amino acids in the predicted protein. Four of the mutations have occurred in more than one patient; two of these involve CpG islands that may be hot spots for mutations. Somatic and gonadal mosaicism has been described in one female, and gonadal mosaicism is presumed in a second family in which two sisters are affected and their mother is phenotypically normal (68). Although the possibility of gonadal mosaicism must be considered in genetic counseling for apparently sporadic CDPX2 cases, there is no conclusive evidence that this phenomenon is more common in CDPX2 than other X-linked disorders.

There are no obvious genotype-phenotype correlations among the 25 patients reported with EBP mutations, a fact that probably reflects determination of the phenotype as much by the pattern of X inactivation in affected tissues as by the nature of the mutation itself. A number of reports have documented the increasing severity of the clinical features of CDPX2 in succeeding generations (135, 138, 172, 182), a phenomenon called anticipation. In analogy with several neurological disorders (32), Traupe et al. (182) speculated that anticipation in familial CDPX2 was caused by expansion of an unstable triplet repeat. With the identification of mutations in the EBP gene in numerous females, such anticipation now appears to result from the skewing of X inactivation, somatic and/or gonadal mosaicism, or decreased reproductive fitness of more severely affected females. However, the possibility remains that the frequently reported anticipation results from an effect of the abnormal sterol metabolism of the mother on her developing CDPX2 embryo and fetus.

Gene Structure and Enzymology

EBP, the original protein designation for human sterol-Δ^8,Δ^7-isomerase, was first identified as a high-affinity binding protein for the antiischemic drug emopamil— hence, the name. As such, EBP is a member of the sigma class of drug-binding proteins whose ligands include numerous pharmacologically active compounds, including chlorpromazine, haloperidol (129), and tamoxifen (27). EBP was later found to possess sterol isomerase activity in mammalian cells and to complement sterol isomerase deficient (*erg2*) mutants of *Saccharomyces cerevisiae* (167). Surprisingly, there is no significant amino acid homology between EBP and the yeast sterol isomerase protein encoded by the *erg2* gene. Another member of the sigma receptor family, sigma-1 receptor (Table 1), has substantial amino acid homology with the erg2 protein but has no measurable isomerase activity (95, 132)

The EBP gene spans approximately 7 kb of genomic DNA, and the genomic sequence for the entire human gene has been assembled (GenBank Accession No. NT_011609). There are five exons in the gene, with the translation start site in exon 2. The EBP protein is widely expressed, with highest levels in tissues involved in cholesterol synthesis and steroidogenesis, such as liver, adrenal gland, intestines, and gonads. It is predicted to be an integral membrane protein with four transmembrane domains and has been localized within the ER [(59); G. Herman, unpublished results]. The C-termini of the mouse and human proteins contain a lysine-rich consensus sequence for retention of proteins within the ER membrane (177). In vitro mutagenesis of the EBP protein identified six residues (His77, Glu81, Glu123, Thr126, Asn194, and Trp197) located in the cytoplasmic halves of several of the predicted transmembrane segments 2–4, which appear to be essential for enzymatic activity and, therefore, probably form part of the protein's catalytic site (131).

Mouse Models of X-linked Dominant, Male-Lethal Skeletal Dyplasias

THE TATTERED (TD) MOUSE   The X-linked tattered (*Td*) mouse is a model for CDPX2. Heterozygous *Td* females are dwarfed and develop hyperkeratotic skin lesions at approximately postnatal day 4–5. The adult coat is striped, following the pattern of X inactivation in the mouse, and subtle cataracts have been detected in *Td* females. Affected male embryos die between 12.5 days post coitum (dpc) and birth, depending on the genetic background (39). Those males that survive until birth are hydropic with a short-limbed skeletal dysplasia, craniofacial abnormalities, cleft palate, and absent mid- and hindgut. *Td* results from a missense mutation in a conserved amino acid within the murine EBP protein (G107R), and *Td* females accumulate 8DHC and 8(9)-cholestenol, similar to human CDPX2 females (39).

THE BARE PATCHES (BPA) MOUSE   Interestingly, although *Td* is now the proven ortholog of human CDPX2, another X-linked dominant, male lethal mouse disorder with a similar phenotype, bare patches (*Bpa*) had long been considered the ortholog of human CDPX2 (5, 61). *Bpa* females are also dwarfed (45, 143, 144), and abnormal deposits of calcium in the tail vertebrae, consistent with epiphyseal stippling, have been reported (67). Over 50% of affected *Bpa* females have asymmetric cataracts, frequently associated with microphthalmia. On postnatal day 5–7, they develop a hyperkeratotic skin eruption, which resolves and leaves bare patches arranged in a horizontal, striped pattern, following the lines of X inactivation. Affected *Bpa* male embryos have not been recovered and die shortly after implantation [(143); B. Cattanach, personal communication]. Several *Bpa* alleles have now been identified, both X-irradiation induced and spontaneous (118). The milder alleles were originally thought to be a different locus called striated (*Str*) (145). *Str* females are normal in size, and affected *Str* male embryos die in midgestation (118). In 1999, mutations in a gene called *Nsdhl* (NADH steroid dehydrogenase-like) were identified in several *Bpa* and *Str* alleles (118). Nsdhl functions as a 3β-hydroxysteroid dehydrogenase in the oxidative decarboxylation of C-4 methyl groups in the conversion of C28, C29, and C30 precursor sterols to 8(9)-cholestenol and related 8(9)-unsaturated sterols (Figure 6). Tissues and skin fibroblasts from *Bpa* females accumulate 4-methyl, 4,4'-dimethyl, and 4-carboxy sterols, consistent with an enzymatic block at this step of the cholesterol biosynthetic pathway [(118); G.E. Herman, R.I. Kelley, unpublished data]. The removal of C-4 methyl groups immediately precedes the sterol-Δ^8,Δ^7-isomerase reaction, explaining the similarity in the phenotype between the *Bpa/Str* and *Td* mutations and the original suggestion that bare patches was orthologous to CDPX2.

NSDHL GENE STRUCTURE AND ENZYMOLOGY   The NSDHL gene in the human and mouse spans approximately 40 kb and contains eight exons, with the translation start site in exon 2 (Table 1). The gene is transcribed from a dual promoter in head-to-head orientation with a gene, caltractin, involved in centrosome function, whose role appears unrelated to that of NSDHL (121). The NSDHL protein is ubiquitously expressed but with higher levels of expression in tissues with higher rates of sterol and steroid biosynthesis. The predicted protein sequence contains an N-terminal NADH cofactor binding site and a single membrane spanning region near the 3' end of the protein (118). NSDHL also contains several conserved Tyr-X-X-X-Lys motifs, at least one of which appears to be involved in the catalytic site of other 3β-hydroxysteroid dehydrogenases (3β-HSDs) (142). NSDHL represents a new subfamily of 3β-HSDs, as it possesses more amino acid sequence identity

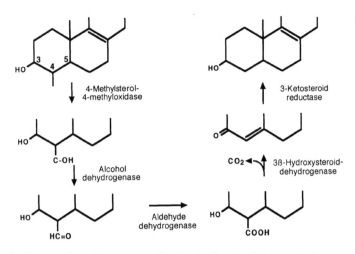

Figure 6   Enzymatic and nonenzymatic steps in the sterol-4-demethylase complex of cholesterol biosynthesis. The first three steps in the oxidative demethylation as studied in yeast appear to be carried out by a single 4-methylsterol oxidase enzyme protein. NSDHL, which is deficient in some cases of CHILD syndrome and the Bare patches mouse, is the 3β-hydroxysteroid dehydrogenase activity indicated at the fourth step.

with similar proteins from lower eukaryotes than with the other known mammalian 3β-HSDs (118).

In *S. cerevisiae*, *erg26* encodes the sterol biosynthetic 3β-HSD. In this organism, the removal of the two C-4 methyl groups from lanosterol (or related 4,4′-dimethyl intermediates) requires the sequential action of a C-4 methyloxidase (encoded by *erg25*), a 3β-HSD (*erg26*), and a keto-reductase (*erg27*) (113). A gene, *erg28*, encoding a regulatory or "scaffolding" protein that tethers the demethylase complex to the ER membrane has also recently been identified (52). Conservation of the majority of the steps in the yeast and mammalian sterol biosynthetic pathways predicts a similar C-4 demethylase complex in mammalian cells. Although a human C-4 methyloxidase gene has been identified (U60205), a mammalian 3-keto reductase gene has not as yet been isolated. A human ortholog for *erg28* has been isolated, although its function is not known (130).

CHILD Syndrome (MIM 308050)

CLINICAL FEATURES AND GENETICS   CHILD syndrome is a disorder with phenotypic similarities to CDPX2 but with a striking unilateral distribution of abnormalities (65, 66, 69). Unilateral ichthyosiform skin lesions (sometimes referred to as an ichthyosiform nevus or inflammatory epidermal nevus) usually are present at birth and, in contrast to the skin lesions of CDPX2, often persist throughout life. Although the lesions may occasionally follow the lines of Blaschko, they usually involve large regions on one side of the body, with a sharp line of demarcation at the midline. Small patches of involved skin may occur on the opposite side, although both sides of the face are spared. Alopecia may occur on the affected side and nail involvement also is common. Ipsilateral limb reduction defects are encountered with epiphyseal stippling, on X rays, during infancy. Internal malformations, including CNS, renal, and cardiac, have been reported, typically occurring on the affected side. Left-sided CHILD syndrome usually is more severe than right-sided, but it is also only half as common. The characteristic skin lesion of CHILD syndrome is a large epidermal plaque or nevus, often with a xanthomatous or even warty appearance. Histologically, there is a thick parakeratotic stratum corneum overlying a psoriasiform, acanthotic epidermis, often with inflammatory infiltrates and lipid-laden histiocytes.

Although CDPX2 and CHILD syndrome have many phenotypic similarities, there are some notable differences. For example, cataracts are not reported in CHILD syndrome, the skin lesions more often persist beyond infancy, and the skeletal anomalies—such as limb reductions or amelia—are more severe. Follicular plugging, which is characteristic of CDPX2, does not occur in classic CHILD syndrome (66, 69, 71). EM findings are similar to those reported in CDPX2 except that crystals within the lipid vacuoles, consistent with cholesterol, may be present (44, 70, 71). Despite the histologic and ultrastructural differences in the skin between CDPX2 and CHILD syndrome, the dermatological features distinguishing CDPX2 and CHILD features are not absolute, as indicated by a rare female with skin lesions typical of CHILD syndrome but with bilateral, symmetrical involvement (48) and by CDPX2 females with persistent diffuse erythroderma (92) or unilateral ichthyosis resembling the skin lesions of CHILD syndrome with only patchy involvement of the other side (30). The presence of cataracts and characteristic skin histology led the authors to classify the latter cases as CDPX2.

Like CDPX2, CHILD syndrome has been presumed to be an X-linked dominant disorder with male lethality since its delineation as a syndrome in 1980 (65). Two 46,XY males with otherwise typical right-sided CHILD syndrome (64, 195) and rare familial cases with mother to daughter transmission have been reported (64, 106). Early somatic mutations may explain the occurrence in males; however, such mutations cannot explain the unusual distribution of skin lesions in CHILD syndrome.

BIOCHEMISTRY AND GENE MUTATIONS   Because of the similarities between CDPX2 and CHILD syndrome, Grange et al. (56) evaluated a four-year-old girl with CHILD syndrome and found her to have the characteristic biochemical sterol abnormalities of CDPX2 and a nonsense mutation in *EBP*. Subsequently, a second patient with CHILD syndrome and a mutation in *EBP* was identified (D.K. Grange, A. Metzenberg, G.E. Herman, & R.I. Kelley, manuscript in preparation). In contrast, following the discovery of *Nsdhl* mutations in *Bpa* mice, Konig et al. reported mutations in the human NSDHL gene in five patients with CHILD syndrome (106), including one male with a normal karyotype who was heterozygous for wild-type and mutant alleles on the affected side. On the unaffected side, only the wild-type allele could be detected, consistent with the unilateral somatic mosaicism associated with a postzygotic mutation. In one family, a mildly affected mother had the same NSDHL mutation as her more severely affected daughter (106), excluding a postzygotic somatic mutation as the cause of CHILD syndrome, at least in that family. We have recently identified three additional CHILD patients with mutations in *NSDHL* (D.K. Grange, A. Metzenberg, G.E. Herman, & R.I. Kelley, manuscript in preparation). To date, all individuals with CHILD syndrome and mutations in the two associated X-linked genes, *EBP* and *NSDHL*, have had right-sided hemidysplasia and skin disease.

The occurrence, as in CHILD syndrome, of mutations in two or more genes in patients with similar or identical phenotypes is not uncommon. More surprising, perhaps, is the striking and unexplained unilateral distribution of lesions in CHILD syndrome. The patterning of skin lesions in CDPX2 and in *Bpa*, *Str*, and *Td* mice follows Blaschko's lines and is consistent with functional mosaicism for X-linked genes (62). However, to explain the midline demarcation and extensive uninterrupted skin lesions in CHILD syndrome, Happle proposed that a clone of midline, early embryonic organizer cells expressing the mutant NSDHL gene could affect the process of X inactivation and, thereby, alter the patterning of a large developmental field on one side of the body (56, 106). However, Happle suggested no specific molecular or biochemical mechanism. The process of X inactivation begins in the extra-embryonic cell lineages, at least in the mouse, at the blastocyst stage (3.5–4.5 dpc) [reviewed in (55)] before laterality is established. However, it is possible that genes, such as Sonic hedgehog, that play a role in laterality determination and that also are influenced by abnormalities in sterol metabolism may provide the link between the two events. Sonic hedgehog is expressed at Hensen's node, the site where asymmetry is first detected in the mammalian embryo (20). Although the expression of Shh in mammals is bilaterally symmetric, Shh-deficient mouse embryos demonstrate laterality defects (20, 25, 128, 183). Moreover, proper Shh signaling is required to prevent left-determining factors from being expressed on the right side of the embryo (128). In CHILD syndrome, Happle's organizer cells expressing the mutant NSDHL or EBP gene could exert their effect on Shh or one of its downstream effectors, such as Patched. Presumably, whether bilateral (CDPX2) or unilateral (CHILD) disease ensues from *EBP* mutations would depend on the pattern of affected vs. unaffected putative organizer cells in the laterality-determining perinodal tissues. Such a model does not explain why unilateral disease is never seen with *Td*, *Str*, or *Bpa* mutations or why unilateral involvement is so common in human *NSDHL* mutations but much less common when *EBP* is mutated. It is possible that species' differences in the timing of X inactivation or in cholesterol metabolism or transport in the developing fetus could explain the phenotypic differences. Unique species-specific features, such as the laterality defects seen in CHILD syndrome but not in the mutant mice, may require studies in in vitro human systems to understand more fully the mechanisms of disease pathogenesis.

HYDROPS–ECTOPIC CALCIFICATION–MOTH-EATEN SKELETAL DYSPLASIA (MIM 215140)

By screening tissues and cells from a skeletal dysplasia repository for evidence of abnormal sterol metabolism, a previously unreported sterol abnormality was found in several fetuses with a lethal short-limbed dwarfism known as Hydrops–ectopic calcification–moth-eaten skeletal dysplasia (HEM) (100). HEM, or Greenberg dysplasia, is a rare skeletal dysplasia characterized by fetal hydrops, short-limbed dwarfism, and severely disorganized chondro-osseous proliferation and mineralization (26, 57). Radiographic abnormalities include moth-eaten appearing, severely shortened long bones, ectopic epiphyseal calcification, laryngeal and tracheal calcification, and platyspondyly. Histologically, there is severe disorganization of cartilaginous tissues, obliteration of the marrow spaces by mesenchyme-like tissue, and extramedullary hematopoiesis. Unlike in SLOS, CDPX2, and NSDHL deficiency, HEM fetuses lack internal malformations, although two unrelated fetuses had postaxial polydactyly of the hands. The genetic characteristics of six known cases include diverse ethnic backgrounds, involvement of both sexes, affected siblings, and parental consanguinity in four of five families, thus making autosomal recessive inheritance almost certain. Possibly because of early in utero lethality—the longest surviving fetus died in utero at 30 weeks—HEM is one of the rarest skeletal dysplasias known. Although not a true chondrodysplasia punctata, HEM has enchondral bone with strikingly disordered calcification that resembles the dense punctate calcifications of severe CDPX2 and that may reflect, in part, a similar process of apoptosis or other premature death of prechondrocytes.

GC/MS of sterols extracted from the cartilage of four HEM fetuses showed increased levels of cholesta-8,14-dien-3β-ol and cholesta-8,14,24-trien-3β-ol, suggesting a deficiency of sterol-Δ^{14}-reductase (100). When grown in cholesterol-depleted culture medium, cultured skin fibroblasts or chondrocytes from two of the patients accumulated very large amounts (>20% of cholesterol) of the same two sterols (R.I. Kelley, unpublished observations). As shown in Figure 2, sterol-Δ^{14}-reductase just precedes the sterol-4-demethylase complex and sterol-Δ^8,Δ^7-isomerase in the normal pathway for cholesterol biosynthesis. The recent discoveries that the nuclear lamin B receptor has intrinsic sterol-Δ^{14}-reductase activity (167) and that at least one 14-dehydrosterol possesses signaling activity as a postfertilization meiosis activator (58) suggest that 14-dehydrosterols may have important effects on DNA replication and nuclear signaling and may explain the especially severe phenotype in HEM dysplasia. Although the yeast sterol-Δ^{14}-reductase was cloned in 1994 (109) and the Δ^{14}-reductase activity of the nuclear lamin B receptor was demonstrated in 1998 (167), the identity of the gene encoding the primary sterol-Δ^{14}-reductase functioning in mammalian cholesterol biosynthesis has not been established (Table 1). Sequencing of all exons of the gene for TM7SF2, an ER protein with widespread tissue distribution and substantial homology with other sterol-Δ^{14}-reductases (76), failed to disclose a mutation in HEM (G.E. Herman, unpublished data). Nevertheless, a block at the level of the sterol-Δ^{14}-reductase involved in sterol biosynthesis remains the most likely cause of HEM.

CONCLUSION

The discovery of new inborn errors of cholesterol metabolism in the past decade has provided geneticists with many new insights in several areas of biology and medicine: normal and abnormal embryogenesis, cholesterol synthesis and nutrition, biochemical genetics, the epidemiology of mutations, and even behavioral genetics. The central importance of cholesterol homeostasis in human biochemistry and development, which has been underscored by the consequences of even subtle genetic deficiencies of cholesterol biosynthesis, raises important questions about the widespread use of drugs for the reduction of blood cholesterol levels and the near elimination of cholesterol from recommended, nutritionally balanced diets.

The study of disorders of cholesterol biosynthesis has brought into focus, among many embryological phenomena, the role of cholesterol nutrition in the developing embryo. The evidence that the maternal supply of cholesterol to the developing embryo can influence the incidence and severity of certain SLOS malformations may be important with regard to the development the same malformations, such as cleft palate and holoprosencephaly, in other syndromes or even in otherwise normal individuals. There may also be important implications of the apparent interaction between cholesterol metabolism and hedgehog proteins in the abnormal morphogenesis of SLOS, if not all of the primary defects of cholesterol biosynthesis. Thus, the creation of mouse models lacking a functional 7DHC reductase gene or other, related genes of sterol biosynthesis will provide geneticists and embryologists with years of work that should bring even closer clinical dysmorphology and biochemical genetics.

Especially important among the emerging connections between sterol metabolism and cellular signaling is the evidence that sterols as well as steroids have nuclear signaling functions and that some of the most potent signaling molecules are normal, trace level intermediates in the cholesterol biosynthetic pathway. Moreover, the intriguing discovery that the lamin B receptor has sterol Δ^{14}-reductase activity—and could even be the primary Δ^{14}-reductase of cholesterol biosynthesis—suggests there may be important effects of disordered cholesterol biosynthesis and metabolism on nuclear signaling during both prenatal and postnatal life. The sequence homology of certain sterol metabolizing enzymes, like NSDHL, with other enzymes apparently dedicated to the synthesis of steroid hormones (118) raises the possibility that phylogenetically ancient sterol hormones evolved to become steroid hormones and took on new roles in interorgan signaling as the size and organizational complexity of organisms increased. Thus, there may be a host of heretofore unsuspected sterol signaling functions at work in the nuclei or cytoplasm of both primitive and advanced eukaryotes, the elucidation of which may shed important light on less well-understood clinical consequences of inborn errors of cholesterol biosynthesis.

ACKNOWLEDGMENTS

GEH was supported in part by a grant from NIH (R01 HD38572) and funds from the Children's Research Institute.

Visit the Annual Reviews home page at www.AnnualReviews.org

LITERATURE CITED

1. Aboushadi N, Engfelt WH, Paton VG, Krisans SK. 1999. Role of peroxisomes in isoprenoid biosynthesis. *J. Histochem. Cytochem.* 47:1127–32

2. Anderson AJ, Stephan MJ, Walker WO, Kelley RI. 1998. Variant RSH/Smith-Lemli-Opitz syndrome with atypical sterol metabolism. *Am. J. Med. Genet.* 78: 413–18

3. Andersson HC, Frentz J, Martinez JE, Tuck-Muller CM, Bellizaire J. 1999. Adrenal insufficiency in Smith-Lemli-Opitz syndrome. *Am. J. Med. Genet.* 82: 382–84

4. Andersson HC, Kratz LE, Kelley RI. 2000. Desmosterolosis presenting with multiple congenital anomalies and profound developmental delay. *J. Inherit. Metab. Dis.* 23:200

5. Angel TA, Faust CJ, Gonzales JC, Kenwrick S, Lewis RA, Herman GE. 1993. Genetic mapping of the X-linked dominant mutations striated (*Str*) and bare patches (*Bpa*) to a 600-kb region of the mouse X chromosome: implications for mapping human disorders in Xq28. *Mamm. Genome* 4:171–76

6. Appelkvist EL, Reinhart M, Fischer R, Billheimer J, Dallner G. 1990. Presence of individual enzymes of cholesterol biosynthesis in rat liver peroxisomes. *Arch. Biochem. Biophys.* 282:318–25

7. Bae SH, Lee JN, Fitzky BU, Seong J, Paik YK. 1999. Cholesterol biosynthesis from lanosterol. Molecular cloning, tissue distribution, expression, chromosomal localization, and regulation of rat 7-dehydrocholesterol reductase, a Smith-Lemli-Opitz syndrome–related protein. *J. Biol. Chem.* 274:14624–31

8. Bae SH, Seong J, Paik YK. 2001. Cholesterol biosynthesis from lanosterol: molecular cloning, chromosomal localization, functional expression and liver-specific gene regulation of rat sterol delta8-isomerase, a cholesterogenic enzyme with multiple functions. *Biochem. J.* 353:689–99

9. Ballabio A, Andria G. 1992. Deletions and translocations involving the distal short arm of the human X chromosome: review and hypotheses. *Hum. Mol. Genet.* 1:221–27

10. Bellknap WM, Dietschy JM. 1988. Sterol synthesis and low-density lipoprotein clearance in vivo in the pregnant rat, placenta, and fetus. *J. Clin. Invest.* 82:2077–85

11. Berger R, Smit GP, Schierbeek H, Bijsterveld K, le Coultre R. 1985. Mevalonic aciduria: an inborn error of cholesterol biosynthesis? *Clin. Chim. Acta* 152:219–22

12. Bialer MG, Penchaszadeh VB, Kahn E, Libes R, Krigsman G, Lesser ML. 1987.

Female external genitalia and mullerian duct derivatives in a 46,XY infant with the Smith-Lemli-Opitz syndrome. *Am. J. Med. Genet.* 28:723–31

13. Biardi L, Krisans SK. 1996. Compartmentalization of cholesterol biosynthesis. Conversion of mevalonate to farnesyl diphosphate occurs in the peroxisomes. *J. Biol. Chem.* 271:1784–88

14. Biardi L, Sreedhar A, Zokaei A, Vartak NB, Bozeat RL, et al. 1994. Mevalonate kinase is predominantly localized in peroxisomes and is defective in patients with peroxisome deficiency disorders. *J. Biol. Chem.* 269:1197–205

15. Bitgood MJ, Shen L, McMahon AP. 1996. Sertoli cell signaling by Desert hedgehog regulates the male germline. *Curr. Biol.* 6:298–304

16. Bourre JM, Clement M, Gerard D, Chaudiere J. 1989. Alterations of cholesterol synthesis precursors (7-dehydrocholesterol, 7-dehydrodesmosterol, desmosterol) in dysmyelinating neurological mutant mouse (quaking, shiverer, and trembler) in the PNS and the CNS. *Biochim. Biophys. Acta* 1004:387–90

17. Braverman N, Lin P, Moebius FF, Obie C, Moser A, et al. 1999. Mutations in the gene encoding 3beta-hydroxysteroid-delta8-delta7 isomerase cause X-linked dominant Conradi-Hunermann syndrome. *Nat. Genet.* 22:291–94

18. Braverman N, Steel G, Obie C, Moser A, Moser H, et al. 1997. Human PEX7 encodes the peroxisomal PTS2 receptor and is responsible for rhizomelic chondrodysplasia punctata. *Nat. Genet.* 15:369–76

19. Brown MS, Goldstein JL. 1980. Multivalent feedback regulation of HMG-CoA reductase, a control mechanism coordinating isoprenoid synthesis and cell growth. *J. Lipid Res.* 21:505–17

20. Capdevila J, Vogan KJ, Tabin CJ, Izpisua Belmonte JC. 2000. Mechanisms of left-right determination in vertebrates. *Cell* 101:9–21

21. Carr BR, Simpson ER. 1981. Lipopro-tein utilization and cholesterol synthesis by the human fetal adrenal gland. *Endocr. Rev.* 2:306–26

22. Carr BR, Simpson ER. 1982. Cholesterol synthesis in human fetal tissues. *J. Clin. Endocrin. Metab.* 55:447–52

23. Chasalow FI, Blethen SL, Taysi K. 1985. Possible abnormalities of steroid secretion in children with Smith-Lemli-Opitz syndrome and their parents. *Steroids* 46:827–43

24. Cherstvoy ED, Lazjuk GI, Lurie IW, Nedzved MK, Usoev SS. 1975. The pathological anatomy of the Smith-Lemli-Opitz syndrome. *Clin. Genet.* 7:382–87

25. Chiang C, Litingtung Y, Lee E, Young KE, Corden JL, et al. 1996. Cyclopia and defective axial patterning in mice lacking *Sonic hedgehog* gene function. *Nature* 383:407–13

26. Chitayat D, Gruber H, Mullen BJ, Pauzner D, Costa T, et al. 1993. Hydrops-ectopic calcification-moth-eaten skeletal dysplasia (Greenberg dysplasia): prenatal diagnosis and further delineation of a rare genetic disorder. *Am. J. Med. Genet.* 47:272–77

27. Cho SY, Kim JH, Paik YK. 1998. Cholesterol biosynthesis from lanosterol: differential inhibition of sterol-delta8 isomerase and other lanosterol-converting enzymes by tamoxifen. *Mol. Cells* 8:233–39

28. Clark RM, Fey MB, Jensen RG, Hill DW. 1983. Desmosterol in human milk. *Lipids* 18:264–66

29. Cooper MK, Porter JA, Young KA, Beachy PA. 1998. Plant-derived and synthetic teratogens inhibit the ability of target tissues to respond to Sonic hedgehog signaling. *Science* 280:1603–7

30. Corbi MR, Conejo-Mir JS, Linares M, Jimenez G, Rodriguez Canas T, Navarrete M. 1998. Conradi-Hunermann syndrome with unilateral distribution. *Pediatr. Dermatol.* 15:299–303

31. Cruz MLA, Wong WW, Mimouni F, Hachey DL, Setchell KDR, et al. 1994.

Effects of infant nutrition on cholesterol synthesis rates. *Pediatr. Res.* 35:135–40

32. Cummings CJ, Zoghbi HY. 2000. Fourteen and counting: unraveling trinucleotide repeat diseases. *Hum. Mol. Genet.* 9:909–16

33. Cunniff C, Kratz LE, Moser A, Natowicz MR, Kelley RI. 1997. Clinical and biochemical spectrum of patients with RSH/Smith-Lemli-Opitz syndrome and abnormal cholesterol metabolism. *Am. J. Med. Genet.* 68:263–69

34. Curless RG. 1983. Dominant chondrodysplasia punctata with neurologic symptoms. *Neurology* 33:1095–97

35. Curry CJ, Carey JC, Holland JS, Chopra D, Fineman R, et al. 1987. Smith-Lemli-Opitz syndrome–type II: multiple congenital anomalies with male pseudohermaphroditism and frequent early lethality. *Am. J. Med. Genet.* 26:45–57

36. Dahl NK, Daunais MA, Liscum L. 1994. A second complementation class of cholesterol transport mutants with a variant Niemann-Pick type C phenotype. *J. Lipid Res.* 35:1839–49

37. Dallaire L, Fraser FC. 1969. The Smith-Lemli-Opitz syndrome of retardation, urogenital and skeletal anomalies in siblings. In *Birth Defects: Original Article Series*, ed. D Bergsma, pp. 180–200. New York: March of Dimes

38. Dehart DB, Lanoue L, Tint GS, Sulik KK. 1997. Pathogenesis of malformations in a rodent model for Smith-Lemli-Opitz syndrome. *Am. J. Med. Genet.* 68:328–37

39. Derry JM, Gormally E, Means GD, Zhao W, Meindl A, et al. 1999. Mutations in a delta8-delta7-sterol isomerase in the tattered mouse and X-linked dominant chondrodysplasia punctata. *Nat. Genet.* 22:286–90

40. Donnai D, Young ID, Owen WG, Clark SA, Miller PF, Knox WF. 1986. The lethal multiple congenital anomaly syndrome of polydactyly, sex reversal, renal hypoplasia, and unilobular lungs. *J. Med. Genet.* 23:64–71

41. Drenth JP, Cuisset L, Grateau G, Vasseur C, van de Velde-Visser SD, et al. 1999. Mutations in the gene encoding mevalonate kinase cause hyper-IgD and periodic fever syndrome. International Hyper-IgD Study Group. *Nat. Genet.* 22:178–81

42. Elias ER, Irons MB, Hurley AD, Tint GS, Salen G. 1997. Clinical effects of cholesterol supplementation in six patients with the Smith-Lemli-Opitz syndrome (SLOS). *Am. J. Med. Genet.* 68:305–10

43. Emami S, Hanley KP, Esterly NB, Daniallinia N, Williams ML. 1994. X-linked dominant ichthyosis with peroxisomal deficiency. An ultrastructural and ultracytochemical study of the Conradi-Hunermann syndrome and its murine homologue, the bare patches mouse. *Arch. Dermatol.* 130:325–36

44. Emami S, Rizzo WB, Hanley KP, Taylor JM, Goldyne ME, Williams ML. 1992. Peroxisomal abnormality in fibroblasts from involved skin of CHILD syndrome. Case study and review of peroxisomal disorders in relation to skin disease. *Arch. Dermatol.* 128:1213–22

45. Evans EP, Phillips RJ. 1975. Inversion heterozygosity and the origin of XO daughters of *Bpa/+* female mice. *Nature* 256:40–41

46. Farese RV Jr, Ruland SL, Flynn LM, Stokowski RP, Young SG. 1995. Knockout of the mouse apolipoprotein B gene results in embryonic lethality in homozygotes and protection against diet-induced hypercholesterolemia in heterozygotes. *Proc. Natl. Acad. Sci. USA* 92:1774–78

47. Fierro M, Martinez AJ, Harbison JW, Hay SH. 1977. Smith-Lemli-Opitz syndrome: neuropathological and ophthalmological observations. *Dev. Med. Child Neurol.* 19:57–62

48. Fink-Puches R, Soyer HP, Pierer G, Kerl H, Happle R. 1997. Systematized inflammatory epidermal nevus with symmetrical involvement: an unusual case of

CHILD syndrome? *J. Am. Acad. Dermatol.* 36:823–26

49. Fitzky BU, Witsch-Baumgartner M, Erdel M, Lee JN, Paik YK, et al. 1998. Mutations in the delta7-sterol reductase gene in patients with the Smith-Lemli-Opitz syndrome. *Proc. Natl. Acad. Sci. USA* 95:8181–86

50. Fitzpatrick DR, Keeling JW, Evans MJ, Kan AE, Bell JE, et al. 1998. Clinical phenotype of desmosterolosis. *Am. J. Med. Genet.* 75:145–52

51. Franco B, Meroni G, Parenti G, Levilliers J, Bernard L, et al. 1995. A cluster of sulfatase genes on Xp22.3: mutations in chondrodysplasia punctata (CDPX) and implications for warfarin embryopathy. *Cell* 81:15–25

52. Gachotte D, Eckstein J, Barbuch R, Hughes T, Roberts C, Bard M. 2001. A novel gene conserved from yeast to humans is involved in sterol biosynthesis. *J. Lipid Res.* 42:150–54

53. Gibson KM, Hoffmann GF, Sweetman L, Buckingham B. 1997. Mevalonate kinase deficiency in a dizygotic twin with mild mevalonic aciduria. *J. Inherit. Metab. Dis.* 20:391–94

54. Goodman P, Dominguez R. 1990. Cervicothoracic myelopathy in Conradi-Hunermann disease: MRI diagnosis. *Magn. Reson. Imaging* 8:647–50

55. Goto T, Monk M. 1998. Regulation of X-chromosome inactivation in development in mice and humans. *Microbiol. Mol. Biol. Rev.* 62:362–78

56. Grange DK, Kratz LE, Braverman NE, Kelley RI. 2000. CHILD syndrome caused by deficiency of 3beta-hydroxysteroid-delta8, delta7-isomerase. *Am. J. Med. Genet.* 90:328–35

57. Greenberg CR, Rimoin DL, Gruber HE, DeSa DJ, Reed M, Lachman RS. 1988. A new autosomal recessive lethal chondrodystrophy with congenital hydrops. *Am. J. Med. Genet.* 29:623–32

58. Grondahl C, Ottesen JL, Lessl M, Faarup P, Murray A, et al. 1998. Meiosis-activating sterol promotes resumption of meiosis in mouse oocytes cultured in vitro in contrast to related oxysterols. *Biol. Reprod.* 58:1297–302

59. Hanner M, Moebius FF, Weber F, Grabner M, Striessnig J, Glossmann H. 1995. Phenylalkylamine Ca^{2+} antagonist binding protein. Molecular cloning, tissue distribution, and heterologous expression. *J. Biol. Chem.* 270:7551–57

60. Happle R. 1979. X-linked dominant chondrodysplasia punctata. Review of literature and report of a case. *Hum. Genet.* 53:65–73

61. Happle R. 1981. Cataracts as a marker of genetic heterogeneity in chondrodysplasia punctata. *Clin. Genet.* 19:64–66

62. Happle R. 1985. Lyonization and the lines of Blaschko. *Hum. Genet.* 70:200–6

63. Happle R. 1995. X-linked dominant chondrodysplasia punctata/ichthyosis/cataract syndrome in males. *Am. J. Med. Genet.* 57:493

64. Happle R, Effendy I, Megahed M, Orlow SJ, Kuster W. 1996. CHILD syndrome in a boy. *Am. J. Med. Genet.* 62:192–94

65. Happle R, Koch H, Lenz W. 1980. The CHILD syndrome. Congenital hemidysplasia with ichthyosiform erythroderma and limb defects. *Eur. J. Pediatr.* 134:27–33

66. Happle R, Mittag H, Kuster W. 1995. The CHILD nevus: a distinct skin disorder. *Dermatology* 191:210–16

67. Happle R, Phillips RJ, Roessner A, Junemann G. 1983. Homologous genes for X-linked chondrodysplasia punctata in man and mouse. *Hum. Genet.* 63:24–27

68. Has C, Bruckner-Tuderman L, Muller D, Floeth M, Folkers E, et al. 2000. The Conradi-Hunermann-Happle syndrome (CDPX2) and emopamil binding protein: novel mutations and somatic and gonadal mosaicism. *Hum. Mol. Genet.* 9:1951–55

69. Hashimoto K, Prada S, Lopez AP, Hoyos JG, Escobar M. 1998. CHILD syndrome with linear eruptions, hypopigmented

bands, and verruciform xanthoma. *Pediatr. Dermatol.* 15:360–66

70. Hashimoto K, Topper S, Sharata H, Edwards M. 1995. CHILD syndrome: analysis of abnormal keratinization and ultrastructure. *Pediatr. Dermatol.* 12:116–29

71. Hebert AA, Esterly NB, Holbrook KA, Hall JC. 1987. The CHILD syndrome. Histologic and ultrastructural studies. *Arch. Dermatol.* 123:503–9

72. Herman TE, Siegel MJ, Lee BC, Dowton SB. 1993. Smith-Lemli-Opitz syndrome type II: report of a case with additional radiographic findings. *Pediatr. Radiol.* 23:37–40

73. Hodge VJ, Gould SJ, Subramani S, Moser HW, Krisans SK. 1991. Normal cholesterol synthesis in human cells requires functional peroxisomes. *Biochem. Biophys. Res. Commun.* 181:537–41

74. Hoffmann G, Gibson KM, Brandt IK, Bader PI, Wappner RS, Sweetman L. 1986. Mevalonic aciduria—an inborn error of cholesterol and nonsterol isoprene biosynthesis. *N. Eng. J. Med.* 314:1610–14

75. Hoffmann GF, Charpentier C, Mayatepek E, Mancini J, Leichsenring M, et al. 1993. Clinical and biochemical phenotype in 11 patients with mevalonic aciduria. *Pediatrics* 91:915–21

76. Holmer L, Pezhman A, Worman HJ. 1998. The human lamin B receptor/sterol reductase multigene family. *Genomics* 54:469–76

77. Honda A, Shefer S, Salen G, Xu G, Batta AK, et al. 1996. Regulation of the last two enzymatic reactions in cholesterol biosynthesis in rats: effects of BM 15.766, cholesterol, cholic acid, lovastatin, and their combinations. *Hepatology* 24:435–39

78. Honda M, Tint GS, Honda A, Nguyen LB, Chen TS, Shefer S. 1998. 7-dehydrocholesterol downregulates cholesterol biosynthesis in cultured Smith-Lemli-Opitz syndrome skin fibroblasts. *J. Lipid Res.* 39:647–57

79. Houten SM, Frenkel J, Kuis W, Wanders RJ, Poll-The BT, Waterham HR. 2000. Molecular basis of classical mevalonic aciduria and the hyperimmunoglobulinaemia D and periodic fever syndrome: high frequency of three mutations in the mevalonate kinase gene. *J. Inherit. Metab. Dis.* 23:367–70

80. Houten SM, Kuis W, Duran M, de Koning TJ, van Royen-Kerkhof A, et al. 1999. Mutations in MVK, encoding mevalonate kinase, cause hyperimmunoglobulinaemia D and periodic fever syndrome. *Nat. Genet.* 22:175–77

81. Houten SM, Wanders RJ, Waterham HR. 2000. Biochemical and genetic aspects of mevalonate kinase and its deficiency. *Biochim. Biophys. Acta* 1529:19–32

82. Ikegawa S, Ohashi H, Ogata T, Honda A, Tsukahara M, et al. 2000. Novel and recurrent EBP mutations in X-linked dominant chondrodysplasia punctata. *Am. J. Med. Genet.* 94:300–5

83. Irons M, Elias ER, Abuelo D, Bull MJ, Greene CL, et al. 1997. Treatment of Smith-Lemli-Opitz syndrome: results of a multicenter trial. *Am. J. Med. Genet.* 68:311–14

84. Irons M, Elias ER, Salen G, Tint GS, Batta AK. 1993. Defective cholesterol biosynthesis in Smith-Lemli-Opitz syndrome. *Lancet* 341:1414

85. Iwasaki M, Le AX, Helms JA. 1997. Expression of Indian hedgehog, bone morphogenetic protein 6, and gli during skeletal morphogenesis. *Mech. Dev.* 69:197–202

86. Jackson SM, Ericsson J, Edwards PA. 1997. Signaling molecules derived from the cholesterol biosynthetic pathway. *Subcell. Biochem.* 28:1–21

87. Jira PE, Wevers RA, de Jong J, Rubio-Gozalbo E, Janssen-Zijlstra FS, et al. 2000. Simvastatin. A new therapeutic approach for Smith-Lemli-Opitz syndrome. *J. Lipid Res.* 41:1339–46

88. Johnson VP. 1975. Smith-Lemli-Opitz

syndrome: review and report of two affected siblings. *Z. Kinderheilkd.* 119:221–34

89. Jones JH. 1997. Regulation of cholesterol biosynthesis by diet in humans. *Am. J. Clin. Nutr.* 66:438–46

90. Jones KL. 1997. *Smith's Recognizable Patterns of Human Malformation*, pp. 112–13. Philadelphia: WB Saunders

91. Joseph DB, Uehling DT, Gilbert E, Laxova R. 1987. Genitourinary abnormalities associated with the Smith-Lemli-Opitz syndrome. *J. Urol.* 137:719–21

92. Kalter DC, Atherton DJ, Clayton PT. 1989. X-linked dominant Conradi-Hunermann syndrome presenting as congenital erythroderma. *J. Am. Acad. Dermatol.* 21:248–56

93. Kandutsch AA, Russell AE. 1960. Preputial gland tumor sterols. III. A metabolic pathway from lanosterol to cholesterol. *J. Biol. Chem.* 235:2256–61

94. Kearsey SE, Edwards J. 1987. Mutations that increase the mitotic stability of minichromosomes in yeast: characterization of RAR1. *Mol. Gen. Genet.* 210:509–17

95. Kekuda R, Prasad PD, Fei YJ, Leibach FH, Ganapathy V. 1996. Cloning and functional expression of the human type 1 sigma receptor (hSigmaR1). *Biochem. Biophys. Res. Commun.* 229:553–58

96. Keller GA, Scallen TJ, Clarke D, Maher PA, Krisans SK, Singer SJ. 1989. Subcellular localization of sterol carrier protein-2 in rat hepatocytes: its primary localization to peroxisomes. *J. Cell Biol.* 108:1353–61

97. Kelley RI. 1983. Review: the cerebrohepatorenal syndrome of Zellweger, morphologic and metabolic aspects. *Am. J. Med. Genet.* 16:503–17

98. Kelley RI. 2000. Inborn errors of cholesterol biosynthesis. *Adv. Pediatr.* 47:1–53

99. Kelley RI, Hennekam RC. 2000. The Smith-Lemli-Opitz syndrome. *J. Med. Genet.* 37:321–35

100. Kelley RI, Kratz LE, Wilcox WG. 2000.

Abnormal metabolism of 14-dehydrosterols in hydrops-ectopic calcification–moth-eaten skeletal dysplasia: evidence for new defect of cholesterol biosynthesis. *Proc. Greenwood Gen. Cen.* 20:116

101. Kelley RI, Roessler E, Hennekam RC, Feldman GL, Kosaki K, et al. 1996. Holoprosencephaly in RSH/Smith-Lemli-Opitz syndrome: does abnormal cholesterol metabolism affect the function of Sonic hedgehog? *Am. J. Med. Genet.* 66:478–84

102. Kelley RI, Wilcox WG, Smith M, Kratz LE, Moser A, Rimoin DS. 1999. Abnormal sterol metabolism in patients with Conradi-Hünermann-Happle syndrome and sporadic lethal chondrodysplasia punctata. *Am. J. Med. Genet.* 83:213–19

103. Kim EH, Boutwell WC. 1985. Smith-Lemli-Opitz syndrome associated with Hirschsprung disease, 46,XY female karyotype, and total anomalous pulmonary venous drainage. *J. Pediatr.* 106:861

104. Kohler HG. 1983. Brief clinical report: familial neonatally lethal syndrome of hypoplastic left heart, absent pulmonary lobation, polydactyly, and talipes, probably Smith-Lemli-Opitz (RSH) syndrome. *Am. J. Med. Genet.* 14:423–28

105. Kolde G, Happle R. 1984. Histologic and ultrastructural features of the ichthyotic skin in X-linked dominant chondrodysplasia punctata. *Acta Derm. Venereol.* 64:389–94

106. Konig A, Happle R, Bornholdt D, Engel H, Grzeschik KH. 2000. Mutations in the NSDHL gene, encoding a 3beta-hydroxysteroid dehydrogenase, cause CHILD syndrome. *Am. J. Med. Genet.* 90:339–46

107. Kratz LE, Kelley RI. 1999. Prenatal diagnosis of the RSH/Smith-Lemli-Opitz syndrome. *Am. J. Med. Genet.* 82:376–81

108. Krisans SK. 1992. The role of peroxisomes in cholesterol metabolism. *Am. J. Respir. Cell Mol. Biol.* 7:358–64

109. Lai MH, Bard M, Pierson CA, Alexander JF, Goebl M, et al. 1994. The identification of a gene family in the *Saccharomyces cerevisiae* ergosterol biosynthesis pathway. *Gene* 140:41–49

110. Lange Y, Echevarria F, Steck TL. 1991. Movement of zymosterol, a precursor of cholesterol, among three membranes in human fibroblasts. *J. Biol. Chem.* 266:21439–43

111. Langius FAA, Waterham HR, Koster J, Dorland L, Duran M, et al. 2000. Smith-Lemli-Opitz syndrome: the mild end of the clinical and biochemical spectrum. *J. Inherit. Metab. Dis.* 23:196

112. Lanoue L, Dehart DB, Hinsdale ME, Maeda N, Tint GS, Sulik KK. 1997. Limb, genital, CNS, and facial malformations result from gene/environment-induced cholesterol deficiency: further evidence for a link to Sonic hedgehog. *Am. J. Med. Genet.* 73:24–31

113. Lees ND, Bard M, Kirsch DR. 1999. Biochemistry and molecular biology of sterol synthesis in *Saccharomyces cerevisiae*. *Crit. Rev. Biochem. Mol. Biol.* 34:33–47

114. Lichtenthaler HK, Schwender J, Disch A, Rohmer M. 1997. Biosynthesis of isoprenoids in higher plant chloroplasts proceeds via a mevalonate-independent pathway. *FEBS Lett.* 400:271–74

115. Lin AE, Ardinger HH, Ardinger RH Jr, Cunniff C, Kelley RI. 1997. Cardiovascular malformations in Smith-Lemli-Opitz syndrome. *Am. J. Med. Genet.* 68:270–78

116. Lin DS, Connor WE, Wolf DP, Neuringer M, Hachey DL. 1993. Unique lipids of primate spermatozoa: desmosterol and docosahexaenoic acid. *J. Lipid Res.* 34:491–99

117. Liscum L, Klansek JJ. 1998. Niemann-Pick disease type C. *Curr. Opin. Lipidol.* 9:131–35

118. Liu XY, Dangel AW, Kelley RI, Zhao W, Denny P, et al. 1999. The gene mutated in bare patches and striated mice encodes a novel 3beta-hydroxysteroid dehydrogenase. *Nat. Genet.* 22:182–87

119. Long SS. 1999. Syndrome of Periodic fever, Aphthous stomatitis, Pharyngitis, and Adenitis (PFAPA)—what it isn't. What is it? *J. Pediatr.* 135:1–5

120. Lowry RB, Yong SL. 1980. Borderline normal intelligence in the Smith-Lemli-Opitz (RSH) syndrome. *Am. J. Med. Genet.* 5:137–43

121. Mallon AM, Platzer M, Bate R, Gloeckner G, Botcherby MR, et al. 2000. Comparative genome sequence analysis of the *Bpa/Str* region in mouse and man. *Genome Res.* 10:758–75

122. Mancini J, Philip N, Chabrol B, Divry P, Rolland MO, Pinsard N. 1993. Mevalonic aciduria in three siblings: a new recognizable metabolic encephalopathy. *Pediatr. Neurol.* 9:243–46

123. Marigo V, Davey RA, Zuo Y, Cunningham JM, Tabin CJ. 1996. Biochemical evidence that patched is the hedgehog receptor. *Nature* 384:176–79

124. Marion RW, Alvarez LA, Marans ZS, Lantos G, Chitayat D. 1987. Computed tomography of the brain in the Smith-Lemli-Opitz syndrome. *J. Child Neurol.* 2:198–200

125. Mayatepek E, Hoffmann GF, Bremer HJ. 1993. Enhanced urinary excretion of leukotriene E4 in patients with mevalonate kinase deficiency. *J. Pediatr.* 123: 96–98

126. McKeever PA, Young ID. 1990. Smith-Lemli-Opitz syndrome II: a disorder of the fetal adrenals? *J. Med. Genet.* 27:465–66

127. Meinecke P, Blunck W, Rodewald A. 1987. Smith-Lemli-Opitz syndrome. *Am. J. Med. Genet.* 28:735–39

128. Meyers EN, Martin GR. 1999. Differences in left-right axis pathways in mouse and chick: functions of FGF8 and Shh. *Science* 285:403–6

129. Moebius FF, Burrows GG, Striessnig J, Glossmann H. 1993. Biochemical characterization of a 22-kDa high affinity antiischemic drug-binding polypeptide in the endoplasmic reticulum of guinea pig liver: potential common target for

antiischemic drug action. *Mol. Pharmacol.* 43:139–48

130. Moebius FF, Fitzky BU, Lee JN, Paik YK, Glossmann H. 1998. Molecular cloning and expression of the human Δ7-sterol reductase. *Proc. Natl. Acad. Sci. USA* 95:1899–902

131. Moebius FF, Soellner KE, Fiechtner B, Huck CW, Bonn G, Glossmann H. 1999. Histidine77, glutamic acid81, glutamic acid123, threonine126, asparagine194, and tryptophan197 of the human emopamil binding protein are required for in vivo sterol delta8-delta7 isomerization. *Biochemistry* 38:1119–27

132. Moebius FF, Striessnig J, Glossmann H. 1997. The mysteries of sigma receptors: new family members reveal a role in cholesterol synthesis. *Trends Pharmacol. Sci.* 18:67–70

133. Morell P, Jurevics H. 1996. Origin of cholesterol in myelin. *Neurochem. Res.* 21:463–70

134. Motley AM, Hettema EH, Hogenhout EM, Brites P, ten Asbroek AL, et al. 1997. Rhizomelic chondrodysplasia punctata is a peroxisomal protein targeting disease caused by a non-functional PTS2 receptor. *Nat. Genet.* 15:377–80

135. Mueller RF, Crowle PM, Jones RA, Davison BC. 1985. X-linked dominant chondrodysplasia punctata: a case report and family studies. *Am. J. Med. Genet.* 20:137–44

136. Ness GC, Lopez D, Borrego O, Gilbert-Barness E. 1997. Increased expression of low-density lipoprotein receptors in a Smith-Lemli-Opitz infant with elevated bilirubin levels. *Am. J. Med. Genet.* 68:294–99

137. Nikolakaki E, Meier J, Simos G, Georgatos SD, Giannakouros T. 1997. Mitotic phosphorylation of the lamin B receptor by a serine/arginine kinase and p34(cdc2). *J. Biol. Chem.* 272:6208–13

138. O'Brien TJ. 1990. Chondrodysplasia punctata (Conradi disease). *Int. J. Dermatol.* 29:472–76

139. Opitz JM, Zellweger H, Shannon WR, Ptacek LJ. 1969. The RSH syndrome. *Birth Defects: Orig. Artic. Ser.* 5(2):43–52

140. Oulmouden A, Karst F. 1990. Isolation of the ERG12 gene of *Saccharomyces cerevisiae* encoding mevalonate kinase. *Gene* 88:253–57

141. Parker CR Jr, Simpson ER, Bilheimer DW, Leveno K, Carr BR, MacDonald PC. 1980. Inverse relation between low-density lipoprotein cholesterol and dehydroisoandrosterone sulfate in human fetal plasma. *Science* 208:512–14

142. Penning TM, Bennett MJ, Smith-Hoog S, Schlegel BP, Jez JM, Lewis M. 1997. Structure and function of 3alpha-hydroxysteroid dehydrogenase. *Steroids* 62:101–11

143. Phillips RJ, Hawker SG, Moseley HJ. 1973. Bare-patches, a new sex-linked gene in the mouse, associated with a high production of XO females. I. A preliminary report of breeding experiments. *Genet. Res.* 22:91–99

144. Phillips RJ, Kaufman MH. 1974. Bare-patches, a new sex-linked gene in the mouse, associated with a high production of XO females. II. Investigations into the nature and mechanism of the XO production. *Genet. Res.* 24:27–41

145. Phillips RJS. 1963. Striated, a new sex-linked gene in the house mouse. *Genet. Res.* 4:151–53

146. Porter JA, von Kessler DP, Ekker SC, Young KE, Lee JJ, et al. 1995. The product of *hedgehog* autoproteolytic cleavage active in local and long-range signaling. *Nature* 374:363–66

147. Porter JA, Young KE, Beachy PA. 1996. Cholesterol modification of Hedgehog signaling proteins in animal development. *Science* 274:255–59

148. Purdue PE, Zhang JW, Skoneczny M, Lazarow PB. 1997. Rhizomelic chondrodysplasia punctata is caused by deficiency of human PEX7, a homologue of the yeast PTS2 receptor. *Nat. Genet.* 15:381–84

149. Raine J, Winter RM, Davey A, Tucker SM. 1989. Unknown syndrome: microcephaly, hypoplastic nose, exophthalmos, gum hyperplasia, cleft palate, low set ears, and osteosclerosis. *J. Med. Genet.* 26:786–88

150. Riou C, Tourte Y, Lacroute F, Karst F. 1994. Isolation and characterization of a cDNA encoding *Arabidopsis thaliana* mevalonate kinase by genetic complementation in yeast. *Gene* 148:293–97

151. Roessler E, Belloni E, Gaudenz K, Jay P, Berta P, et al. 1996. Mutations in the human *Sonic hedgehog* gene cause holoprosencephaly. *Nat. Genet.* 14:357–60

152. Rohmer M, Knani M, Simonin P, Sutter B, Sahm H. 1993. Isoprenoid biosynthesis in bacteria: a novel pathway for the early steps leading to isopentenyl diphosphate. *Biochem. J.* 295:517–24

153. Rosser ZH, Zerjal T, Hurles ME, Adojaan M, Alavantic D, et al. 2000. Y-chromosomal diversity in Europe is clinal and influenced primarily by geography, rather than by language. *Am. J. Hum. Genet.* 67:1526–43

154. Roux C, Aubry MM. 1966. Action tératogène chez le rat d'un inhibiteur de la synthèse du cholesterol, le AY 9944. *C. R. Soc. Biol.* 160:1353–57

155. Roux C, Dupuis R, Horvath C, Giroud A. 1979. Interpretation of isolated agenesis of the pituitary. *Teratology* 19:39–43

156. Roux C, Dupuis R, Horvath C, Talbot JN. 1980. Teratogenic effect of an inhibitor of cholesterol synthesis (AY 9944) in rats: correlation with maternal cholesterolemia. *J. Nutr.* 110:2310–12

157. Roux C, Wolf C, Mulliez N, Gaoua W, Cormier V, et al. 2000. Role of cholesterol in embryonic development. *Am. J. Clin. Nutr.* 71:1270S–9S

158. Ryan AK, Bartlett K, Clayton P, Eaton S, Mills L, et al. 1998. Smith-Lemli-Opitz syndrome: a variable clinical and biochemical phenotype. *J. Med. Genet.* 35:558–65

159. Schafer BL, Bishop RW, Kratunis VJ, Kalinowski SS, Mosley ST, et al. 1992. Molecular cloning of human mevalonate kinase and identification of a missense mutation in the genetic disease mevalonic aciduria. *J. Biol. Chem.* 267:13229–38

160. Setchell KA, O'Connell NC. 2000. Disorders of bile acid synthesis and metabolism. In *Pediatric Gastrointestinal Disease*, ed. WA Walker, PR Durie, JR Hamilton, JA Walker-Smith, JB Watkins, pp. 1138–71. Hamilton, Ontario: BC Decker

161. Shackleton CH, Roitman E, Kelley R. 1999. Neonatal urinary steroids in Smith-Lemli-Opitz syndrome associated with 7-dehydrocholesterol reductase deficiency. *Steroids* 64:481–90

162. Shackleton CH, Roitman E, Kratz LE, Kelley RI. 1999. Equine-type estrogens produced by a pregnant woman carrying a Smith-Lemli-Opitz syndrome fetus. *J. Clin. Endocrinol. Metab.* 84:1157–59

163. Shackleton CH, Roitman E, Kratz LE, Kelley RI. 1999. Midgestational maternal urine steroid markers of fetal Smith-Lemli-Opitz (SLO) syndrome (7-dehydrocholesterol 7-reductase deficiency). *Steroids* 64:446–52

164. Shefer S, Salen G, Batta AK, Honda A, Tint GS, et al. 1995. Markedly inhibited 7-dehydrocholesterol-delta7-reductase activity in liver microsomes from Smith-Lemli-Opitz homozygotes. *J. Clin. Invest.* 96:1779–85

165. Shefer S, Salen G, Batta AK, Tint GS, Irons M, Elias ER. 1994. Reduced 7-dehydrocholesterol reductase activity in Smith-Lemli-Opitz syndrome. *Am. J. Med. Genet.* 50:326–38

166. Shefer S, Salen G, Honda A, Batta AK, Nguyen LB, et al. 1998. Regulation of rat hepatic 3beta-hydroxysterol delta7-reductase: substrate specificity, competitive and non-competitive inhibition, and phosphorylation/dephosphorylation. *J. Lipid Res.* 39:2471–76

167. Silve S, Dupuy PH, Ferrara P, Loison G. 1998. Human lamin B receptor

exhibits sterol C14-reductase activity in *Saccharomyces cerevisiae. Biochim. Biophys. Acta* 1392:233–44

168. Smith DW, Lemli L, Opitz JM. 1964. A newly recognized syndrome of multiple congenital anomalies. *J. Pediatr.* 64:210–17

169. Soullam B, Worman HJ. 1995. Signals and structural features involved in integral membrane protein targeting to the inner nuclear membrane. *J. Cell Biol.* 130:15–27

170. Stamellos KD, Shackelford JE, Shechter I, Jiang G, Conrad D, et al. 1993. Subcellular localization of squalene synthase in rat hepatic cells. Biochemical and immunochemical evidence. *J. Biol. Chem.* 268:12825–36

171. Stamellos KD, Shackelford JE, Tanaka RD, Krisans SK. 1992. Mevalonate kinase is localized in rat liver peroxisomes. *J. Biol. Chem.* 267:5560–68

172. Sutphen R, Amar MJ, Kousseff BG, Toomey KE. 1995. XXY male with X-linked dominant chondrodysplasia punctata (Happle syndrome). *Am. J. Med. Genet.* 57:489–92

173. Sybert VP, ed. 1997. *Genetic Skin Disorders.* Oxford: Oxford Univ. Press. 105 pp.

174. Tanaka RD, Lee LY, Schafer BL, Kratunis VJ, Mohler WA, et al. 1990. Molecular cloning of mevalonate kinase and regulation of its mRNA levels in rat liver. *Proc. Natl. Acad. Sci. USA* 87:2872–76

175. Tanaka Y, Saitoh A, Taniguchi H, Oba K, Kitaoka T, Amemiya T. 1999. Conradi-Hunermann syndrome with ocular anomalies. *Ophthalmic Genet.* 20:271–74

176. Taybi H, Lachman RS. 1996. *Radiology of Syndromes, Metabolic Disorders, and Skeletal Dysplasias*, pp. 776–86. St. Louis, MO: Mosby

177. Teasdale RD, Jackson MR. 1996. Signal-mediated sorting of membrane proteins between the endoplasmic reticulum and the golgi apparatus. *Annu. Rev. Cell Dev. Biol.* 12:27–54

178. Tierney E, Nwokoro NA, Porter FD, Fre-

und LS, Ghuman JK, Kelley RI. 2001. Behavior phenotype in the RSH/Smith-Lemli-Opitz syndrome. *Am. J. Med. Genet.* 98:191–200

179. Tint GS, Irons M, Elias ER, Batta AK, Frieden R, et al. 1994. Defective cholesterol biosynthesis associated with the Smith-Lemli-Opitz syndrome. *N. Engl. J. Med.* 330:107–13

180. Tint GS, Seller M, Hughes-Benzie R, Batta AK, Shefer S, et al. 1995. Markedly increased tissue concentrations of 7-dehydrocholesterol combined with low levels of cholesterol are characteristic of the Smith-Lemli-Opitz syndrome. *J. Lipid Res.* 36:89–95

181. Traupe H. 1999. Functional X-chromosomal mosaicism of the skin: Rudolf Happle and the lines of Alfred Blaschko. *Am. J. Med. Genet.* 85:324–29

182. Traupe H, Muller D, Atherton D, Kalter DC, Cremers FP, et al. 1992. Exclusion mapping of the X-linked dominant chondrodysplasia punctata/ichthyosis/cataract/short stature (Happle) syndrome: possible involvement of an unstable premutation. *Hum. Genet.* 89:659–65

183. Tsukui T, Capdevila J, Tamura K, Ruiz-Lozano P, Rodriguez-Esteban C, et al. 1999. Multiple left-right asymmetry defects in Shh(−/−) mutant mice unveil a convergence of the shh and retinoic acid pathways in the control of Lefty-1. *Proc. Natl. Acad. Sci. USA* 96:11376–81

184. Ullrich K, Koch HG, Meschede D, Flotmann U, Seedorf U. 1996. Smith-Lemli-Opitz syndrome: treatment with cholesterol and bile acids. *Neuropediatrics* 27:111–12

185. Wanders RJ, Jansen G, van Roermund CW, Denis S, Schutgens RB, Jakobs BS. 1996. Metabolic aspects of peroxisomal disorders. *Ann. NY Acad. Sci.* 804:450–60

186. Wanders RJ, Romeijn GJ. 1998. Cholesterol biosynthesis, peroxisomes and peroxisomal disorders: mevalonate kinase is not only deficient in Zellweger syndrome

but also in rhizomelic chondrodysplasia punctata. *J. Inherit. Metab. Dis.* 21:309–12

187. Wanders RJ, Romeijn GJ. 1998. Differential deficiency of mevalonate kinase and phosphomevalonate kinase in patients with distinct defects in peroxisome biogenesis: evidence for a major role of peroxisomes in cholesterol biosynthesis. *Biochem. Biophys. Res. Commun.* 247:663–67

188. Wassif CA, Maslen C, Kachilele-Linjewile S, Lin D, Linck LM, et al. 1998. Mutations in the human sterol delta7-reductase gene at 11q12-13 cause Smith-Lemli-Opitz syndrome. *Am. J. Hum. Genet.* 63:55–62

189. Waterham HR, Koster J, Romeijn GJ, Vreken P, Hennekam RC, et al. 2001. Mutations in the *DHCR24* gene encoding 3β-hydroxysteroid Δ24-reductase cause autosomal recessive desmosterolosis. Submitted

190. Waterham HR, Wijburg FA, Hennekam RC, Vreken P, Poll-The BT, et al. 1998. Smith-Lemli-Opitz syndrome is caused by mutations in the 7-dehydrocholesterol reductase gene. *Am. J. Hum. Genet.* 63:329–38

191. Willnow TE, Hilpert J, Armstrong SA, Rohlmann A, Hammer RE, et al. 1996. Defective forebrain development in mice lacking gp330/megalin. *Proc. Natl. Acad. Sci. USA* 93:8460–64

192. Witsch-Baumgartner M, Fitzky BU, Ogorelkova M, Kraft HG, Moebius FF, et al. 2000. Mutational spectrum in the delta7-sterol reductase gene and genotype-phenotype correlation in 84 patients with Smith-Lemli-Opitz syndrome. *Am. J. Hum. Genet.* 66:402–12

193. Yu H, Lee MH, Starck L, Elias ER, Irons M, et al. 2000. Spectrum of delta7-dehydrocholesterol reductase mutations in patients with the Smith-Lemli-Opitz (RSH) syndrome. *Hum. Mol. Genet.* 9:1385–91

194. Yu H, Tint GS, Salen G, Patel SB. 2000. Detection of a common mutation in the RSH or Smith-Lemli-Opitz syndrome by a PCR-RFLP assay: IVS8–1G>C is found in over 60% of U.S. propositi. *Am. J. Med. Genet.* 90:347–50

195. Zellweger H, Uehlinger E. 1948. Ein Fall von halbseitiger Knochenchondromatose (Ollier) mit Naevus ichthyosiformis. *Helv. Paediatr. Acta* 2:153–63

196. Zizka J, Maresova J, Kerekes Z, Nozicka Z, Juttnerova V, Balicek P. 1983. Intestinal aganglionosis in the Smith-Lemli-Opitz syndrome. *Acta Paediatr. Scand.* 72:141–43

Annu. Rev. Genomics Hum. Genet. 2001. 2:343–72

A NEW APPROACH TO DECODING LIFE:
Systems Biology

Trey Ideker[1,2], Timothy Galitski[1], and Leroy Hood[1,2,3,4,5]
Institute for Systems Biology[1], Seattle, Washington 98105; Departments of
Molecular Biotechnology[2], Immunology[3], Bioengineering[4], and Computer
Science and Engineering[5], University of Washington, Seattle, Washington 98195;
e-mail: tideker@systemsbiology.org, tgalitski@systemsbiology.org,
lhood@systemsbiology.org

Key Words   biological information, discovery sciences, genome, proteome

■ **Abstract**   Systems biology studies biological systems by systematically perturbing them (biologically, genetically, or chemically); monitoring the gene, protein, and informational pathway responses; integrating these data; and ultimately, formulating mathematical models that describe the structure of the system and its response to individual perturbations. The emergence of systems biology is described, as are several examples of specific systems approaches.

INTRODUCTION

Perhaps the most important consequence of the Human Genome Project is that it is pushing scientists toward a new view of biology—what we call the systems approach. Systems biology does not investigate individual genes or proteins one at a time, as has been the highly successful mode of biology for the past 30 years. Rather, it investigates the behavior and relationships of all of the elements in a particular biological system while it is functioning. These data can then be integrated, graphically displayed, and ultimately modeled computationally. How has the Human Genome Project moved us to this new view? It has done so by catalyzing a new scientific approach to biology, termed discovery science; by defining a genetic parts list of human and many model organisms; by strengthening the view that biology is an informational science; by providing us with powerful new high-throughput tools for systematically perturbing and monitoring biological systems; and by stimulating the creation of new computational methods.

Discovery Science

The Human Genome Project was one of the first modern biological endeavors to practice discovery science. The objective of discovery science is to define all of the elements in a system and to create a database containing that information. For

1527-8204/01/0728-0343$14.00

example, discovery approaches are providing the complete sequences of the 24 different human chromosomes and of the 20 distinct mouse chromosomes. The transcriptomes and proteomes of individual cell types (e.g., quantitative measurements of all of the mRNAs and protein species) also represent discovery projects. Discovery science lies in contrast to hypothesis-driven science, which creates hypotheses and attempts to distinguish among them experimentally. The integration of these two approaches, discovery and hypothesis-driven science, is one of the mandates of systems biology.

Genomic Sequences in Humans and Model Organisms

The complete genomic sequences of human (78, 125), nematode (121), fly (2), arabadopsis (81), yeast (54), *Escherichia coli* (19), and a host of microbes and parasites are now available; others, including mouse, are in the pipeline. These sequences offer a number of powerful opportunities.

GENETIC PARTS LIST   Software and global experimental techniques are now becoming available to identify the gene locations, and even coding regions, embedded in a sequenced genome (110). Comparative analysis of these coding regions reveals a lexicon of motifs and functional domains (essential to solving the protein-folding and structure/function problems). Moreover, genomic sequence provides access to the adjacent regulatory sequences—a vital component to solving the regulatory code (34)—and opens access to polymorphisms, some of which are responsible for differences in physiology and disease predisposition. Combined, these components make up the elements in the periodic table of life. With these components now in hand, the immediate challenge is to place them in the context of their informational pathways and networks.

MODEL ORGANISMS ARE THE ROSETTA STONES FOR DECIPHERING BIOLOGICAL SYSTEMS   The genomic sequences of humans and model organisms have elegantly confirmed a basic unity in the strategy of life. Informational pathways in yeast are remarkably similar to those in fly, worm, and humans, and many orthologous genes can be identified across these species. Thus, it is feasible to use genetically and biologically facile model organisms (yeast, fly, worm) to infer the function of human genes and to place these genes in the context of their informational pathways. Alternatively, comparison of different model genomes offers the possibility of comparing and contrasting the logic of life between organisms. Differences in logic provide fundamental insights into the mechanisms of evolution, development, and physiology.

Biology is an Informational Science

The Human Genome Project has propelled us toward the view that biological systems are fundamentally composed of two types of information: genes, encoding the molecular machines that execute the functions of life, and networks of regulatory interactions, specifying how genes are expressed. All of this information is hierarchical in nature: DNA → mRNA → protein → protein interactions → informational pathways → informational networks → cells → tissues or networks of cells → an organism → populations → ecologies. Of course, other macromolecules and small molecules also participate in these information hierarchies, but the process is driven by genes and interactions between genes and their environments. The central task of systems biology is (*a*) to comprehensively gather information from each of these distinct levels for individual biological systems and (*b*) to integrate these data to generate predictive mathematical models of the system.

Biological information has several important features:

- It operates on multiple hierarchical levels of organization.
- It is processed in complex networks.
- These information networks are typically robust, such that many single perturbations will not greatly effect them.
- There are key nodes in the network where perturbations may have profound effects; these offer powerful targets for the understanding and manipulation of the system.

Perturbation of Biological Systems

The development of systems biology has been driven by a number of recent advances in our ability to perturb biological systems systematically. Three technological trends have emerged in this respect. First, techniques for genetic manipulation have become more high-throughput, automated, and standardized by several orders of magnitude. Second, the availability of complete genomic sequences has stimulated the development of several systematic mutagenesis projects to complement more traditional efforts involving random mutagenesis. Third, technologies for disrupting genes *in trans* allow the application of genetic perturbations to a wide range of eukaryotic organisms.

HIGH-THROUGHPUT GENETIC MANIPULATION   A number of recent and ongoing technological developments are making it possible to rapidly and systematically manipulate genomic material. To illustrate these developments, consider some of the tools available for the budding yeast *Saccharomyces*. Expanding yeast's already-formidable genetic toolkit is a series of plasmids that has greatly facilitated PCR-based gene replacement (83, 127). These versatile vectors render the gene-insertion process simple, standardized, and applicable to any gene or genomic region. First, forward and reverse PCR primers are synthesized with ~40 bp of DNA homologous to a gene of interest and another ~20 bp designed to flank a plasmid-encoded module. These primers are then used to PCR-amplify the module from the plasmid template. PCR products are directly transformed into yeast cells, and homologous recombination occurs with the desired gene at high efficiency. A wide variety of readymade sequence modules are available to disrupt, replace, or modify essentially any genomic sequence using this technique. These include modules for gene knockouts, promoter fusions, protein fusions, and epitope tags, and because the 20-bp sequences flanking the modules are standard, the same primers can be used for multiple constructions involving several of these different module types. In order to select for successful recombination events while minimizing the formation of undesired recombinant types, modules usually include a marker gene that is non-native to yeast.

The demands of manipulating many genes via many different constructs, then observing the results of these manipulations in many strains simultaneously, are driving the development of ever-more facile and standardized plasmid-construction systems. For example, the GATEWAY recombinational-cloning system [Life Technologies (128)] allows for automated, high-throughput generation of an unlimited array of constructs derived from genes of interest. This procedure exploits the advantages of the bacteriophage lambda integration/excision reaction to transfer genes or other sequences of interest to a virtually unlimited number of clones in separate but identical in vitro reactions. PCR-based cloning into a single entry vector allows transfer of the gene to a variety of destination vectors without the need for restriction endonucleases. This procedure is efficient, standardized, precise, and directional and involves minimal investment in clone isolation or confirmation.

SYSTEMATIC GENE MUTATIONS   Presently, there is a transition in gene mutation methods from the random generation of mutant alleles to a systematic approach in which genes are specifically targeted for mutation. The systematic approach has several distinct advantages. With random mutageneses, one employs a selection process or a screen to visually identify mutants of interest. Typically, some genes are identified multiple times, whereas others are not found. In contrast, with a systematic approach, the response of all generated genotypes is documented and the coverage of the genome is unambiguous.

For instance, one recent development is the completion of a collection of deletions of essentially all yeast genes (109, 134). This collection allows the systematic characterization of all yeast gene knockouts for phenotypes that one can assay in a high-throughput screen. In addition, this mutant collection uniquely identifies each deletion mutant genotype with a 20 base-pair "barcode" that can be used to quantify the relative numbers of each genotype in a pooled population of mutant strains (using a microarray of probes against the barcodes). Thus, one can quantitatively assess the fitness of each deletion strain in a given condition in a single experiment. With all single-gene deletion strains now available, ongoing and future research is attempting to characterize phenotypes among genes in combination through the use of double-deletion strains.

GENE DISRUPTION *IN TRANS*   Many of the powerful genetic approaches available to yeast and other model organisms are not practicable in higher eukaryotes. However, recent studies suggest that genes in these higher organisms may be perturbed *in trans* using antisense inhibitors of mRNA translation or technologies based on

RNA-mediated interference (RNAi). Gene disruption *in trans* is inducible, thus allowing transient interrogation of practically any gene.

For example, modified oligonucleotides have proven very effective as antisense inhibitors of mRNA translation (44, 91). The modifications (e.g., morpholino groups or phosphoramidate linkages) render the oligonucleotides resistant to nucleases, and the modified oligonucleotides can be delivered by transfection or microinjection.

RNA-mediated interference (RNAi) is another recently discovered mechanism to silence genes in organisms ranging from mice to trypanosomes [recently reviewed by Bass (14)]. The introduction of double-stranded RNA (dsRNA) corresponding to a particular mRNA results in the specific and rapid destruction of that mRNA in cells. The current model (14) proposes that dsRNA is cleaved into 21-25 bp dsRNA molecules that then serve as a template for an RNA helicase that exchanges the sense strand for the mRNA, which is followed by cleavage of the mRNA. The cleavage destroys the mRNA and regenerates the 21-25 bp dsRNA. RNAi was discovered in the nematode worm *C. elegans* (46); systematic investigations of gene function in worm development have used libraries of bacterial clones (fed to the worms) expressing dsRNA (48, 55). These RNAi screens have multiplied the numbers of genes with functional assignments.

Quantitative High-Throughput Biological Tools

Just as the Human Genome Project has led to improvements in our ability to systematically perturb cells, it has also provided us with new technologies for systematically characterizing their cellular response: DNA sequencers, microarrays, and high-throughput proteomics. Because these tools can carry out global (or nearly global) analyses, they become the methods of choice for rapid and comprehensive assessment of biological system properties and dynamics. Typically, the development of these tools goes through three distinct stages: (*a*) proof-of-principle, (*b*) development of a robust instrument, and (*c*) the creation of an automated production line. For example, the automated DNA sequencer was first demonstrated in 1986 (112), made feasible by about 1990 (67), and has enjoyed widespread use in genome sequencing centers since about 1999 (98). The current production-line instrument includes 96 capillary sequencers with a front-end interface of automated sample preparation and a back-end process for tracking the data. A single 96-capillary sequencer can produce ~500,000 base pairs of raw DNA sequence data per day. From 1985 to the present, there has been a 2000-fold increase in the throughput of sequencing, with corresponding increases in the quality of sequence information and simultaneous decreases in cost—all achieved by incremental improvements in chemistry, engineering, and software. High-throughput DNA sequencing production lines may analyze genomic DNA and cDNAs as well as identify and type polymorphisms [either simple sequence repeats or single nucleotide polymorphisms (SNPs)]. Powerful new applications of sequencing technology are also emerging. For example, the biotechnology company Lynx

Therapeutics, Inc. has developed a technique that allows up to 500,000 different sequences to be determined simultaneously for 16 to 20 residues (22). Moreover, this is a powerful discovery approach for determining complete transcriptomes in individual cell types from organisms whose genome has been sequenced.

DNA arrays represent a second kind of powerful discovery tool. Two types of arrays are in common use: cDNA microarrays (102) and oligonucleotide arrays (47, 66). cDNA microarrays consist of double-stranded cDNA or PCR products spotted on a glass slide. If, indeed, the human genome only contains 30,000 to 40,000 genes (78, 125), this approach will easily allow interrogation of complete human transcriptomes. Oligonucleotide arrays are synthesized (66) or spotted (58) on glass slides at densities that can exceed 50,000 spots/slide. In principle, they are more specific than the cDNA microarray and make it possible to distinguish single-nucleotide differences. Using the oligonucleotide array, the mRNAs from individual members of multigene families can be distinguished, alternatively spliced genes can be characterized, alternative forms of SNPs can be identified and typed, and whole stretches of DNA can be resequenced. Clearly, DNA array technology is less mature than sequencing; although the technology is now robust, it is just entering the production-line stage at some companies.

Proteomics, the characterization of the many proteins within a cell type, involves analysis of different types of information corresponding to each protein species: protein identity, abundance, processing, chemical modifications, interactions, compartmentalization, turnover time, etc. Perhaps the major challenge of proteomics is to deal with the enormous dynamic range of protein abundances found in a single cell type—from 1 to 10^6 copies or greater.

For organisms whose genome has been sequenced, mass spectrometry is an especially powerful tool for identifying and quantifying large numbers of proteins (42), identifying and typing SNPs, and analyzing protein modifications. For example, Dr. Ruedi Aebersold and colleagues have recently developed a technique, termed isotope coded affinity tags (ICAT), for measuring the relative expression levels of proteins between two different cell populations (59). In brief, the ICAT reagent is a molecule with three functions: a biotin tag, a linker sequence containing either eight deuterium atoms (heavy reagent) or eight hydrogen atoms (light reagent), and a group reactive to cysteine residues. Proteins from the first cell population are labeled with the heavy reagent, whereas those from the second cell population are labeled with the light reagent. Equal quantities of each protein sample are combined and digested with trypsin, and cysteine-labeled peptides are isolated with an avidin column. Mass spectrometry is used to analyze the paired atomic masses for each peptide (light vs. heavy peptides differ by eight mass units) and, after further fragmentation, to determine their amino-acid sequences. Thus, peptides can be quantitated (to an accuracy of ~20%) and the corresponding genes identified. Aebersold and colleagues have created an automated high-throughput production line for this procedure, capable of analyzing 1000 proteins per day; they are currently developing a next-generation facility to analyze up to 1 million proteins per day.

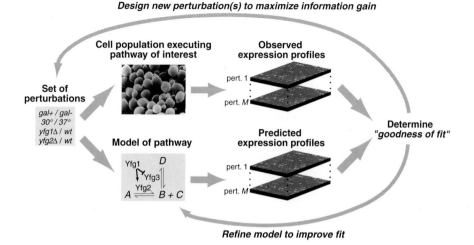

Design new perturbation(s) to maximize information gain

Refine model to improve fit

Figure 1 Overview of the systems biology approach, involving pathway verification and refinement through systematic, successive perturbations. The pathway of interest is perturbed genetically by gene deletion or overexpression and/or biologically by modulation of metabolite levels, temperature, or other pathway components. Gene expression profiles measured in response to each perturbation, obtained using microarrays or related technologies, are compared to those predicted by a model of the pathway mechanism. Perturbations are initially selected to target known pathway components and are thereafter chosen to distinguish between alternative models that are consistent with the present set of observations. All aspects of the process are amenable to automation (laboratory or computational), including model refinement and choice of perturbations.

Figure 2 (see figure page C-2) A *cis*-regulatory network at the sea urchin *endo16* gene. (*A1–A6*) A developmental time course of *endo16* in situ expression patterns in sea urchin. The gene is expressed early in the vegetal plate (*A1*), although not in the early blastula (*A2*) nor at ingression of skeletogenic cells (*A3*). After gastrulation, *endo16* expression is observed throughout the archenteron (*A4*). Subsequently, expression is shut down in the foregut, the secondary mesenchyme (*A5*), and the hindgut (*A6*) but remains in the midgut. (*B*) A map of protein-DNA interactions in the 2300bp *endo16 cis*-regulatory sequences. Different colors represent different proteins. Repeated sites are marked with symbols below the line. Distinct modules with identifiable roles in *endo16* expression patterns are indicated and annotated. (*C*) Control logic model for modules A and B. Binding sites are indicated above the line. Below the line, circles indicate logical operations. Effects exerted by module A are indicated in red; those of module B are in blue. Interactions that can be modeled as boolean inputs are indicated as dashed lines, scalars as thin solid lines, and time-dependent quantitative inputs as heavy lines. Outputs indicated with an arrowhead exert positive effects; perpendicular bars represent negative effects. As an example, CY and CB1 interact synergistically to promote the output of the module B spatial-temporal control element U1. Originals reprinted with permission from Davidson (34). Copyright 2001 Academic Press.

See legend page C-1

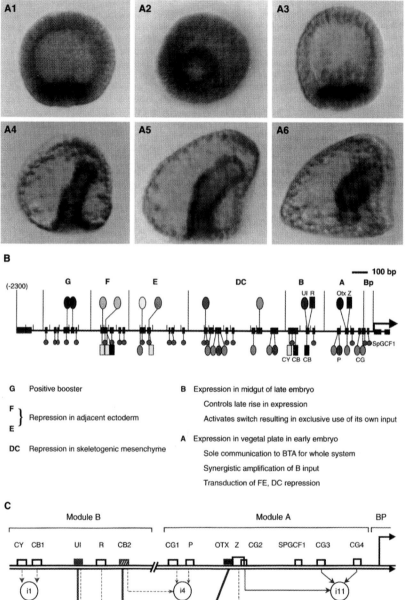

B

G   Positive booster

F
 } Repression in adjacent ectoderm
E

DC   Repression in skeletogenic mesenchyme

B   Expression in midgut of late embryo

Controls late rise in expression

Activates switch resulting in exclusive use of its own input

A   Expression in vegetal plate in early embryo

Sole communication to BTA for whole system

Synergistic amplification of B input

Transduction of FE, DC repression

C

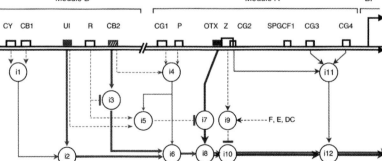

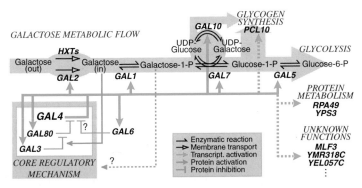

Figure 3 The galactose system. Yeast metabolize galactose through a series of steps involving the *GAL2* transporter and enzymes produced by *GAL1, 7, 10*, and *5*. These genes are transcriptionally regulated by a mechanism consisting primarily of *GAL4, 80*, and *3*. *GAL6* produces another regulatory factor thought to repress the GAL enzymes in a manner similar to *GAL80*. Dotted interactions denote model refinements supported by our systems approach. Reprinted with permission from Ideker et al. (68). Copyright 2001 American Association for the Advancement of Science.

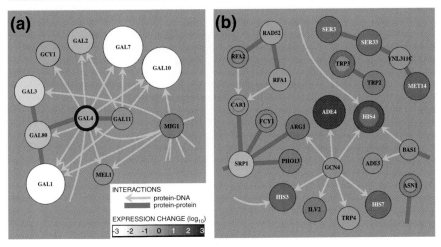

Figure 4 Sample regions of the integrated physical-interaction network, corresponding to (*a*) galactose utilization and (*b*) amino-acid biosynthesis. Each node represents a gene, a yellow arrow directed from one node to another represents a protein-DNA interaction, and a blue line between nodes represents a protein-protein interaction. The intensity of each node indicates the change in mRNA expression of the corresponding gene, with medium-gray representing no change and darker or lighter shades representing an increase or decrease in expression, respectively (node diameter also scales with the magnitude of change). Nodes for which protein data are also available (panel *b*) contain two distinct regions: an outer circle, or ring, representing the change in mRNA expression and an inner circle representing the change in protein expression. To signify that the expression level of *GAL4* has been perturbed by external means (panel *a*), it is highlighted with a red border. Reprinted with permission from Ideker et al. (68). Copyright 2001 American Association for the Advancement of Science.

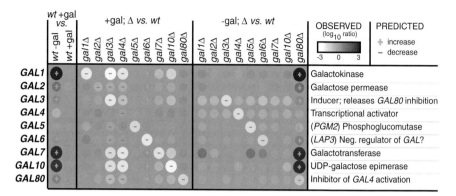

Figure 5 Matrix of observed vs. predicted gene-expression responses. DNA microarrays were used to measure the mRNA-expression responses of yeast cells undergoing steady-state growth in the presence of each of 20 perturbations to the galactose-utilization pathway. Each spot in the matrix represents the quantitative change in expression observed for a GAL gene (*rows*) in one of the perturbations (*columns*), according to the intensity scale shown at upper right. Superimposed on each spot are the corresponding (qualitative) predictions of the network model as shown in Figure 3, with the symbols + vs. − indicating a predicted increase vs. decrease in expression, respectively. Unannotated spots represent genes for which no expression change is predicted. Reprinted with permission from Ideker et al. (68). Copyright 2001 American Association for the Advancement of Science.

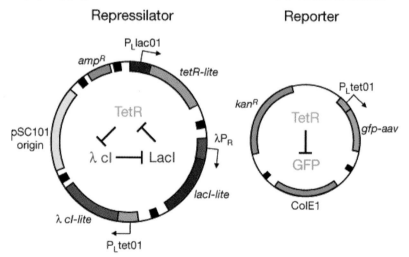

Figure 7 A synthetic, three-protein oscillatory network. The "repressilator" plasmid encodes three synthetic proteins, each fused to a heterologous promoter such that one protein represses the transcription of the next in a closed negative-feedback loop (TetR represses cI, cI represses LacI, and LacI represses TetR). A reporter plasmid is used to track oscillations in TetR concentration: it contains the TetR-binding sequence upstream of the gene encoding for green fluorescent protein. Reprinted with permission from Elowitz & Leibler (40). Copyright 2001 Nature.

Because of their ability to separate many different cells and cell types at high speed, multiparameter cell sorters are another technology critical to systems biology. Although microarray and proteomics experiments typically measure average levels of mRNA or protein within a cell population, in reality these levels can vary from cell to cell; this distribution of expression levels contains important information about the underlying control mechanisms and regulatory network structure. Dr. Ger van den Engh and colleagues have developed a new cell sorter capable of separating 30,000 elements per second against 32 different parameters.

Computation for Systems Biology: Databases and Models

Biology is unique among the natural sciences in that it has a digital code at its core. Together with colleagues in computer science, mathematics, and statistics, biologists are developing the necessary tools to acquire, store, analyze, graphically display, model, and distribute this information. An enormous challenge for the future is how to integrate the different levels of information pertaining to genes, mRNAs, proteins, and pathways.

THE INCREASING IMPORTANCE OF COMPUTER DATABASES   Computer databases first rose to prominence in molecular biology as central repositories for the plethora of data generated by large-scale sequencing projects. Although databases of nucleic-acid and amino-acid sequences are still the largest, most utilized, and best maintained, there has been a sudden explosion of interest in databases to store other types of molecular data. Such interest is primarily in response to demands placed by functional genomics and other emerging systems approaches. For instance, the Database of Interacting Proteins (137), BIND (10), and MIPS (90a) contain searchable indices of known protein-protein interactions; TRANSFAC (133) and SCPD (142) catalog interactions between proteins and DNA (i.e., transcription-factor interactions), and databases of metabolic pathways have also recently been established [e.g., EcoCyc (73), KEGG (92), and WIT (106)]. A growing number of databases are also under development for storing the now sizeable number of mRNA-expression data sets (1, 43, 63, 96, 116); companies, such as Affymetrix, Rosetta, Spotfire, Informax, Incyte, Gene Logic, and Silicon Genetics, market gene-expression databases commercially. A comprehensive review of recent developments in the molecular biology databases is available elsewhere (15).

This recent explosion, in both the variety and amount of information of interest, poses two challenges to database users and developers alike. First, the information must be maintained systematically in a format that is compatible with both single queries and global searches. Often, the desired information is present in the database but is not annotated consistently for all entries: For instance, an EST sequence or expression profile may have been derived from cancer cells, but in the absence of an enforced annotation style, this information may be recorded using many different keywords (e.g., cancer, tumor, metastatic, carcinoma, etc.). Alternately, the information recorded in the databases may be incomplete: For instance, a protein-interaction database may correctly document an interaction between two proteins but may fail to include related, highly informative data, such as the strength of binding or the result of interaction on the functional activity of each protein.

The second challenge, and perhaps the more difficult one, is keeping the databases updated against the ever-increasing body of biological knowledge. In this regard, computer scientists working in the field of natural-language processing have made promising advances in computer programs that can parse textual passages, extract the key concepts, and catalog these concepts systematically (6, 18, 33, 38, 95, 117, 122). Thus in the near future, there is hope of updating biological databases automatically with the thousands of relevant results published each month in the primary biological literature, reducing the dependence on humans to perform this tedious, error-prone, and time-consuming task.

Why are databases so important to the future of systems biology? Although individual researchers may amass a great deal of knowledge about the genes, molecular interactions, and other biological information underlying one particular pathway, no single biologist can be familiar with the extremely large and complex number of interactions in an entire cell. The databases track all of these, provided that the analytical approaches are available to help the biologist to access, display, and interpret the information. In the end, however, biology cannot be done solely in silico. Biologists must employ their insights to bring coherence to the massive data sets.

THE INCREASING IMPORTANCE OF GLOBAL ANALYSES   Given the recent accumulation of expression profiles, molecular interactions, and a variety of other global data in the biological databases, the immediate task is to develop powerful analyses and experimental strategies to integrate and analyze these data to make biological discoveries. To date, methods to analyze patterns of gene expression have received the most attention. In the most straightforward approach, gene-expression data are used to identify genes involved in a particular biological process, by selecting genes with clear changes in expression over different biological conditions or over time. Depending on the experiment, genes of interest have been implicated in cancer (129), development (131), aging (79, 107), or a specific cellular response (28, 29, 35, 50, 71, 114). In most of these cases, expression levels of tens, hundreds, or even thousands of genes changed over the conditions examined, often expanding the known number of changes by an order of magnitude.

Genes with similar responses over multiple conditions are often clustered together to form functional groups or to reveal coordinated patterns of expression. Several clustering methods have been proposed: Most are excellent and have associated software packages that are publicly available (4, 23, 25, 36, 39, 57, 62, 118). Some analyses achieve more specific and/or accurate functional predictions by integrating gene-expression clusters with complementary types of global data: for example, searching for shared regulatory sequences in the promoters of co-expressed genes (29, 89, 97, 111, 114, 119). Identification of these regulatory sequences provides evidence that the associated genes are under the control of common transcription factor(s). Alternatively, gene-expression data may be combined with information on protein-protein interactions and protein phylogenetic profiles (85) or augmented with the genomic location of each gene to find shared regulatory elements (32). Finally, deciding whether an expression level changes significantly can be a difficult problem and is also an active area of research (27, 57, 65, 69, 84). Methods for analyzing expression data, including gene-expression clustering and its extensions, have been extensively reviewed elsewhere (21, 30, 64, 108, 141).

Where do we go from here? Although these analyses have certainly been informative, global data sets undoubtedly provide additional information that remains untapped. Ultimately, it would be highly desirable to analyze expression levels and other global measurements in a way that validates our current knowledge of a cellular process and isolates discrepancies between expected and observed levels. To achieve this level of analysis, we believe that it will be necessary to compare and incorporate global data with a well-defined model of the biological process of interest.

THE INCREASING IMPORTANCE OF COMPUTER MODELS   Conventionally, a biological model begins in the mind of the individual researcher, as a proposed mechanism to account for some experimental observations. Often, the researcher represents their ideas by sketching a diagram using pen and paper. This diagram is a tremendous aid in thinking clearly about the model, in predicting possible experimental outcomes, and in conveying the model to others. Not surprisingly, diagrammatic representations form the basis of the vast majority of models discussed in journal articles, textbooks, and lectures.

Despite the useful simplicity of these conventional models, advances in systems biology are prompting some biologists to forego "mental" models, or pen-and-paper diagrams, for more sophisticated computer representations. Although the notion of modeling a biological process computationally is almost as old as the electronic computer itself [e.g., see biological models proposed by Turing (124)]; such models are gaining in importance for several reasons. First, it is now apparent that the magnitude and complexity of interactions in a cell are simply too vast for an unaided human mind to process and organize (132). Second, as DNA microarrays, sequencers, and other large-scale technologies begin to generate vast amounts of quantitative biological data, a paradigm shift is occurring in biology away from a descriptive science and toward a predictive one (52, 77). Computer systems are required to store, catalogue, and condense the rapidly accumulating mass of data, and automated tools are needed that, by assimilating these data into a network model, can predict network behaviors and outcomes that may be tested experimentally. It is encouraging that recent computer simulations of partial or whole genetic networks have demonstrated network behaviors, commonly called systems properties or emergent properties, that were not apparent from examination of a few isolated interactions alone (5, 16, 74).

Computer modeling tools have already achieved widespread acceptance within the engineering and physical sciences. For example, computer-aided design packages, such as SPICE, VHDL, or Prolog, are heavily used to simulate and test electronic circuitry (93). In contrast, a relative paucity of software and methods exists for analyzing biological circuits. Several useful tools are available for simulating small networks of chemical reactions [e.g., Gepasi (90) or the Chemical Reaction Network Toolbox (45)], but larger-scale simulations are still emerging. For example, ongoing projects, such as E-CELL (123) and the Virtual Cell (100, 101), attempt to model all molecular interactions in the cell as an integrated, computational process. Other efforts, such as BioJake (99) and a collection of guidelines set forth by Kohn (75), are working to define a standard graphical environment in which biologists may interactively define and simulate genetic circuit models. A widespread, standard notation (and/or software environment) is attractive because systems biologists working on diverse systems and at different institutions would be able to directly exchange their fully detailed models.

TYPES OF COMPUTER MODELS   A wide variety of cellular models have been proposed, each of differing complexity and abstraction. For example, chemical kinetic models attempt to represent a cellular process as a system of distinct chemical reactions. In this case, the network state is defined by the instantaneous quantity (or concentration) of each molecular species of interest in the cell, and molecular species may interact via one or more reactions. Often, each reaction is represented by a differential equation relating the quantity of reactants to the quantity of postreaction products, according to a reaction rate and other parameters. This system of differential equations is usually too complex to be solved explicitly, but given an initial network state, the quantity of each gene product or other molecular species can be simulated to produce a state transition path or trajectory, i.e., the succession of states adopted by the network over time. A variety of biological systems have been modeled in this way, including the networks controlling bacterial chemotaxis (5, 20), developmental patterning in *Drosophila* (24, 86), and infection of *E. coli* by lambda phage (88). Recently, it has been pointed out that transcription, translation, and other cellular processes may not behave deterministically but instead are better modeled as random events (87). Models have been investigated that address this concern by abandoning differential equations in favor of stochastic relations to describe each chemical reaction (8, 53).

In contrast to models involving systems of chemical reactions, another popular approach has been to model a genetic network as a simplified discrete circuit. Much like a neural network, this approach represents the network as a graph with nodes and arrows (i.e., directed edges), where a node represents the quantity or level of a distinct molecular species and an edge directed from one node to another represents the effect of the first node's level on that of the second. Also required is a function for each node, describing how all of the incoming effects should be combined to determine its level. Typically, nodes may assume one of two discrete levels, signifying whether the molecule is present or absent or whether a gene is turned on or off. Given a starting state of levels for all nodes, the next level of each node may be determined directly from its function. In this way, the network state over all nodes evolves over a series of discrete time steps, where the state of the next step is computed from the current state.

Discrete circuit models have been investigated extensively (74, 113, 136), and simulation software is available (135). Clearly, such models are greatly simplified compared to a kinetic model. Proponents of discrete circuit models argue that they preserve the essential features of the underlying biology while greatly reducing network complexity and simulation time. A major criticism has been that they require the model to update simultaneously for all nodes, whereas molecular interactions within the cell are not synchronous. Also, a two-level representation of molecular species may not always be sufficient to capture the underlying biological behavior of the network.

CHOICE OF MODEL DETAIL   Formulation of a model involves important choices about which genes, gene products, and other molecular species should be included in the network state. Genes may be regulated at the level of transcription or translation, and once translated, a protein may exist in one of several modified forms. Through alternative splicing, a single gene may encode several distinct mRNAs, and nongenetic molecules, such as metabolites, may also affect the network. Furthermore, some interactions are restricted to the nucleus, cell membrane, golgi apparatus, or other organelles. A complete model would therefore have to include molecular species, such as alternatively-spliced mRNA and modified protein products, and would have to restrict interactions between species located in different cellular compartments.

Increasing levels of detail are not always desirable, however, and deciding which information to include can be a difficult task. In general, one identifies the types of properties or behaviors that the model should be able to predict (e.g., mRNA levels, protein activation states, or growth rates) and includes only the components that impact these properties. The number of model parameters must be compatible with the amount and type of available data: If only mRNA-expression levels are measured, for instance, then detailed information about protein structures or compartmentalization may overload the model with too many hidden variables.

INFERENCE OF MODELS FROM GLOBAL PATTERNS OF GENE EXPRESSION   Several methods have been proposed for inference of a genetic network from a measured time series of mRNA-expression profiles. These methods try to infer a discrete circuit model by looking for statistical correlations between expression levels (7), by training a neural network (130), or by information theoretic methods (80). Also under development are methods for inferring models from steady-state expression profiles, e.g., recorded over a battery of biological conditions or gene deletions (3, 70). In the future, several or all of these methods will almost certainly be expanded to take advantage of other types of global data, such as protein-expression levels, protein-modification states, or metabolite concentrations.

A FRAMEWORK FOR SYSTEMS BIOLOGY

Ultimately, one wishes to understand the underlying interactions, molecular or otherwise, that are responsible for the global changes observed in a system. In order to most directly address this goal, we argue that it will be necessary to integrate the various levels of global measurements together and with a mathematical model of the biological system of interest. Although these model-driven approaches may differ in the particulars of implementation, all follow a fundamental framework involving several distinct steps (as shown in Figure 1, see color insert):

1. **Define all of the components of the system.** Use these components, along with prior biochemical and genetic knowledge, to formulate an initial model. Ideally, a global approach is the most powerful (i.e., defining all genes in the genome, all mRNAs and proteins expressed in a particular condition, or all protein-protein interactions occurring in the cell) because it does not require any prior assumptions about system components. Constructing a model by interrogating these components will ultimately accomplish two objectives: (*a*) to describe the structure of the interactions that govern the systems behavior and (*b*) to predict accurately relevant properties of the system given specified perturbations. If prior knowledge about the system is limited, the initial model may be rough and may involve purely hypothetical interactions.

2. **Systematically perturb and monitor components of the system.** Specific perturbations may be genetic (e.g., gene deletions, gene overexpressions, or undirected mutations) or environmental (e.g., changes in growth conditions, temperature, or stimulation by hormones or drugs). The corresponding response to each perturbation is measured using large-scale discovery tools to capture changes at relevant levels of biological information (e.g., mRNA expression, protein expression, protein activation state, overall pathway function). Once observed, data from all levels are integrated with each other and with the current model of the system. As in step 1, an approach in which all components are systematically perturbed and globally monitored is the most desirable.

3. **Reconcile the experimentally observed responses with those predicted by the model.** Refine the model such that its predictions most closely agree with experimental observations. Agreement between the observed and predicted responses is evaluated qualitatively and/or quantitatively using a goodness-of-fit measure. When predictions and observations disagree, alternative hypotheses are proposed to alleviate the discrepancies (maximize the good-ness-of-fit), resulting in a refined model for each competing hypothesis. If the initial model is largely incomplete or is altogether unavailable, the observed responses may be used to directly infer the particular components required for system function and, among these, the components most likely to interact. If the model is relatively well defined, its predictions may already

be in good qualitative agreement with the observations, differing only in the extent of their predicted changes.

4. **Design and perform new perturbation experiments to distinguish between multiple or competing model hypotheses.** Even for a moderate number of observations, the proposed refinements may result in several distinct models whose predictions fit equally well with the observations. These models are indistinguishable by the current data set, requiring new perturbations and measurements to discriminate among them. New perturbations are informative only if they elicit different systems responses between models, with the most desirable perturbations resulting in model predictions that are most dissimilar from one another. After choosing the set of new perturbations, repeat steps 2 through 4, thereby expanding and refining the model continually, over successive iterations. The idea is to bring the theoretical predictions and experimental data into close apposition by repeated iterations of this process so that the model predictions reflect biological reality.

Thus, systems biology requires that all of the elements of a system be studied (at multiple levels of the information hierarchy and in the context of their responses to perturbations), that these data be integrated and graphically displayed, and finally, that these responses be modeled mathematically to predict the structure and behavior of the informational pathway. Moreover, systems biology involves an iterative, strategic interplay between discovery- and hypothesis-driven science. Global observations (discoveries) are matched against model predictions (hypotheses) in an iterative manner, leading to the formation of new models, new predictions, and new experiments to test them.

EXAMPLES OF SYSTEMS BIOLOGY

A large number of recent and ongoing efforts are putting this systems biology framework into practice (as illustrated in Table 1). We now examine in detail four such studies, representing four distinct types of biological networks: (*a*) a *cis*-regulatory network, (*b*) a *trans*-regulatory network, (*c*) a signal-transduction network, and (*d*) a synthetic regulatory system engineered according to a predetermined network model.

Cis Gene Regulation in the Sea Urchin

Gene-regulatory networks are defined by *trans* and *cis* logic (34). *Trans* logic defines the interactions between protein transcription factors and the batteries of genes they control (e.g., other transcription factors as well as genes in the network periphery). Conversely, *cis* logic defines the precise relationships among promoter elements (DNA sequences) whose states (i.e., bound vs. unbound by transcription factors) are combined to produce the temporal and spatial patterns of expression for a particular gene. Both of these types of regulatory networks

TABLE 1 A sampling of systems-biology approaches

Model systems	Organisms	Approaches	References
Viral infection of E. coli	phage λ	Computer simulations via a mixed model (discrete and continuous); stochastic simulations	(8, 87, 88)
	phage T7	Time-differential equations	(41)
Bacterial chemotaxis	E. coli	Time-differential equations; stochastic simulations	(5, 20, 31, 138)
Embryo patterning or development	Sea urchin	Identification of *cis*-regulatory elements and interactions; computational modeling of the proposed *cis* network	(9, 139)
	Drosophila	Simulation via time-differential equations	(24, 86, 126)
Cross-talk between signaling pathways	Mammals	Simulation via time-differential equations	(16, 132)
Sugar metabolism	S. cerevisiae	Integrated physical-interaction network, Bayesian networks	(61, 68, 120)
Protein networks	S. cerevisiae	Directed-graph models	(68, 94, 105)
General metabolism	E. coli	Flux-balance analyses	(37, 103, 104)
Whole-cell model	E. coli, neuron	Simulation via time-differential equations	(100, 123)
Synthetic and circadian oscillators	E. coli, Drosophila, Neurospora, mice	Synthesis of a multigene network in vivo using a computer model as blueprint	(12, 40)
Synthetic flip-flop	E. coli	" "	(51)
Cell cycle	Mammals, yeast	Molecular-interaction maps; Bayesian networks	(49, 75)

have input and output. For instance, network inputs may arise from exogenous signals (e.g., sperm penetrating the egg, steroid hormones, etc.) or from signal-transduction pathways. The output of the network, concentration of nuclear RNA, exhibits many possible levels of posttranscriptional control (e.g., RNA processing, alternative RNA splicing, protein processing, protein chemical modification, etc.).

The sea urchin is a powerful model for studying *cis* and *trans* regulation because its development is relatively simple (34) (the embryo has only 12 different cell types); enormous numbers of eggs can be obtained in a single summer (30 billion); the eggs can be fertilized synchronously and development stopped at any stage; and many transcription factors can be readily isolated and characterized, and their genes may be cloned using affinity chromatography, conventional protein chemistry, protein microsequencing, DNA probe synthesis, or library screening (60). A great deal of developmental biology has been carried out on the sea urchin, and a modest genome effort has defined the general features of its genome (26).

The sea urchin *endo16* gene has the most completely defined *cis*-regulatory system to date (34, 139, 140). Strikingly, this system is highly analogous to a computer or other electronic circuit; multiple inputs from a wide range of transcription factors are integrated to send a signal to the RNA-synthesizing machine (the basal transcription apparatus) as to whether and how much transcript to synthesize. Endo16 is expressed in the endoderm of the embryo: It appears first in the vegetal plate (which gives rise to endodermal and mesodermal cell types), emerges later in the archenteron, and finally, intensifies in the midgut while diminishing in the fore- and hindgut (Figure 2*A*, see color insert). Thus, the *cis*-regulatory apparatus must turn on gene expression in the appropriate cells, establish sharp boundaries of expression (expressed in endodermal but not mesodermal cells), and specify the cells of terminal expression.

A 2.3 kb sequence of genomic DNA contains all of the control elements necessary for normal endo16 expression patterns. Figure 2*B* (see color insert) depicts the 34 binding sites spread across this region, together with the 13 different transcription factors that bind them. The binding sites fall into seven regions of DNA sequence: six functional regions (modules A–G) and the basal promoter region where the transcriptional apparatus assembles. Each of these was defined by mutating one or more binding sites, attaching the resulting sequence to a reporter construct, placing this construct into transgenic sea urchins, then measuring the spatial and temporal gene-expression output. For example, this approach revealed that the module G is a positive booster, whereas the F and D modules are responsible for repressing gene expression in the adjacent ectoderm. Module A is the sole means of communication between the six functional regions and the basal transcription apparatus, integrating the positive or negative inputs from the G, F, E, DC, and B modules. In addition, module A mediates expression in the vegetal plate of the early embryo.

From these studies, a logic model was constructed delineating all the operations executed by these modules and their interactions (Figure 2*C*, see color insert). This model clearly indicates that the output of module B runs through module A (which amplifies it) on the way to instructing the basal transcription apparatus. These interactions can be boolean (*dotted lines*), scalar (*thin solid lines*), or time-varying quantitative (*heavy solid lines*) inputs (Figure 2*C*). The important point is that the *cis*-control region behaves as a series of integrated electronic circuits (modules), each combining their environmental inputs (quantitatively changing levels of transcription factors) to send signals through module A to set the overall circuit output. Moreover, the model was derived according to the systems biology framework described earlier, having been developed after many iterative cycles of perturbation and gene-expression measurements.

The *cis*-regulatory regions of other genes and other organisms employ a very similar logic, although they certainly differ in detail (34). Thus, the responses of an organism, both developmental and physiological, are hardwired into its *cis*-regulatory circuitry. Almost certainly, the central driver of evolution is not changes in individual genes, but rather changes in this circuitry. Clearly one of the major challenges for biology in the twenty-first century will be coming to understand the nature of the *cis-* and *trans*-regulatory networks that control an organism's development, physiological responses, and even its trajectory of evolution.

A Network Controlling Galactose Utilization in Yeast

As another example, we (T. Ideker & L. Hood) have recently used a systems approach to explore, expand, and refine the understanding of galactose utilization (GAL) in yeast (68). Like the previous example of sea urchin development, a regulatory network model is used to predict changes in gene expression resulting from a battery of directed perturbation experiments. Unlike the sea urchin example in which the components of the network are the *cis*-regulatory sequences controlling a particular gene, the network controlling galactose utilization consists of a large number of genes and gene products interacting in *trans*. Although the *trans* model includes less detail about the regulation of individual genes, it provides new information on how groups of genes interact to control a cellular process.

As shown in Figure 3 (see color insert), the yeast galactose-utilization system employs at least nine genes. Four encode the enzymes that catalyze the conversion of galactose to glucose-6-phosphate (*GAL1*, *5*, *7*, and *10*), whereas a fifth (*GAL2*) encodes a transporter molecule that sets the state of the system. If galactose is present in the yeast cell, the system is turned on; if galactose is absent, the system is turned off. A number of transcription factors regulate this on/off switch, including *GAL3*, *4*, *80*, and possibly *GAL6*.

GENETIC AND ENVIRONMENTAL PERTURBATION OF THE NETWORK   We wished to determine whether the molecular interactions in the galactose network were sufficient to account for changes in gene expression resulting from extensive perturbations to the GAL pathway. Toward this goal, we constructed nine genetically perturbed yeast strains, each with a deletion of a different GAL gene (see above). These strains, along with wild-type yeast (no genes deleted), were grown to steady state in the presence (+gal) or absence (−gal) of 2% galactose. For each of these 20 perturbation conditions (10 strains × 2 media types), we used a whole-yeast-genome microarray to monitor changes in mRNA expression relative to wild type (+gal).

Nine hundred ninety-seven mRNAs (out of ~6200) showed statistically significant concentration changes during one or more of these perturbations. The corresponding perturbed genes could be divided into 16 different clusters, where the genes within a cluster behaved in a similar manner through all perturbations. The striking observation was that the genes encoding various metabolic, cellular, and synthetic pathways tended to fall in individual clusters—thus beginning to reveal the network of interconnected informational pathways within the yeast cell.

CONSTRUCTION AND VISUAL DISPLAY OF THE NETWORK MODEL   To reveal the nature of these informational pathways, we constructed a model of the known molecular interactions connecting galactose utilization with other metabolic processes in yeast. For this purpose, we compiled a list of 3026 previously observed physical interactions, using all available entries from publicly available databases of protein-protein (105) and protein-DNA interactions (133, 142). The interactions in these databases came from several different sources relying on a variety of experimental approaches: Most were derived from biochemical association studies reported in the literature or through large-scale experiments such as the two-hybrid screen.

The interactions in these databases define a model of the molecular-interaction network, shown in Figure 4 (see color insert), for two small regions. Each node in the network represents a gene and is labeled with its corresponding gene name. An arrow directed from one node to another signifies that the protein encoded by the first gene can influence the transcription of the second by DNA binding (a protein → DNA interaction), whereas an undirected line between two nodes signifies that the proteins encoded by each gene can physically interact (a protein-protein interaction).

Expression data from each perturbation can be visually superimposed on the network. For example, Figure 4a shows the result of the $gal4\Delta$ deletion in the presence of galactose. In the figure, the grayscale intensity of each node represents the change in mRNA expression of its corresponding gene. When other types of information are available, they too can be superimposed on the network display. For example, we also measured changes in protein abundance for wild-type cells grown in the presence vs. absence of galactose (68). Using a procedure based on isotope coded affinity tags (ICAT) and tandem mass spectrometry (59), we detected a total of 289 proteins and quantified their expression-level changes between these two conditions. Strikingly, 30 proteins showed significant concentration changes, ~15 of which showed no changes at the mRNA level. The implication is clear—these 15 proteins are regulated by posttranscriptional mechanisms—a compelling argument for the need to integrate both mRNA- and protein-expression changes to understand eukaryotic gene regulation. Figure 4b illustrates the addition of protein-abundance information to the visual display, focusing on the region of the network corresponding to amino-acid biosynthesis. By comparing the mRNA- and protein-expression responses displayed on each node, one can visually assess whether the mRNA and protein data are correlated and quickly spot genes for which they are remarkably discordant.

COMPARISON OF OBSERVED AND PREDICTED RESPONSES   In keeping with the systems-biology framework outlined previously, we wished to determine if the expression changes observed across the 20 perturbations were consistent with changes as predicted by the molecular-interaction network. By itself, the network model makes only very coarse predictions. For example, a protein-DNA interaction involving protein A and gene B predicts that a change in expression of A could result in a change in expression of B. If A is additionally involved in a protein-protein interaction with C (C—A→B), then a change in expression of C could also elicit a change at B, by first altering the activity of A. However, the network does not dictate whether these interactions activate or repress transcription or, in the case that multiple interactions affect a gene, how these interactions should be combined to produce an overall change in expression. Similarly, the network does not specify whether a protein-protein interaction results in the formation of a functional protein complex or if, instead, one protein transiently modifies the other. Because none of these levels of information are encoded in the protein-DNA or protein-protein databases, they are also absent from the network model and the graphical display.

Interestingly, much of this information is known outside of the databases: In the case of the GAL genes, classic genetic and biochemical experiments have determined that Gal4p is a strong transcriptional activator and that Gal80p can bind to Gal4p to repress this function [see reviews by Johnston et al. (72) and Lohr et al. (82)]. Thus, we have supplemented the model with these and other results from the literature that address the effect of perturbing GAL genes on the expression of other genes in the cell. For example, when such information is known, we can indicate whether each relevant protein-DNA interaction serves to activate or repress gene expression. Likewise, we can indicate whether each protein-protein interaction can alter the activity of either protein and whether this change is positive or negative. Once incorporated into the model, these added levels of detail greatly increase its predictive power. Note that it is not necessary to integrate all previous evidence into the model, just evidence that bears on gene expression.

Figure 5 (see color insert) compares the observed to the predicted expression responses of the GAL genes, for each of the 20 perturbations. Although the observed response is obviously more complex than the predicted one, the two responses agree in many of their salient features. Not shown in the figure are the approximately 990 additional genes, outside of the core GAL pathway, whose mRNA expression levels were affected in at least one perturbation. Interestingly, very little is known about the molecular interactions that determine how the galactose-utilization pathway may influence these other genes. The majority of these expression changes, therefore, are not yet addressed by the model and will call for the addition of new interactions.

REFINING THE MODEL THROUGH ADDITIONAL PERTURBATIONS   We used discrepancies between the predicted and observed expression responses to suggest possible refinements to the model. For example, the current model predicts that in galactose, perturbations to GAL enzymes (i.e., $gal1\Delta$, 7Δ, or 10Δ) should not affect expression levels of other GAL genes. Although this is largely true for $gal1\Delta$, the $gal7\Delta$ and $gal10\Delta$ deletions clearly affect expression levels of GAL1, 2, 3, 7, 10, and 80 (see Figure 5). Because both $gal7\Delta$ or $gal10\Delta$ deletions block the conversion of galactose-1-phosphate (Gal-1-P), leading to increased levels of this and other metabolites (76), one hypothesis is that one of these metabolites exerts control over GAL-gene expression. To address this hypothesis, we examined expression changes in a $gal1gal10\Delta$ double deletion strain grown in galactose. Although deletion of GAL10 blocks the conversion of Gal-1-P, deletion of GAL1 blocks a preceding step in the galactose-utilization pathway such that Gal-1-P levels are greatly reduced. Thus, if metabolites are the cause of the change in GAL-gene expression observed in a *gal10Δ* mutant, these changes are predicted to disappear in the *gal1Δgal10Δ* strain. In fact, when we measured the gene-expression profile of this double deletion using a microarray, this is exactly what happened, lending support for this revised model.

Thus, the systems approach to galactose utilization has given us new insights into how the pathway is regulated (e.g., it has generated many new, testable hypotheses) and how it is interconnected with other informational pathways in the yeast cell. Accordingly, this approach may be very powerful in elucidating the network of informational pathways in other biological systems and, ultimately, the interconnected networks of cells in metazoan organisms.

Bacterial Chemotaxis: A Robust Signal-Transduction Network

Bacterial chemotaxis, the process by which bacteria move toward or away from a chemical source, has intrigued researchers for 120 years. The last few decades of research (largely in *E. coli*) have focused on elucidating the molecular interactions responsible for the chemotaxic response, resulting in a large body of genetic, structural, physiological, and biochemical data [see recent reviews (17, 56)]. These large data sets have led to a number of recent attempts to model the chemotaxis network using systems approaches.

BEHAVIOR AND MOLECULAR BIOLOGY OF CHEMOTAXIS   The physiology of the chemotactic response has been relatively well characterized (17). The bacterium moves in a chemical gradient by means of a biased random walk, alternating episodes of swimming straight (running) and random reorientation (tumbling). To run, the bacterium turns its flagellar motors counter-clockwise; a tumble ensues when the motors are reversed. As it moves, the bacterium senses gradients of chemical attractant (e.g., aspartate) or repellent (e.g., hydrogen peroxide) as changes in concentration over time. Increasing attractant or decreasing repellent results in an increase in the duration of runs in the desired direction.

As shown in Figure 6, the molecular biology of the chemotaxis system is also known in considerable detail. Five transmembrane attractant receptors, known as methyl-accepting chemotaxis proteins (MCPs), have multiple methylation sites whose modification state governs signal transduction via a phospho-relay system. This signaling network modulates the output of flagellar motors. It also adapts the sensitivity of the network to changing concentrations of attractant or repellent.

MODELING "ROBUSTNESS" AS A SYSTEMS PROPERTY   The molecular interactions responsible for chemotaxis have been studied quantitatively. For instance, Spiro et al. (115) modeled the transitions of the various chemotactic signaling molecules among different states of ligand occupancy, phosphorylation, and methylation. Using the equations of mass-action kinetics, they succeeded in recapitulating the

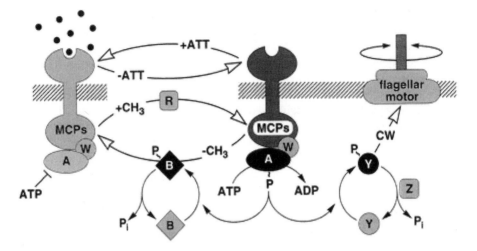

Figure 6   The bacterial chemotaxis system. Transmembrane attractant receptors, known as methyl-accepting chemotaxis proteins (MCPs), form a signaling complex with CheW and a kinase, CheA, that autophosphorylates and transfers phosphates to a response regulator, CheY. Phospho-CheY interacts with flagellar motor proteins to induce clockwise rotation (tumbling). The CheZ protein promotes CheY dephosphorylation. CheA transfers phosphates also to the CheB MCP methylesterase. The activated phospho-CheB demethylates MCPs; this demethylation diminishes the kinase activity of the MCP-CheW-CheA complex as part of the adaptation mechanism. CheR is a constitutive MCP methyltransferase. In the presence of increasing attractant, CheA autophosphorylation is inhibited, counterclockwise flagellar rotation (running) is extended, and subsequent MCP methylation allows adaptation to the higher attractant concentrations. When moving through decreasing attractant, CheA autophosphorylation is stimulated. This, in turn, promotes phospho-CheY induced tumbling and then adaptation through demethylation of MCPs. Figure from Spiro et al. (115).

response of these molecules to attractant gradients, step increases, and saturation. In doing so, Spiro et al. moved beyond the study of isolated components toward a quantitative and integrative reproduction of many interactions simultaneously. This complex web of interactions makes it possible to observe, model, and predict system properties (i.e., properties that are observed behaviorally but are not readily understood by studying any individual component of the system).

One interesting systems property of the chemotaxis network is its robustness of adaptation to different attractant concentrations (5, 11, 138). In this context, robustness means that the output of the system is insensitive to particular choices of its inputs or biochemical parameters (e.g., enzyme levels and rate constants). In particular, robust adaptation means that the system output responds to changes in the input, without depending on the overall input magnitude. Thus, *E. coli* cells respond to changes in attractant concentration (system input) by changing their tumbling frequency (output). However, owing to robust adaptation, a homogenous solution of attractant will always result in the same tumbling frequency, regardless of the attractant concentration.

PRECISION OF ADAPTATION   Modeling work of Barkai & Leibler (11) demonstrated that the robust adaptation of chemotaxis is a result of the structure of the underlying molecular signaling network. Assuming a simple two-state model in which the MCP-CheW-CheA protein complex is either active or inactive, they were able to capture a wide variety of behaviors that had been previously observed by experiment. In addition, they were able to make several striking predictions regarding the properties of adaptation. They defined the precision of adaptation of the chemotaxis network to mean its ratio of steady-state output before vs. after a change in input. The network model predicted that this ratio was always equal to one, thus indicating that precision of adaptation is a robust property; that is, the tumbling frequency is only transiently affected by an increase in the level of attractant and will eventually return to its initial steady-state value. In fact, one can change rate constants over several orders of magnitude and still maintain precise adaptation, regardless of whether Michaelis-Menten or cooperative kinetics are used for the model simulations. Interestingly, many other properties of chemotaxis were predicted to be nonrobust; for example, neither the adaptation time (the interval between the input change and the re-establishment of steady-state output) nor the tumbling frequency (which depends on enzyme levels) were robust properties under the model.

In a related study, Alon et al. tested these predictions directly, by varying enzyme concentrations over two orders of magnitude and observing the response of *E. coli* to the addition of saturating attractant (5). They observed steady-state tumbling frequency, adaptation time, and precision of adaptation. As predicted by the model, tumbling frequency and adaptation time were highly sensitive to changes in enzyme level, whereas precision of adaptation remained relatively constant (within the bounds of experimental error).

Returning to the model, the key structural feature responsible for this robust behavior appears to be a feedback-control loop involving modification of the MCP-CheW-CheA protein complex. Because the modification rate depends on the activity of the protein complex and not on the concentrations of its various modified forms, system activity tends to return to an initial steady state following a change in input.

Forward Engineering of Biological Networks

All of the examples discussed thus far have involved constructing a mechanistic model of a naturally occurring biological system. However, the systems-biology framework outlined above can equally be used to construct a synthetic system according to a predetermined model, with one key difference: In the former scenario, one alters the model to best fit the biological system (i.e., reverse engineering), whereas in the latter scenario, one alters the biological system to fit the model (i.e., forward engineering).

Although the idea of engineering biological systems to have desired properties or particular functions is not new, a series of ongoing research projects are putting these ideas into practice. Spearheading these efforts is work by Elowitz et al. (40), who constructed a gene-regulatory system in *E. coli* that functions as a synthetic oscillator, and work by Gardner et al. (51), who demonstrated a genetic toggle switch based on similar principles.

The basic network configuration of Elowitz et al. involves three transcriptional repressor proteins organized into a negative-feedback loop (Figure 7, see color insert). To explore the potential oscillatory behavior of this configuration, the group constructed quantitative models (both kinetic and stochastic) describing the change in protein concentration over time for each of the three genes in the system. These models involved a number of biochemical parameters, including the overall rate of translation, the rates of mRNA and protein degradation, and the dependence of transcription rate on the concentration of the corresponding protein repressor. In simulations, a high protein-degradation rate (relative to mRNA degradation) tended to produce the desired oscillatory behavior. These simulations prompted the group to insert a carboxy-terminal tag at the $3'$ end of each of the three repressor genes: These tags increased the degradation rate of each protein by targeting them for destruction by cellular proteases. In this way, parameters of the biological system were adjusted to match the desired parameters of the model.

To explore the behavior of the network in vivo, *E. coli* cells were transformed with two plasmids: one encoding the three repressor proteins and another containing green-fluorescent protein (GFP) under the transcriptional control of one of the repressors. By monitoring levels of fluorescence over time, the group showed that, as desired, individual cells exhibited oscillations with an average period of 150 minutes (three times that of the typical cell cycle, although there was some difficulty in synchronizing oscillations over an entire *E. coli* population).

Research efforts such as these in which novel biological networks are designed from a model will eventually converge and couple with efforts to study existing biological systems. In this scenario, one would not only possess predictive models but would also have the power to use these models to re-engineer cells. A range of potential modifications could be rigorously evaluated through model simulations then later verified directly in the biological system. This dualistic approach is one of the "holy grails" of biology and medicine in which a predictive model of a complex disease pathway is used to design and test cellular modifications that can, ultimately, ameliorate the disease response.

SUMMARY

In conclusion, what are the most striking challenges arising from systems biology?

- ▪ The inclusion of nongenetic molecules, small and large, into the systems picture. The cell contains thousands of distinct metabolic substrates and other small and large molecules, a variety of which exert influence on gene expression (through direct interactions with proteins or DNA) and on allosteric enzymes. Methods to systematically measure levels of such molecules would be of enormous benefit.

- Further development of theoretical frameworks and tools for integrating the various levels of biological information, displaying them graphically, and, finally, mathematical modeling and simulation of biological systems.

- Systematic and detailed annotations of information in the public databases. As the databases become more advanced, so will our models of cellular processes. For instance, rather than simply provide a list of interactions, physical-interaction databases should specify if, and how, each interaction affects cell state.

- Education of cross-disciplinary scientists. Cross-disciplinary scientists should have a deep understanding of biology (their contributions will be proportional to their understanding). We believe that the solution to this problem is to teach biology as an informational science. This approach is conceptual, hierarchical, economical, and in the future mainstream of education in biology.

- The integration of technology, biology, and computation. Integration (also see points discussed below) presents one of the most striking challenges for systems biology, both for academia and industry.

In addition to these general challenges, the development and practice of systems biology involves a number of requirements that will pose particular difficulties for academic institutions. Among these requirements are:

- high-throughput facilities for global technologies, such as DNA sequencing, DNA arrays, genotyping, proteomics, and protein interactions;

- integration of different levels of biological information generated at each of these facilities;

- the integration of excellent biology with a strong computational infrastructure and analytic tools;

- the formation of teams of biologists, technologists, and computational scientists to attack the iterative challenges of systems biology;

- the integration of discovery- and hypothesis-driven science; and

- the development of diverse partnerships with academia and industry. Academia will provide new systems for exploration; industry will provide new technologies and resources to take on demanding problems.

These six challenges pose difficulties for most academic institutions—if they are to provide their biologists the opportunities to practice systems biology. Most academic institutions do not have the diverse scientific talent, funds, or space to initiate a self-sustaining systems biology effort. Although these resources may be available through industrial partnerships, it is difficult for academic laboratories to form such partnerships, particularly when intellectual property is involved. Owing to severe salary constrains, recruiting scientists from high-demand fields, such as bioinformatics, proteomics, and engineering, can be equally problematic. Within an academic institution, individual departments often provide barriers for cross-disciplinary science—in geographical isolation, in the training of students, and in the constraints of what is expected from faculty (research projects, teaching, etc.). Finally, the demands of tenure force young faculty to carry out safe projects independently—at potentially the most creative phase of their careers—and penalize them for research performed as part of a team. Of course, some academic institutions may circumvent many of these limitations by the creation of special centers. In other cases, independent, nonprofit research institutes, such as our Institute for Systems Biology, can be fashioned to take advantage of these opportunities.

Regardless of these initial hurdles, it is clear that systems biology will necessarily be a leading academic and industrial thrust in the years to come. Its impact on medicine, agriculture, biological energy production, and many other areas will make biotechnology a powerful driving force as we move into the century of biology.

Visit the Annual Reviews home page at www.AnnualReviews.org

LITERATURE CITED

1. Aach J, Rindone W, Church GM. 2000. Systematic management and analysis of yeast gene expression data. *Genome Res.* 10:431–45

2. Adams MD, Celniker SE, Holt RA, Evans CA, Gocayne JD, et al. 2000. The genome sequence of *Drosophila melanogaster. Science* 287:2185–95

3. Akutsu T, Kuhara S, Maruyama O, Miyano S. 1998. Identification of gene regulatory networks by strategic gene disruptions and gene overexpressions. *Proc. ACM-SIAM Symp. Discrete Algorithms, 9th.* New York: ACM Press

4. Alon U, Barkai N, Notterman DA, Gish K, Ybarra S, et al. 1999. Broad patterns of gene expression revealed by clustering analysis of tumor and normal colon tissues probed by oligonucleotide arrays. *Proc. Natl. Acad. Sci. USA* 96:6745–50

5. Alon U, Surette MG, Barkai N, Leibler S. 1999. Robustness in bacterial chemotaxis. *Nature* 397:168–71

6. Andrade MA, Valencia A. 1998. Automatic extraction of keywords from scientific text: application to the knowledge domain of protein families. *Bioinformatics* 14:600–7

7. Arkin A, Ross J. 1995. Statistical construction of chemical reaction mechanisms from measured time series. *J. Phys. Chem.* 99:970

8. Arkin A, Ross J, McAdams HH. 1998. Stochastic kinetic analysis of developmental pathway bifurcation in phage lambda-infected Escherichia coli cells. *Genetics* 149:1633–48

9. Arnone MI, Davidson EH. 1997. The hardwiring of development: organization and function of genomic regulatory systems. *Development* 124:1851–64

10. Bader GD, Donaldson I, Wolting C, Ouellette BF, Pawson T, et al. 2001. BIND—The biomolecular interaction network database. *Nucleic Acids Res.* 29: 242–45

11. Barkai N, Leibler S. 1997. Robustness in simple biochemical networks. *Nature* 387:913–17

12. Barkai N, Leibler S. 2000. Circadian clocks limited by noise. *Nature* 403:267–68

13. Deleted in proof

14. Bass BL. 2000. Double-stranded RNA as a template for gene silencing. *Cell* 101:235–38

15. Baxevanis AD. 2001. The molecular biology database collection: an updated compilation of biological database resources. *Nucleic Acids Res.* 29:1–10

16. Bhalla US, Iyengar R. 1999. Emergent properties of networks of biological signaling pathways. *Science* 283:381–87

17. Blair DF. 1995. How bacteria sense and swim. *Annu. Rev. Microbiol.* 49:489–522

18. Blaschke C, Andrade MA, Ouzounis C, Valencia A. 1999. Automatic extraction of biological information from scientific text: protein-protein interactions. *Proc. Int. Conf. Intell. Syst. Mol. Biol., 7th, Heidelberg, 1999*, pp. 60–67. Menlo Park, CA: AAAI Press

19. Blattner FR, Plunkett G, Bloch CA, Perna NT, Burland V, et al. 1997. The complete genome sequence of *Escherichia coli* K-12. *Science* 277:1453–74

20. Bray D, Levin MD, Morton-Firth CJ. 1998. Receptor clustering as a cellular mechanism to control sensitivity. *Nature* 393:85–88

21. Brazma A, Vilo J. 2000. Gene expression data analysis. *FEBS Lett.* 480:17–24

22. Brenner S, Johnson M, Bridgham J, Golda G, Lloyd DH, et al. 2000. Gene expression analysis by massively parallel signature sequencing (MPSS) on microbead arrays. *Nat. Biotechnol.* 18:630–34

23. Brown MP, Grundy WN, Lin D, Cristianini N, Sugnet CW, et al. 2000. Knowledge-based analysis of microarray gene expression data by using support vector machines. *Proc. Natl. Acad. Sci. USA* 97:262–67

24. Burstein Z. 1995. A network model of developmental gene hierarchy. *J. Theor. Biol.* 174:1–11

25. Butte AJ, Kohane IS. 2000. Mutual information relevance networks: functional genomic clustering using pairwise entropy measurements. *Pac. Symp. Biocomput.* 5:418–29

26. Cameron RA, Mahairas G, Rast JP, Martinez P, Biondi TR, et al. 2000. A sea urchin genome project: sequence scan, virtual map, and additional resources. *Proc. Natl. Acad. Sci. USA* 97:9514–18

27. Chen Y, Dougherty E, Bittner M. 1997. Ratio-based decisions and the quantitative analysis of cDNA microarray images. *J. Biomed. Optics* 2:364–74

28. Cho RJ, Campbell MJ, Winzeler EA, Steinmetz L, Conway A, et al. 1998. A genome-wide transcriptional analysis of the mitotic cell cycle. *Mol. Cell* 2:65–73

29. Chu S, Derisi J, Eisen M, Mulholland J, Botstein D, et al. 1998. The transcriptional program of sporulation in budding yeast. *Science* 282:699–705

30. Claverie JM. 1999. Computational methods for the identification of differential and coordinated gene expression. *Hum. Mol. Genet.* 8:1821–32

31. Cluzel P, Surette M, Leibler S. 2000. An ultrasensitive bacterial motor revealed by monitoring signaling proteins in single cells. *Science* 287:1652–55

32. Cohen BA, Mitra RD, Hughes JD, Church GM. 2000. A computational analysis of whole-genome expression data reveals chromosomal domains of gene expression. *Nat. Genet.* 26:183–86

33. Craven M, Kumlien J. 1999. Constructing biological knowledge bases by extracting information from text sources. *Proc. Int. Conf. Intell. Syst. Mol. Biol., 7th, Heidelberg, 1999*, pp. 77–86. Menlo Park, CA: AAAI Press

34. Davidson EH. 2001. *Genomic Regulatory Systems: Development and Evolution*. San Diego, CA: Academic

35. Derisi JL, Iyer VR, Brown PO. 1997. Exploring the metabolic and genetic control

of gene expression on a genomic scale. *Science* 278:680–86

36. Dysvik B, Jonassen I. 2001. Exploring gene expression data using Java. *Bioinformatics* 17:369–70

37. Edwards JS, Palsson BO. 2000. Robustness analysis of the *Escherichia coli* metabolic network. *Biotechnol. Prog.* 16: 927–39

38. Eilbeck K, Brass A, Paton N, Hodgman C. 1999. INTERACT: an object oriented protein-protein interaction database. *Proc. Int. Conf. Intell. Syst. Mol. Biol., 7th, Heidelberg, 1999*, pp. 87–94. Menlo Park: AAAI Press

39. Eisen MB, Spellman PT, Brown PO, Botstein D. 1998. Cluster analysis and display of genome-wide expression patterns. *Proc. Natl. Acad. Sci. USA* 95:14863–68

40. Elowitz MB, Leibler S. 2000. A synthetic oscillatory network of transcriptional regulators. *Nature* 403:335–38

41. Endy D, You L, Yin J, Molineux IJ. 2000. Computation, prediction, and experimental tests of fitness for bacteriophage T7 mutants with permuted genomes. *Proc. Natl. Acad. Sci. USA* 97:5375–80

42. Eng JK, McCormack AL, Yates JRI. 1994. An approach to correlate tandem mass spectral data of peptides with amino acid sequences in a protein database. *J. Am. Soc. Mass. Spectrom.* 5:976–89

43. Ermolaeva O, Rastogi M, Pruitt KD, Schuler GD, Bittner ML, et al. 1998. Data management and analysis for gene expression arrays. *Nat. Genet.* 20:19–23

44. Faria M, Spiller DG, Dubertret C, Nelson JS, White MRH, et al. 2001. Phosphoramidate oligonucleotides as potent antisense molecules in cells and in vivo. *Nat. Biotechnol.* 19:40–44

45. Feinberg M. 1995. *The Chemical Reaction Network Toolbox*, Version 1.02, ftp.che.rochester.edu/pub/feinberg/. Rochester, NY

46. Fire A, Xu S, Montgomery MK, Kostas SA, Driver SE, et al. 1998. Potent and specific genetic interference by double-stranded RNA in *Caenorhabditis elegans*. *Nature* 391:806–11

47. Fodor SPA, Dower WJ, Ekins RP, Flynn GC, Houghten RA, et al. 1991. Light-directed, spatially addressable parallel chemical synthesis. *Science* 251:767–73

48. Fraser AG, Kamath RS, Zipperlen P, Martinez-Campos M, Sohrmann M, et al. 2000. Functional genomic analysis of cell division in *C. elegans* chromosome 1 by systematic RNA inteference. *Nature* 408:325–30

49. Friedman N, Linial M, Nachman I, Pe'er D. 2000. Using Bayesian networks to analyze expression data. *J. Comput. Biol.* 7:601–20

50. Galitski T, Saldanha AJ, Styles CA, Lander ES, Fink GR. 1999. Ploidy regulation of gene expression. *Science* 285:251–54

51. Gardner TS, Cantor CR, Collins JJ. 2000. Construction of a genetic toggle switch in *Escherichia coli*. *Nature* 403:339–42

52. Gilbert W. 1991. Towards a paradigm shift in biology. *Nature* 349:99

53. Gillespie DT. 1976. A general method for numerically simulating the stochastic time evolution of coupled chemical reactions. *J. Comput. Physics* 22:403–34

54. Goffeau A, Barrell BG, Bussey H, Davis RW, Dujon B, et al. 1996. Life with 6000 genes. *Science* 274:546, 563–67

55. Gonczy P, Echeverri G, Oegema K, Coulson A, Jones SJ, et al. 2000. Functional genomic analysis of cell division in *C. elegans* using RNAi of genes on chromosome III. *Nature* 408:331–36

56. Grebe TW, Stock J. 1998. Bacterial chemotaxis: the five sensors of a bacterium. *Curr. Biol.* 8:R154–57

57. Greller LD, Tobin FL. 1999. Detecting selective expression of genes and proteins. *Genome Res.* 9:282–96

58. Guo Z, Guilfoyle RA, Thiel AJ, Wang R, Smith LM. 1994. Direct fluorescence analysis of genetic polymorphisms by hybridization with oligonucleotide arrays on glass supports. *Nucleic Acids Res.* 22: 5456–65

59. Gygi SP, Rist B, Gerber SA, Turecek F, Gelb MH, et al. 1999. Quantitative analysis of complex protein mixtures using isotope-coded affinity tags. *Nat. Biotechnol.* 17:994–99

60. Harrington MG, Coffman JA, Calzone FJ, Hood LE, Britten RJ, et al. 1992. Complexity of sea urchin embryo nuclear proteins that contain basic domains. *Proc. Natl. Acad. Sci. USA* 89:6252–56

61. Hartemink AJ, Gifford DK, Jaakkola TS, Young RA. 2001. Using graphical models and genomic expression data to statistically validate models of genetic regulatory networks. *Pac. Symp. Biocomput.* 6:422–33

62. Hartuv E, Schmitt A, Lange J, Meier-Ewert S, Lehrach H, et al. 1998. *An algorithm for clustering cDNAs for gene expression analysis.* Presented at Hum. Genome Meet. 1998. Torino, Italy

63. Hawkins V, Doll D, Bumgarner R, Smith T, Abajian C, et al. 1998. PEDB: the Prostate Expression Database. *Nucleic Acids Res.* 27:204–8

64. Heyer LJ, Kruglyak S, Yooseph S. 1999. Exploring expression data: identification and analysis of coexpressed genes. *Genome Res.* 9:1106–15

65. Hilsenbeck SG, Friedrichs WE, Schiff R, O'Connell P, Hansen RK, et al. 1999. Statistical analysis of array expression data as applied to the problem of tamoxifen resistance. *J. Natl. Cancer Inst.* 91:453–59

66. Hughes TR, Mao M, Jones AR, Burchard J, Marton MJ, et al. 2001. Expression profiling using microarrays fabricated by an ink-jet oligonucleotide synthesizer. *Nat. Biotechnol.* 19:342–47

67. Hunkapiller MW. 1991. Advances in DNA sequencing technology. *Curr. Opin. Genet. Dev.* 1:88–92

68. Ideker T, Thorsson V, Ranish J, Christmas R, Buhler J, et al. 2001. Integrated genomic and proteomic analyses of a systematically perturbed metabolic network. *Science* 292:929–934

69. Ideker T, Thorsson V, Siegel A, Hood L. 2000. Testing for differentially expressed genes by maximum likelihood analysis of microarray data. *J. Comput. Biol.* 7:805–17

70. Ideker TE, Thorsson V, Karp RM. 2000. Discovery of regulatory interactions through perturbation: inference and experimental design. *Pac. Symp. Biocomput.* 5:305–16

71. Iyer VR, Eisen MB, Ross DT, Schuler G, Moore T, et al. 1999. The transcriptional program in the response of human fibroblasts to serum. *Science* 283:83–87

72. Johnston M, Carlson M. Regulation of carbon and phosphate utilization. 1992. *The Molecular and Cellular Biology of the Yeast Saccharomyces*, ed. E Jones, J Pringle, J Broach, Vol. 2. New York: Cold Spring Harbor Lab. Press

73. Karp PD, Riley M, Saier M, Paulsen IT, Paley SM, et al. 2000. The EcoCyc and MetaCyc databases. *Nucleic Acids Res.* 28:56–59

74. Kauffman SA. 1993. *The Origins of Order: Self Organization and Selection in Evolution.* New York: Oxford Univ. Press

75. Kohn KW. 1999. Molecular interaction map of the mammalian cell cycle control and DNA repair systems. *Mol. Biol. Cell* 10:2703–34

76. Lai K, Elsas LJ. 2000. Overexpression of human UDP-glucose pyrophosphorylase rescues galactose-1-phosphate uridyltransferase-deficient yeast. *Biochem. Biophys. Res. Commun.* 271:392–400

77. Lander ES. 1996. The new genomics: global views of biology. *Science* 274:536–39

78. Lander ES, Linton LM, Birren B, Nusbaum C, Zody MC, et al. 2001. Initial sequencing and analysis of the human genome. *Nature* 409:860–921

79. Lee CK, Klopp RG, Weindruch R, Prolla TA. 1999. Gene expression profile of aging and its retardation by caloric restriction. *Science* 285:1390–93

80. Liang S, Fuhrman S, Somogyi S. 1998. REVEAL, a general reverse engineering

algorithm for inference of genetic network architectures. *Pac. Symp. Biocomput.* 3:18–29

81. Lin X, Kaul S, Rounsley S, Shea TP, Benito MI, et al. 1999. Sequence and analysis of chromosome 2 of the plant *Arabidopsis thaliana*. *Nature* 402:761–68

82. Lohr D, Venkov P, Zlatanova J. 1995. Transcriptional regulation in the yeast GAL gene family: a complex genetic network. *FASEB J.* 9:777–87

83. Longtine MS, McKenzie A, Demarini DJ, Shah NG, Wach A, et al. 1998. Additional modules for versatile and economical PCR-based gene deletion and modification in *Saccharomyces cerevisiae*. *Yeast* 14:953–61

84. Manduchi E, Grant GR, McKenzie SE, Overton GC, Surrey S, et al. 2000. Generation of patterns from gene expression data by assigning confidence to differentially expressed genes. *Bioinformatics* 16:685–98

85. Marcotte EM, Pellegrini M, Thompson MJ, Yeates TO, Eisenberg D. 1999. A combined algorithm for genome-wide prediction of protein function. *Nature* 402:83–86

86. Marnellos G, Mjolsness E. 1998. A gene network approach to modeling early neurogenesis in *Drosophila*. *Pac. Symp. Biocomput.* 3:30–41

87. McAdams HH, Arkin A. 1997. Stochastic mechanisms in gene expression. *Proc. Natl. Acad. Sci. USA* 94:814–19

88. McAdams HH, Shapiro L. 1995. Circuit simulation of genetic networks. *Science* 269:650–56

89. McGuire AM, Hughes JD, Church GM. 2000. Conservation of DNA regulatory motifs and discovery of new motifs in microbial genomes. *Genome Res.* 10:744–57

90. Mendes P. 1997. Biochemistry by numbers: simulation of biochemical pathways with Gepasi 3. *Trends Biochem. Sci.* 22:361–63

90a. Mewes HW, Heumann K, Kaps A, Mayer K, Pfeiffer F, et al. 1999. MIPS: a database for genomes and protein sequences. *Nucleic Acids Res.* 27:44–48

91. Nasevicius A, Ekker SC. 2000. Effective targeted gene *knockdown* in zebrafish. *Nat. Genet.* 26:216–20

92. Ogata H, Goto S, Sato K, Fujibuchi W, Bono H, et al. 1999. KEGG: Kyoto Encyclopedia of Genes and Genomes. *Nucleic Acids Res.* 27:29–34

93. Pillage L. 1994. *Electronic Circuit and System Simulation Methods.* New York: McGraw Hill

94. Pirson I, Fortemaison N, Jacobs C, Dremier S, Dumont JE, et al. 2000. The visual display of regulatory information and networks. *Trends Cell Biol.* 10:404–8

95. Pulavarthi P, Chiang R, Altman RB. 2000. Generating interactive molecular documentaries using a library of graphical actions. *Pac. Symp. Biocomput.* 5:266–77

96. Ringwald M, Eppig JT, Kadin JA, Richardson JE. 2000. GXD: a Gene Expression Database for the laboratory mouse: current status and recent enhancements. The Gene Expression Database Group. *Nucleic Acids Res.* 28:115–19

97. Roth FP, Hughes JD, Estep PW, Church GM. 1998. Finding DNA regulatory motifs within unaligned noncoding sequences clustered by whole-genome mRNA quantitation. *Nat. Biotechnol.* 16:939–45

98. Rowen L, Lasky S, Hood L. 1999. Deciphering genomes through automated large-scale sequencing. In *Methods in Microbiology*, ed. AG Craig, JD Hoheisel, pp. 155–91. San Diego, CA: Academic

99. Salamonsen W, Mok KY, Kolatkar P, Subbiah S. 1999. BioJAKE: a tool for the creation, visualization and manipulation of metabolic pathways. *Pac. Symp. Biocomput.* 4:392–400

100. Schaff J, Fink CC, Slepchenko B, Carson JH, Loew LM. 1997. A general computational framework for modeling cellular structure and function. *Biophys. J.* 73:1135–46

101. Schaff JC, Slepchenko BM, Loew LM.

2000. Physiological modeling with virtual cell framework. *Methods Enzymol.* 321:1–23

102. Schena M, Shalon D, Davis RW, Brown PO. 1995. Quantitative monitoring of gene expression patterns with a complementary DNA microarray. *Science* 270:467–70

103. Schilling CH, Letscher D, Palsson BO. 2000. Theory for the systemic definition of metabolic pathways and their use in interpreting metabolic function from a pathway-oriented perspective. *J. Theor. Biol.* 203:229–48

104. Schilling CH, Palsson BO. 1998. The underlying pathway structure of biochemical reaction networks. *Proc. Natl. Acad. Sci. USA* 95:4193–98

105. Schwikowski B, Uetz P, Fields S. 2000. A network of protein-protein interactions in yeast. *Nat. Biotechnol.* 181257–61

106. Selkov E Jr, Grechkin Y, Mikhailova N, Selkov E. 1998. MPW: the Metabolic Pathways Database. *Nucleic Acids Res.* 26:43–45

107. Shelton DN, Chang E, Whittier PS, Choi D, Funk WD. 1999. Microarray analysis of replicative senescence. *Curr. Biol.* 9:939–45

108. Sherlock G. 2000. Analysis of large-scale gene expression data. *Curr. Opin. Immunol.* 12:201–5

109. Shoemaker DD, Lashkari DA, Morris D, Mittmann M, Davis RW. 1996. Quantitative phenotypic analysis of yeast deletion mutants using a highly parallel molecular bar-coding strategy. *Nat. Genet.* 14:450–56

110. Shoemaker DD, Schadt EE, Armour CD, He YD, Garrett-Engele P, et al. 2001. Experimental annotation of the human genome using microarray technology. *Nature* 409:922–27

111. Sinha S, Tompa M. 2000. A statistical method for finding transcription factor binding sites. *Proc. Int. Conf. Intell. Syst. Mol. Biol., 8th, La Jolla, California, 2000*, 8:344–54. Menlo Park: AAAI Press

112. Smith LM, Sanders JZ, Kaiser RJ, Hughes P, Dodd C, et al. 1986. Fluorescence detection in automated DNA sequence analysis. *Nature* 321:674–79

113. Somogyi R, Sniegoski C. 1996. Modeling the complexity of genetic networks: understanding multigenic and pleiotropic regulation. *Complexity* 1:45–63

114. Spellman PT, Sherlock G, Zhang MQ, Iyer VR, Anders K, et al. 1998. Comprehensive identification of cell cycle–regulated genes of the yeast *Saccharomyces cerevisiae* by microarray hybridization. *Mol. Biol. Cell* 9:3273–97

115. Spiro PA, Parkinson JS, Othmer HG. 1997. A model of excitation and adaptation in bacterial chemotaxis. *Proc. Natl. Acad. Sci. USA* 94:7263–68

116. Stoeckert CJ Jr, Salas F, Brunk B, Overton GC. 1999. EpoDB: a prototype database for the analysis of genes expressed during vertebrate erythropoiesis. *Nucleic Acids Res.* 27:200–3

117. Tamames J, Ouzounis C, Casari G, Sander C, Valencia A. 1998. EUCLID: automatic classification of proteins in functional classes by their database annotations. *Bioinformatics* 14:542–43

118. Tamayo P, Slonim D, Mesirov J, Zhu Q, Kitareewan S, et al. 1999. Interpreting patterns of gene expression with self-organizing maps: methods and application to hematopoietic differentiation. *Proc. Natl. Acad. Sci. USA* 96:2907–12

119. Tavazoie S, Hughes JD, Campbell MJ, Cho RJ, Church GM. 1999. Systematic determination of genetic network architecture. *Nat. Genet.* 22:281–85

120. Termonia Y, Ross J. 1998. Oscillations and control features in glycolysis: numerical analysis of a comprehensive model. *Proc. Natl. Acad. Sci. USA* 78:2952–56

121. The *C. elegans* Sequencing Consortium. 1998. Genome sequence of the nematode *C. elegans*: a platform for investigating biology. *Science* 282:2012–18

122. Thomas J, Milward D, Ouzounis C, Pulman S, Carroll M. 2000. Automatic

extraction of protein interactions from scientific abstracts. *Pac. Symp. Biocomput.* 5:541–52

123. Tomita M, Hashimoto K, Takahashi K, Shimizu TS, Matsuzaki Y, et al. 1999. E-CELL: software environment for whole-cell simulation. *Bioinformatics* 15:72–84

124. Turing AM. 1952. The chemical basis of morphogenesis. *Proc. R. Philos. Soc. B* 237:37–72

125. Venter JC, Adams MD, Myers EW, Li PW, Mural RJ, et al. 2001. The sequence of the human genome. *Science* 291:1304–51

126. Von Dassow G, Meir E, Munro EM, Odell GM. 2000. The segment polarity network is a robust developmental module. *Nature* 406:188–92

127. Wach A, Brachat A, Alberti-Segui C, Rebischung C, Philippsen P. 1997. Heterologous HIS3 marker and GFP reporter modules for PCR-targeting in *Saccharomyces cerevisiae*. *Yeast* 13:1065–75

128. Walhout AJ, Temple GF, Brasch MA, Hartley JL, Lorson MA, et al. 2000. GATEWAY recombinatorial cloning: application to the cloning of large numbers of open reading frames or ORFs. *Methods Enzymol.* 328:575–92

129. Wang K, Gan L, Jeffrey E, Gayle M, Gown AM, et al. 1999. Monitoring gene expression profile changes in ovarian carcinomas using a cDNA microarray. *Gene* 229:101–8

130. Weaver DC, Workman CT, Stormo GD. 1999. Modeling regulatory networks with weight matrices. *Pac. Symp. Biocomput.* 4:112–23

131. Wen X, Fuhrman S, Michaels GS, Carr DB, Smith S, et al. 1998. Large-scale temporal gene expression mapping of central nervous system development. *Proc. Natl. Acad. Sci. USA* 95:334–39

132. Weng G, Bhalla US, Iyengar R. 1999. Complexity in biological signaling systems. *Science* 284:92–96

133. Wingender E, Chen X, Hehl R, Karas H, Liebich I, et al. 2000. TRANSFAC: an integrated system for gene expression regulation. *Nucleic Acids Res.* 28:316–19

134. Winzeler EA, Shoemaker DD, Astromoff A, Liang H, Anderson K, et al. 1999. Functional characterization of the *S. cerevisiae* genome by gene deletion and parallel analysis. *Science* 285:901–6

135. Wuensche A. 1995. Discrete Dynamics Laboratory (DDLAB). http://www.ddlab.com/. Santa Fe, NM: Discrete Dynamics, Inc.

136. Wuensche A. 1998. Genomic regulation modeled as a network with basins of attraction. *Pac. Symp. Biocomput.* 3:89–102

137. Xenarios I, Fernandez E, Salwinski L, Duan XJ, Thompson MJ, et al. 2001. DIP: the database of interacting proteins. *Update. Nucleic Acids Res.* 29:239–41

138. Yi TM, Huang Y, Simon MI, Doyle J. 2000. Robust perfect adaptation in bacterial chemotaxis through integral feedback control. *Proc. Natl. Acad. Sci. USA* 97:4649–53

139. Yuh CH, Bolouri H, Davidson EH. 1998. Genomic *cis*-regulatory logic: experimental and computational analysis of a sea urchin gene. *Science* 279:1896–902

140. Yuh CH, Bolouri H, Davidson EH. 2001. *Cis*-regulatory logic in the *endo16* gene: switching from a specification to a differentiation mode of control. *Development* 128:617–29

141. Zhang MQ. 1999. Large-scale gene expression data analysis: a new challenge to computational biologists. *Genome Res.* 9:681–88

142. Zhu J, Zhang MQ. 1999. SCPD: a promoter database of the yeast *Saccharomyces cerevisiae*. *Bioinformatics* 15:607–11

Annu. Rev. Genomics Hum. Genet. 2001. 2:373–400

THE GENOMICS AND GENETICS OF HUMAN INFECTIOUS DISEASE SUSCEPTIBILITY

Adrian V.S. Hill

Wellcome Trust Centre for Human Genetics, University of Oxford, Oxford OX3 7BN, United Kingdom; e-mail: adrian.hill@well.ox.ac.uk

■ **Abstract** A genetic basis for interindividual variation in susceptibility to human infectious diseases has been indicated by twin, adoptee, pedigree, and candidate gene studies. This has led to the identification of a small number of strong genetic associations with common variants for malaria, HIV infection, and infectious prion diseases. Numerous other genes have shown less strong associations with these and some other infectious diseases, such as tuberculosis, leprosy, and persistent hepatitis viral infections. Many immunogenetic loci influence susceptibility to several infectious pathogens. Recent genetic linkage analyses of measures of infection as well as of infectious disease, including some genome-wide scans, have found convincing evidence of genetic linkage to chromosomal regions wherein susceptibility genes have yet to be identified. These studies indicate a highly polygenic basis for susceptibility to many common infectious diseases, with some emerging examples of interaction between variants of specific polymorphic host and pathogen genes.

INTRODUCTION

Analysis of the genetic basis of susceptibility to major infectious diseases is potentially the most complex area in the genetics of complex disease. Not only are these highly polygenic diseases with important, if not overwhelming, genetic components, but there is well documented interpopulation heterogeneity; and in all cases, one essential required environmental factor with, almost always, its own genome in play. Nonetheless, steady progress is being made in untangling the complex interplay of host genes and microorganism that results in some striking interindividual variation in susceptibility.

This is one of the oldest areas of complex disease genetics in humans, with one major susceptibility locus for malaria analyzed almost 50 years ago (9). Since the 1930s, several twin studies have supported a substantial role for host genetics in variable susceptibility to tuberculosis (35, 43, 78), leprosy (33), *Helicobacter pylori* infection (93), and hepatitis B virus persistence (88). Early reports in which malaria (71) and tuberculosis microbes (69) were deliberately or accidentally administered to large numbers of nonimmunes have documented clear variation in susceptibility to these pathogens. A large adoptee study has also supported the importance of host genetic factors in susceptibility to fatal infectious diseases in Northern Europeans (152). However, good estimates of the increase in risk to siblings of affected individuals compared to the general population (the λ_s value) are generally lacking for infectious diseases; and where available, it may be difficult to dissect the genetic from the environmental contributors to this value.

In the 1980s, HLA analysis was added to hematologic candidate gene studies of malaria; and in the 1990s, more and more non-MHC candidate genes began to be assessed. In recent years, the first genome-wide linkage studies have been reported, and the utility of this approach is now clear. This review begins with consideration of these genetic linkage studies before tackling the large number of reports of association studies with candidate genes. The field has now expanded to the point where no review article of this size can hope to be comprehensive. Priority is given to papers published in the last three years; but even among these, there is inevitably some personal selection. Also excluded from consideration is the large literature on monogenic disorders that give rise to immunodeficiency and infectious disease susceptibility. These disorders, like studies of gene knockout mice and susceptibility gene mapping in other species, may identify important candidate genes for analysis in common infectious diseases; but as these disorders are invariably rare, the analytic approaches involved are different. To date, there is surprisingly little overlap between the loci involved in monogenic immunodeficiencies and those implicated as common susceptibility loci.

Analysis of infectious disease susceptibility has long been of interest to evolutionary biologists, and the debate over the extent to which MHC polymorphism may have been driven by infectious pathogens is well rehearsed. Two further objectives have become more prominent in recent years. The first, in common with much of genomics, is the identification on new pathways of pathogenesis or resistance that may eventually lead to new prophylactic of therapeutic agents for these infections. For examples, industrial interest in blockers of the chemokine receptor, CCR5, has been encouraged by some striking genetic findings, and the design of investigational vaccines for malaria has been influenced by HLA association studies. The second is the potential for genetics studies to facilitate targeting of therapeutic or prophylactic interventions. For example, IL-10 genotypes might be of value in choosing which patients with chronic hepatitis should receive α-interferon therapy (46), or mannose-binding lectin-deficient individuals could be prioritized for pneumococcal vaccination. However, for a more useful assessment of risk profile, many loci will likely need to be evaluated.

GENETIC LINKAGE STUDIES

Some of the first family studies of the genetics of complex disease searched for linkage and association of the HLA region with the mycobacterial diseases leprosy and tuberculosis. Whole genome scans have more recently been undertaken, initially for the phenotype of parasite burden and subsequently for disease manifestations.

Schistosomiasis

Complex segregation analysis of Brazilian families with *Schistosoma mansoni* infection provided evidence that a major gene could determine susceptibility to high parasite burdens in this region. Careful estimation of the major environmental risk factor, water contact, for this infection allowed this variable to be included in the genetic analysis. A whole genome scan of 11 families with 246 microsatellite markers found a single region of linkage on chromosome 5q31-q33 (96). A parametric linkage analysis in these families revealed a multipoint lod score in excess of 4.5, mainly accounted for by two extended pedigrees (95). This region of chromosome 5q contains a cluster of cytokine and other immunologically important loci, particularly those for the TH2-type cytokines IL-4, IL-9, and IL-13 and also the colony stimulating factor-1 receptor. Immunological studies have suggested a key protective role of TH2-type immune responses and IgE antibodies in protection against this parasitic disease. Further evidence for a role for this chromosomal region in susceptibility to *S. mansoni* was found in a study of West African families, suggesting that the relevant susceptibility loci are not limited to a small number of Brazilian pedigrees (117).

The Dessein and Abel group have also studied disease manifestations resulting from *S. mansoni* infection (42). An analysis of hepatitic periportal fibrosis in Sudanese families employed initially complex segregation analysis that provided evidence of a codominant major gene with a susceptibility allele frequency of 0.16. Subsequent linkage analysis of a subset of these pedigrees assessed four candidate chromosomal regions for evidence of linkage. A lod score of 3.12 was reported for the 6q22-q23 region that includes the interferon-γ-receptor-1 gene. The antifibrogenic role of γ interferon suggests a potential mechanism for the linkage, but association with this gene has not been reported.

Malaria

The success of case-control studies in identifying numerous loci associated with resistance to malaria suggests that this disease is highly polygenic and that genome-wide studies for new susceptibility loci should be worthwhile. Jepson et al. found evidence of linkage of clinical malaria to the MHC in a study of Gambian children (75), and her twin studies indicated a major role for non-MHC loci in determining some cellular immune responses to the malaria parasite (74). Studies in Burkina-Faso, West Africa, have shown consistent parasitological, clinical, and immunological differences between ethnic groups in *Plasmodium falciparum* infection rates, malaria morbidity, and prevalence and levels of antibodies to various *P. falciparum* antigens, and these could not be ascribed to known susceptibility loci (114, 115). Yet, no genome-wide linkage analyses of malarial disease have been reported. This in part reflects operational difficulties. Although hospital-based studies of severe malaria cases and matched controls have frequently been undertaken in regions of hyperendemic malaria, recruitment of affected sibling pairs with severe malaria is more challenging for an acute infectious disease. Often hospital records will not allow reliable ascertainment of another sibling who has had strictly defined severe malaria in earlier years. Recruitment of sibling pairs with nonsevere clinical malaria is more feasible; but in general, clearer genetic associations have been observed with severe malaria than with the much more prevalent phenotype of clinical malaria.

However, progress has been made in analysis of another malarial phenotype, that of *P. falciparum* parasite density. In highly endemic regions, with age individuals acquire substantial anti-disease immunity to *P. falciparum* that allows them to tolerate high peripheral blood parasite densities with often minimal or no clinical manifestations. Studies have now addressed the genetic regulation of this level of parasite density by frequent sampling of populations in endemic areas and family segregation and genetic linkage analysis. An early complex segregation analysis of 42 Cameroonian families suggested a single strong susceptibility locus (1), but this was not apparent in a larger study of families from Burkina Faso where a marked age effect was noted (131). Linkage analysis of four candidate regions in nine Cameroonian families showed some weak evidence of linkage to the 5q31-33 chromosome regions (53). Analysis of 154 sibs from 34 Burkina Faso families for 5q31-33 markers using nonparametric analysis showed stronger evidence of linkage (P < 0.001), supporting the view that a gene or genes in this region may influence parasite density. Together with evidence for linkage to this interval in asthma and atopy (173), the reported linkage studies of both schistosome and malarial parasite density justify further detailed analysis of the chromosome 5q11-13 region to search for associated and potentially causative loci.

Mycobacterial Diseases

Mycobacterial diseases were among the first to show evidence of both linkage and association with the HLA region, and early examples of the use of a variant of the increasingly popular transmission disequilibrium test may be found in these studies (40, 148, 162). More recently, family linkage analysis was used in the identification of the interferon-γ-receptor-1 gene as the susceptibility locus for rare familial susceptibility to usually nonpathogenic atypical mycobacteria and the BCG vaccine (77, 121). Mutations in the IL-12 receptor beta-1 gene have more recently been causally associated with this phenotype and with monogenic susceptibility to *Salmonella* infections (11, 76). In a study of an extended Canadian aboriginal family with unusually marked susceptibility to tuberculosis, linkage to the chromosome 2q35 region was identified (59). However, linkage analysis of this region in families from other regions with other mycobacterial phenotypes has provided mixed and generally negative results (see below) (3, 7, 18, 26, 134, 144, 167).

The first whole genome scan for an infectious disease, tuberculosis, was reported by Bellamy and colleagues (18, 23). In the first stage of this two-stage study,

92 African affected sibling pairs with tuberculosis were studied, mainly families from Gambia (23). Several chromososomal regions, including the MHC, initially showed weak evidence of linkage, and these were reassessed in a second-stage study. In this stage, 81 predominantly South African sibling pairs were studied because of the limited availability of more West African sibling pairs. Nonetheless, two of these chromosomal regions, around bands 15q11 and Xq27, again showed evidence of linkage, with overall lod scores of the order of 2.0 (18). The location of susceptibility genes at these chromosomal regions was further supported by an independent analysis, employing common ancestry using microsatellites mapping. In this variant of homozygosity mapping, chromosomal regions are assessed for increased homozygosity in cases that are compared to controls using Goldstein's genetic distance as a sensitive measure of inbreeding at each locus (18). Several positional candidate genes in these regions are under investigation and linkage disequilibrium mapping is being used to search for the putative susceptibility loci. The X-chromosome linkage may relate to the observation in Africa and other continents that clinical tuberculosis is more frequently found in males than females.

Negative findings in this genome-wide analysis are also of interest. Despite speculation on the potential role of NRAMP1 as a major susceptibility locus for tuberculosis in humans, no support has been found for this possibility in the linkage analysis of these African families. Although there is a clear association of clinical tuberculosis with the HLA class II region in several Asian studies, significant linkage to the MHC was not found in the final analysis of these African families. Although this lack of linkage does not exclude susceptibility genes in these chromosomal regions, it does limit the potential magnitude of their effects. Finally, placed alongside the estimates of the magnitude of a host genetic effect suggested for tuberculosis by early twin studies (35, 43, 78), the lack of a clear major locus for tuberculosis in these African families suggests that much or most of the genetic component, at least in Africans, may be dispersed among many loci, with no locus or chromosomal region sufficiently important to show clear linkage. This inference, together with the evidence from candidate gene studies of malaria (63) that show that susceptibility to this disease is also highly polygenic, raised the possibility that, in general, susceptibility to major infectious disease might be too polygenic for major loci to be mapped convincingly using the available genome-wide linkage strategies. Fortunately, other data soon contradicted this negative view.

Leprosy is one of the infectious diseases most amenable to genetic linkage analysis. In a study of approximately 250 affected sibling pairs from South India, it was possible to recruit almost all the parents in these families, and a genome scan using almost 400 microsatellite markers was undertaken (146). One area of strong linkage with a lod score over 4.0 was identified on chromosome 10p13. In contrast, the MHC showed only weak evidence of linkage, despite evidence of association with HLA-DR2 within these families. There are several positional candidate genes within the chromosome 10 region of linkage, and further fine mapping studies are in progress. Estimation of the locus-specific sibling risk in these families suggested that this region may account for a substantial proportion of the overall genetic component in this geographic region. The latter was estimated indirectly in a different South Indian population (60, 132). Nonetheless, this study provides the clearest indication yet that genome scans can provide a useful approach to major gene identification in a major infectious disease, despite the polygenic nature of these diseases.

Viral Diseases

HIV infection is now frequently investigated by human geneticists, but the rarity of multicase families has generally prevented useful linkage analyses. An exception is an HLA study of 95 HIV-infected hemophiliac brother-pairs where HLA concordant sibling pairs and those sharing one but not zero HLA haplotypes were significantly concordant in their rate of CD4 T-cell number decline (85).

One of the clearest dichotomies in the response to an infectious pathogen is the ability of most, but not all, individuals infected by the hepatitis B virus to clear this infection. In most populations, 5%–20% of individuals fail to clear the virus and develop a chronic carrier state that substantially increases their risk of chronic liver disease and hepatocellular carcinoma. A small Taiwanese twin study provided some evidence of a host genetic influence of viral clearance (88). Whole genome scans to examine the phenotype of persistent HBV infection have now been undertaken in both Gambian and Italian populations. In the European families, there was evidence of linkage to chromosome 6q (50). The West African study assessed almost 200 Gambian sibling pairs and identified a linkage (lod score > 3.5) to a region on chromosome 21 that encodes numerous cytokine receptor genes (L. Zhang, A. Frodsham, U. Dumpis, S. Best, A. Hall, H. Whittle, B. Hennig, S. Hellier, M. Thursz, H. Thomas, & A. Hill, unpublished data). Preliminary evidence of association with a variant of one of these genes has been found in these families, and the availability of a full sequence of this chromosome should facilitate further analysis. Interestingly, it has been recognized for over 30 years that individuals with trisomy 21 have a higher prevalence of chronic HBV infection (28), suggesting that the same gene or genes on chromosome 21 may underlie this finding and the linkage result.

CANDIDATE GENE STUDIES

Human Leukocyte Antigens

Studies of HLA and malaria in Gambia helped to establish the view that natural selection by infectious disease has contributed to the maintenance of the remarkable allelic diversity of HLA class I and II loci (64, 65). Gilbert et al. (57) have extended these studies to assess association of malaria parasite variants with HLA class I type. When parasites were divided into strains according to allelic types of an HLA class I restricted epitope in the major coat protein of the sporozoite, an association was observed between parasite type and the relevant HLA class I molecule, HLA-B*35. A remarkable nonrandom distribution of parasite allelic type, termed cohabitation, was documented in which parasite types that could mutually antagonize CD8 T-cell responses in primary (129) and effector T-cell assays (57) were found together more frequently than expected in mixed parasite infections. A mathematical model was employed that supported the inference that the immunological mechanisms documented in vitro may be maintaining the nonrandom distribution of parasite types through HLA class I restricted T-cell antiparasite responses. This study of coevolution provides insight into the potentially powerful influences that HLA restricted responses may exert on pathogen diversity and population structure and suggests that further analyses of host HLA and pathogen diversity in the same sample sets should be instructive.

Some further evidence of HLA association with the rate of progress of HIV infection has been provided by several recent studies. HLA-B*5701 was found in 11 of 13 long-term nonprogressors with low viral loads, but only 10% of controls (111). In a study of large U.S. cohorts, HLA-B*35 and Cw*04 were associated with rapid progression to AIDS-defining illnesses (32). In a powerful study of 75 rapid progressors and 200 long-term nonprogressors, representing the extremes of this spectrum, a variety of alleles were associated with protection and susceptibility (60a). HLA-A29 and -B22 were significantly associated with rapid progression; whereas B14, C8, and, though less strongly, B27, B57, and C14 were protective. Interestingly, in contrast to other chronic viral diseases, such as persistent hepatitis B and C infection, HLA class II associations have been less evident in these studies of HIV/AIDS. In contrast, in a rare study of susceptibility to HIV/AIDS in Africans (92), HLA-A2 related subtypes were associated with resistance to disease progression, and the class II type HLA-DR1 was associated with resistance to HIV infection in Nairobi commercial sex workers.

Further studies of Europeans (156, 166) have supported previous findings (10, 113, 126, 175) that the linked HLA class II alleles, HLA-DRB1*11 and HLA-DQB1*0301, are associated with resistance to persistent HCV infection; this is now probably the most consistently documented HLA association with a viral disease. In a cohort study of infection with the HTLV-1 retrovirus, HLA-A2 was strongly associated with a reduced risk of developing HTLV-associated myelopathy, and viral load also reduced in individuals with this class I type (73). Other types, HLA-B*54, Cw*08, and DRB1*0101, showed less strong associations (72). In this study and the Carrington et al. study of HIV infection in American cohorts (32), the influence of HLA heterozygosity was examined. A graded protective effect of HLA-A, -B, and -C heterozygosity against disease progression to AIDS and death was found with heterozygotes at all three loci showing the slowest progression (32). In the study of HTLV-1 infection, individuals heterozygous at all three HLA class I loci had significantly reduced viral load compared to other genotypes (72). Together with an association of HLA class II heterozygosity in HBV infection (157), these data provide important support for the proposal (44) that heterozygote advantage against viral infectious disease plays some part in maintaining the polymorphism of HLA loci.

Two studies have analyzed further the well-established association of HLA-DR2 with susceptibility to tuberculosis as well as to leprosy in Indian populations (107, 130). Analysis of HLA-DR2 subtypes showed that the HLA-DRB1*1501, but not the *1502 allele, is associated with susceptibility to tuberculosis, but the HLA-DQB1*0601 allele in strong linkage disequilibrium with HLA-DRB1*1501 is also strongly associated. In a small study of tuberculosis in Vietnam, a susceptibility association with the rare HLA-DQB1*0503 allele was reported (58). Intriguingly, an HLA-DR2 association has now been reported with a third mycobacterial disease. In North American AIDS patients, HLA-DRB1*1501 was associated with an accelerated onset of disseminated *Mycobacterium avium* complex disease (86), suggesting a common mechanism underlying these three mycobacterial disease associations.

Cytokines and Their Receptors

Since the initial report of a TNF polymorphism association with cerebral malaria by Kwiatkowski & colleagues (105), there have been several studies of polymorphism in cytokines and infectious diseases. The promoter variant allele at position -308 has now been associated with cerebral malaria, mucocutaneous leishmaniasis, leprosy type, and scarring trachoma in various populations (31, 36, 105, 138). Some, but not all, of these studies analyzed flanking HLA polymorphisms to allow independent assessment of the relevance of the TNF promoter variant. Some functional analyses have found evidence of increased transcription (170) by the rarer TNF2 allele at this -308 position, but this remains controversial (4). Analysis of the other common clinical presentation of severe malaria in African children, severe malarial anemia, has demonstrated an association of the variant allele at position -238 of the TNF promoter with this phenotype (106), suggesting that these different complications of malaria infection are influenced by separate genetic factors near the TNF gene. The -238 variant has also been associated with chronic hepatitis B virus infection in Europeans (67).

Another TNF promoter variant has recently been associated with cerebral malaria (80). In this case, the polymorphism at position -376 altered binding of a transcription factor, identified as OCT-1, to that region of the promoter and resulted in altered gene expression in a human monocytic cell line. Although relatively uncommon, this variant was associated with a fourfold increase in risk of cerebral malaria after allowing for flanking polymorphisms.

Interleukin (IL)-1 genetic variation has been investigated in several autoimmune diseases, and evidence of its relevance to infectious disease susceptibility has been recently reported. A twin study has demonstrated that host genetic factors influence susceptibility to *H. pylori* infection (93), and a study of IL-1 beta polymorphisms has now found association with complications of this chronic gastric infection. Two polymorphisms in near-complete linkage disequilibrium were associated with both *H. pylori*–induced hypochlorhydria and increased risk of gastric cancer, and one of these variants is a TATA box promoter polymorphism that was found to alter DNA-protein interactions (47). Two studies have also reported preliminary evidence that polymorphism in the IL-1 beta and the flanking IL-1 receptor antagonist genes may affect either risk or clinical presentation of tuberculosis (20, 169).

Three single nucleotide polymorphisms (SNPs) in the promoter of the IL-10 gene and two flanking microsatellite polymorphisms have been investigated in several autoimmune disorders. The variants at position -1082, -819, and -592 are G to A changes, and assays of IL-10 production suggest that the GGA haplotype is associated with higher IL-10 levels (37, 159). The variants associated with higher IL-10 production are associated with clearance, rather than persistence, of hepatitis B virus in both West Africans and Europeans (L. Zhang, A. Frodsham, S. Knapp, H. Thomas, M. Thursz, & A. Hill, unpublished). A study of response to α-interferon therapy in hepatitis C infected patients also suggested that the A allele at -592, associated with lower IL-10 production, is a marker of good response to this intervention (46). In American HIV-infected cohorts, the presence of the -592 A allele was associated with more rapid disease progression, particularly late in the course of infection (145).

An SNP at position -589 of the IL-4 promoter has been investigated in several diseases, in view of its potential relevance to TH1-TH2 switching of the cellular immune response. In a study of Japanese HIV-infected individuals, homozygotes for the T allele were more likely to develop syncytium-inducing strains of HIV than other genotypes (120). These strains use the CXCR4 coreceptor, appear later in HIV infection, and their emergence represents a marker of more rapid disease progression. In this study, no association of this viral phenotype with RANTES or IL-10 promoter variants was observed.

There have been fewer reported studies of variation in cytokine receptors than cytokine genes in common infectious diseases. However, it is clear that rare inactivating mutations in the interferon-γ-receptor-1 and the IL-12Rβ1 genes are associated with susceptibility to usually nonpathogenic mycobacteria and *Salmonella* species (11, 76, 77, 121). There is, to date, no evidence from population studies that common variants in the interferon-γ-receptor genes affect susceptibility to tuberculosis, although several groups have addressed this question.

Chemokines and Their Receptors

The discovery in 1996 of the resistance to HIV infection of Caucasian homozygotes for a 32-base pair deletion in CC chemokine receptor 5 (CCR5) (41, 90, 140) has led to numerous studies of chemokine receptors, and more recently chemokine polymorphisms, in HIV infection and disease. It soon became apparent that this variant is not significantly protective against infection in the heterozygous state, but that heterozygotes manifest slower disease progression to AIDS and death, at least among homosexuals but possibly not among hemophiliacs (39, 41, 68, 109, 177).

Rare homozygotes infected by HIV-1 have been described so protection is not absolute (16). Sadly, in regions of Africa and Asia where the epidemic is most marked, the variant is essentially absent (100).

A valine-to-isoleucine change in the first transmembrane region of the flanking CCR2 gene is also associated with delayed disease progression but not altered susceptibility to HIV infection (151). This protective effect has now been confirmed in another U.S. cohort and in Swiss and British studies (45, 81, 133). In a large Texas-based study, this association was observed in African Americans, but not Caucasians, with HIV infection (118). The lack of association in another published study (110) may be due to analyses of a seroprevalent rather than a seroincident cohort (150), as this CCR2 polymorphism may be most relevant early in HIV infection. In a study of the genetics of susceptibility to HIV or AIDS in Africans, Anzala et al. found an increased frequency of the protective CCR2 allele in Kenyan long-term nonprogressors (14). A possible molecular mechanism underlying this association was suggested by the finding that in vitro CCR2 can heterodimerize with CCR5 but that the CCR2 isoleucine variant, unlike the wild type, cannot heterodimerize with CXCR4 (108).

Several studies have now found that variants in the promoter region of CCR5 also influence the rate of disease progression. An A to G change at position 59029 [in the numbering of Genbank clone U95626, equivalent to position 303 as numbered in the study by Martin & colleagues (99) as well as 2459 as numbered in the report by An et al. (13)] of the CCR5 promoter region was associated with reduced promoter activity in Jurkat cells (104). G/G homozygotes progressed to AIDS more slowly than A/A homozygotes, particularly in the absence of the 32-bp CCR5 deletion and CCR2-isoleucine variants (104). In parallel, haplotypes of eight other SNPs in the CCR5 promoter region were defined, and one of these, termed P1, was significantly associated with accelerated progression to AIDS (99). The P1 haplotype and the 59029 A allele are in very strong linkage disequilibrium in both Caucasisans and African Americans (13). An Australian study found that 59029 A/A homozygotes were less frequent among long-term nonprogressors (34), consistent with the U.S. datasets. Among African Americans, the 59029 A association with rapid progression was also observed, but in contrast to Caucasians, the effect was dominant rather than recessive (13). Evidence of further complexity in the CCR5 promotor and possible population differences in haplotypic associations was provided by an evolutionary analysis of extended haplotypes among a Texas cohort (119). Analysis of another promoter variant, a C to T change at position 590356 that is more prevalent in African Americans, showed increased perinatal transmission of HIV-1 from mothers to 21 African American offspring who were homozygous for the variant T allele (82). In a study of French HIV-infected individuals who were homozygotes for a coding change at position 280 of the fractalkine receptor, CX_3CR1, showed increased rates of progress to AIDS (48). However, analysis of U.S. cohorts failed to replicate this association and instead suggested that heterozygotes for this change might have a reduced rate of disease progression (103).

Variation in the regulatory regions of two chemokine genes has now been associated with HIV infection or disease progression. The promoter of the gene for RANTES, one of the ligands for CCR5, was sequenced in Japanese subjects and two promoter variants at positions -28 and -403 were identified. The G allele at position -28 showed increased expression in transfection assays and was associated with delayed disease progression in Japanese (89). The -403 position change has also been found to be of functional significance in in vitro assays and has been associated with atopic dermatitis, atopy, and asthma (51, 122). In the U.S. Multicenter AIDS Cohort Study, the A allele at -403 of the RANTES promoter was associated with an increased risk of HIV infection but with a slower rate of disease progression to AIDS (102). Clearly, further studies of these RANTES promoter variants are required to address this complexity. More heterogeneous results have been reported for a 3'-UTR variant in the gene for stromal-derived factor-1, SDF-1, the principal ligand for CXCR4. Winkler et al. initially reported that homozygotes for a variant A SNP were highly significantly protected from disease progression to AIDS (172). Five studies have now failed to replicate this strong protective effect, and four of them find evidence of greater disease susceptibility of 3'-UTR A/A homozygotes at this locus (15, 29, 45, 118, 163). A London cohort study showed no SDF association (45), but a Texas cohort showed that A/A homozygotes had faster progression to death (118). Balotta et al. (15) associated A/A homozygosity with low CD4 T-cell counts, Brambilla et al. (29) observed a more rapid late progression of A/A homozygotes; and in a Dutch cohort (163), homozygotes progressed more rapidly to AIDS but not to death.

NRAMP1

Genetic linkage studies in mice led to the mapping of a gene, initially termed *Lsh/Ity/Bcg*, that influences early resistance to several intramacrophage pathogens, *Leishmania donovani*, *Salmonella typhimurium*, and some strains of *M. bovis* BCG (25). This gene was positionally cloned by Gros and colleagues in Montreal and termed Nramp1 (natural-resistance–associated macrophage protein-1) (165). Cellular and molecular studies have now indicated that Nramp1 is expressed in both macrophages and neutrophils, is a transporter of divalent cations, and is localized to the phagolysosomal membrane (149). Recent studies from the Gros laboratory have suggested that Nramp1 can pump manganese ions out of the phagolysosomal space in a pH-dependent manner (70), although inward pumping has also been advocated (27). Iron and other ions may also be pumped out (17), thereby perhaps modifying mycobacterial viability.

The susceptibility allele of Nramp1 in mice bears a glycine-to-arginine substitution at position 105, leading effectively to a null phenotype (164). The clear effect of this change on susceptibility to BCG Montreal infection (149), together with some complex segregation analysis of human mycobacterial disease that suggested a major gene effect (2), led to speculation that the human homologue, NRAMP1, might be a major gene for human mycobacterial disease. It is now increasingly clear that this speculation is generally incorrect.

Recent linkage studies in mice have shown that the major gene effect evident in studies of BCG strains is not observed in challenge with *M. tuberculosis* (123), and other loci have now been mapped that do affect tuberculosis susceptibility in mice (84). Whole genome scans for tuberculosis in West and South Africans (18) and for leprosy in India (146) have not found significant linkage to the chromosome 2 region encoding the human NRAMP1 gene. However, linkage to leprosy has been reported in a small number of Vietnamese families (3) and in analyses of skin test responses (Mitsuda reactions) to leprosy antigens in Vietnamese (7). No evidence of association with NRAMP1 variants was reported in these studies. Furthermore, in a large aboriginal Canadian family with multiple affected individuals significant linkage to the NRAMP1 locus has been found (59). Interpretation of such positive linkage data is complicated by the finding that in mice a locus termed Sst1 has been mapped near to the NRAMP1 locus on chromosome 1 as a tuberculosis susceptibility locus (84).

However, several studies have now found evidence for association of NRAMP1 polymorphisms with mycobacterial disease. The largest study, undertaken in Gambians, found association with several variants, in particular a 4-bp insertion-deletion polymorphism in the 3' untranslated region (21). Recently the same variants have been associated with tuberculosis in independent studies of Koreans (139) and Japanese (52). In Bengal (136) and Mali (107a), no association was found with leprosy per se; but in the latter study, there was evidence of a possible association with leprosy type.

How may these apparently heterogeneous reports be reconciled? With the exception of the Canadian aboriginal family (59), there appears to be no evidence of a major effect of NRAMP1 or a flanking gene on general tuberculosis susceptibility. Evidence of genetic association in several tuberculosis studies exists, but the magnitude of these effects is modest and compatible with the absence of significant linkage in family studies. Indeed, the man-mouse difference may be more apparent than real, in that small differences in susceptibility to tuberculosis between the susceptible and resistant mouse strains might be missed in linkage studies. It remains possible that the associations observed, like the positive linkage data, result from linkage disequilibrium with variation in some flanking gene (97). However, similar allelic associations in Japanese and Africans and the known function of NRAMP1 make this gene still the most parsimonious culprit. A different issue is whether the associations result from a primary effect of the NRAMP1 gene on *M. tuberculosis* susceptibility. Variable degrees of exposure to environmental mycobacteria may underlie the variable efficacy of BCG vaccine against tuberculosis in different populations, and a primary effect of NRAMP1 variation on other mycobacterial infections might conceivably result in altered tuberculosis susceptibility.

Recently, associations have been reported with other diseases, including HIV infection in Columbians (98), juvenile arthritis in Latvians (142), and sarcoidosis in African Americans (94).

Vitamin D and Other Receptors

Polymorphism in the vitamin D receptor gene was initially extensively investigated in osteoporosis and other bone disorders (116). Although there are many apparently conflicting reports of associations with SNPs in the 3' region of this gene, overall it appears that, at least in some populations, variation at a Taq I site in codon 352 (with alleles denoted T and t) and at flanking sites is associated with susceptibility to reduced bone mineral density. Several studies have now suggested that this and perhaps other variants of the VDR gene may be associated with susceptibility to various infectious diseases. In a case-control study of pulmonary tuberculosis in Gambia, homozygotes for the rarer tt genotype were reduced in frequency in the cases, suggesting a protective effect (19). Some support for this conclusion was provided by a study of a Gujarati tuberculosis patient in West London where both the T allele and deficiency of 25-hydroxycholecalciferol were associated with tuberculosis (168). In vitro studies have reported that dihydroxy vitamin D is one of the few mediators identified that can lead to a reduction in the growth of *M. tuberculosis* in human macrophages (135).

However, this gene may also, or alternatively, influence the type of cellular immune response evoked by pathogens. In a case-control study of leprosy in Bengalis, the TT and tt genotypes were associated with the two polar forms of leprosy, lepromatous and tuberculoid (136). As the latter form is associated with a stronger cellular immune response to *M. tuberculosis*, the tt genotype here and in the tuberculosis studies may be modulating the predominant type of cellular immune response evoked. 1,25-dihydroxy vitamin D has been found to affect IL-12 production by macrophages and to modulate dendritic cell maturation (38, 127). In Gambians, the tt genotype was also associated with a greater rate of viral clearance in those infected by the hepatitis B virus (19), again consistent with a stronger cellular immune response. However, vitamin D stores are influenced by both diet and sunlight exposure, providing opportunities for important gene-environment interactions and suggesting that there may be substantial heterogeneity in VDR infectious disease associations between populations. Variation in the VDR gene has also been associated with susceptibility to *Mycobacterium malmoense* pulmonary disease (56), localized early-onset peridontitis (61), and, which is intriguing, to Crohn's disease (147), a granulomatous disease of the intestine for which a possible mycobacterial or other infectious etiology has been mooted.

The immunoglobulin receptor Fcγ RIIa, CD32, has a common dimorphism at position 131 where the variant with histidine has higher opsonic activity for IgG2 antibodies than the arginine 131 variant. In 1994, small clinical studies by the van de Winkel group suggested that children homozygous for the arginine variant may be at increased risk of recurrent bacterial infection (141) and meningococcal septic shock (30). An association with meningococcal disease has now been reported in a small study of Slavic children over 5 years of age (128). A possible increased risk of bacteremic pneumococcal disease associated with homozygosity for the arginine variant has also been suggested (174; S. Segal, K. Knox, D. Crook, &

A. Hill, unpublished data). If this genotype is associated with susceptibility to disease caused by several encapsulated bacteria, this would suggest that some other fairly strong positive selective pressure has been maintaining the arginine variant at high frequencies.

A proposed role for the class B scavenger receptor CD36 in the downregulation of dendritic cell activation following malaria parasite clearance raised the possibility that some common deficiency variants of this gene observed in Africans might have been selected through enhanced malaria resistance (161). However, African case-control study data have not, so far, supported this idea (6).

Mannose-Binding Lectin

Mannose-binding lectin (MBL) (also known as mannose-binding protein) is a serum collagenous lectin that has a remarkably high prevalence of alleles and genotypes that produce little or no protein as a result of mutations in codons 52, 54, or 57 of the gene (160). Heterozygotes for one or other of these variant alleles are found at frequencies in the order of 0.33 in major population groups. Homozygotes or compound heterozygotes for the variant alleles (collectively sometimes termed functional mutant homozygotes) produce very little or no MBL and, thus, have impaired opsonization of some pathogens and lack the ability to activate complement through MBL-associated serine proteases. This manifestly functional genetic variation has led to searches for MBL association with many infectious diseases.

Although initially it was proposed that MBL deficiency led to susceptibility to recurrent infections in young children, evidence supporting this is weak. Studies of children in London suggested that both heterozygotes and homozygotes may be susceptible to a variety of infections and to menigococcal disease (62, 153), but the ethnic complexity of these study populations and the unusual distribution of specific genotypes in the groups raises the possibility of significant confounding by population stratification (154). A Danish study suggested that functional mutant homozygotes show increased susceptibility to HIV infection and progress to death more rapidly following diagnosis of AIDS (55). A difficulty with these and some other studies in the literature is the surprisingly low frequency of functional mutant homozygotes in the control groups employed, sometimes as low as 1%, when the expectation from the frequency of heterozygotes is about 5%. However, a Finnish study also reported a higher frequency of MBL homozygotes in HIV-infected individuals than controls (125), and a small study of Italian children found an association of codon 54 heterozygotes with an increased rate of progression to AIDS but not with infection (12). A study in London of HIV disease progression failed to find any genotypic association (101). Finally, in an Amsterdam cohort, there was a suggestion that MBL heterozygotes might progress more slowly to AIDS and death. Overall, this heterogeneous literature fails to provide convincing evidence, as yet, that MBL genotype really influences any of these phenotypes.

Several groups have studied MBL genotypes and tuberculosis, following a suggestion that MBL deficiency might have been maintained evolutionarily by a reduced capacity of mycobacteria to invade macrophages in the absence of MBL, leading to resistance to tuberculosis (54). Small studies in India and South Africa suggested that homozygotes may be susceptible to tuberculosis (143) and that codon 54 heterozygotes may be protected from tuberculous meningitis (66), but a larger study in Gambia (22) found no genotypic association. Recently, a Mexican study of surfactant genes expressing collectins that are evolutionarily and functionally related to MBL has suggested that variation in these genes may influence tuberculosis susceptibility (49).

Evidence from a study in London that MBL codon 52 heterozygotes may be susceptible to hepatitis B virus persistence (155) was not supported by data from Gambia (22), and a weak association of MBL heterozygotes with severe malaria susceptibility seen in Gabon (91) was not found in a larger study of Gambians (22).

In an Oxford-based study of Caucasian hospital patients with invasive pneumococcal disease, we have recently found a significantly increased frequency of MBL functional mutant homozygotes in cases, suggesting a protective role for MBL in this disease (S. Roy, K. Knox, D. Crook, & A. Hill, unpublished). But protective associations with MBL deficiency genotypes have yet to be identified. Thus, the enigma of why multiple MBL deficiency alleles are so prevalent remains.

Hemoglobins and Blood Groups

There has been evidence for many years that heterozygotes for sickle hemoglobin and for β thalassemia enjoy protection from severe malaria, but data on the other common hemoglobin variant in Africa, hemoglobin C, have been lacking. This is distributed more focally than hemoglobin S in West Africa, and a Dogon population from Mali has now been studied (5). Substantial protection against severe cerebral malaria was documented with an odds ratio of 0.14, suggesting that the level of protection afforded by hemoglobin C may be very substantial.

Lack of the Duffy blood group on red blood cells, manifest as the FY (a-b-) phenotype, is associated with complete protection against *Plasmodium vivax* malaria (112). This chemokine receptor gene that encodes the Duffy antigen has a point mutation in its promoter, which prevents erythroid expression (158), and this FY*A null allele is present at frequencies of almost 100% in most sub-Saharan African populations but is rare or absent in Caucasians. Zimmerman et al. (178) have now identified a FY*A null allele at low frequencies in Papua New Guinea and present preliminary evidence that the rate of *P. vivax* infection may be lower in heterozygotes for this variant.

The inability to secrete blood groups' substances into saliva and at other mucosal surfaces was one of the earliest human genetic markers studied and is determined by null alleles of the fucosyltransferase-2 gene (79). Nonsecretors make up 15%–25% of major population groups and may be at increased risk of bacterial urinary tract infections. A study of Senegalese commercial sex workers has found that nonsecretors were at lower risk of HIV-1 infection (8), supporting a previous study of a heterosexual HIV transmission (24).

Prion Protein Gene

One of the strongest genetic associations described with an infectious disease is for the prion protein gene (PRNP). About 50% of Caucasians are heterozygous for methionine and valine at position 129 of this gene, and these individuals are strongly protected against sporadic Creutzfeld-Jacob disease (CJD) (124) and new-variant CJD (176), which resulted from the bovine spongiform encephalopathy epidemic. Methionine homozygotes are at greater risk of sporadic CJD than valine homozygotes (171), and these two genotypes are associated with subtly different clinicopathologic phenotypes (83). Recently, it has been possible to genotype kuru cases and matched controls from the Fore tribe of the New Guinea highlands where the original epidemic of this transmissible spongiform encephalopathy was described. Methionine homozygotes show the highest risk of disease and manifest a shorter incubation period than other genotypes, raising the possibility that more heterozygotes will be identified in the current British new variant CJD epidemic as time progresses (87).

CONCLUDING REMARKS

There has been a clear upturn in the amount of activity in this field in recent years, driven by the greater ease of genotyping and new analytical approaches. Larger and more realistic sample sizes are becoming common because of technical improvements. Although, as ever, there are apparently inconsistent reports from different laboratories, usually studying very different populations, some important areas of consensus are clear. There are some clear and repeatable genetic associations with particular diseases. The existence of HLA associations with several infectious diseases now appears beyond dispute, particularly with leprosy, tuberculosis, persistent hepatitis, HIV and HTLV-1, and malaria. Evidence for the importance of heterozygote advantage in maintaining HLA polymorphism is growing. The associations of variation in the chemokine receptor genes, CCR5 and CCR2, and altered rate of HIV disease progression are now very well supported, but evidence of chemokine associations is more preliminary. Several other loci are credibly associated with malaria, tuberculosis, and pneumococcal disease manifestations. But the great majority of these associations to date are of modest effect.

The exceptions are worth noting. Hemoglobin S and possibly hemoglobin C provide a very substantial reduction in risk of severe malaria in the heterozygous state. The absence of the Duffy blood group on red blood cells is associated with complete resistance to vivax malaria. Homozygosity for the CCR5 32-bp deletion is very substantially protective against HIV infection, and heterozygotes at position 129 of the prion protein gene hardly ever develop Creutzfeld-Jacob disease. Interestingly, these strong associations are all disease specific, whereas the many immunogenetic loci that have smaller more modulatory effects (e.g., HLA,

VDR, CD32) are often associated with multiple infectious diseases. All four of these loci were identified as candidate genes, but increasing use of genome-wide linkage analysis of multicase families suggests that new major susceptibility loci should soon emerge from this approach. Although this is clearly a more demanding approach than candidate gene analysis, progress, so far, with linkage studies of several infectious diseases has been encouraging.

In the near future, two new approaches will become of increasing importance. The first will be the rise of detailed linkage disequilibrium mapping, resulting from the availability of huge numbers of new SNPs in all areas of the genome. This will initially allow much more precise mapping of known associations and linkages and eventually lead to genome-wide association studies. The potential of the latter has been extensively discussed in the field of complex disease in general, and the apparently highly polygenic nature of common infectious disease suggests that this approach may be particularly fruitful in this arena. The other approach that is already attracting more interest, that of combined host-parasite genetic analysis, is also fuelled by genomic information. Viral genome sequences have been available for some years, bacterial genomes, such as that of *M. tuberculosis* and the meningococcus, are newly available, and those of larger parasites, such as *P. falciparum*, will be available in the near future. In diseases where it is readily possible to sample the genomes of both host and pathogen simulataneously, such as for HIV, malaria, and many other infections, this should lead to new combined analytical approaches that may reveal much about the nature of evolutionary driving forces for host and parasite genetic diversity.

Visit the Annual Reviews home page at www.AnnualReviews.org

LITERATURE CITED

1. Abel L, Cot M, Mulder L, Carnevale P, Feingold J. 1992. Segregation analysis detects a major gene controlling blood infection levels in human malaria. *Am. J. Hum. Genet.* 50:1308–17

2. Abel L, Demenais F. 1988. Detection of major genes for susceptibility to leprosy and its subtypes in a Caribbean island: Desirade island. *Am. J. Hum. Genet.* 42:256–66

3. Abel L, Sanchez FO, Oberti J, Thuc NV, Hoa LV, et al. 1998. Susceptibility to leprosy is linked to the human NRAMP1 gene. *J. Infect. Dis.* 177:133–45

4. Abraham LJ, Kroeger KM. 1999. Impact of the -308 TNF promoter polymorphism on the transcriptional regulation of the TNF gene: relevance to disease. *J. Leukocyte Biol.* 66:562–66

5. Agarwal A, Guindo A, Cissoko Y, Taylor JG, Coulibaly D, et al. 2000. Hemoglobin C associated with protection from severe malaria in the Dogon of Mali, a West African population with a low prevalence of hemoglobin S. *Blood* 96:2358–63

6. Aitman TJ, Cooper LD, Norsworthy PJ, Wahid FN, Gray JK, et al. 2000. Malaria susceptibility and CD36 mutation. *Nature* 405:1015–16

7. Alcais A, Sanchez FO, Thuc NV, Lap VD, Oberti J, et al. 2000. Granulomatous reaction to intradermal injection of lepromin (Mitsuda reaction) is linked to

the human NRAMP1 gene in Vietnamese leprosy sibships. *J. Infect. Dis.* 181:302–8

8. Ali S, Niang MA, N'Doye I, Critchlow CW, Hawes SE, et al. 2000. Secretor polymorphism and human immunodeficiency virus infection in Senegalese women. *J. Infect. Dis.* 181:737–39

9. Allison AC. 1954. Protection afforded by sickle-cell trait against subtertain malarial infection. *Br. Med. J.* i:290–94

10. Alric L, Fort M, Izopet J, Vinel JP, Charlet JP, et al. 1997. Genes of the major histocompatibility complex class II influence the outcome of hepatitis C virus infection. *Gastroenterology* 113:1675–81

11. Altare F, Durandy A, Lammas D, Emile JF, Lamhamedi S, et al. 1998. Impairment of mycobacterial immunity in human interleukin-12 receptor deficiency. *Science* 280:1432–35

12. Amoroso A, Berrino M, Boniotto M, Crovella S, Palomba E, et al. 1999. Polymorphism at codon 54 of mannose-binding protein gene influences AIDS progression but not HIV infection in exposed children. *AIDS* 13:863–64

13. An P, Martin MP, Nelson GW, Carrington M, Smith MW, et al. 2000. Influence of CCR5 promoter haplotypes on AIDS progression in African Americans. *AIDS* 14:2117–22

14. Anzala AO, Ball TB, Rostron T, O'Brien SJ, Plummer FA, Rowland-Jones SL. 1998. CCR2-64I allele and genotype association with delayed AIDS progression in African women. University of Nairobi Collaboration for HIV Research. *Lancet* 351:1632–33

15. Balotta C, Bagnarelli P, Corvasce S, Mazzucchelli R, Colombo MC, et al. 1999. Identification of two distinct subsets of long-term nonprogressors with divergent viral activity by stromal-derived factor-1 chemokine gene polymorphism analysis. *J. Infect. Dis.* 180:285–89

16. Balotta C, Bagnarelli P, Violin M, Ridolfo AL, Zhou D, et al. 1997. Homozygous delta 32 deletion of the CCR-5 chemokine receptor gene in an HIV-1–infected patient. *AIDS* 11:F67–71

17. Barton CH, Biggs TE, Baker ST, Bowen H, Atkinson PG. 1999. Nramp1: a link between intracellular iron transport and innate resistance to intracellular pathogens. *J. Leukocyte Biol.* 66:757–62

18. Bellamy R, Beyers N, McAdam KP, Ruwende C, Gie R, et al. 2000. Genetic susceptibility to tuberculosis in Africans: a genome-wide scan. *Proc. Natl. Acad. Sci. USA* 97:8005–9

19. Bellamy R, Ruwende C, Corrah T, McAdam KP, Thursz M, et al. 1999. Tuberculosis and chronic hepatitis B virus infection in Africans and variation in the vitamin D receptor gene. *J. Infect. Dis.* 179:721–24

20. Bellamy R, Ruwende C, Corrah T, McAdam KP, Whittle HC, Hill AV. 1998. Assessment of the interleukin 1 gene cluster and other candidate gene polymorphisms in host susceptibility to tuberculosis. *Tuberc. Lung Dis.* 79:83–89

21. Bellamy R, Ruwende C, Corrah T, McAdam KP, Whittle HC, Hill AV. 1998. Variations in the NRAMP1 gene and susceptibility to tuberculosis in West Africans. *N. Engl. J. Med.* 338:640–44

22. Bellamy R, Ruwende C, McAdam KP, Thursz M, Sumiya M, et al. 1998. Mannose-binding protein deficiency is not associated with malaria, hepatitis B carriage nor tuberculosis in Africans. *Q. J. Med.* 91:13–18

23. Bellamy RJ, Hill AV. 1998. Host genetic susceptibility to human tuberculosis. *Novartis Found. Symp.* 217:3–13

24. Blackwell CC, James VS, Davidson S, Wyld R, Brettle RP, et al. 1991. Secretor status and heterosexual transmission of HIV. *Br. Med. J.* 303:825–26

25. Blackwell JM, Barton CH, White JK, Roach TI, Shaw MA, et al. 1994. Genetic regulation of leishmanial and

mycobacterial infections: the Lsh/Ity/ Bcg gene story continues. *Immunol. Lett.* 43:99–107

26. Blackwell JM, Black GF, Peacock CS, Miller EN, Sibthorpe D, et al. 1997. Immunogenetics of leishmanial and mycobacterial infections: the Belem Family Study. *Philos. Trans. R. Soc. London Ser. B Biol. Sci.* 352:1331–45

27. Blackwell JM, Searle S, Goswami T, Miller EN. 2000. Understanding the multiple functions of Nramp1. *Microbes Infect.* 2:317–21

28. Blumberg BS, Gerstley BJ, Hungerford DA, London WT, Sutnick AI. 1967. A serum antigen (Australia antigen) in Down's syndrome, leukemia, and hepatitis. *Ann. Intern. Med.* 66:924–31

29. Brambilla A, Villa C, Rizzardi G, Veglia F, Ghezzi S, et al. 2000. Shorter survival of SDF1-3′A/3′A homozygotes linked to CD4$^+$ T-cell decrease in advanced human immunodeficiency virus type 1 infection. *J. Infect. Dis.* 182:311–15

30. Bredius RG, Derkx BH, Fijen CA, de Wit TP, de Haas M, et al. 1994. Fc gamma receptor IIa (CD32) polymorphism in fulminant meningococcal septic shock in children. *J. Infect. Dis.* 170:848–53

31. Cabrera M, Shaw M-A, Sharples C, Williams H, Castes M, et al. 1995. Polymorphism in tumor necrosis factor genes associated with mucocutaneous leishmaniasis. *J. Exp. Med.* 182:1259–64

32. Carrington M, Nelson GW, Martin MP, Kissner T, Vlahov D, et al. 1999. HLA and HIV-1: heterozygote advantage and B*35-Cw*04 disadvantage. *Science* 283:1748–52

33. Chakravarti MR, Vogel F. 1973. *A Twin Study on Leprosy.* Stuttgart: Thieme

34. Clegg AO, Ashton LJ, Biti RA, Badhwar P, Williamson P, et al. 2000. CCR5 promoter polymorphisms, CCR5 59029A and CCR5 59353C, are under represented in HIV-1–infected long-term non-progressors. The Australian Long-

Term Non-Progressor Study Group. *AIDS* 14:103–8

35. Comstock GW. 1978. Tuberculosis in twins: a re-analysis of the Prophit survey. *Am. Rev. Respir. Dis.* 117:621–24

36. Conway DJ, Holland MJ, Bailey RL, Campbell AE, Mahdi OS, et al. 1997. Scarring trachoma is associated with polymorphism in the tumor necrosis factor alpha (TNF-alpha) gene promoter and with elevated TNF-alpha levels in tear fluid. *Infect. Immun.* 65:1003–6

37. Crawley E, Kay R, Sillibourne J, Patel P, Hutchinson I, Woo P. 1999. Polymorphic haplotypes of the interleukin-10 5′ flanking region determine variable interleukin-10 transcription and are associated with particular phenotypes of juvenile rheumatoid arthritis. *Arthritis Rheum.* 42:1101–8

38. D'Ambrosio D, Cippitelli M, Cocciolo MG, Mazzeo D, Di Lucia P, et al. 1998. Inhibition of IL-12 production by 1,25-dihydroxy vitamin D3. Involvement of NF-kappaB downregulation in transcriptional repression of the p40 gene. *J. Clin. Invest.* 101:252–62

39. de Roda Husman AM, Koot M, Cornelissen M, Keet IP, Brouwer M, et al. 1997. Association between CCR5 genotype and the clinical course of HIV-1 infection. *Ann. Intern. Med.* 127:882–90

40. de Vries RR, Fat RF, Nijenhuis LE, van Rood JJ. 1976. HLA linked genetic control of host response to *Mycobacterium leprae. Lancet* 2:1328–30

41. Dean M, Carrington M, Winkler C, Huttley GA, Smith MW, et al. 1996. Genetic restriction of HIV-1 infection and progression to AIDS by a deletion allele of the CKR5 structural gene. Hemophilia Growth and Development Study, Multicenter AIDS Cohort Study, Multicenter Hemophilia Cohort Study, San Francisco City Cohort, ALIVE Study. *Science.* 273:1856–62. Erratum. 1996. *Science* 274(5290):1069

42. Dessein AJ, Hillaire D, Elwali NE, Marquet S, Mohamed-Ali Q, et al. 1999.

Severe hepatic fibrosis in *Schistosoma mansoni* infection is controlled by a major locus that is closely linked to the interferon-γ-receptor gene. *Am. J. Hum. Genet.* 65:709–21

43. Diehl K, Von Verscheur O. 1936. *Der Erbeinfluss bei den Tuberkulose.* Jena: Gustav Fischer

44. Doherty PC, Zinkernagel RM. 1975. A biological role for the major histocompatibility antigens. *Lancet* 1:1406–9

45. Easterbrook PJ, Rostron T, Ives N, Troop M, Gazzard BG, Rowland-Jones SL. 1999. Chemokine receptor polymorphisms and human immunodeficiency virus disease progression. *J. Infect. Dis.* 180:1096–105

46. Edwards-Smith CJ, Jonsson JR, Purdie DM, Bansal A, Shorthouse C, Powell EE. 1999. Interleukin-10 promoter polymorphism predicts initial response of chronic hepatitis C to interferon α. *Hepatology* 30:526–30

47. El-Omar EM, Carrington M, Chow WH, McColl KE, Bream JH, et al. 2000. Interleukin-1 polymorphisms associated with increased risk of gastric cancer. *Nature* 404:398–402

48. Faure S, Meyer L, Costagliola D, Vaneensberghe C, Genin E, et al. 2000. Rapid progression to AIDS in HIV$^+$ individuals with a structural variant of the chemokine receptor CX3CR1. *Science* 287:2274–77

49. Floros J, Lin HM, Garcia A, Salazar MA, Guo X, et al. 2000. Surfactant protein genetic marker alleles identify a subgroup of tuberculosis in a Mexican population. *J. Infect. Dis.* 182:1473–78

50. Frodsham AJ. 2000. *The genetics of susceptibility to chronic hepatitis B infection.* PhD thesis. Oxford Univ.

51. Fryer AA, Spiteri MA, Bianco A, Hepple M, Jones PW, et al. 2000. The -403 G-A promoter polymorphism in the RANTES gene is associated with atopy and asthma. *Genes Immun.* 1:509–14

52. Gao PS, Fujishima S, Mao XQ, Remus N, Kanda M, et al. 2000. Genetic variants of NRAMP1 and active tuberculosis in Japanese populations. *Clin. Genet.* 58:74–76

53. Garcia A, Marquet S, Bucheton B, Hillaire D, Cot M, et al. 1998. Linkage analysis of blood *Plasmodium falciparum* levels: interest of the 5q31-q33 chromosome region. *Am. J. Trop. Med. Hyg.* 58:705–9

54. Garred P, Harboe M, Oettinger T, Koch C, Svejgaard A. 1994. Dual role of mannan-binding protein in infections: another case of heterosis? *Eur. J. Immunogenet.* 21:125–31

55. Garred P, Madsen HO, Balslev U, Hofmann B, Pedersen C, et al. 1997. Susceptibility to HIV infection and progression of AIDS in relation to variant alleles of mannose-binding lectin. *Lancet* 349:236–40

56. Gelder CM, Hart KW, Williams OM, Lyons E, Welsh KI, et al. 2000. Vitamin D receptor gene polymorphisms and susceptibility to *Mycobacterium malmoense* pulmonary disease. *J. Infect. Dis.* 181:2099–102

57. Gilbert SC, Plebanski M, Gupta S, Morris J, Cox M, et al. 1998. Association of malaria parasite population structure, HLA, and immunological antagonism. *Science* 279:1173–77

58. Goldfeld AE, Delgado JC, Thim S, Bozon MV, Uglialoro AM, et al. 1998. Association of an HLA-DQ allele with clinical tuberculosis. *JAMA* 279:226–28

59. Greenwood CM, Fujiwara TM, Boothroyd LJ, Miller MA, Frappier D, et al. 2000. Linkage of tuberculosis to chromosome 2q35 loci, including NRAMP1, in a large aboriginal Canadian family. *Am. J. Hum. Genet.* 67:405–16

60. Haile RW, Iselius L, Fine PE, Morton NE. 1985. Segregation and linkage analyses of 72 leprosy pedigrees. *Hum. Hered.* 35:43–52

60a. Hendel H, Caillat-Zucman S, Lebuanec H, Carrington M. O'Brien S, et al. 1999.

New class I and II HLA alleles strongly associated with opposite patterns of progression to AIDS. *J. Immunol.* 162:6942–45

61. Hennig BJ, Parkhill JM, Chapple IL, Heasman PA, Taylor JJ. 1999. Association of a vitamin D receptor gene polymorphism with localized early-onset periodontal diseases. *J. Periodontol.* 70:1032–38

62. Hibberd ML, Sumiya M, Summerfield JA, Booy R, Levin M. 1999. Association of variants of the gene for mannose-binding lectin with susceptibility to meningococcal disease. Meningococcal Research Group. *Lancet* 353:1049–53

63. Hill AV. 1996. Genetic susceptibility to malaria and other infectious diseases: from the MHC to the whole genome. *Parasitology* 112:S75–84

64. Hill AV, Allsopp CE, Kwiatkowski D, Anstey NM, Twumasi P, et al. 1991. Common West African HLA antigens are associated with protection from severe malaria. *Nature* 352:595–600

65. Hill AV, Elvin J, Willis AC, Aidoo M, Allsopp CE, et al. 1992. Molecular analysis of the association of HLA-B53 and resistance to severe malaria. *Nature* 360:434–39

66. Hoal-Van Helden EG, Epstein J, Victor TC, Hon D, Lewis LA, et al. 1999. Mannose-binding protein B allele confers protection against tuberculous meningitis. *Pediatr. Res.* 45:459–64

67. Hohler T, Kruger A, Gerken G, Schneider PM, Meyer zum Buschenefelde KH, Rittner C. 1998. A tumor necrosis factor-alpha (TNF-alpha) promoter polymorphism is associated with chronic hepatitis B infection. *Clin. Exp. Immunol.* 111:579–82

68. Huang Y, Paxton WA, Wolinsky SM, Neumann AU, Zhang L, et al. 1996. The role of a mutant CCR5 allele in HIV-1 transmission and disease progression. *Nat. Med.* 2:1240–43

69. Huebner R. 1996. Bacillus of Calmette and Guerin (BCG) vaccine. In *Tuberculosis*, ed. WN Rom, SM Gray, pp. 893–904. Boston: Little, Brown

70. Jabado N, Jankowski A, Dougaparsad S, Picard V, Grinstein S, Gros P. 2000. Natural resistance to intracellular infections. Natural resistance–associated macrophage protein 1 (nramp1) functions as a pH-dependent manganese transporter at the phagosomal membrane. *J. Exp. Med.* 192:1237–48

71. James SP, Nicol WD, Shute PG. 1932. A study of induced malignant tertian malaria. *Proc. R. Soc. Med.* 25:1153–86

72. Jeffery KJ, Siddiqui AA, Bunce M, Lloyd AL, Vine AM, et al. 2000. The influence of HLA class I alleles and heterozygosity on the outcome of human T-cell lymphotropic virus type I infection. *J. Immunol.* 165:7278–84

73. Jeffery KJ, Usuku K, Hall SE, Matsumoto W, Taylor GP, et al. 1999. HLA alleles determine human T-lymphotropic virus-I (HTLV-I) proviral load and the risk of HTLV-I–associated myelopathy. *Proc. Natl. Acad. Sci. USA* 96:3848–53

74. Jepson A, Banya W, Sisay-Joof F, Hassan-King M, Nunes C, et al. 1997. Quantification of the relative contribution of major histocompatibility complex (MHC) and non-MHC genes to human immune responses to foreign antigens. *Infect. Immun.* 65:872–76

75. Jepson A, Sisay-Joof F, Banya W, Hassan-King M, Frodsham A, et al. 1997. Genetic linkage of mild malaria to the major histocompatibility complex in Gambian children: study of affected sibling pairs. *Br. Med. J.* 315:96–97

76. Jong R, Altare F, Haagen IA, Elferink DG, Boer T, et al. 1998. Severe mycobacterial and Salmonella infections in interleukin-12 receptor–deficient patients. *Science* 280:1435–38

77. Jouanguy E, Altare F, Lamhamedi S, Revy P, Emile JF, et al. 1996. Interferon-γ-receptor deficiency in an infant with

fatal bacille Calmette-Guerin infection. *N. Engl. J. Med.* 335:1956–61

78. Kallmann FJ, Reisner D. 1942. Twin studies on the significance of genetic factors in tuberculosis. *Am. Rev. Tuberc.* 47:549–74

79. Kelly RJ, Rouquier S, Giorgi D, Lennon GG, Lowe JB. 1995. Sequence and expression of a candidate for the human secretor blood group alpha(1,2)fucosyltransferase gene (FUT2). Homozygosity for an enzyme-inactivating nonsense mutation commonly correlates with the non-secretor phenotype. *J. Biol. Chem.* 270:4640–49

80. Knight JC, Udalova I, Hill AV, Greenwood BM, Peshu N, et al. 1999. A polymorphism that affects OCT-1 binding to the TNF promoter region is associated with severe malaria. *Nat. Genet.* 22:145–50

81. Kostrikis LG, Huang Y, Moore JP, Wolinsky SM, Zhang L, et al. 1998. A chemokine receptor CCR2 allele delays HIV-1 disease progression and is associated with a CCR5 promoter mutation. *Nat. Med.* 4:350–53

82. Kostrikis LG, Neumann AU, Thomson B, Korber BT, McHardy P, et al. 1999. A polymorphism in the regulatory region of the CC-chemokine receptor 5 gene influences perinatal transmission of human immunodeficiency virus type 1 to African-American infants. *J. Virol.* 73:10264–71

83. Kovacs GG, Head MW, Bunn T, Laszlo L, Will RG, Ironside JW. 2000. Clinicopathological phenotype of codon 129 valine homozygote sporadic Creutzfeldt-Jakob disease. *Neuropathol. Appl. Neurobiol.* 26:463–72

84. Kramnik I, Dietrich WF, Demant P, Bloom BR. 2000. Genetic control of resistance to experimental infection with virulent *Mycobacterium tuberculosis. Proc. Natl. Acad. Sci. USA* 97:8560–65

85. Kroner BL, Goedert JJ, Blattner WA, Wilson SE, Carrington MN, Mann DL. 1995. Concordance of human leukocyte antigen haplotype sharing, CD4 decline and AIDS in hemophilic siblings. Multicenter Hemophilia Cohort and Hemophilia Growth and Development Studies. *AIDS* 9:275–80

86. LeBlanc SB, Naik EG, Jacobson L, Kaslow RA. 2000. Association of DRB1*1501 with disseminated *Mycobacterium avium* complex infection in North American AIDS patients. *Tissue Antigens* 55:17–23

87. Lee HS, Brown P, Cervenakova L, Garruto RM, Alpers MP, et al. 2001. Increased susceptibility to kuru of carriers of the PRNP 129 methionine/methionine genotype. *J. Infect. Dis.* 183:192–96

88. Lin TM, Chen CJ, Wu MM, Yang CS, Chen JS, et al. 1989. Hepatitis B virus markers in Chinese twins. *Anticancer Res.* 9:737–41

89. Liu H, Chao D, Nakayama EE, Taguchi H, Goto M, et al. 1999. Polymorphism in RANTES chemokine promoter affects HIV-1 disease progression. *Proc. Natl. Acad. Sci. USA* 96:4581–85

90. Liu R, Paxton WA, Choe S, Ceradini D, Martin SR, et al. 1996. Homozygous defect in HIV-1 coreceptor accounts for resistance of some multiply-exposed individuals to HIV-1 infection. *Cell* 86:367–77

91. Luty AJ, Kun JF, Kremsner PG. 1998. Mannose-binding lectin plasma levels and gene polymorphisms in *Plasmodium falciparum* malaria. *J. Infect. Dis.* 178:1221–24

92. MacDonald KS, Fowke KR, Kimani J, Dunand VA, Nagelkerke NJ, et al. 2000. Influence of HLA supertypes on susceptibility and resistance to human immunodeficiency virus type 1 infection. *J. Infect. Dis.* 181:1581–89

93. Malaty HM, Engstrand L, Pedersen NL, Graham DY. 1994. *Helicobacter pylori* infection: genetic and environmental influences. A study of twins. *Ann. Intern. Med.* 120:982–86

94. Maliarik MJ, Chen KM, Sheffer RG, Rybicki BA, Major ML, et al. 2000. The natural resistance–associated macrophage protein gene in African Americans with sarcoidosis. *Am. J. Respir. Cell Mol. Biol.* 22:672–75

95. Marquet S, Abel L, Hillaire D, Dessein A. 1999. Full results of the genome-wide scan which localises a locus controlling the intensity of infection by *Schistosoma mansoni* on chromosome 5q31-q33. *Eur. J .Hum. Genet.* 7:88–97

96. Marquet S, Abel L, Hillaire D, Dessein H, Kalil J, et al. 1996. Genetic localization of a locus controlling the intensity of infection by *Schistosoma mansoni* on chromosome 5q31-q33. *Nat. Genet.* 14:181–84

97. Marquet S, Lepage P, Hudson TJ, Musser JM, Schurr E. 2000. Complete nucleotide sequence and genomic structure of the human NRAMP1 gene region on chromosome region 2q35. *Mamm. Genome* 11:755–62

98. Marquet S, Sanchez FO, Arias M, Rodriguez J, Paris SC, et al. 1999. Variants of the human NRAMP1 gene and altered human immunodeficiency virus infection susceptibility. *J. Infect. Dis.* 180:1521–25

99. Martin MP, Dean M, Smith MW, Winkler C, Gerrard B, et al. 1998. Genetic acceleration of AIDS progression by a promoter variant of CCR5. *Science* 282:1907–11

100. Martinson JJ, Chapman NH, Rees DC, Liu YT, Clegg JB. 1997. Global distribution of the CCR5 gene 32-basepair deletion. *Nat. Genet.* 16:100–3

101. McBride MO, Fischer PB, Sumiya M, McClure MO, Turner MW, et al. 1998. Mannose-binding protein in HIV-seropositive patients does not contribute to disease progression or bacterial infections. *Int. J. Stud. AIDS* 9:683–88

102. McDermott DH, Beecroft MJ, Kleeberger CA, Al-Sharif FM, Ollier WE, et al. 2000. Chemokine RANTES promoter polymorphism affects risk of both HIV infection and disease progression in the Multicenter AIDS Cohort Study. *AIDS* 14:2671–78

103. McDermott DH, Colla JS, Kleeberger CA, Plankey M, Rosenberg PS, et al. 2000. Genetic polymorphism in CX3CR1 and risk of HIV disease. *Science* 290:2031a

104. McDermott DH, Zimmerman PA, Guignard F, Kleeberger CA, Leitman SF, Murphy PM. 1998. CCR5 promoter polymorphism and HIV-1 disease progression. Multicenter AIDS Cohort Study (MACS). *Lancet* 352:866–70

105. McGuire W, Hill AV, Allsopp CE, Greenwood BM, Kwiatkowski D. 1994. Variation in the TNF-alpha promoter region associated with susceptibility to cerebral malaria. *Nature* 371:508–10

106. McGuire W, Knight JC, Hill AV, Allsopp CE, Greenwood BM, Kwiatkowski D. 1999. Severe malarial anemia and cerebral malaria are associated with different tumor necrosis factor promoter alleles. *J. Infect. Dis.* 179:287–90

107. Mehra NK, Rajalingam R, Mitra DK, Taneja V, Giphart MJ. 1995. Variants of HLA-DR2/DR51 group haplotypes and susceptibility to tuberculoid leprosy and pulmonary tuberculosis in Asian Indians. *Int. J. Lepr. Other Mycobact. Dis.* 63:241–48

107a. Meisner SJ, Mucklow S, Warner G, Sow SO, Lienhardt C, Hill AVS. 2001. Association of *NRAMP1* polymorphism with leprosy type but not susceptibility to leprosy per se in West Africans. *Am. J. Trop. Med. Hyg.* In press

108. Mellado M, Rodriguez-Frade JM, Vila-Coro AJ, de Ana AM, Martinez AC. 1999. Chemokine control of HIV-1 infection. *Nature* 400:723–24

109. Michael NL, Chang G, Louie LG, Mascola JR, Dondero D, et al. 1997. The role of viral phenotype and CCR-5 gene defects in HIV-1 transmission and disease progression. *Nat. Med.* 3:338–40

110. Michael NL, Louie LG, Rohrbaugh AL, Schultz KA, Dayhoff DE, et al. 1997. The role of CCR5 and CCR2 polymorphisms in HIV-1 transmission and disease progression. *Nat. Med.* 3:1160–62

111. Migueles SA, Sabbaghian MS, Shupert WL, Bettinotti MP, Marincola FM, et al. 2000. HLA-B*5701 is highly associated with restriction of virus replication in a subgroup of HIV-infected long-term nonprogressors. *Proc. Natl. Acad. Sci. USA* 97:2709–14

112. Miller LH, Mason SJ, Clyde DF, McGinniss MH. 1976. The resistance factor to *Plasmodium vivax* in blacks. The Duffy-blood-group genotype, FyFy. *N. Engl. J. Med.* 295:302–4

113. Minton EJ, Smillie D, Neal KR, Irving WL, Underwood JC, James V. 1998. Association between MHC class II alleles and clearance of circulating hepatitis C virus. Members of the Trent Hepatitis C Virus Study Group. *J. Infect. Dis.* 178:39–44

114. Modiano D, Chiucchiuini A, Petrarca V, Sirima BS, Luoni G, et al. 1999. Interethnic differences in the humoral response to non-repetitive regions of the *Plasmodium falciparum* circumsporozoite protein. *Am. J. Trop. Med. Hyg.* 61:663–67

115. Modiano D, Petrarca V, Sirima BS, Nebie I, Diallo D, et al. 1996. Different response to *Plasmodium falciparum* malaria in West African sympatric ethnic groups. *Proc. Natl. Acad. Sci. USA* 93:13206–11

116. Morrison NA, Qi JC, Tokita A, Kelly PJ, Crofts L, et al. 1994. Prediction of bone density from vitamin D receptor alleles. *Nature* 367:284–87. Erratum. 1997. *Nature* 387(6628):106

117. Muller-Myhsok B, Stelma FF, Guisse-Sow F, Muntau B, Thye T, et al. 1997. Further evidence suggesting the presence of a locus, on human chromosome 5q31-q33, influencing the intensity of infection with *Schistosoma mansoni. Am. J. Hum. Genet.* 61:452–54

118. Mummidi S, Ahuja SS, Gonzalez E, Anderson SA, Santiago EN, et al. 1998. Genealogy of the CCR5 locus and chemokine system gene variants associated with altered rates of HIV-1 disease progression. *Nat. Med.* 4:786–93

119. Mummidi S, Bamshad M, Ahuja SS, Gonzalez E, Feuillet PM, et al. 2000. Evolution of human and non-human primate CC chemokine receptor 5 gene and mRNA. Potential roles for haplotype and mRNA diversity, differential haplotype-specific transcriptional activity, and altered transcription factor binding to polymorphic nucleotides in the pathogenesis of HIV-1 and simian immunodeficiency virus. *J. Biol. Chem.* 275:18946–61

120. Nakayama EE, Hoshino Y, Xin X, Liu H, Goto M, et al. 2000. Polymorphism in the interleukin-4 promoter affects acquisition of human immunodeficiency virus type 1 syncytium-inducing phenotype. *J. Virol.* 74:5452–59

121. Newport MJ, Huxley CM, Huston S, Hawrylowicz CM, Oostra BA, et al. 1996. A mutation in the interferon-γ receptor gene and susceptibility to mycobacterial infection. *N. Engl. J. Med.* 335:1941–49

122. Nickel RG, Casolaro V, Wahn U, Beyer K, Barnes KC, et al. 2000. Atopic dermatitis is associated with a functional mutation in the promoter of the C-C chemokine RANTES. *J. Immunol.* 164:1612–16

123. North RJ, LaCourse R, Ryan L, Gros P. 1999. Consequence of Nramp1 deletion to *Mycobacterium tuberculosis* infection in mice. *Infect. Immun.* 67:5811–14

124. Palmer MS, Dryden AJ, Hughes JT, Collinge J. 1991. Homozygous prion protein genotype predisposes to sporadic Creutzfeldt-Jakob disease. *Nature.* 352:340–42. Erratum. 1991. *Nature* 352:547

125. Pastinen T, Liitsola K, Niini P, Salminen M, Syvanen AC. 1998. Contribution of the CCR5 and MBL genes to susceptibility to HIV type 1 infection in the Finnish population. *AIDS Res. Hum. Retrovir.* 14:695–98

126. Peano G, Menardi G, Ponzetto A, Fenoglio LM, Czaja AJ, et al. 1994. HLA-DR5 antigen. A genetic factor influencing the outcome of hepatitis C virus infection? Significance of human leukocyte antigens DR3 and DR4 in chronic viral hepatitis. *Arch. Intern. Med.* 154:2733–36

127. Penna G, Adorini L. 2000. 1α,25-dihydroxyvitamin D3 inhibits differentiation, maturation, activation, and survival of dendritic cells leading to impaired alloreactive T-cell activation. *J. Immunol.* 164:2405–11

128. Platonov AE, Shipulin GA, Vershinina IV, Dankert J, van de Winkel JG, Kuijper EJ. 1998. Association of human Fc gamma RIIa (CD32) polymorphism with susceptibility to and severity of meningococcal disease. *Clin. Infect. Dis.* 27:746–50

129. Plebanski M, Lee EA, Hannan CM, Flanagan KL, Gilbert SC, et al. 1999. Altered peptide ligands narrow the repertoire of cellular immune responses by interfering with T-cell priming. *Nat. Med.* 5:565–71

130. Ravikumar M, Dheenadhayalan V, Rajaram K, Lakshmi SS, Kumaran PP, et al. 1999. Associations of HLA-DRB1, DQB1 and DPB1 alleles with pulmonary tuberculosis in South India. *Tuberc. Lung Dis.* 79:309–17

131. Rihet P, Abel L, Traore Y, Traore-Leroux T, Aucan C, Fumoux F. 1998. Human malaria: segregation analysis of blood infection levels in a suburban area and a rural area in Burkina Faso. *Genet. Epidemiol.* 15:435–50

132. Risch N. 1987. Assessing the role of HLA linked and unlinked determinants of disease. *Am. J. Hum. Genet.* 40:1–14

133. Rizzardi GP, Morawetz RA, Vicenzi E, Ghezzi S, Poli G, et al. 1998. CCR2 polymorphism and HIV disease. Swiss HIV Cohort. *Nat. Med.* 4:252–53

134. Roger M, Levee G, Chanteau S, Gicquel B, Schurr E. 1997. No evidence for linkage between leprosy susceptibility and the human natural resistance–associated macrophage protein 1 (NRAMP1) gene in French Polynesia. *Int. J. Lepr. Other Mycobact. Dis.* 65:197–202

135. Rook GA, Steele J, Fraher L, Barker S, Karmali R, et al. 1986. Vitamin D3, γ interferon, and control of proliferation of *Mycobacterium tuberculosis* by human monocytes. *Immunology* 57:159–63

136. Roy S, Frodsham A, Saha B, Hazra SK, Mascie-Taylor CG, Hill AV. 1999. Association of vitamin D receptor genotype with leprosy type. *J. Infect. Dis.* 179:187–91

137. Deleted in proof

138. Roy S, McGuire W, Mascie-Taylor CG, Saha B, Hazra SK, et al. 1997. Tumor necrosis factor promoter polymorphism and susceptibility to lepromatous leprosy. *J. Infect. Dis.* 176:530–32

139. Ryu S, Park YK, Bai GH, Kim SJ, Park SN, Kang S. 2000. 3′ UTR polymorphisms in the NRAMP1 gene are associated with susceptibility to tuberculosis in Koreans. *Int. J. Tuberc. Lung Dis.* 4:577–80

140. Samson M, Libert F, Doranz BJ, Rucker J, Liesnard C, et al. 1996. Resistance to HIV-1 infection in Caucasian individuals bearing mutant alleles of the CCR-5 chemokine receptor gene. *Nature* 382:722–25

141. Sanders LA, van de Winkel JG, Rijkers GT, Voorhorst Ogink MM, de Haas M, et al. 1994. Fc gamma receptor IIa (CD32) heterogeneity in patients with recurrent bacterial respiratory tract infections. *J. Infect. Dis.* 170:854–61

142. Sanjeevi CB, Miller EN, Dabadghao P, Rumba I, Shtauvere A, et al. 2000. Polymorphism at NRAMP1 and D2S1471

loci associated with juvenile rheumatoid arthritis. *Arthritis Rheum.* 43:1397–404

143. Selvaraj P, Narayanan PR, Reetha AM. 1999. Association of functional mutant homozygotes of the mannose-binding protein gene with susceptibility to pulmonary tuberculosis in India. *Tuberc. Lung Dis.* 79:221–27

144. Shaw MA, Collins A, Peacock CS, Miller EN, Black GF, et al. 1997. Evidence that genetic susceptibility to *Mycobacterium tuberculosis* in a Brazilian population is under oligogenic control: linkage study of the candidate genes NRAMP1 and TNFA. *Tuberc. Lung Dis.* 78:35–45

145. Shin HD, Winkler C, Stephens JC, Bream J, Young H, et al. 2000. Genetic restriction of HIV-1 pathogenesis to AIDS by promoter alleles of IL10. *Proc. Natl. Acad. Sci. USA* 97:14467–72

146. Siddiqui MR, Meisner S, Tosh K, Balakrishnan K, Ghei S, et al. 2001. A major susceptibility locus for leprosy in India maps to chromosome 10p13. *Nat. Genet.* 27:439–41

147. Simmons JD, Mullighan C, Welsh KI, Jewell DP. 2000. Vitamin D receptor gene polymorphism: association with Crohn's disease susceptibility. *Gut* 47:211–14

148. Singh SP, Mehra NK, Dingley HB, Pande JN, Vaidya MC. 1983. Human leukocyte antigen (HLA) linked control of susceptibility to pulmonary tuberculosis and association with HLA-DR types. *J. Infect. Dis.* 148:676–81

149. Skamene E, Schurr E, Gros P. 1998. Infection genomics: Nramp1 as a major determinant of natural resistance to intracellular infections. *Annu. Rev. Med.* 49:275–87

150. Smith MW, Carrington M, Winkler C, Lomb D, Dean M, et al. 1997. CCR2 chemokine receptor and AIDS progression. *Nat. Med.* 3:1052

151. Smith MW, Dean M, Carrington M, Winkler C, Huttley GA, et al. 1997. Contrasting genetic influence of CCR2 and CCR5 variants on HIV-1 infection and disease progression. Hemophilia Growth and Development Study (HGDS), Multicenter AIDS Cohort Study (MACS), Multicenter Hemophilia Cohort Study (MHCS), San Francisco City Cohort (SFCC), ALIVE Study. *Science* 277:959–65

152. Sorensen TI, Nielsen GG, Andersen PK, Teasdale TW. 1988. Genetic and environmental influences on premature death in adult adoptees. *N. Engl. J. Med.* 318:727–32

153. Summerfield JA, Sumiya M, Levin M, Turner MW. 1997. Association of mutations in mannose-binding protein gene with childhood infection in consecutive hospital series. *Br. Med. J.* 314:1229–32

154. Tang C, Kwiatkowski D. 1999. Mannose-binding lectin and meningococcal disease. *Lancet* 354:336

155. Thomas HC, Foster GR, Sumiya M, McIntosh D, Jack DL, et al. 1996. Mutation of gene of mannose-binding protein associated with chronic hepatitis B viral infection. *Lancet* 348:1417–19

156. Thursz M, Yallop R, Goldin R, Trepo C, Thomas HC. 1999. Influence of MHC class II genotype on outcome of infection with hepatitis C virus. The HENCORE Group. Hepatitis C European Network for Cooperative Research. *Lancet* 354:2119–24

157. Thursz MR, Thomas HC, Greenwood BM, Hill AV. 1997. Heterozygote advantage for HLA class II type in hepatitis B virus infection. *Nat. Genet.* 17:11–12

158. Tournamille C, Colin Y, Cartron JP, Le Van Kim C. 1995. Disruption of a GATA motif in the Duffy gene promoter abolishes erythroid gene expression in Duffy-negative individuals. *Nat. Genet.* 10:224–28

159. Turner DM, Williams DM, Sankaran D, Lazarus M, Sinnott PJ, Hutchinson IV. 1997. An investigation of polymorphism

in the interleukin-10 gene promoter. *Eur. J. Immunogenet.* 24:1–8

160. Turner MW. 1998. Mannose-binding lectin (MBL) in health and disease. *Immunobiology* 199:327–39

161. Urban BC, Ferguson DJ, Pain A, Willcox N, Plebanski M, et al. 1999. *Plasmodium falciparum*–infected erythrocytes modulate the maturation of dendritic cells. *Nature* 400:73–77

162. van Eden W, de Vries RR, Mehra NK, Vaidya MC, D'Amaro J, van Rood JJ. 1980. HLA segregation of tuberculoid leprosy: confirmation of the DR2 marker. *J. Infect. Dis.* 141:693–701

163. van Rij RP, Broersen S, Goudsmit J, Coutinho RA, Schuitemaker H. 1998. The role of a stromal cell–derived factor-1 chemokine gene variant in the clinical course of HIV-1 infection. *AIDS* 12:F85–90

164. Vidal S, Tremblay ML, Govoni G, Gauthier S, Sebastiani G, et al. 1995. The Ity/Lsh/Bcg locus: natural resistance to infection with intracellular parasites is abrogated by disruption of the Nramp1 gene. *J. Exp. Med.* 182:655–66

165. Vidal SM, Malo D, Vogan K, Skamene E, Gros P. 1993. Natural resistance to infection with intracellular parasites: isolation of a candidate for Bcg. *Cell* 73:469–85

166. Wawrzynowicz-Syczewska M, Underhill JA, Clare MA, Boron-Kaczmarska A, McFarlane IG, Donaldson PT. 2000. HLA class II genotypes associated with chronic hepatitis C virus infection and response to α-interferon treatment in Poland. *Liver* 20:234–39

167. White JK, Shaw MA, Barton CH, Cerretti DP, Williams H, et al. 1994. Genetic and physical mapping of 2q35 in the region of the NRAMP and IL8R genes: identification of a polymorphic repeat in exon 2 of NRAMP. *Genomics* 24:295–302

168. Wilkinson RJ, Llewelyn M, Toossi Z, Patel P, Pasvol G, et al. 2000. Influence of vitamin D deficiency and vitamin D receptor polymorphisms on tuberculosis among Gujarati Asians in West London: a case-control study. *Lancet* 355:618–21

169. Wilkinson RJ, Patel P, Llewelyn M, Hirsch CS, Pasvol G, et al. 1999. Influence of polymorphism in the genes for the interleukin (IL)-1 receptor antagonist and IL-1beta on tuberculosis. *J. Exp. Med.* 189:1863–74

170. Wilson AG, Symons JA, McDowell TL, McDevitt HO, Duff GW. 1997. Effects of a polymorphism in the human tumor necrosis factor alpha promoter on transcriptional activation. *Proc. Natl. Acad. Sci. USA* 94:3195–99

171. Windl O, Dempster M, Estibeiro JP, Lathe R, de Silva R, et al. 1996. Genetic basis of Creutzfeldt-Jakob disease in the United Kingdom: a systematic analysis of predisposing mutations and allelic variation in the PRNP gene. *Hum. Genet.* 98:259–64

172. Winkler C, Modi W, Smith MW, Nelson GW, Wu X, et al. 1998. Genetic restriction of AIDS pathogenesis by an SDF-1 chemokine gene variant. ALIVE Study, Hemophilia Growth and Development Study (HGDS), Multicenter AIDS Cohort Study (MACS), Multicenter Hemophilia Cohort Study (MHCS), San Francisco City Cohort (SFCC). *Science* 279:389–93

173. Xu J, Postma DS, Howard TD, Koppelman GH, Zheng SL, et al. 2000. Major genes regulating total serum immunoglobulin E levels in families with asthma. *Am. J. Hum. Genet.* 67:1163–73

174. Yee AM, Phan HM, Zuniga R, Salmon JE, Musher DM. 2000. Association between Fc gamma RIIa-R131 allotype and bacteremic pneumococcal pneumonia. *Clin. Infect. Dis.* 30:25–28

175. Zavaglia C, Martinetti M, Silini E, Bottelli R, Daielli C, et al. 1998. Association between HLA class II alleles and protection from or susceptibility to chronic hepatitis C. *J. Hepatol.* 28:1–7

176. Zeidler M, Stewart G, Cousens SN, Estibeiro K, Will RG. 1997. Codon 129 genotype and new variant CJD. *Lancet* 350:668

177. Zimmerman PA, Buckler-White A, Alkhatib G, Spalding T, Kubofcik J, et al. 1997. Inherited resistance to HIV-1 conferred by an inactivating mutation in CC chemokine receptor 5: studies in populations with contrasting clinical phenotypes, defined racial background, and quantified risk. *Mol. Med.* 3:23–36

178. Zimmerman PA, Woolley I, Masinde GL, Miller SM, McNamara DT, et al. 1999. Emergence of FY*A(null) in a *Plasmodium vivax*–endemic region of Papua New Guinea. *Proc. Natl. Acad. Sci. USA* 96:13973–77

Annu. Rev. Genomics Hum. Genet. 2001. 2:401–33

PRIVACY AND CONFIDENTIALITY OF GENETIC INFORMATION: What Rules for the New Science?

Mary R. Anderlik and Mark A. Rothstein

Institute for Bioethics, Health Policy, and Law, University of Louisville School of Medicine, Louisville, Kentucky 40292; e-mail: mrande02@gwise.louisville.edu, mark.rothstein@louisville.edu

Key Words   genetic discrimination, legal issues, ethical issues, policy issues

■ **Abstract**   This review covers the ethical, legal, and policy issues associated with the generation and dissemination of genetic information. First, conceptual issues, such as the definition of terms and the description of two modes of analysis, are addressed. Research findings on public attitudes toward privacy and genetics and other factors relevant to policy making are also reviewed. Second, the example of genetic research is used to highlight the importance of attention to the intrinsic harms associated with violations of genetic privacy. Subtopics include national databases and biobanks, gene brokers, and pharmacogenomics. Third, the example of insurer access to genetic information is used to highlight the importance of attention to discrimination and other instrumental harms associated with failures of regulation. Fourth, a summary of the preceding sections leads into an outline of a program for realizing the benefits of the new science in a manner that affirms rather than erodes privacy and other important values.

INTRODUCTION

Developments in genetics present significant challenges in the areas of privacy and confidentiality. This review of the many issues to be considered by scientists and policy makers is organized into four sections. First, we address some of the major conceptual issues, such as distinctions among privacy, confidentiality, and security, and different ways of framing arguments concerning privacy protections. In particular, we describe two complementary approaches to the ethical and legal problems that are associated with the generation and dissemination of genetic information: One focuses on the intrinsic value of privacy and the intrinsic harms that may result from violations of genetic privacy, and the other focuses on the instrumental value of privacy and the negative consequences of failures to protect genetic privacy. We conclude the first section with a summary of research findings on public attitudes toward privacy and genetics and other factors relevant to policy making in this area.

   Second, we use the example of genetic research to highlight the importance of attention to the intrinsic value of privacy. We look at debates over the creation

1527-8204/01/0728-0401$14.00

of national databases and biobanks to facilitate genetic research in Iceland and elsewhere, and at private alternatives to these national efforts, and we analyze the evolving rules for exchanges of genetic information and for research uses of genetic information. We conclude with a discussion of privacy and related issues in the context of pharmacogenomics.

Third, we focus on instrumental harms. Insurer access to genetic information serves as an entry point for a more in-depth examination of the negative consequences of failures of regulation. We begin by distinguishing between irrational and rational discrimination in insurance; although both are problematic, the justifications for regulation of these two forms of discrimination are slightly different. We then undertake separate analyses of the policy considerations relevant to health insurance, life and disability insurance, and long-term care insurance. Finally, we present a synopsis of the key points in the preceding sections and a proposal for a future in which the benefits of the new science are realized in a manner that affirms rather than erodes privacy and other important individual and social values.

CONCEPTUAL ISSUES

Defining Privacy, Confidentiality, and Security

Clear understanding of terminology is one of the keys to productive dialogue between scientists and policy makers. Accordingly, we begin with definitions. Although the terms privacy and confidentiality are sometimes used interchangeably, in a policy analysis the two should be distinguished. Privacy is a broad concept that subsumes at least four categories: access to persons and to personal spaces; access to information by third parties; third-party interference with personal choices, especially in intimate spheres such as procreation; and ownership of materials and information derived from persons (1). Privacy is also a term with deep emotional resonance. When people voice fears about an erosion or invasion of privacy, they are speaking of a serious threat to their sense of well-being. Confidentiality describes the duties that accompany the disclosure of nonpublic information to a third party within a professional, fiduciary, or contractual relationship. By law, social norm, or contract, and usually by some combination of these, the third party entrusted with the information is prohibited from redisclosing it or discussing it outside the confines of the relationship except under very restricted circumstances.

A third term, security, is related to privacy and confidentiality. It refers to the measures taken to prevent unauthorized access to persons, places, or information. The measures used to achieve the goal of security will vary according to the context and the state of technology. For example, the measures required to maintain the security of data in the virtual world are very different from the measures required to maintain the security of data in the physical world. In ancient Greece, a physician's pledge to keep his patient's secret was largely self-executing, a matter of a single individual resisting the temptation to speak. In contemporary health care, an assurance of confidentiality is nearly meaningless without attention to the potential for accidental release through system malfunction or access by hackers as well as a lapse of vigilance on the part of a range of persons, including data processors and researchers as well as health care professionals.

Arguments in Favor of Protecting Privacy

Philosophers engaged in social ethics typically distinguish between intrinsic value and instrumental value. Something has intrinsic value if it is linked to our most basic beliefs about humanity—protected, not because its protection will produce good consequences but because a good or meaningful human life would be impossible without it. Something has instrumental value if its protection is justified by its potential to produce outcomes we value. Empirical evidence is extremely relevant in a dispute about the degree of protection warranted for something that has only instrumental value. The "right" answer will emerge from a review of the good and bad consequences associated with each of the possible courses of action. Matters are much less clear when the dispute involves something with intrinsic value. Empirical evidence may be relevant, but that evidence will likely concern beliefs rather than benefits and costs.

Privacy has both intrinsic and instrumental value. The argument for intrinsic value is linked to the ethical principle of autonomy or individual self-governance. In the West, this principle came to prominence with the Enlightenment, and it has become perhaps the dominant idea in American law and ethics. In brief, the idea is that authentic humanity requires autonomy, the capacity and opportunity to make important life choices for oneself. Autonomy is linked to privacy because autonomy is only possible where there is a sphere of privacy. "Personal autonomy carries over the idea of having a domain or territory of sovereignty for the self and a right to protect it" (8).

In the approach to ethics most closely associated with the philosopher Immanuel Kant, each person has a duty to respect the autonomy of others. This means, first, that it is wrong to treat a fellow human being solely as a means to an end, even if that end is something noble like the advancement of science or the cure of disease and the prevention of suffering (8). Second, the rules that govern society should be consistent with the principle of respect for autonomy. Other related ethical principles include nonmaleficence, not harming others, and beneficence, taking affirmative steps to further the well-being of others. However, these are supplements to, rather than substitutes for, the principle of respect for autonomy. Using another human being as a mere means—a thing with instrumental rather than intrinsic value—is never appropriate nor is the adoption of rules that violate the principle of respect for autonomy. Although there may be debate concerning the contours of the sphere of privacy necessary for autonomy, the erosion of privacy should never be watched with complacency. The researchers involved in the notorious Tuskegee syphilis experiment were not monsters. They believed they were benefiting society. Their error lay in disregarding the intrinsic harm of enlisting human beings in research without consent and treating individuals as mere means to scientific ends.

Because genetic information is connected to personal and group identity, protecting the privacy of genetic information is an important individual and social priority. Genetic privacy has intrinsic value as a facet of autonomy, and respect for autonomy implies a duty to respect the genetic privacy of others. Within a legal framework, genetic privacy must be considered a fundamental right, and individuals should be able to block or seek redress for invasions of their genetic privacy by other people and by the government. In the next section, we look at the intrinsic value of genetic privacy and the intrinsic harms associated with violations of genetic privacy, using genetic research as an example.

Privacy also may have instrumental value. The instrumental approach to ethics is usually termed utilitarianism or consequentialism. This approach is associated with a group of British philosophers, most notably Jeremy Bentham and John Stuart Mill, who urged a fresh appraisal of social norms and legal rules in light of empirical investigations of their effects on social welfare. The winner in any contest between rules or policies would be the proposal that would eventuate in "the greatest good for the greatest number," the greatest net benefit. Even those who are not strict utilitarians may recognize assessment of outcomes as one dimension of a thorough ethical analysis. Discussions of the ethical, legal, and social implications of the Human Genome Project (HGP) often focus on harms associated with the generation and dissemination of genetic information. These include the erosion of the idea of unconditional acceptance as a foundation of the parent-child relationship and the violation of a child's right to an "open future," the exacerbation of tensions in families and in the courtroom as genetic information is transformed into a weapon to be deployed against a spouse or adversary in a child custody or personal injury lawsuit, discrimination in insurance and employment, and a lessening of social solidarity. Accompanying each of these more specific harms is anxiety about loss of control over the production and sharing of intimate knowledge and the harms associated with the dissemination of unwanted information. Further, because genetic information usually has implications for the family and for the group, harms extend beyond the individual or individuals immediately affected to those with whom they share genetic ties.

Rules protecting the privacy of genetic information are intended to prevent, lessen, or eliminate negative consequences of the new genetics. For example, recent policy statements from the American Society of Human Genetics and other organizations have addressed the appropriate use of predictive genetic testing of children. These statements discourage testing of children for adult-onset disorders and disorders that cannot be ameliorated or cured. Concern is especially great in the adoption context, where children may be labeled on the basis of testing in a way that harms their chances for adoption and their self-esteem (2, 3). The happiness and well-being of children are also at risk when beliefs about parentage are unsettled by an unexpected finding in the course of clinical genetic testing, or through widely marketed DNA-based paternity testing, or when parents who are facing off in a custody battle attempt to force collection or disclosure of information about disease risk.

In adopting rules that restrict access to genetic information by third parties, we greatly reduce the potential for genetic discrimination in insurance and employment and a range of negative outcomes, including the destabilization of the insurance market, the side-lining of productive employees, and an erosion of social solidarity, culminating in the creation of a "genetic underclass" of uninsurables and unemployables. As one of us (M. Rothstein) has argued on many occasions, prohibitions on genetic discrimination are virtually meaningless without limitations on access to genetic information (62). Concerns about genetic discrimination may be used to justify laws that restrict access to persons for purposes of creating genetic information, i.e., restrictions on genetic testing as well as laws that address the confidentiality and security of genetic material and information. Later, we use the example of insurance to explore the instrumental harms associated with violations of genetic privacy.

A related justification for privacy protection is concern for public health. Increasingly, genetic testing will present opportunities to reduce morbidity and mortality. A positive test for a mutation that is associated with a heightened risk of disease may lead to an aggressive surveillance and prevention regimen that includes lifestyle changes and chemoprevention. Pharmacogenomic research, and the identification of genetic variations affecting drug response, should lead to better targeting of treatments and minimization of side-effects. Yet people may forego beneficial testing because they fear that genetic information can and will be used against them. On this point, evidence about public perceptions will be more significant than evidence about the actual prevalence of genetic discrimination, if perceptions of risk influence behavior. In the mid-1990s, researchers looked at the utilization of BRCA1 testing in families with BRCA1-linked hereditary breast-ovarian cancer. They found that subjects without health insurance were nearly four times less likely to request testing than individuals with health insurance, and 34% of subjects said that the possibility of losing health insurance was an important risk. Fear may also inhibit information sharing with health care providers and discourage use of potentially beneficial preventive interventions (35).

Policy makers and advisors across the political spectrum may find the public health rationale for regulation persuasive. Newt Gingrich, currently a senior fellow at the conservative American Enterprise Institute and former Speaker of the U.S. House of Representatives, recently commented, "We want people to know whether they have a propensity for breast cancer, for example, or a particular kind of disease that they ought to be aware of and treating even before symptoms appear. People won't actively seek that knowledge if it could mean that they get fired or dropped by their insurance company. My guess is that we'll eventually adopt a blanket law that prohibits this kind of information from being used for either employment or insurance purposes" (14). On the other hand, there may be costs as well as benefits associated with the regulation of genetic information. Public health officials, health plans, and researchers argue that laws restricting their access to health information hamper disease surveillance efforts and impede the implementation of programs to improve population health. Administrators of publicly financed health care programs also want access to health information in order to discover and investigate fraud and abuse.

The costs of privacy protection are not limited to the health sector. Law enforcement officials resist restrictions that affect public safety. They have, for example, objected strongly to a new regulation from the U.S. Department of Health and Human Services that would require some form of legal process as a precondition for police access to medical records under most circumstances. The consequences of this requirement could include delays in apprehending suspects and, potentially, missed arrests. It is important that the public understand the "price tag" for privacy as well as the instrumental and intrinsic harms associated with gaps in privacy protection. Policy makers need to weigh the costs and benefits of privacy protection in particular contexts.

Public Views on Privacy and Genetics

Even if one accepts that privacy is a fundamental value, it does not follow that privacy is an absolute value. And, as noted above, there will always be costs as well as benefits associated with privacy protection. In recent years, the views of a number of groups concerning the privacy of genetic information, or more broadly, health information, have been solicited. Surveys of public opinion provide insight into evolving social norms. Those norms will, in turn, inform the legal rules that govern the new genetics and the "omics" spawned by the HGP.

In a national survey conducted in August 2000, over three quarters of respondents said it was very important that their medical records be kept confidential (22). The term confidential was defined as, "no one can see it without your permission." Large majorities opposed access to medical records without permission by either employers or insurers (84% and 82%, respectively). Medical researchers fared only slightly better, with 67% of respondents opposing access without permission. The concluding question was, "Should medical and government researchers be allowed to study your genetic information (for example, to identify genes thought to be associated with various medical conditions) without first obtaining your permission, or do you feel they should first obtain your permission." Over nine in ten respondents said that permission should be obtained. Unfortunately, there were no questions regarding trade-offs in terms of cost, convenience, or quality of care. Even if the results overstate the resistance to access without permission, they suggest that privacy is a significant public concern. Researchers and others seeking health information will need to make a strong case if they want access unimpeded by authorization requirements.

The results of a series of focus groups, or mini-hearings, commissioned by the U.S. National Bioethics Advisory Commission (NBAC) are also illuminating (49). Participant demographics varied by location: in Richmond, Virginia, the target population consisted of educated baby boomers; in Honolulu and Mililani, Hawaii, members of neighborhood boards; in San Francisco, California, students and young adults; in Cleveland, Ohio, African-Americans; in Boston, Massachusetts, senior citizens; and in Miami, Florida, Jewish women. Although the samples were admittedly unrepresentative, the format allowed for a more nuanced exploration of attitudes toward privacy and research.

Focus group participants were generally supportive of research, although they varied in their willingness to trade privacy for the benefits of research. For example, most groups lacked consensus on the desirability of a consent requirement for research with stored tissue. Groups were also split between those who would find a general one-time consent acceptable and those who questioned the meaning of consent in the absence of specificity about potential uses. Few participants were troubled by anonymous research. In this area, the chief concern was that restrictions would impede valuable research. Generally, participants in the Cleveland and Miami groups were the most emphatic about privacy protection. Majorities across sites were concerned about insurance company access to research results. Most participants did not object to research that links demographic information to stored tissue samples and were only slightly more concerned about links to medical histories. Indeed, the response to a question about notification in the event of discovery of medically useful information suggests a preference for storing tissue in a manner that permits identification of the source. Many participants across sites believed that in the contemporary health care environment there can be no guarantee of confidentiality. Participants in the Miami group were particularly skeptical about privacy protection in "the age of computers."

Although many participants were not disturbed by the prospect of research that involves specific ethnic or racial groups, in some regions the issue generated disagreement between those who focused on risks and those who focused on benefits to the targeted groups. For example, in Cleveland, "[o]ne participant felt there would be negative consequences from research that publicized that disease belongs to certain groups," whereas "[a]nother person said that information gained from such research is valuable and that ethnic and racial groups are better off when more is known about diseases that may affect them." Regarding sponsorship of research, most participants felt positively about research whether carried out or funded by academic medical centers, drug or biotechnology companies, or the government. Most participants in the Richmond, Mililani, San Francisco, and Cleveland groups were not upset by the profit potential of research that uses stored tissue. Others had mixed feelings about profit and profit-sharing with tissue donors. Most participants across groups favored a requirement that all studies be approved by a committee appointed to oversee the ethics of research.

These results suggest a public that is supportive of research but unwilling to trust researchers to conduct ethical research or protect subject privacy and confidentiality without oversight. A similar study in Britain found the same positive attitude toward research, although genetic research was regarded as mysterious and even sinister (77). In addition, people appear to be realistic about the possibility (or impossibility) of perfect data security, but they are unwilling to compromise demands for the best data protection attainable. Owing to these tensions, a scandal that involves carelessness on the part of researchers has the potential to undermine public support for research, just as efforts to educate the public about specific safeguards implemented to protect privacy have the potential to diminish anxiety about research risks. There also appears to be fear bordering on fatalism about the erosion of privacy due to computerization. A growing sense of insecurity may explain a survey finding from January 1999 that pertains to the public health rationale for privacy regulation. Questioned about privacy-protective actions, one sixth of respondents reported providing inaccurate information, changing physicians, avoiding care, or taking some other action to safeguard their privacy (30).

Another theme in the report of focus group results is the variation among individuals in areas such as perception of and aversion to risk. The difference can be great even among individuals who belong to the same demographic group. Members of some groups do have heightened concerns about privacy, however. For example, in the Gallup survey cited earlier, women and older adults felt more strongly about the confidentiality of medical records than other groups (22). In the NBAC–commissioned study, participants identified as African Americans or Jews—groups injured by research abuses in the past—were the most concerned about privacy protection (49). Reaching out to these groups presents unique challenges and opportunities. One participant in the Cleveland group expressed skepticism, if not distrust, of researchers, commenting that the Tuskegee experiment was not so long ago. Another participant responded that researchers today are different. Researchers have a responsibility to demonstrate as well as assert that they can be trusted with sensitive information.

Other Policy Considerations

As suggested by focus group participants' comments about insurers and computers, policy development at the intersection of genetic science and privacy regulation is affected by social context. History, the social safety net, and technology are among the most important contextual factors. Commentators note that Europeans tend to view privacy protection within a civil rights or civil liberties framework (66, 67). Although Americans may treat privacy as a basic right, especially in the negative sense of "a right to be left alone," there are also instances in which privacy (or personal information) is treated as a commodity to be dealt with according to standard rules of property law. History may be partly responsible for this divergence. Nazi success in segregating and transporting Jews and seizing their assets has been attributed, in part, to the existence of extensive repositories of personal data from public and private sector sources (66, 67). In the United States, government data sources were used to locate Japanese-Americans for internment during World War II, but public awareness of this episode is limited.

Gaps in the social safety net also register in discussions of genetic discrimination. In the United States, the absence of universal access to health care has led the public and policy makers to focus on discrimination in health insurance. In countries with national health insurance, the potential for genetic discrimination in that arena is lessened, if not eliminated. To be sure, private health insurance may be an important supplement to a public health care system, and access to genetic information generated through the public health care system remains a matter of grave concern insofar as this information can serve as a basis for discrimination in other areas. Still, given the different social context, in European countries there has been greater attention than in the United States to the issue of genetic discrimination in life insurance.

As for technology, Scott McNealy, President and Chief Executive Officer of computer giant Sun Microsystems, reportedly said, "You've got zero privacy now. Get over it" (70). Some privacy advocates are concerned that such hyperbole may trigger a downward spiral in privacy protection. To the extent that regulators and the courts use reasonable expectations of privacy to set the bar, the guardians of individual rights and public order may end up facilitating rather than impeding the erosion of privacy. "The more intrusive surveillance technology becomes, the less reasonable is any expectation that individuals will have privacy, and as a consequence, the less privacy the law will recognize" (66). If privacy is to be preserved, rules must be fashioned that do more than simply enshrine the status quo. In the remaining sections of this chapter, we consider the harms that may result from disregard for privacy on the part of researchers and insurers, and we look at the prospects for consensus on rules that safeguard privacy without unduly restricting the flow of genetic information for beneficial purposes.

ISSUES SURROUNDING GENETIC RESEARCH

In the previous section, we reviewed evidence concerning public attitudes toward research. We also touched on some controversies that are likely to arise in genetic research: the nature and value of consent, including the acceptability of research without consent or of blanket consents to future research; the appropriate role of community representatives in research; and the desirability of commercialization with or without benefit sharing. In each case, the debate is less about consequences than it is about the meaning of respect for autonomy and about fundamental human rights. The announcement of plans for the creation of a database containing genetic and genealogical information and medical records for a substantial portion of the population of Iceland and the licensing of that data to a private, for-profit firm served as a catalyst for a lively public discussion of these and related issues.

A Contemporary Icelandic Saga

As the focus of research shifts to complex disorders, family linkage studies may have diminishing scientific value. With computerization, it has become possible to amass and link large quantities of medical, lifestyle, and environmental data for selected populations and to use sophisticated algorithms to analyze or "mine" the data for patterns of association between genetic and other variations, manifestations of disease, and presumably, dimensions of health, such as unusual longevity.

Iceland has much to recommend it as a site for this new research strategy. Its assets include: a relatively homogenous population, positive public attitudes toward medical research, a national health system with extensive stores of health data and tissue, and unusually detailed genealogical records. In terms of commercial potential, it may help that the population is Caucasian (26).

deCODE Genetics, Inc. (deCODE), a private for-profit company with ties to Iceland and the United States, has spearheaded the drive to develop Iceland's genetic resources. The Iceland project attracted world attention, in part because a public-private partnership was ratified by national law. In December 1998, the Icelandic parliament passed the Act on a Health Sector Database (Database Act) (31a). The Database Act provides for the aggregation of health system and other data in a single large database under license. Among other things, the Database Act requires the licensee to adopt security measures consistent with the requirements of the national data protection commission and to provide a regularly updated copy of the database to the health ministry. It also gives the ministry free access to statistical information from the database for health planning and other projects. Further, the licensee is not permitted to grant direct access to the database to third parties. The licensee must process information in a manner that precludes linking output to individuals. On January 21, 2000, a license for the Icelandic Health Care Database was formally awarded to deCODE (26).

The decision to seek ratification of the partnership through legislation has allowed database proponents to claim the legitimacy of the democratic process. It has also exposed the project to an unusual degree of public scrutiny, resulting in some alterations in design. These alterations have not, however, mollified the opposition. The most controversial feature of the database is its opt-out structure. The Database Act allows individuals to request that information from their health records not be entered into the database, but the default rule is inclusion. The commerical aspect of the database project has also been much criticized.

In making the project a matter of national policy, deCODE secured advantages not typically available to private sector entities. It also ensured that certain benefits would accrue to the citizens of Iceland. Most immediately and directly, the licensing arrangement is a mechanism for funding the computerization of the Icelandic health system. In addition, the law affirms the status of the information shared under license as a public resource that continues to be available for public purposes. Because the work is to be conducted in Iceland, the project is a potential engine of economic and scientific progress for the country. Also, in a five-year collaborative agreement between deCODE and pharmaceutical giant Roche Holdings, Roche has reportedly agreed to provide any products resulting from deCODE's work to Icelanders without charge, at least during the period of patent protection. The Database Act does not itself require this form of benefit-sharing. As of February 2001, deCODE and Roche were claiming credit for reaching "research milestones" in the mapping of genes linked to osteoporosis, osteoarthritis, stroke, Alzheimer's disease, and schizophrenia (15).

Collection of genetic samples for analysis is an independent operation not covered under the Database Act. Therefore, this aspect of the deCODE business plan was, at least initially, subject to a more general law that requires consent as a precondition to participation in biomedical research. However, on May 13, 2000, the Icelandic parliament passed an Act on Biobanks (Biobank Act) (31b). It defines a biobank as a collection of biological samples, organic material from human beings living or dead, that is permanently preserved. The law does not apply to the "temporary keeping" of samples for clinical testing, treatment, or a specific research study for a period of up to five years, so long as arrangements are made for destruction at completion. Materials stored for assisted reproduction or organ transplantation are also excluded. Anyone seeking to establish or operate a biobank must obtain a license from the minister of health. Conditions for approval of a license include submission of a description of the bank's objectives, the basis of operation, and the conditions of storage as well as the protocols for collaboration with foreign parties and an evaluation of security measures for consistency with rules issued by the national data protection commission. Location in Iceland is also required, and another provision of the Biobank Act prohibits any export of biological samples without approval from the national data protection and bioethics bodies.

The Biobank Act requires affirmative informed consent for the collection of samples, with one major exception. The law provides that a sample collected for a clinical test or treatment may be stored in a biobank if written information of this possibility was available to the source or "assumed donor." Individuals can register a standing objection to the banking of their samples, and a donor is permitted to withdraw consent at any time, triggering destruction of any stored samples. However, "material that has been produced from a biological sample by performance of a study," and the results of studies already carried out, will not be destroyed. Before a researcher can gain access to samples in a biobank, the national data protection commission must grant its permission, and the specific research protocol must be approved by the national bioethics committee or an institutional ethics committee. Significantly, the board of a biobank, with approval from the national data protection and bioethics bodies, may authorize the use of a sample for purposes other than those for which it was collected, "provided that important interests are at stake, and that the potential benefit outweighs any potential inconvenience to the donor of a biological sample or other parties." By its terms, the Biobank Act takes effect on January 1, 2001. However, legal challenges have already been filed. In particular, opponents assert that the qualification to the right of withdrawal, a late addition to the law, violates Article 71 of the Icelandic Constitution and Article 8 of the European Convention for the Protection of Human Rights and Fundamental Freedoms (78).

With the passage of time, it has become possible to speak of the lessons of the Iceland experiment. The strongest and most enduring criticisms of the database have been directed at its opt-out structure. Many commentators believe this structure guts or trivializes the principle of informed consent. Henry Greely has written one of the most balanced assessments of the project. Greely calls the opt-out structure a form of conscription, and he does not believe an adequate case has been made for this departure from respect for autonomy. "Against the unusual breadth of the information to be gathered, the potential social benefits of the information are entitled to some weight. But the medical benefits are speculative and the commercial benefits would accrue mainly to a private corporation. More importantly, there seems no special reason to believe that informed consent would be unusually difficult or expensive to obtain" (26). To be sure, scientists may fear that less than universal participation will bias the results of their research, if those who fail to respond share some relevant characteristic. The case for "presumed consent" is strongest where scientists can point to some factor, other than objection to participation in the research, that will hamper response. Given claims of broad public support and Iceland's well-educated, affluent population, the bias argument is weak. Indeed, the negative publicity generated by the consent controversy may cause some individuals who would have given consent under an opt-in structure to opt-out. The opposition group Mannvernd reports that as of February 28, 2001, over 19,675 individuals had registered with the government for exclusion from the database (41).

In the twentieth century, consent to research has become a legal issue as well as an ethical one. The Nuremberg Code, drafted by the judges who tried German physicians for crimes in the course of medical research, endorses the principle of informed consent and describes consent as an affirmative decision by the experimental subject. The World Medical Association's Declaration of Helsinki and the Council of Europe's Convention on Human Rights and Biomedicine also stress the importance of informed consent as a fundamental human right. Arguably, these pronouncements are sufficient to give informed consent for participation in research the status of customary international law. Database proponents argue that it is common practice to conduct medical records research and research with stored tissues without consent, provided that identifiers are removed. This distinction is, for example, incorporated in the European Union Directive 95/46/EC, which requires subject consent before personal data are included in a database. Greely believes the Database Act is incoherent, treating a one-way coding system as a means of rendering data nonidentifiable, yet contemplating that data from different sources will be linked within the database and will be continually updated. It appears possible that at least one of the parties to the operation will be able to link database information to a personal identifier. According to Greely, that makes the data personally identifiable under the law of Iceland and the European Union (26). In going forward with its own national DNA database, Estonia plans to use an opt-in structure in which consent is a prerequisite to inclusion (20).

The charge of commericalization of the genome through creation of the Icelandic database is partly false, and somewhat misleading if it implies a commerce-free alternative. Genes are not being sold, although there is a clear acknowledgment in the Database Act that the data flowing to the licensee will be used to generate profit. A process of development that would not involve commercial entities at some stage is difficult to envision. The question is where to draw the line, considering factors such as comparative advantage and the appropriate allocation of responsibilities and anticipated rewards (26). Estonia plans to vest control over its database in a nonprofit state-controlled foundation, a joint venture of the country's health ministry and a private entity, the Estonian Genome Foundation. However, investors are being solicited to finance the start-up, and a for-profit subsidiary of the foundation will have the power to sell access and information (20). Benefit-sharing is another facet of the issue. Commercial entities may accept benefit-sharing as a matter of justice, beneficence, or public relations. Organizers of the Estonian database have announced that they will permit donors to access their own genetic data. They have also said that public researchers will be able to access data for free or a small fee. Whether that access will be restricted to accommodate the biotechnology or pharmaceutical companies that pay (presumably) substantial fees to mine the database for information on specific conditions is unclear.

Privacy considerations do counsel strongly against auctioning off data rights to the highest bidder. In a world of imperfect security, trust and trustworthiness are important. In response to insistent questioning from an interviewer on the topic of security, deCODE's Kari Stefansson replied, "We have never claimed that the protection of privacy cannot be broken. The principal element here is trust. This is crucial because it would be silly for people to say you can't put together a database like this unless it can never be violated. It's like saying you can't build or drive cars because they are the cause of deaths on the road. It's not going to be impossible to violate the database. But it's going to be a lot harder than in any other place where people are doing research involving humans. And unlike any other place, it will be a criminal offense" (43). Studies of opposition to genetically modified foods suggest that the level of trust in institutions is the key factor affecting public perceptions of risk (33). In England, a national database project is a joint venture of a public agency and a private foundation. The Medical Research Council and the Wellcome Trust are collaborating to create a U.K. Population Biomedical Collection, containing samples, linked medical records, and family histories from up to 500,000 people. Access by researchers will be limited to data stripped of identifiers, all protocols will be peer-reviewed, and no one will have exclusive rights to data. Commercial entities will compete for access on the basis of scientific merit rather than price. The planned concentration on cancer and cardiovascular disease is also likely to lessen controversy, although some may complain about a narrow focus on the disorders of the affluent.

Community involvement has not figured prominently in the Icelandic database debate, perhaps because Icelanders have played a major role in the initiation, ratification, and implementation of the database project. The issue has been significant where researchers and subjects have had little in common. For example, charges of biopiracy, vampirism, and helicoptor genetics have been leveled in connection with a blood collection expedition carried out by U.S. researchers in Newfoundland (5, 27, 71). In the Newfoundland case, a major source of dissatisfaction was the researchers' failure to share findings with the people who contributed samples and information. Reports suggest that Newfoundlanders are not angered by the notion that others may reap financial rewards from research involving their DNA (5, 27, 71). The greater the cultural and socioeconomic distance between researchers and research subjects, the greater the potential for misunderstanding as well as the perception, if not the reality, of exploitation for ends that are not shared. The label of biocolonialism has been affixed to genetic research conducted by investigators from elite Western institutions and allied commercial interests in China, in other developing countries, and among Native American populations (16, 68). The Human Genome Diversity Project has also been tagged with this negative label (51).

A related issue, intertwined with concerns about commercialization, is what to do about the expectations created in people, often culturally as well as genetically isolated, who see biobanking as their economic salvation. Although money factored in the push to create the Icelandic database, Iceland is a relatively wealthy country, and the fractiousness of the debate is an indicator of an absence of economic desperation as well as deep ethical disagreement. Observers of the Estonian database project attribute the lack of opposition, at least in part, to "eagerness to become a player in the world economy" (21, 39). According to news reports, a growing number of villages in the Cilento region of southern Italy see their future in the creation of a genetic park. The mayor of Gioi Cilento is quoted as saying, "Many of our children have gone, it's mostly old people, which means our communities are dying. This has given us hope for the future. It is a chance to create tourism" (12). The most eagerly awaited members of the advance team are the experts who will offer workshops on how to set up bed and breakfasts, not the geneticists. In this global equivalent of the game of Monopoly, other biotechnology companies are collecting island nations in the Pacific (52). It is somewhat ironic that China plans to establish its own genetic park (50).

Amid all this activity, it is difficult to discern a consensus on basic principles, although agreement on some key points, such as informed consent, may be emerging. The Ethics Committee of the Human Genome Organization has already published statements on topics such as benefit sharing (31). Pharmaceutical companies worry that inconsistent guidance at the regional and national levels on matters such as sample identification, consent for future research, and length of storage may adversely affect the international enterprise of genetic research (24).

The Emergence of Private Gene Brokers

In the United States, genetic heterogeneity and the multiplicity of legal regimes may be barriers to the creation of a public database or biobank for genetic research. It is difficult to imagine a scenario that would lead to the creation of a U.S. Population Biomedical Collection. Large collections of tissue and information do exist. For the most part though, these are special-purpose databases and banks governed by policies that limit research and commercial development. Examples include the Defense Department's DNA Specimen Repository for Remains Identification, identified as the largest DNA bank in the world, and the forensic DNA banks maintained by the Federal Bureau of Investigation and the states. A survey of storage sites for biological materials in the United States arrived at conservative totals of 176.5 million cases and more than 282 million specimens, with accrual of new cases at the rate of 20 million a year (17). Some specimens are well-characterized or accompanied by identifiers that permit linkage to medical and other records, but little uniformity exists in this area. Certain groups have engaged in extended collaborations with geneticists and genetic epidemiologists interested in particular disorders. The best-known examples are the Amish, Ashkenazi Jews, Mormons, and Native Americans. In some cases, state or federal public health agencies have been involved in the research. However, so far as we know, there have been no cases of direct involvement by a political unit in creating a large-scale database or bank for the broad purpose of facilitating a data mining approach to genetic research.

It is perhaps unsurprising, then, that the United States has become the incubator for database and biobanking projects initiated, funded, and operated solely within the private sector. Pioneers include First Genetic Trust, Inc. (First Genetic), based in Chicago, Illinois; The Gene Trust, a project of DNA Sciences Inc., a privately held company with headquarters in Mountain View, California; and Genomics Collaborative Inc. (GCI), a privately held company based in Cambridge, Massachusetts.

The business idea behind First Genetic is that an intermediary will address the privacy concerns thought to limit participation in research. Through a web site, the company will offer information on clinical trials, plus decision-making algorithms and contact information for genetic counselors. Individuals will also be able to deposit their genetic information and then register to participate in clinical trials through online accounts. First Genetic has stated that any information provided to external researchers will be de-identified. Although promoters suggest that the long-term goal is to facilitate personalized medicine, at least initially the aim will be to facilitate patient recruitment for research conducted by pharmaceutical companies (18).

Like First Genetic, DNA Sciences uses a web site to solicit participation. Individuals are asked to provide contact information and a personal and family health history. If they match certain profiles, a phlebotomist collects a blood sample. In the 3-month period after project launch, DNA Sciences reportedly recruited nearly 5000 volunteers via the Internet (58). Like First Genetic, DNA Sciences emphasizes privacy. Three of the six guiding principles for the Gene Trust concern privacy and security, pledging the company to always obtain "the express, written permission of the individual" before selling or releasing individual identifying information, to "work for laws that protect individuals from discriminatory, inappropriate, or unfair use of information about their genetic makeup," and to store health and genetic data in "restricted databases accessible only by authorized individuals" (16a). Revenues are to be generated through development of intellectual property, alliances with health care service providers, and sales of aggregated health and genetic information to third party researchers. Volunteers are promised updates from medical journals and free access to any diagnostic tests developed by the company for their conditions (56).

GCI uses a feeder network of 200–300 physicians (or physician groups) in the United States, as well as physicians in Eastern and Western Europe, North Africa, India, and Southeast Asia, to recruit patients on its behalf and handle the informed consent process. The president of GCI has said that the company is enrolling approximately 7500 patients a month, with 55% of subjects coming from the United States (58). The company also contracts with Ameripath (a supplier of pathology services to hospitals) to provide tissue, blood samples, and clinical information to a GCI bank for genetic studies. According to news reports, in contrast to First Genetic and DNA Sciences, GCI has not promised information or free services to research participants (19).

The privacy problem looms so large in research because anonymous research has deficiencies from nearly every perspective. As suggested by the responses from focus groups, many individuals would like to be notified of research results that have medical relevance, and this weighs against the complete removal of identifiers from genetic information. Further, the ability to link genetic information derived from biological materials with medical records and other sources of personal information may be extremely important for genetic epidemiology. "Tissues without a link are of limited use because of the inability to update outcomes data; to link with related clinical, outcome, or utilization data; or to solicit subjects for more information or for participation in follow-up studies" (45). Gene brokers purport to solve the privacy problem. However, their trustworthiness has yet to be established, and the rules for this activity are unclear.

What Rules for Exchanges of Genetic Information?

Within the United States, there is no definitive legal framework for brokering of genetic information. A few state laws obliquely address the relationship between property law and genetic information. For example, a section of the Oregon genetic privacy act provides that "an individual's genetic information and DNA sample are the property of the individual except when the information or sample is used in anonymous research" (54). However, the law adds that this provision "does not apply to any law, contract, or other arrangement that determines a person's rights to compensation relating to substances or information derived from a sample of an individual from which genetic information has been obtained." Further, a 1999 amendment to the Oregon law, effective until January 1, 2002, deems as anonymous any research conducted in accordance with federal regulations for the protection of human subjects. Several states have laws that contain general declarations that genetic information is the unique or exclusive property of the person tested. Clearly, such statements are intended to send a message about the intrinsic value of genetic information and its relationship to individual identity, but their practical import is unclear. In a widely cited case, *Moore v. Regents of University of California* (46), a court rejected a patient's argument that commercial development of a cell-line derived from his excised cells amounted to theft, but it granted patients a right to information concerning their physicians' economic and research interests.

Even if DNA, and the information derived from it, is best regarded as property, it must be a special kind, in view of the intrinsic harms associated with careless handling. One legal scholar has made the case for applying rules developed for intellectual property to transactions involving personal information (66). Significant features of intellectual property law include a default rule that licenses are nonexclusive and nontransferable (i.e., redisclosure is prohibited unless specifically authorized) and that any restrictions on use travel with the information. If courts instead apply traditional property law, taking no account of the special nature of genetic information, then serious violations of privacy are likely. For example, if an intermediary files for or is forced into bankruptcy, the contents of a bank may be sold to the highest bidder (72). Until a guarantor of the obligations of DNA banks and databases appears, legal enforcement of privacy policies may provide some protection for the public.

What Rules for Research Uses of Genetic Information?

In the United States, as in many other countries, a special body of law imposes additional obligations on researchers conducting studies that involve human subjects. At the national level, current standards are derived from the Policy for the Protection of Human Subjects (often referred to as the Common Rule) (76a). Technically, the Common Rule only applies to research conducted, supported, or otherwise subject to regulation by a department or agency of the U.S. government. Also, because the definition of the term human subject refers to living individuals, biological materials and data derived from deceased persons are not covered. Another provision of the Common Rule limits coverage to research involving intervention or interaction with the individual, or identifiable private information, meaning that the identity of the subject is or may be readily either ascertained by the researcher or associated with the information. In the past, the federal oversight body has taken the position that information is identifiable where codes can be broken with the cooperation of others (48). Some research, though covered, may be eligible for an exemption from regulatory requirements. The categories of exempt research include research involving the collection or study of existing data or specimens, if these sources are publicly available or if the information is recorded by the investigator in such a manner that subjects cannot be identified directly or indirectly through identifiers.

If research is within the scope of the federal regulations and nonexempt, there are two basic requirements: Approval must be given by an institutional review board (IRB) and informed consent must be obtained from participants. Privacy has not been a major component of IRB review, although the regulations provide that when appropriate, adequate provision to protect the privacy of subjects and maintain the confidentiality of data is a condition of approval. Privacy and other autonomy-related concerns have become linked to the informed consent requirement. As the law stands, informed consent for research can be altered or waived if an IRB finds and documents that (*a*) the research involves no more than minimal risk to the subjects, (*b*) the waiver or alteration will not adversely affect the rights and welfare of the subjects, (*c*) the research could not be practicably carried out without the waiver or alteration, and (*d*) whenever appropriate, the subjects will be provided with additional pertinent information after participation. Historically, tissues, blood, and other biological materials used for research have been stored and studied without explicit consent. With the advent of genetic research, some have argued that the risk associated with the use of identified or identifiable data or specimens is no longer minimal and therefore consent should be obtained, particularly where protocols propose sharing information with the source or a third party (13).

In August 1999, NBAC issued recommendations for the interpretation of federal regulations relating to the use of human biological materials in research (48). NBAC concluded that use of specimens for purposes other than the purpose for which they were originally collected raises the strongest privacy concerns. It recommended that any research involving identified or coded (identifiable) samples be considered greater than minimal risk and, therefore, ineligible for a consent waiver, unless an IRB were to find that a particular study adequately protected the confidentiality of personal information and incorporated an appropriate plan for whether and how to reveal findings to the donors or their physicians. For nonwaiver research, NBAC expressed support for the menu model of informed consent. An alternative proposal from Greely would mandate IRB approval for all research with human information or materials, end the practice of making protections contingent upon increasingly arbitrary distinctions concerning identifiability, and allow for informed consent to be replaced by informed permission for unforeseen future research uses, at least under some circumstances (25).

Individual states sometimes add their own rules for research. Research is usually exempt from genetic privacy laws if it is anonymous. Depending on the language of the law, this may entail that the genetic information be unidentifiable, that identifiers be replaced with a code, or that identifiers not be released to researchers. In the pre-HGP, pre-managed care era, medical records research was considered entirely benign and was conducted without subject knowledge, let alone consent. Hence, state medical records laws generally permit access by researchers, so long as they maintain the confidentiality of identifiable information. Minnesota created a considerable stir when it passed a law imposing strict requirements on external researchers (44). Forensic databanks are perhaps the most heavily regulated repositories in the United States. Typically, disclosure of stored information is permitted only for law enforcement–related purposes.

In the United States, the handling and use of unidentified or de-identified samples and information is almost completely unregulated. Research involving biological material or information that has been stripped of individual identifiers is very unlikely to generate information that will result in immediate and direct harm to particular donors. It may, however, reveal sensitive information about the groups to which the donors belong, and policymakers are only now beginning to pay attention to this problem. The problem is especially acute where third parties can compel the production of research data. A 1998 federal law makes data generated through federal grants to nonprofit organizations publicly available through the Freedom of Information Act procedures (42, 73). The law provides for individual identifiers to be removed before release, but this step has not allayed the concerns of researchers and privacy advocates. With expanded capacities to manipulate data, redaction to eliminate obvious identifiers may not adequately protect the privacy of individuals; and it certainly does not protect the privacy of groups. Under the Public Health Service Act, certificates of confidentiality may be obtained for sensitive research, including genetic research (75). However, certificates of confidentiality protect against forced disclosure of identifying information about individuals, not forced disclosure of aggregate data.

Pharmacogenomics

As in other areas of genetic research involving human subjects, pharmacogenomics-related clinical trials raise important questions of informed consent and protection of privacy and confidentiality. Current notions of autonomy require that subjects agree to enter into research based on adequate information regarding the risks and benefits of participation. Genotyping appropriate to pharmacogenomic research may not produce information regarding susceptibility to disease or early death, but it may reveal evidence of a genetic variation or characteristic that would lead to a classification as difficult to treat, less profitable to treat, or more expensive to treat (64). Also, genetic tests performed to determine drug response or susceptibility to side effects could reveal information about susceptibility to disease. For example, scientists have found that a SNP pattern in a cholesterol-metabolizing gene affects response to the cholesterol-lowering drug pravastatin. However, the SNP pattern that serves as an indicator for treatment with pravastatin is also a marker for a bad prognosis if cholesterol levels are not lowered. As Francis Collins has remarked, "You won't be able to separate the reason you did the test from the response to it" (34).

In addition, the risk of stigmatization exists where clinical trials target members of specific population subgroups. Genetic variations of pharmacological significance may occur with greater or lesser frequency in different ethnic groups. The meaning of this finding can be debated. Nevertheless, ethnic origin must be considered in pharmacogenetic studies and in pharmacotherapy. Researchers need to account for ethnicity where scientifically appropriate, while respecting the autonomy of individuals and taking account of the potential for group harms. Additional human subject protections for research involving specific subgroups may be necessary, not only as a safeguard against undue risk but also as a means of educating the communities that may be involved in research about the absence of undue risk (64). These additional protections would likely include community consultation.

Philosopher Allen Buchanan has distinguished among three different rationales for community consultation: abandonment of the assumption of individual agency as an ethical universal, respect for local customs of collective decision making or consultation with honored representatives such as community elders, and reduction of the potential for group-based harms. Only the first rationale would justify a community veto over the individual, and this would not accord with respect for autonomy. Buchanan believes that the second and third rationales may be compelling in particular circumstances, supporting a requirement of community consultation. The difficulties that may attend attempts at community consultation include identifying the relevant community; handling the risks of coercion of the individual by the community or a particular segment of the community; preserving neutrality in contests for control within communities; and avoiding complicity in racial, ethnic, religious, caste, or gender oppression. Buchanan concludes, "Whether these risks are worth taking will depend largely on three factors: 1. whether there is a significant risk of group-based harms (rather than a mere possibility of them), 2. whether other protections against the group-based harms in question are likely to be adequate, and 3. whether a process of consultation can be devised that is not likely to reinforce oppressive inequalities within the group or to become an arena for political entrepreneurship by would-be leaders of the group" (10, 32, 69).

Some researchers attempt to minimize misuse—or the perceived danger of misuse—of genotypic information by withholding all information generated in the course of research from participants. This strategy may be appropriate if the research will not reveal information of medical relevance to participants and participants are informed before entering into the research that no findings will be disclosed. As a general approach to the problem of privacy in research, it is gravely flawed. Incorporating an information restriction in an informed consent form may adhere to technical requirements for human subject protection, but depriving individuals of the opportunity to request or decline significant personal information in the possession of a third party reduces rather than enlarges autonomy.

INSURER ACCESS TO GENETIC INFORMATION

Any rational individual contemplating genetic testing or participation in genetic research will weigh the benefits and risks. The possible benefits of predictive genetic testing include relief from anxiety, increased knowledge and ability to plan (depending on the nature of the result), and better decisions about the prevention and treatment of disease. The risks include the cost of testing, negative psychological and emotional reactions to test results, familial and social disruption, and discrimination by third parties with access to test results. Most of these have a common source: the gap between our ability to identify genetic risk and our ability to ameliorate it. Nearly two thirds of respondents in a 1997 survey reported that they would not undergo genetic testing if health insurers and employers would have access to the results, and a 1995 survey found that over 85% of respondents were very or somewhat concerned about access to and use of genetic information by insurers and employers (76c). These data and the data concerning utilization of genetic testing suggest that, lacking legal protections against genetic discrimination, many individuals will refuse testing and will fail to take advantage of available interventions that might lower the morbidity and mortality associated with genetic disorders. Presumably, they will also decline to participate in research if they fear that research data could find their way to an insurer or employer. The individual and societal costs of this behavior are likely to be enormous.

Two Types of Concerns: Rational and Irrational Discrimination

In assessing the social costs and benefits of regulation of insurers' access to genetic information, distinguishing between rational and irrational discrimination is important. In insurance, irrational discrimination results from decision making on the basis of faulty or incomplete data, misunderstandings of genetic science, or misreadings of the implications of genetic test results for morbidity and mortality. At present, the predictive value of most genetic tests is limited. A positive result does not necessarily mean that the person undergoing the test will develop a disease, and a negative result does not necessarily mean that the person will escape it. The reliability of genetic tests in predicting mortality from complex disorders is even more questionable (60). For new genetic tests, the generation of good mortality figures will take years. Further, changes in individual behavior triggered by testing may alter risk significantly. A positive result may motivate the person undergoing the test to take specific steps to prevent the likelihood of a premature death. For example, a person with a genetic predisposition to breast cancer may have regular mammograms and conduct frequent self-examinations, thereby detecting cancer in its earliest stages. On the other hand, a negative result may give the person a false sense of security and reduce the incentive for precautionary measures, leading to premature death.

Few defenders of irrational discrimination exist, and many states prohibit insurers from charging nonstandard premiums or denying an application for insurance unless these determinations can be justified by sound actuarial data or reasonably anticipated experience. Such laws imply that rational, scientifically sound and empirically supported discrimination is permissible. Indeed, in the insurance context, it has been common for issuers of health, life, disability, and long-term care insurance to engage in risk classification based on characteristics such as age, individual and family health history, health status, occupation, serum cholesterol, and alcohol and tobacco use. Insurers view genetic information as simply one additional factor to be evaluated (55). Discrimination among risks is considered ethically problematic only where there is no sound actuarial basis for the manner in which risks are classified, or individuals of the same risk class are treated differently. Hence, the more information available to insurers the better—the more precise the discriminations and the greater the actuarial fairness of the system.

The pledge to actuarial fairness is often combined with an expression of concern about the financial well-being of the industry should access to information be restricted or denied: "When there is suppression of information and insurers are forced to charge the same rate to insured persons who have different expected costs, then the insurance pricing might not only be viewed as unfairly discriminatory to the group of lower expected cost persons, but will also encourage moral hazard and adverse selection against the insurer. The financial effects of these adverse incentives ultimately drive rates upward and may even threaten the solvency of insurers" (9).

Moral hazard refers to the lessening of incentives to exercise care once an individual has purchased insurance. Adverse selection refers to the idea that where individuals have knowledge of a characteristic that increases the likelihood of a claim and insurers lack it, they will exploit the informational asymmetry and load up on insurance at bargain prices. As the proportion of higher-risk individuals in an insurance pool increases, payouts will increase; and as payouts increase, premiums for all policyholders will go up. In a voluntary system of insurance, premium increases can be expected to drive out lower-risk individuals, resulting in a further increase in the proportion of higher-risk individuals and, in the nightmare vision of representatives of the insurance industry, total market collapse.

A recent Utah study found little evidence that confidentiality protections for predictive genetic testing lead to adverse selection in life insurance (79). Insurers already cover many people who have genetic disorders or predispositions to disease (38). The drive for greater precision may have more to do with ideology than economics; if the outcome is increased premiums for some existing insureds and decreased premiums for other existing insureds, it is difficult to justify the administrative costs associated with genetic underwriting in purely economic terms. The economic analysis will likely vary with the potential for loading up behavior, and that will vary by type of insurance.

A consequentialist analysis will inevitably involve reflection on the moral meaning of the different types of insurance. In the past, there was little need to choose between the two major rationales for insurance, individual risk-diversification and social risk–spreading. Predictive genetic testing will tend to drive a wedge between the two if testing becomes widely available, and there are few restrictions on use. Individuals who do well in the genetic lottery will be able to pinpoint their few areas of risk and may target their purchases of insurance accordingly, or they may drop out of the insurance market altogether. The case for regulation of insurers is based on convictions about social solidarity, the principles of beneficence and nonmaleficence, and the public health rationale for restricting practices that will discourage individuals from undergoing beneficial genetic testing.

Coerced collection of genetic information also implicates intrinsic values. When third parties demand genetic testing or access to genetic information as a condition to the receipt of essential goods, such as a job or health care, individuals effectively lose control over access to their bodies. More importantly, they lose control over the generation and dissemination of personal information. Of course, this happens every time an insurer demands a cholesterol test. It is difficult to argue that genetic tests vary from cholesterol tests in kind. Genetic tests are, however, at the far end of the spectrum in terms of the sensitiveness of the information, its potential for misinterpretation, and its relevance to family members. An insurer that requires a young woman with a family history of Huntington disease to undergo genetic testing effectively requires her to know, many years in advance, that she will (in all likelihood) die at an early age of a terrible disease or that she has escaped this fate, a fact that may alienate her from other family members. Finally, unless classification on the basis of genetic information is prohibited, the person who tests positive for a disease mutation will likely find that his or her autonomy and range of opportunities have been sharply restricted.

Skeptics rightly point out that evidence of genetic discrimination has been anecdotal or derived from studies with serious methodological flaws. Hence, a recent study combining in-person interviews with health insurers and a direct market test has attracted considerable attention (29). To the surprise of some, the investigators found that a person with a serious genetic condition but asymptomatic for disease would have little or no difficulty obtaining individual health insurance under current market conditions. The investigators also concluded that there was no significant association between the degree of difficulty in obtaining insurance and the existence or absence of a state law regulating the use of genetic information. Although this study may reassure individuals worried about losing their health insurance, particularly if they are considering testing for the BRCA1 or BRCA2 mutations, other risks are not addressed, such as the risk of discrimination in life insurance or employment. Further, the finding that there is no widespread genetic discrimination in health insurance at present does not necessarily undermine the case for regulation. The amount of genetic information in the average individual's medical file is likely to increase greatly. In addition, in the interviews, a general sense of legal and social disapproval emerged as an important consideration in insurers' decisions not to inquire about or request genetic testing.

Considerations Relating to Health Insurance

In the United States, many people affirm the importance of health care as a basic individual and social good (8). At the same time, Americans tolerate a system, or nonsystem, in which millions of people lack health insurance and access to high-quality basic health care. Any treatment of health insurance in economic or ethical terms is complicated by the variety of health insurance arrangements. In 1996, 43.1% of the population relied on health insurance paid for by private-sector employers, 34.2% had publicly funded insurance, 7.1% purchased their own coverage, and 15.6% were uninsured (11). Medical underwriting for individuals is unlawful in the Medicare and Medicaid programs and, under the Health Insurance Portability and Accountability Act (HIPAA) (74), in employer-sponsored group health plans. In addition, HIPAA precludes the use of genetic information to establish a pre-existing condition in the absence of a diagnosis.

Those who purchase their own coverage in the individual market are the most vulnerable to discrimination. Insurers may gain access to genetic information through the underwriting process, in connection with applications for insurance, and through claims data. The Medical Information Bureau allows insurers to share information with one another. Insurers clearly use information derived from family histories in underwriting, e.g., denying insurance to people with a family history of Huntington disease.

Because many employers in the United States either fund their employees' health care directly by self-insuring or purchase and subsidize employees' health insurance, they are motivated to seek genetic information about job applicants and employees in order to control their expenditures. In any given year, 5% of health care claimants consume 50% of health care resources, and 10% of claimants consume 70% of resources (36). At present, there is little risk that employers will conduct genetic testing themselves because it is not cost-effective. There is considerable risk that employers will gain access to the results of genetic tests in their role as insurer or purchaser of insurance (65). The Americans with Disabilities Act (ADA) (76) regulates the timing and scope of medical examinations conducted by employers. The ADA also requires that employers keep information obtained from medical examinations confidential. However, benefit records, which contain reimbursement information for specific medical conditions, are usually not considered medical records and hence may be stored in benefits or personnel files that are more widely accessible. Also, the ADA does not prohibit employers from requesting blanket releases of personal medical information or collecting information through voluntary wellness programs or health surveys.

The majority of states have enacted a genetic nondiscrimination law. The tremendous variation in these laws testifies to the dispute between insurers and consumer groups over the economic consequences of regulation and basic notions of fairness. Most of these laws prohibit genetic testing or disclosure of genetic information to a third party without prior informed consent, with exceptions such as anonymous research and law enforcement. They also contain targeted provisions prohibiting health insurers from requiring genetic testing as a condition of insurance and from using genetic information in a discriminatory fashion. Privacy advocates argue that these protections are fairly meaningless so long as insurers can obtain information from other sources or persuade or pressure individuals into submitting to genetic testing, sharing genetic information, or signing a blanket release of medical records. Once a third party has possession of information, its use is very difficult to police (62). To address the problem of compelled disclosure, some states prohibit covered insurers from requesting genetic testing or genetic information. Others prohibit insurers from seeking genetic information for a nontherapeutic purpose or make it unlawful for employers to intentionally collect or seek to obtain genetic information unless it can be demonstrated that the information is job-related and consistent with business necessity or sought in connection with a bona fide employee welfare or benefit plan. Whether these laws effectively deter employers from using information from health insurance claims in employment decisions is open to question.

State laws also reflect divisions between those who regard genetic information as a subcategory of health information, favoring a uniform set of rules for all such information, and those who emphasize the unique or special character of genetic information and DNA. There are good ethical reasons for protecting the privacy of genetic information, including minimizing opportunities for discrimination. However, considerations of justice weigh against the treatment of genetic discrimination as different from related forms of health-based discrimination. In addition, practical problems bedevil attempts to separate genetic information from other health-related information. For example, defining "genetic" has become increasingly difficult. Information concerning inherited genetic disorders can be derived from many sources. Yet state laws frequently protect only the information derived from direct tests of DNA, while allowing unrestricted access to and use of the same or similar information derived from tests analyzing proteins and other gene products or family history. Problems of definition will be exacerbated as genetic research turns to complex disorders. With the increasing attention to complex or multifactorial disorders in both research and clinical settings, a DNA-based definition of genetic is demonstrably underinclusive, but a more comprehensive definition would include virtually all medical conditions.

Even if a defensible definition of genetic is crafted, the separation of genetic information from other information is simply not feasible. In most medical records, information about family history and similar matters is interspersed with other kinds of information. Editing or otherwise expunging genetic information from the patient's central medical record would be burdensome and impractical. (By comparison, the results of HIV testing are fairly easy to isolate in the medical record.) Attempting to isolate genetic information could also compromise the quality of patient care, by impeding the access of health care professionals to this clinically significant information. Finally, separate treatment of genetic information increases the stigma attached to genetic conditions. People may believe that because genetic conditions are singled out for protection, they must be particularly shameful. Separate treatment may also foster genetic reductionism—the attribution of all traits, health problems, and behaviors to genes without regard to other factors—and genetic determinism—the belief that an individual's future "is defined and predicted by genetic make up and cannot be changed" (59).

The most important factor behind the focus on genetic discrimination may be political reality. In the United States, efforts to pass more comprehensive legislation protecting the privacy of health information and eliminating the potential for discrimination on the basis of health status have failed, so far. Although genetic discrimination may be the more manageable target, enactment of such legislation may give the misleading impression that health privacy issues have been adequately addressed, thereby further delaying enactment of more meaningful reforms. Internationally, most countries regulate the collection, storage, and use of genetic information through omnibus data protection legislation.

Considerations Relating to Life and Disability Insurance

For reasons reviewed above, in the United States, the bioethics and policy communities have focused on the relationship of genetics to health insurance. Scholars who examine life insurance typically concentrate on the business aspects of risk classification. They tend to reject regulation where life insurance is involved on the premise that life insurance is a less necessary product than health insurance or that the risk of adverse selection is greater, because it is possible to purchase large amounts of life insurance (28). The importance of life insurance must not be underestimated, however, because for many individuals it is the primary means of financial planning (23). Moreover, in contrast to health insurers, insurers that issue life and disability policies have a greater stake in obtaining predictive genetic information that could alert them to a person's likely health status many years in the future. Also, these policies tend to be sold in the individual market, heightening insurers' interest in individual health risks.

Insurance-related restrictions in state genetic privacy laws are almost always limited to health insurance. Separate legislation for life insurance has been enacted by a number of states, but the protections afforded applicants for life insurance under these laws are fairly minimal. Typically, these laws require only that informed consent be obtained prior to performance of any genetic testing and/or that any use of genetic information in medical underwriting meet standards of actuarial fairness.

An approach that has been considered, though not adopted by any of the states, is a prohibition on the use of genetic information for policies below a certain dollar amount. This approach has found support in various European countries and in Canada. The Netherlands prohibits any use of genetic information for ordinary life insurance policies under 200,000 guilders. The Canadian Privacy Commission has recommended that no medical underwriting be allowed for life insurance policies under $100,000 (61). In Great Britain, the Association of British Insurers (ABI) took a position in early 1997 that the results of genetic tests would not be considered, if detrimental, for policies under £100,000 that were linked to a new mortgage on a principal residence. This cap is believed to be consistent with actuarial calculations, which indicate that this is the point at which adverse selection from at-risk individuals occurs (4). Although the ABI code of conduct remains in place, on October 23, 2000, Britain became the first country to approve insurers' use of results from a predictive genetic test in underwriting. The Genetics and Insurance Committee, an advisory body under the health department, found the genetic test for Huntington disease technically reliable. Decisions are pending on tests for hereditary breast cancer and Alzheimer's disease. The finding means that, apart from the mortgage context, insurers will be permitted to deny life insurance to individuals who refuse to reveal existing genetic test results and to void policies in cases of deception (6). In Australia, a trade organization representing life insurers recently asked a national commission to approve a policy banning genetic testing in life insurance. The commission initially vetoed the policy as anticompetitive, but it later reversed its position after receiving protests from the Australian Medical Association and other groups (57).

Regardless of the specific approach adopted, a uniform set of rules is important for two reasons. First, consistency in the handling of cases contributes to fair treatment of individuals. Second, consistency across insurers decreases the potential for adverse selection.

Considerations Relating to Long-Term Care Insurance

The social importance of long-term care is likely to increase dramatically as the population ages. At the same time, pressures exist to rein in public spending. In the United States, a large portion of long-term care in institutions is currently financed through the Medicaid program. Over time, policymakers will try to shift some of these long-term care costs to the private sector, meaning individuals (and their families) will have to draw on personal assets or purchase long-term care insurance. Because the cost for a year in a nursing home can range from $30,000 to $100,000, depending on the location and services provided, few people will be able to finance their own care or the care of a dependent for an extended period of time (37). Hence, the demand for long-term care insurance is likely to increase. Indeed, in the United States, the long-term care insurance market has grown at a rate of about 21% a year over the past decade (63). As with life insurance, most long-term care insurance is sold in the individual market.

Whereas life insurers worry about mortality and the possibility of premature death, long-term care insurers are most concerned about the prospects for morbidity without mortality. In particular, long-term care insurers have an economic interest in finding out whether applicants have a genetic predisposition to Alzheimer's disease and other chronic, degenerative conditions affecting independence. Researchers are optimistic about new drugs that may slow the onset of Alzheimer's, but a cure is unlikely in the foreseeable future. Given this reality, at a minimum, rules are required to ensure fair treatment for consumers and a level playing field for insurers. An argument for greater protections would stress the commonalities between long-term care insurance and health insurance. Like health insurance, long-term care insurance is directly tied to the provision of necessary health care. On the other side, the premium structure for long-term care insurance is similar to life and disability insurance, as it is based on mortality risk. Legislators have been slow to address the problem of medical underwriting of long-term care insurance. At present, only two states regulate the use of genetic information in the issuance of long-term care insurance (63).

LOOKING TO THE FUTURE

Although the rules for the new science are not yet fixed, building on the preceding discussion, we find some areas of consensus. First, privacy is too large an issue to be solely the responsibility of geneticists, or any other group. The involvement of ethicists, social scientists, lawyers, and representatives of affected communities in appropriate cases is an important protection against the errors of judgment that may result from narrowness of perspective. Rather than regarding this sort of consultation as a threat to their projects, scientists should embrace the opportunity to broaden their consideration of public concerns and to educate the public concerning their work. The alternative, keeping science to the scientists, is strategically unsound as well as ethically problematic. Without public participation, there is a very real risk that the public will turn against genetic research if projects come to light that violate public expectations of protection of privacy and autonomy.

Second, individuals and organizations working in the field of genetics should add privacy protection to the checklist of items to be reviewed at each stage of a project, from conception through ongoing monitoring. In particular, a privacy review should take place before the initiation of any research project involving human subjects, regardless of funding source or geographic location. Further, if the privacy review reveals deficiencies in certain key areas, the project should not be allowed to proceed. For example, there appears to be widespread agreement that an opt-out structure for participation in databases and biobanks is unacceptable in moral and legal terms. No project involving research with human subjects, especially research with identifiable samples or information, should go forward without affirmative consent from participants. Although anonymous research lessens the potential for violations of confidentiality, and so has been judged permissible without consent in the past, sample identification may be preferable for scientific and ethical reasons. Genetic epidemiologists want the ability to correlate clinical, outcome, and utilization data and, in longitudinal studies, to update data over time. Many individuals want to be notified of research results that have medical implications for them and their families. Further, as the quantity of data stored and the sophistication of data analysis increases, the label anonymous is becoming less and less meaningful. A formal requirement of community consultation is also likely in the near future. Researchers should begin to experiment with different formats now, while they have some flexibility, so that policy making is informed by evidence concerning what works and what does not.

Areas of continuing dispute include the degree of specificity required in a consent for future research and the advisability of mandating destruction of samples or data at the end of a specified period or upon completion of the purpose for which samples or data were collected. In these two areas, a compromise is likely that facilitates use for future projects that defy projection based on current knowledge but prevents use for projects that pose risks to participants that are different from those identified at the time of consent, are controversial, or are of questionable social benefit. As noted above, prominent scholars such as H. Greely have proposed accommodations such as the substitution of informed permission for informed consent in the case of unforeseen future research uses. Greely's proposal attempts to honor the principle of respect for autonomy by retaining an opt-in structure. At the same time, Greely avoids blurring the concept of informed consent by applying it to situations in which its strict requirements cannot be met. The safeguard against the perversion of permission, the use of samples or information for ends that participants would likely reject, is the review of all research projects by an IRB or IRB-like entity for level of risk, social benefit, and compliance with other criteria specified in the consent (25).

Third, although laws protecting the privacy of health information and prohibiting genetic discrimination are in place in most jurisdictions, there are gaps in these laws and in the social safety net. Public fears of irrational and rational discrimination in insurance are not unjustified, and scientists eager to recruit subjects for genetic research will have to address these fears. In the third section of this chapter, we describe the general features of existing state laws affecting genetic discrimination. We note that legal protections are usually incomplete in a number of respects. For example, even in the area of health insurance, family histories are often not protected, and individuals who are diagnosed with a genetic disorder or display symptoms of disease or lose coverage under a group policy are especially vulnerable to discrimination. Protections in the areas of life and disability and long-term care are virtually nonexistent. Health information privacy laws also have significant deficiencies in scope, and they typically fail to address the problem of coerced consent to release of medical records.

We began by citing evidence of growing support for a federal law that would prohibit discrimination on the basis of genetic information. There is every reason to expect that a federal nondiscrimination law would mirror the deficiencies in state nondiscrimination laws. Without a gathering of political will to enact comprehensive health care reform in the United States, the consequences of the genetic revolution are likely to be harsh for many. The picture for privacy protection at the federal level is only slightly more encouraging. On December 28, 2000, the U.S. Department of Health and Human Services issued a final rule (76b) offering limited privacy protections for health information. The rule contains a set of federal requirements for the use and disclosure of individually identifiable health information, without preempting stronger privacy protections at the state level. The basic principle is that covered entities may not use or disclose individually identifiable health information unless authorized by the individual or specifically permitted under the rule. Covered entities are permitted but not required to disclose protected information without authorization for certain activities or purposes, including public health and law enforcement. Covered entities are also permitted to disclose information for a number of purposes, such as listing in a facility directory and marketing, so long as individuals are given the opportunity to opt out. In all other cases, a use or disclosure is only permissible with individual authorization.

Among the major deficiencies in the federal privacy rule is its limited scope. This means that researchers and employers, among others, may escape regulation. Researchers are not regulated directly unless they fit the definition of health care provider. The key provision affecting research permits covered entities to use or disclose protected health information for research purposes without authorization. There are, however, limited safeguards for research subjects. Before information can be used or disclosed in this manner, the covered entity must obtain written documentation of a waiver of authorization from an IRB or a privacy board. This requirement applies to all research for which a covered entity serves as a conduit of protected health information, regardless of funding source. Like an IRB, a privacy board must have members with varying backgrounds and appropriate professional competency, including at least one member with no affiliation to the entity conducting the research. In connection with each research protocol, an IRB or privacy board must review a request for a waiver of consent.

With regard to employers, the rule directs a hybrid entity that performs multiple functions to ensure that protected health information does not flow freely from the health care component to other components. This provision has implications for employers that self-insure. However, because an employer qua employer is not a covered entity, the rule does not appear to prohibit an employer from conditioning employment on the provision of an authorization permitting the employer to access health records. Another serious gap is the absence of a private right of action to redress violations. These deficiencies in the rule, and many others, can be traced to limitations in HIPAA. The enormous loophole opened up by permitting disclosure of protected health information for marketing without prior authorization, on the other hand, cannot be blamed on Congress. The prospects for correction of any of these problems by Congress are dim.

Scientists, and others involved in the enterprise of genetics, can improve the situation by engaging in vigorous advocacy for enhanced legal protections and by conscientiously adhering to guidelines contained in consensus documents, whether or not implemented as a matter of law. Unfortunately, so long as legal protections remain imperfect, one of the principal tasks for researchers committed to ethical conduct will be educating potential subjects about the harms that may be associated with participation in genetic research. Before educating potential subjects, however, genetic researchers need to consider the societal aspects of their research. For better or worse, privacy will be as important to genetic researchers as pedigrees, polymorphisms, and proteomics.

Visit the Annual Reviews home page at www.AnnualReviews.org

LITERATURE CITED

1. Allen AL. 1997. Genetic privacy: emerging concepts and values. See Ref. 61a pp. 31–59

2. Am. Soc. Hum. Genet. Am. Coll. Med. Genet. 1995. Ethical, legal and psychosocial implications of genetic testing of children and adolescents. *Am. J. Hum. Genet.* 57:1233–41

3. Am. Soc. Hum. Genet. Am. Coll. Med. Genet. 2000. Genetic testing in adoption: joint statement of the American Society of Human Genetics and American College of Medical Genetics. *Am. J. Hum. Genet.* 66:761–67

4. Assoc. British Insurers. 1997. *Information Sheet: Life Insurance and Genetics.* London: ABI

5. Atkinson B. 2000. DNA mother lode: like Iceland, Newfoundland's unique gene pool is a priceless commodity that could change the fortunes of the island. *Globe Mail* (Toronto), Jan. 25, p. R8

6. BBC News Online. 2000. *Genetic Test First for UK*. Oct. 12, http://news.bbc.co.uk

7. Deleted in proof

8. Beauchamp TL, Childress JF. 1989. *Principles of Biomedical Ethics*. New York: Oxford Univ. Press

9. Brockett PL, Tankersley SE. 1997. The genetics revolution, economics, ethics and insurance. *J. Bus. Ethics* 16:1661–76

10. Buchanan A. 2000. An ethical framework for biological samples policy. In *Research Involving Human Biological Materials: Ethical Issues and Policy Guidance*, 2:B1–31. Rockville, MD: Natl. Bioethics Advis. Comm.

11. Carrasquillo O, Himmelstein DU, Woolhandler S, Bor DH. 1999. A reappraisal of private employers' role in providing health insurance. *N. Engl. J. Med.* 340:109–14

12. Carroll R. 2000. Meet the people of genetic park: centuries of isolation have turned the inhabitants of remote Italian villages into a living laboratory. *The Guard.*, Oct. 30, p. 3

13. Clayton EW, Steinberg KK, Khoury MJ, Thomson E, Andrews L, et al. 1995. Consensus statement: informed consent for genetic research on stored tissue samples. *JAMA* 274:1786–92

14. Cunningham R. 2000. Two old hands and the new new thing. *Health Aff.* 19:33–40

15. deCODE Genet. 2001. *Press Releases*. http://www.decode.com/news/releases

16. Dembner A. 2000. Harvard-affiliated gene studies in China face federal inquiry. *Boston Globe*, Aug. 1, p. A1

16a. DNA Sciences Inc. *About Us*. http://www.dna.com/sectionHome/sectionHome.jsp?link = DNASciencesInc.html

17. Eiseman E. 2000. Stored tissue samples: an inventory of sources in the United States. In *Research Involving Human Biological Materials: Ethical Issues and Policy Guidance*, 2:D1–52. Rockville, MD: Natl. Bioethics Advis. Comm.

18. Fodor K. 2000. Company plans to manage genetic information. *Reuters*. http://www.reutershealth.com/frame2/arch.html

19. Foubister V. 2000. Genetics companies seeking to expand patient base. *Am. Med. News*, Oct. 9, p. 42

20. Frank L. 2000. Estonia prepares for national DNA database. *Science* 290:31

21. Frank L. 2000. Give and take—Estonia's new model for a national gene bank. *Genome News Netw.* http://www.celera.com/genomics/news/articles/10_00/Estonias_genebank.cfm

22. Gallup Org. Inst. Health Freedom. 2000. *Public Attitudes Toward Medical Privacy*. Princeton, NJ: Gallup Org.

23. Genetic Test. Work. Group. 1996. *Report of the Genetic Testing Working Group to the Life Insurance Committee*. Kansas City, MO: Natl. Assoc. Insur. Comm.

24. Glaxo Wellcome. 2000. *Memorandum by Glaxo Wellcome to the Select Committee on Science and Technology of the House of Lords*. http://www.parliament.uk/

25. Greely HT. 1999. Breaking the stalemate: a prospective regulatory framework for unforeseen research uses of human tissue samples and health information. *Wake Forest Law Rev.* 34:737–66

26. Greely HT. 2000. Iceland's plan for genomics research: facts and implications. *Jurimetrics* 40:153–91

27. Greenwood J. 2000. Newfoundland hopes to reap the benefits after its genetic heritage has helped decode the human genome. *Natl. Post*, June 24, p. D1

28. Hall MA. 1996. Insurers' use of genetic information. *Jurimetrics* 7:13–22

29. Hall MA, Rich SS. 2000. Laws restricting health insurers' use of genetic information: impact on genetic discrimination. *Am. J. Hum. Genet.* 66:293–307

30. Health Priv. Work. Group. 1999. *Best Principles for Health Privacy*. Washington, DC: Georgetown Univ. Inst. Health Care Res. Policy

31. HUGO Ethics Comm. 2000. Statement on benefit sharing. http://www.gene.ucl.ac.uk/hugo/benefit.html

31a. Icelandic Parlem., the Althing. 1998. *Act on a Health Sector Database*, No. 139. http://brunnur.stjr.is/interpro/htr/htr.nsf/pages/gagngr-log-ensk

31b. Icelandic Parlem., the Althing. 2000. *Act Biobanks*, No. 110. http://www. mannvernd.is/english/laws/Act.Biobanks.html

32. Juengst ET. 2000. Commentary: what "community review" can and cannot do. *J. Law Med. Ethics* 28:52–54

33. Kaye J, Martin P. 2000. Safeguards for research using large scale DNA collections. *Br. Med. J.* 321:1146–49

34. Kolata G. 1999. Using gene tests to customize medical treatment. *NY Times*, Dec. 20, p. A1:3

35. Lerman C, Narod S, Schulman K, Hughes C, Gomez-Caminero A, et al. 1996. BRCA1 testing in families with hereditary breast-ovarian cancer. *JAMA* 275:1885–92

36. Light DW. 1992. The practice and ethics of risk-rated health insurance. *JAMA* 267:2503–8

37. Lisko EA. 1998. *Genetic Information and Long-term Care Insurance*. http://www.law.uh.edu/healthlawperspectives/Managed/980629LongTerm.html

38. Lowden AJ. 1998. The current state of genetic testing in life insurance. In *Genetic Testing: Implications for Insurance, Symp. Proc.*, p. 19. Schaumburg, IL: Actuarial Found.

39. Maheshwari V. 2000. Estonia touts its genetic credentials. *Financ. Times*, Oct. 31, p. 3

40. Deleted in proof

41. Mannvernd. 2001. Opt outs from Icelandic health sector database. http://www.mannvernd.is/english/optout.html

42. Marshal E. 2000. Epidemiologists wary of opening up their data. *Science* 290:28–29

43. Masood E. 2000. Gene warrior. *N. Sci.* www.newscientist.co.uk/opinion/opinion_224722.html

44. Melton J. 1997. The threat to medical-records research. *N. Engl. J. Med.* 337:1466–70

45. Merz JF, Sankar P, Taube SE, Livolsi V. 1997. Use of human tissues in research: clarifying clinician and researcher roles and information flows. *J. Investig. Med.* 45:252–57

46. *Moore v. Regents of University of California*, 793 P.2d 479 (Cal. 1989) cert. denied, 499 US 936 (1991)

47. Deleted in proof

48. Natl. Bioeth. Advis. Comm. 1999. *Research Involving Human Biological Materials: Ethical Issues and Policy Guidance*, Vol. 1. Rockville, MD: Natl. Bioeth. Advis. Comm.

49. Natl. Bioeth. Advis. Comm. 2000. *Research Involving Human Biological Materials: Ethical Issues and Policy Guidance*, Vol. II. Rockville, MD: Natl. Bioeth. Advis. Comm.

50. Niiler E. 2000. China's efforts to lure biotech to bio-island criticized. *Nat. Biotech* 18:708

51. N. Am. Reg. Comm. Hum. Genome Divers. Proj. 1997. Model ethical protocol for collecting DNA samples. *Houston Law Rev.* 33:1431–73

52. Nowak R. 2000. Gene sale. *N. Sci.* www.newscientist.com

53. Deleted in proof

54. Oregon Legis. 1999. *Oregon Laws, Chapter 921* (S.B. 937)

55. Pokorski RJ. 1994. Use of genetic information by private insurers. In *Justice and the Human Genome Project*, ed. TF Murphy, MA Lappé, pp. 91–109. Berkeley, CA: Univ. Calif. Press

56. Pollack A. 2000. Company seeking donors of DNA for a "Gene Trust." *NY Times*, Aug. 1, p. A1

57. Reuters Med. News. 2000. *Australian Insurers Banned from Genetic Testing.* Nov. 29. http://www.reutershealth.com

58. Rosenberg R. 2000. A study in data collection: genomics companies go abroad to obtain samples, citing obstacles in the United States. *Boston Globe*, Nov. 1, p. D4

59. Rothenberg KH. 1997. Breast cancer, the

genetic 'quick fix,' and the Jewish community. *Health Matrix* 7:97–124

60. Rothstein MA. 1993. Genetics, insurance, and the ethics of genetic counseling. In *Molecular Genetic Medicine*, ed. T Friedman, 3:200. New York: Academic

61. Rothstein MA. 1996. The challenge of new genetic information for the law of health and life insurance. In *The Human Genome Project: Legal Aspects*, 3:55–56. Madrid: Fund. BBV

61a. Rothstein MA, ed. 1997. *Genetic Secrets: Protecting Privacy and Confidentiality in the Genetic Era.* New Haven, CT: Yale Univ. Press

62. Rothstein MA. 1998. Genetic privacy and confidentiality: why they are so hard to protect. *Am. J. Law Med. Ethics* 26:198–204

63. Rothstein MA. 2001. Predictive genetic testing for Alzheimer's disease in long-term care insurance. *Ga. Law Rev.* 35:707–33

64. Rothstein MA, Epps PG. 2001. Ethical and legal implications of pharmacogenomics. *Nat. Rev. Genet.* 2:228–31

65. Saltus R. 2000. Activists renew calls for genetic privacy law. *Boston Globe*, June 28, p. A01

66. Samuelson P. 2000. Privacy as intellectual property? *Stand. Law Rev.* 52:1125–73

67. Schwartz PM. 1997. European data protection law and medical privacy. See Ref. 61a, pp. 393–417

68. Selden R. 2000. Native American activists zero in on bioengineering efforts. *Knight-Ridder Tribut. Bus. News*, p. WL 27468129

69. Sharp RR, Foster MW. 2000. Involving study populations in the review of genetic research. *J. Law Med. Ethics* 28:41–51

70. Sprenger P. 1999. Sun on privacy: "get over it." *Wired News.* http://www. wired.

com/news/politics/0,1283,17538,00.html

71. Staples S. 2000. Newfoundland's 300-year-old genetic legacy has triggered a gold rush. *Globe Mail* (Toronto), Aug. 25, p. 116

72. Stellin S. 2000. Dot-com liquidations put consumer data in limbo. *NY Times*, Dec. 4, p. C4

73. US Congress. 1998. *Omnibus Consolidated and Emergency Supplemental Appropriations Act.* US PL. 105–277, 112 Stat. 2681

74. US Congress. 2000. *Health Insurance Portability and Accountability Act of 1996.* 42 USC Sect. 300gg–2

75. US Congress. 2000. *Public Health Service Act of 1944, as amended.* 42 USC Sect. 241(d)

76. US Congress. 2000. *Americans with Disabilities Act of 1990.* 42 USC Sec. 12101–13

76a. US Dep. Health Hum. Serv. 1991. Federal Guidelines for Research Involving Human Subjects. *Code Fed. Regul.* Title 45 Pt 46. Washington, DC: GPO

76b. US Dep. Health Hum. Serv. 2000. Standards for privacy of individually identifiable health information. *Fed. Regist.*, Dec. 28. Washington, DC: GPO

76c. US Dep. Labor. 1998. *Genetic Information and the Workplace.* Washington, DC: US Dep. Labor

77. Wellcome Trust Med. Res. Counc. 2000. *Qualitative Research to Explore Public Perceptions of Human Biological Samples.* London: Cragg Ross Dawson

78. Winickoff D. 2000. Correspondence: the Icelandic healthcare database. *N. Engl. J. Med.* 343:1734

79. Zick CD, Smith KR, Mayer RN, Botkin JR. 2000. Genetic testing, adverse selection, and the demand for life insurance. *Am. J. Med. Genet.* 93:29–39

Annu. Rev. Genomics Hum. Genet. 2001. 2:435–62

THE GENETICS OF AGING

Caleb E. Finch
Andrus Gerontology Center and Department Biological Sciences, University of Southern California, Los Angeles, California 90089-0191; e-mail: cefinch@usc.edu

Gary Ruvkun
Department of Genetics, Harvard Medical School, Department of Molecular Biology, Massachusetts General Hospital, Boston, Massachusetts 02114; e-mail: ruvkun@molbio.mgh.harvard.edu

Key Words   longevity, metabolism, phylogeny, polymorphism, senescence

■ **Abstract**   The genetic analysis of life span has only begun in mammals, invertebrates, such as *Caenorhabditis elegans* and *Drosophila*, and yeast. Even at this primitive stage of the genetic analysis of aging, the physiological observations that rate of metabolism is intimately tied to life span is supported. In many examples from mice to worms to flies to yeast, genetic variants that affect life span also modify metabolism. Insulin signaling regulates life span coordinately with reproduction, metabolism, and free radical protective gene regulation in *C. elegans*. This may be related to the findings that caloric restriction also regulates mammalian aging, perhaps via the modulation of insulin-like signaling pathways. The nervous system has been implicated as a key tissue where insulin-like signaling and free radical protective pathways regulate life span in *C. elegans* and *Drosophila*. Genes that determine the life span could act in neuroendocrine cells in diverse animals. The involvement of insulin-like hormones suggests that the plasticity in life spans evident in animal phylogeny may be due to variation in the timing of release of hormones that control vitality and mortality as well as variation in the response to those hormones. Pedigree analysis of human aging may reveal variations in the orthologs of the insulin pathway genes and coupled pathways that regulate invertebrate aging. Thus, genetic approaches may identify a set of circuits that was established in ancestral metazoans to regulate their longevity.

INTRODUCTION

Life spans have a remarkable range between species, approaching 1,000,000-fold across all phyla and 10- to 50-fold within groups of the same grade of organization (13, 26, 28). In mammals, bowhead whales hold the record with a life span of more than 200 years, according to estimates from racemization of lens proteins (28, 33). Laboratory rodents are atypically short-lived at 2–3 years, by comparison with many other rodents of similar size that live 5–10 years (13, 26, 28). Even the simplest invertebrates and plants have similar major species differences. Immortality of some cell lineages is possible, as evident in the endless chain of being that has emanated from the germlines of eukaryotes over the past billion years. Clearly, cells can produce viable progeny ad infinitum, but in the soma, a program of senescence has evolved that is highly plastic and enables the massive variation in animal life spans evident in phylogeny.

The regulatory tool kit that evolved a billion years ago in eukaryotes allowed the construction of organisms with virtually any life span. Specification of the life span may be sought at the level of the physiological architecture (31), which determines the degree of somatic maintenance through cellular protection, replacement, and regeneration as well as the risks associated with reproduction (55, 93).

Although no one can doubt that these species differences in life span are a result of genetic sequence differences, it is instructive that social insects use the same genome to build adults with very different life spans, e.g., honey bee queens live up to five years, whereas workers born in the summer live only a few months. Interestingly, the long-lived queen is extraordinarily reproductive, producing a brood that can number in the millions, while eating copiously (26). This challenges the general rule that reproduction and high metabolism are anticorrelated with longevity, as well as broad generalizations about caloric restriction. Evidently, there are programs that can decouple high metabolism from short life span.

Recent developments on the genetics of aging can be seen as several streams of effort. In general, humans show a relatively modest (<50%) heritability of life spans (results obtained from twin studies discussed below). The apoE polymorphisms are remarkable for their influence on both cardiovascular disease and Alzheimer disease. In contrast, rare mutant genes with high penetrance cause these same diseases but with early onset and a major shortening of the life span. Short-lived laboratory models (fruit flies, nematodes, mice) are yielding rapid advances, with the discovery of mutants that increase life spans in association with altered metabolism, which leads to questions on the physiological organization of aging processes. Although these early findings do not show that a conserved genetic program actually controls aging processes across animal phylogeny, it is striking how frequently findings of metabolic rate, insulin signaling, and free radicals have emerged from very different approaches to aging in nematodes and mammals, for example. These findings hint that the genetic control of life span was already developed in the common ancestor of modern animals so that subsequent evolution of life spans was mediated by quantitative changes in the control of metabolism through insulin and the production of free radicals.

GENETICS OF LIFE SPAN IN HUMANS

Most studies of human twins agree that the heritability of life span is less than 50% (45, 68). Of particular interest is an ongoing study of aging in Swedish twins that includes a large group of adopted twins who were reared separately. Ljungquist et al. (68) concluded that "a maximum of one-third the variance in integrated mortality risk is attributable to genetic factors and that almost all of the remaining variance is due to nonshared, individually unique environmental factors." Moreover, this heritability declined with age and was negligible after the age of 85 in men and 90 in women.

Nonetheless, some strong familial trends for great longevity are found (40, 44 88). For example, siblings of centenarians have a fourfold greater survival to >85 years of age than sibs of those who died by age 73 (88). In Iceland, familial effects on survival to the 95th percentile decrease monotonically with meiotic distance, either within the same generation (sibs and cousins) or between generations (children and grandchildren); the data do not distinguish between the models of a few genes enhancing longevity versus multiple genes with additive (independent) effects (40). Linkage analysis of the centenarian phenotype (L. Kunkel & T. Perls, personal communication) should soon identify probable long-lived offspring of centenarians. The presence of rare genes that favor longevity is, of course, not inconsistent with its modest heritability in the general population.

Only one gene, apolipoprotein E (apoE), is showing indications of having general effects on longevity. Centenarians, as first reported by Schachter et al. (95), tend to show a higher prevalence of the apoE e2 allele, relative to e4. More generally, the apoE e4 allele is remarkable, as its presence is the major susceptibility factor for elevated blood cholesterol, coronary artery disease, and Alzheimer disease (but not for cancer or diabetes). Although allele dose susceptibility varies between populations, and is not significant in some (19, 101, 104), no other public allele has yet approached this level of impact on the pathology of aging. By the age of 90, the risk of Alzheimer disease from apoE e4 reaches a plateau (72). Some e4/e4 centenarians are cognitively normal—we do not know if this is owing to protective effects of other genes or aspects of good luck. A new calculation of mortality risk in Danish centenarians showed that, relative to the apoE e3 allele, the mortality risk from e4 from age 40 to 100 was ∼12% higher, whereas that of the rarer e2 allele was 8% lower (35). This differential mortality results in the progressive enrichment of e2 at the expense of e4, so that by 100 years, ∼40% of survivors are apoE e2. Gerdes et al. (35) argue that apoE should be considered as a frailty gene rather than a longevity gene. Other frailty (or longevity) gene candidates include MHC haplotypes, methylenetetrahydrofolate reductase, and angiotensin converting enzyme (32, 44, 95).

We briefly comment on rare mutations that shorten life span through the early onset of diseases that are increasingly common during aging in the general population, e.g., familial forms of Alzheimer, breast cancer, coronary artery disease, type II diabetes, etc. The later onset forms of these diseases are associated with causes of death at later ages. A major question is what role the more common allelic variants of these same genes have in "normal aging". Although examination of this huge emerging topic goes beyond the present discussion, we may consider the example of Werner's syndrome, a rare autosomal recessive that causes adult onset progeria with a high incidence of cancer and atherosclerosis (70). The absence of Alzheimer-type dementia in Werner's syndrome illustrates the "segmental" nature of this and other progerias (70). Thus, heritable shortening of life span should not be considered as a simple acceleration of general aging processes.

The Werner's lesion maps to a defective gene encoding a helicase and exonuclease, which also has several polymorphisms. In Japan, 1367Arg was associated with a lower risk of myocardial infarction (70), although it was not associated with longevity in Finland (14). In general, we know little of the genetic factors involved in frailty and morbidity at later ages, which are important to the gene-environment interactions implied in the major longevity increase seen during the twentieth century.

The heritability of menopause is of potential importance to longevity in women because estrogen deficits after menopause are strongly associated with osteoporotic fractures, which, in turn, increase mortality risk. (Less resolved and more controversial are associations of estrogen deficits/replacements with cardiovascular disease and Alzheimer disease.) From twin studies, the heritability of age at menopause is determined to be $\sim$30%–60% (20, 102). Approximately 20% of monozygous twins differ in age at menopause by 5 or more years (102), which may be attributed to chance events during development that lead to differing numbers of ovarian oocytes (29). Genetic effects on menopause could include influences on the initial oocyte pool and its rate of loss through atresia during aging, as well as on hypothalamic controls that may be responsive to environmental effects, such as stress and nutrition (31, 117). The age of menarche and menopause are not correlated in twins (102), suggesting that these are governed by distinct neural and endocrine mechanisms.

Another association of ovarian functions with longevity is that female centenarians (1986 birth cohort) had a fourfold greater likelihood of having children after the age of 40 versus those who died by the age of 73 (87). On the other hand, premature ovarian failure before 40 sharply increases mortality (52) and osteoporotic fractures (29). Environmental factors that are implied in the major increases of life span during the twentieth century have not modified the age at menopause as far as is known, e.g., the 30-year increase in the life expectancy of women in developed countries during the twentieth century was not associated with increased fertility after age 50 (9).

Consistent with the findings that indicate neuroendocrine regulation of reproductive life span, one of the unifying features noted by Perls in his pedigrees of centenarians is an optimistic attitude (88a). Although it is hardly a quantitative trait at this point, this observation suggests that mood may be co-regulated with life span. For example, a neuroendocrine signal that triggers the slowing of the life span may also regulate feelings of satiety and satisfaction. On the other hand, optimism may also be the result of surviving longer than anyone else. Still, longevity determination may occur at a point above simple metabolic and reproduction regulation, and closer to the pacemakers of general mood.

OTHER MAMMALS

Whereas most mutations in mammals shorten the life span, a few in mice have the opposite effect. The best documented are those associated with dwarfism at three different loci that increased life span by 30%–75%, depending on strain and gender. The Ames dwarf (*df/df* at *Prop-1*) and Snell dwarf (*dw/dw* at *Pit-1*) have growth deficiencies owing to the absence of pituitary growth hormone (GH) and thyroid stimulating hormone (TSH) (6, 11, 75). Dwarfism can also result from dysfunctional GH receptors (*GHR/BP*) (16). GH deficiency causes chronically lower blood levels of glucose and insulin-like growth factor-1 (IGF-1), as in caloric restriction (6, 16). However, unlike caloric restricted mice, pituitary dwarfs become obese during middle age (77). As discussed below, one of the *C. elegans* mutants with increased longevity has sequence similarities to the insulin receptor, suggesting some common features of life span regulation between mammals and invertebrates.

Inverse correlations of size and longevity of about 0.6–0.7 were found post hoc in two other paradigms of size selection. Mice selected for different postnatal growth rates proved to have 2.7-fold differences in life spans (11–31 months), a range that was larger than the twofold difference in adult size (76). Similarly, domestic dogs have a 35-fold range in body sizes (chihuahua to wolfhound) that differ by up to 4 years in life span (75). Although some breeds differ in IGF-1 production, the hormones that regulate this variation are not fully characterized (22a). Because artificial selection for body or organ size often exceeds the initial extreme value of a quantitative trait (24), greater size selection may further extend longevity. Curiously, these intraspecies differences are opposite to the classic allometric relationships in which body size is positively associated with greater longevity in several vertebrate orders (26).

Several mutations in genes that modulate apoptosis also influence life span. Ablations of p66shc showed gene dose effects on resistance to oxidative stress and also increased life span by up to 30% (74). (p66shc is a splice variant of p52shc/p46shc, which binds Grb2 and may be involved in Ras activation.) Another longevity gene candidate is Werner's syndrome *WRN*. The WRN protein enhances transcription of p53, which activates an apoptotic pathway in response to DNA damage and which is attenuated in Werner's cells (8). A helicase-null mouse appears normal, with no signs of progeria during life spans of at least two years, although it is unclear if this is the Werner's gene (69).

To identify other gene candidates in the laboratory mouse genome, several groups are studying gene influences on life span in common mouse strains by analysis of quantitative trait loci (QTL analysis). De Haan & Van Zant (18) have found a region of chromosome 11 with QTLs that influence life span, which overlapped with a QTL associated with the rate of cell cycling in hematopoietic progenitor cells. The short-lived DBA/2 (mean of 592 days) had a threefold higher cell cycling rate than the somewhat longer-lived C57BL/6 (765 days), which those authors discuss as a possible in vivo example of the Hayflick limit of in vitro replicative potential in relation to organismal longevity. Concurrent with these studies of recombinant inbred mice, Miller and colleagues (76) are examining two of these strains in fourway crosses: (BALB/c × C57BL/6)F1 dams and (C3H × DBA/2)F1 sires. The progeny of this cross are genetically equivalent to sibs in that each has a random sample of half of its alleles with the others. Several segregating loci distinct from those above show large effects on life span, diseases, and biomarkers of aging. Preliminary analysis indicates complex pleiotropic effects with gender differences on different spontaneous diseases of aging (77). Of course, the genetic diversity available in inbred mouse strains may underrepresent the rates in natural populations (see studies on *Drosophila* below).

INSULIN CONTROL OF *C. ELEGANS* METABOLISM, DEVELOPMENT, AND LONGEVITY

During the last decade, major progress on the genetics of life span has been realized through the study of long-lived mutants identified in the nematode *C. elegans*. The biochemical functions of many of these genes is now known, and because they are related to processes (e.g., metabolism, free radical production) implicated in aging of vertebrates, they have potential general significance to aging. The clearest example of such a biochemical convergence is the finding that an insulin-like signaling pathway regulates longevity and metabolism in *C. elegans* (54). *daf-2*, *age-1*, and *pdk-1* mutants (see below) that constitute components of the *C. elegans* insulin signaling pathway live 2 to 3 times longer than wild type (21, 53, 63) (Figure 1).

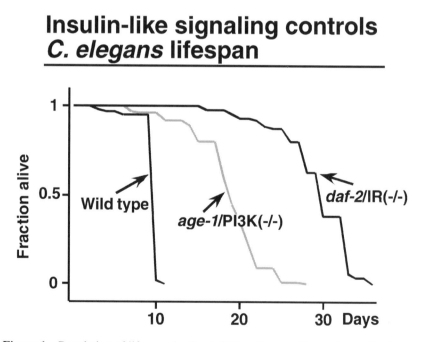

Figure 1   Regulation of life span by the *daf-2* insulin signaling pathway. *C. elegans* animals bearing reduction of function mutations in the insulin-like receptor gene *daf-2* or the downstream phosphotidylinositol kinase 3 gene *age-1* live much longer than wild type.

This insulin-like signaling pathway is part of a global endocrine system that controls whether the animals grow reproductively or arrest at the dauer diapause stage.

Dauer arrest is normally regulated by a combination of a high dauer pheromone (an unidentified fatty acid), high temperature, and low bacterial food (36). Genes that regulate the function of this neuroendocrine pathway were identified by two general classes of mutants: dauer defective and dauer constitutive mutants (36, 37, 96, 110). For example, *daf-2* dauer constitutive mutant animals form dauers in the absence of high pheromone levels. Conversely, *daf-16* dauer defective mutants do not form dauers under normal dauer pheromone induction conditions, and they suppress the dauer constitutive phenotype induced by *daf-2* mutations (37). Based on genetic epistasis and synergistic interactions, most of the dauer defective and dauer constitutive genes have been ordered into a genetic pathway (Figure 1). The most important conclusion from this genetic analysis is that the *daf* genes constitute multiple parallel signaling pathways that converge to regulate *C. elegans* diapause. The *daf-2/age-1/pdk-1/daf-18/akt-1/akt-2/daf-16* subpathway corresponds to an insulin-like signaling pathway (54, 79, 82, 83, 84, 84a) (Figure 2), and the *daf-7/daf-1/daf-4/daf-8/daf-14/daf-3* subpathway corresponds to a TGF-beta–like neuroendocrine signaling pathway (23, 34, 86, 89). Even though these pathways conspire to regulate metabolism and dauer arrest, only the *daf-2* insulin-like

Genetic pathways for dauer arrest

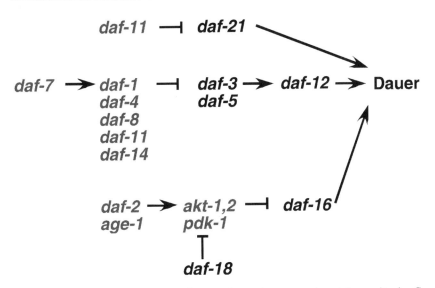

Figure 2 The genetic pathway that regulates dauer arrest and longevity in *C. elegans*.

pathway has effects on the longevity of reproductively growing adults. This is surprising because *daf-7* pathway dauers have increased longevity.

daf-2 encodes the worm ortholog of the insulin/IGF-I receptor gene and is necessary for reproductive development and metabolism (54). A large number of *C. elegans* insulin superfamily genes have been found (88b). Many members of the *C. elegans ins* gene family are organized into clusters, suggesting that these genes arose relatively recently by gene duplication. Clusters of linked insulin superfamily genes are also found in humans and *Drosophila* (see below). GFP fusions to the regulatory regions for 14 of the *ins* genes have shown that they are expressed primarily in subsets of sensory neurons, but also in tissues, such as the intestine and somatic gonad (88b). The INS-1 protein is the most closely related to human insulin. High gene dosage of *ins-1* acts antagonistically to DAF-2, enhancing dauer arrest in wild type or a weak *daf-2* mutant. Human insulin expressed using the *ins-1* regulatory region also antagonizes DAF-2 (88b). Although it is surprising that insulin, an agonist of its receptor, is an antagonist of the worm insulin receptor ortholog, it is important that these insulins engage this pathway. Because all of the INS proteins are predicted to adopt a similar tertiary structure, they may all bind to DAF-2, the only member of this receptor superfamily in the worm genome. Some INS proteins may be DAF-2 agonists, whereas others may be antagonists.

age-1 acts at the same point in the genetic pathway as *daf-2* and encodes the worm ortholog of mammalian phosphatidylinositol 3-kinase (PI 3-kinase) p110 catalytic subunit (37, 79). Reduction of function mutations in *age-1* cause a twofold increase in life span (79). PI 3-kinases generate a membrane-localized signaling molecule, phosphatidylinositol inositol phosphate P3 (PIP3), which binds to the pleckstrin homology domain of mammalian Akt/PKB, which are required for PIP3 activation (1). There are two Akt/PKB orthologs in *C. elegans*. Simultaneous inhibition of both *akt-1* and *akt-2* activities using the technique of RNA interference causes nearly 100% arrest at the dauer stage, whereas inactivation of either gene alone does not (84). One of the kinases that phosphorylates Akt/PKB and is required for its activation is 3-phosphoinositide-dependent kinase-1 (PDK1) (2). A loss of function mutation in *pdk-1* increases *C. elegans* life span almost twofold, similar to a mutation in *age-1* (84).

The dauer arrest, fat accumulation, and longevity phenotypes of *daf-2* and *age-1* mutants are suppressed by *daf-18* mutations. *daf-18* encodes the *C. elegans* ortholog of mammalian PTEN lipid phosphatase gene (83). By genetic epistasis experiments, *daf-18* was shown to act downstream of the AGE-1 PI3K, but upstream of AKT-1 and AKT-2, in this signaling cascade. The DAF-18 lipid phosphatase may normally decrease the level of PIP3 signals, perhaps to insulate signals that emanate from the DAF-2/AGE-1 signaling complex from other PIP3 signals in the cell or to resolve insulin-like signaling episodes. Genetic analysis has not revealed whether DAF-18/PTEN activity is regulated during insulin-like or other signaling.

Mutations in *daf-16* also completely suppress the dauer arrest and metabolic shift of animals bearing *daf-2*, *age-1*, or *pdk-1* mutations or RNAi inhibited *akt-1* and *akt-2* activity (37, 82, 84, 84a). *daf-16* mutations also suppress the increase in longevity caused by decreased *daf-2, age-1,* or *pdk-1* signaling (21, 63, 84, 84a, 111a). Thus, DAF-16 is active in the absence of these upstream inputs and acts to increase life span. *daf-16* encodes two proteins with forkhead DNA binding domains. The mammalian orthologs to DAF-16 are human FKHR, FKHRL1, and AFX (82). DAF-16 contains four consensus sites for phosphorylation by Akt/PKB, and three of these sites are conserved in the human DAF-16 homologs AFX, FKHR, and FKHRL1. Mammalian FKHR, FKHRL1, and AFX activities are regulated by AKT phosphorylation; these transcription factors are nuclearly localized only when insulin-like signaling (AKT activity) is low (22, 38, 41, 42, 80, 90, 98, 107, 112). Consistent with the activity of *C. elegans akt-1* and *akt-2* that lie upstream of and inhibit *daf-16* activity, mutation of those Akt sites in FKHRL1 cause nuclear localization in the presence of insulin or IGF-I signaling (22, 41, 80, 90, 107). The nuclear localization of a functional DAF-16/GFP fusion protein is similarly controlled by upstream pathway activity (R. Lee & G. Ruvkun, personal communication) (Figure 3, see color insert; Table 1).

The mammalian DAF-16 orthologs regulate the expression of target genes, such as the metabolic genes PEPCK and glucose 6 phosphatase (22, 38, 41, 42, 80, 90, 98, 107, 112). DAF-16 binds to this same insulin response sequence in vitro (81). *C. elegans* DAF-16 may regulate the *C. elegans* homologs of these and other metabolic genes.

The molecular genetic pathway suggests that DAF-2, AGE-1, DAF-18, AKT-1, AKT-2, and DAF-16 act in the same cells to regulate *C. elegans* metabolism and longevity. Under reproductive growth conditions, high DAF-2 receptor signaling activates AGE-1 and PDK-1 to, in turn, activate the AKT-1 and AKT-2 kinases, which negatively regulate DAF-16 activity. Phosphorylated DAF-16 is excluded from the nucleus and, therefore, does not activate the genes necessary for dauer

TABLE 1

Gene	Reduction of function phenotype	Ortholog
daf-2	Long life span, dauer arrest	Insulin/IGF receptor
age-1	Long life span, dauer arrest	Phosphatidylinositol 3-kinase
pdk-1	Long life span, dauer arrest	PDK1 kinase
akt-1, akt-2	? life span, dauer arrest	AKT/PKB kinase
daf-18	Short life span, dauer defective	PTEN lipid phosphatase
daf-16	Short life span, dauer defective	FKHR, FKHRL1, AFX transcription factors
daf-7	Normal life span, dauer arrest	GDF-8 (myostatin), GDF-11
daf-3	Normal life span, dauer defective	DPC-4 Smad protein
daf-11	Normal life span, dauer arrest	Guanyl cyclase
Cilia mutants	Slightly longer life span, dauer defective	Kinesin, etc.

arrest and long life span or repress the genes that inhibit reproductive growth and short life span. Under dauer inducing conditions, these kinase cascades are inactive, and DAF-16 is active and nuclear. Active DAF-16 represses genes required for reproductive growth and short life span, and/or it activates genes necessary for dauer arrest and long life span.

The *C. elegans* insulin pathway regulates the expression of key free radical detoxifying enzymes, consistent with free radical theories of aging. *ctl-1* catalase and *sod-3* Mn superoxide dismutase genes are expressed at higher levels in a *daf-2* mutant than in a *daf-2; daf-16* double mutant (47, 62, 108). These enzymes convert the toxic superoxide radicals and peroxides to less reactive products. Supporting an important function for this regulation, a mutation in *ctl-1* reverses the longevity increase of a *daf-2* mutant (108) (Figure 4).

The *C. elegans* insulin-like signaling pathway regulates metabolism as well as free radical protection. These mutants accumulate much larger stores of fat (54). In addition, the rate of CO_2 production, a measure of metabolic rate, is reduced in a *daf-2* mutant to ~30% that of wild type (114). The fat accumulation and the decline in metabolic rate as well as the longevity increase are fully suppressed by a mutation in *daf-16* (82, 114). Similar but less severe declines in metabolism are observed

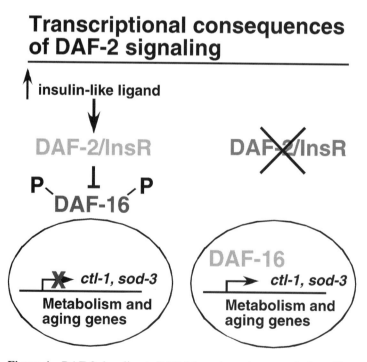

Figure 4   DAF-2 signaling to DAF-16 regulates the transcription of free radical protective enzyme genes. Regulation of free radical protective enzyme genes by the DAF-16 transcription factor.

An insulin-like receptor to PI-3 kinase, PTEN, PDK-1, AKT-1, AKT-2, to DAF-16 signaling cascade controls C. elegans aging and metabolism

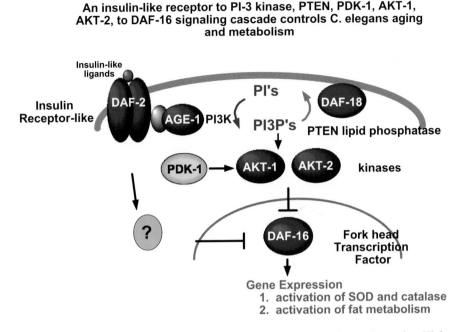

Figure 3 A molecular model for insulin like regulation of *C. elegans* longevity. High insulin signaling activates the DAF-2 receptor which in turn activates the AGE-1 PI 3-kinase to in turn activate the kinases PDK-1, AKT-1, and AKT-2. The AKT-1 and AKT-2 kinases phosphorylate the DAF-16 transcription factor causing cytoplasmic localization to in turn cause a change in transcription of target genes.

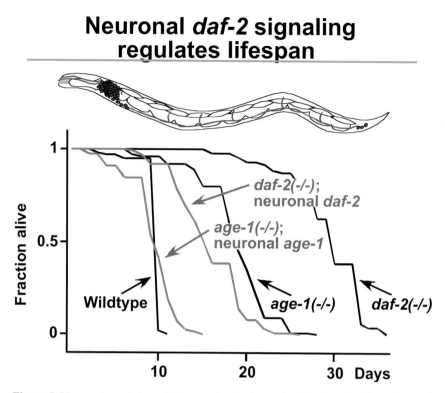

Figure 5 Neuronal regulation of life span by *C. elegans* insulin-like signaling. Neuronal expression of the *C. elegans* insulin pathway is sufficient to rescue the lifespan increase caused by deficits in insulin signaling.

with a weak *age-1* allele (114). The decline in metabolic rate of insulin-like signaling pathway mutants is supported by a 30% decline in the rate of egg laying and in total fecundity of *daf-2* and *age-1* mutants (111a, 114). Conflicting data regarding the metabolic control by *daf-2* pathway mutants also exists; however, because the other studies used a free radical surrogate measure of metabolism rate and because they were done under nonoptimal growth conditions, they must be viewed with caution (113).

THE CONNECTION TO CALORIC RESTRICTION

The increase in longevity associated with decreased DAF-2 signaling is analogous to mammalian longevity increases associated with caloric restriction (103). The involvement of an insulin signaling pathway in worm aging may be mechanistically related to the longevity increase caused by caloric restriction in mammals (26, 103). Insulin secretion by the pancreas is regulated by nutritional and autonomic neural inputs, and this endocrine signal of metabolic status is detected by target tissues to regulate the activities of metabolic enzymes that synthesize or breakdown glucose, amino acids, fat, etc. Like *daf-2* mutants, life span is dramatically increased in dwarf mice with defects in growth hormone signaling and with decreased IGF-I signaling (see above).

Mammalian orthologs of the *C. elegans* insulin signaling pathway may be components of a mammalian longevity determining pathway. Caloric restriction in mammals may cause a decline in insulin-like signaling that induces a partial diapause state (109), like that induced in weak *daf-2* and *age-1* mutants. The induction of diapause-like states, or changes in the mode and tempo of metabolism itself, may affect postreproductive longevity (109), as in *C. elegans*. This association of metabolic rate with longevity is also consistent with the correlation of free radical generation to aging (103).

Life span of *C. elegans* is also coupled to the rate of feeding (61, 105). Perhaps like caloric restriction, some *eat* mutants that ingest bacteria less efficiently than wild type live up to 50% longer. The increase in life span of the *eat* mutants is not fully suppressed by *daf-16* mutations, suggesting that the caloric restriction pathway engages some pathway besides the *daf-2* insulin-like signaling pathway (61).

THE AGING TRANSCRIPTOME

Given that the major output of worm insulin-like signaling is transcriptional and the apparent homology between life span extension by caloric restriction in mammals and *daf-2* pathway mutations in *C. elegans*, the monitoring of the changes in gene expression in calorically restricted animals seems especially attractive. This has been done in mouse skeletal muscle, which shows declines in mass and other functions as animals age. Comparison of one muscle between 5-month-old and 30-month-old adult mice revealed a more than twofold change in 113 of 6347 genes (65). These genes could be classified into three major groups: stress response, energy metabolism, and neuronal signaling. When muscle gene expression was compared between calorically restricted animals, 30% of the 113 major aging response genes no longer showed the twofold or greater change during the aging period, suggesting that they are coupled to the aging process and its modulation by caloric restriction. Caloric restriction also caused an induction (1.8 fold or greater) of 51 and repression (1.6 fold or more) of 57 muscle genes in 30-month-old mice that were not induced in well-fed 30-month-old mice. These genes were also classified into energy metabolism, protein synthesis, and stress response pathways. Many genes not yet represented in EST databases are not present on current gene arrays.

Thus, approximately 1% of the genome that was tested (10% to 30% of total mouse genes, depending on how many genes there are in a mouse) is transcriptionally regulated in a particular tissue during aging. The induction of a stress response pathway could be a result of the accumulated damaged macromolecules, as predicted from the involvement of free radicals in aging processes. The induction of neuronal growth factor signaling pathways could be the response to a decline in muscle targets. The decrease in energy metabolism gene expression is consistent with the known decline in mitochondrial function during aging.

Global gene array searches based on whole organ RNA may be encyclopedic in their documentation of gene expression changes. However, cause and effect in aging can only be resolved by perturbing life spans by varying the expression or activity of these candidate genes. For example, the detection of mutations in these genes in pedigrees with life span variation, or engineering changes in life span by disrupting or misexpressing key regulatory genes, will probably prove valuable in identifying causal chains and, ultimately, possible primary causes of aging. In addition, without determining the exact cell types that regulate aging, the monitoring of gene expression in a particular tissue could miss the regulators. For example, analyses in *C. elegans* have revealed that aging of the entire animal is controlled hormonally from the nervous system (see below). Merely monitoring gene expression in aging muscle would not identify such triggers (though the response pathway could emerge if its transcription was modulated). Thus, the gene screens are the first stage in an exacting process, similar to that begun long ago by developmental biologists that has led to the present molecular understanding of cell fate determination and organogenesis.

NEURONAL CONTROL OF AGING

Insulin-like signaling may directly regulate metabolism and free radical production in aging skin and muscle or in signaling centers that then coordinately control the senescence of the entire organism. *pdk-1*, *daf-18*, *akt-1*, and *daf-16* are all expressed throughout much of the animal, consistent with their function in either signaling cells or target tissues. Studies of *daf-2* genetic mosaic animals showed that *daf-2* can act nonautonomously to regulate life span but did not assign *daf-2* longevity control to particular cell types (3).

In animals that express normal insulin signaling only in the nervous system, life span is normal, whereas insulin signaling in muscles or gut does not confer normal life span. In these studies, *daf-2* pathway function was restricted to particular cell types by using distinct promoters to express *daf-2* or *age-1* in the neurons, intestine, or muscle cells of a *daf-2* or *age-1* mutant. The long life span of *daf-2* and *age-1* mutants is rescued by neuronal expression of *daf-2* or *age-1*, respectively, using a pan-neuronal promoter (118) (Figure 5, see color insert). Restoration of *daf-2* pathway activity to muscles is not sufficient to rescue the long life span of *daf-2* or *age-1* mutants. Similarly, expression of *daf-2* or *age-1* in the intestine, the major site of fat storage, does not rescue life span as efficiently as neural expression of these genes.

These data argue that the key tissue where *daf-2* insulin-like signaling regulates aging is from the nervous system. Precedents for insulin signaling in the mammalian nervous system exist. Although the target tissue responses to insulin are better known, feeding and metabolic responses to insulin are also present in the mammalian brain (99). In addition, insulin receptor signaling defects in the neurosecretory beta cells of the mouse pancreas or only in the nervous system cause metabolic as well as reproductive defects, also suggestive of a role for insulin signaling in neuronal tissues (12, 59).

Neurons may be particularly sensitive to free radical damage during aging. In fact, overexpression of Cu/Zn superoxide dismutase (SOD) in just the motorneurons can extend *Drosophila* life span by 48% (85). And perhaps mechanistically related, motor neuron degeneration in amyotrophic lateral sclerosis is caused by mutations in Cu/Zn SOD (119). It is striking that aging in two different organisms can be controlled by neurons and is correlated with increased free radical protection in those neurons. If this model is correct, neuronal *daf-2* signaling regulates an organism's life span by controlling the integrity of specific neurons that secrete neuroendocrine signals, some of which may regulate the life span of target tissues in the organism (Figure 6). These results, together with those from *Drosophila*, suggest that oxidative damage to neurons may be a primary determinant of life span.

Given that insulin signaling in the *C. elegans* nervous system is implicated in the control of longevity, it is intriguing that several brain regions of aging mice show, via array technology, selective changes in mRNA (66). This study compared the neocortex (a recent mammalian invention) and the cerebellum in 5- and 30-month-old mice, which were fed to 90% and 60% of ad libitum intake. During aging, ~1% of the genes were increased (~twofold or more) and ~1% decreased (~twofold or more) in expression. These numbers are consistent with analysis of brain poly(A)mRNA population sequence diversity by hybridization kinetics, which did not detect age changes within a statistical upper limit of 5%

Models for neuronal control of lifespan

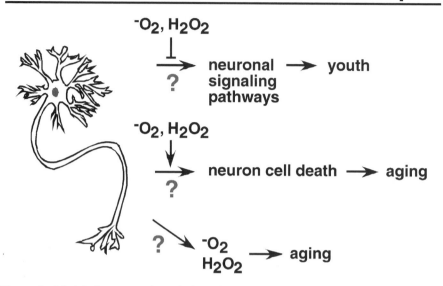

Figure 6 Models for neuronal regulation of life span. Three models for how free radicals produced by neurons could regulate whole animal aging: (*a*) The free radicals could compromise the production of hormonal signals produced by neurons. (*b*) The free radicals could cause cell death of neurons that normally send "youth" hormone signals. (*c*) The free radicals could be hormonal signals, analogous to NO, for example, that are received by target tissues, such as muscle and skin.

(15). Approximately 40% of the mRNAs showing increased expression were in the inflammatory and stress response pathways, whereas ~15% of the genes in the decreased expression class were growth factors. A subset of inflammatory mRNAs show parallel changes in both regions (30). The growth factor mRNA subset has tantalizing implications for age changes in neural plasticity, which is based on the elaboration of neurons and connections. Caloric restriction attenuated the aging responses in approximately half of those genes, with a complete attenuation of a few of them (66). In terms of the intersection between worm genetics and mammalian caloric restriction, it is noteworthy that IRS-3, a transduction component of insulin signaling, is downregulated ~1.7-fold in calorically restricted mice—perhaps a decrease in insulin signaling caused by caloric restriction in turn causes a decrease in IRS-3 expression in an autocrine fashion.

For both *C. elegans* and mammals, particular neurons, rather than the entire nervous system, may prove to be key in aging. In the case of *C. elegans*, it is fairly simple to activate the insulin signaling pathway in particular neural types and then monitor life span, as was done for nervous system vs. muscle, etc. (118). Promoters exist for restricting expression to particular classes of neurons (cholinergic, sero-tonergic, motor, etc.). If one class of neurons is then shown to be key for aging control, then discerning the homologous class of neuron from the mammalian brain might be possible. In such a case, gene array monitoring of the transcripts of purified or dissected neurons of this subtype over the life span of an animal might identify mRNAs critical for aging. For example, the hypothalamus is a plausible location for such regulation of the life span and life cycle because it regulates metabolism and the reproductive schedule (puberty, menopause), which are targets of selection in the evolution of life histories (31). In some examples of animals that die after first spawning (semelparous life history), reproduction and death can be postponed by neuroendocrine manipulations (27, 31). It is interesting that a worm thermoregulatory circuit bearing antagonistic warm and cold regulatory loci (single neurons) may be homologous to the vertebrate hypothalamus (46). Thus, there may be a future genetics of neuronal aging pacemakers.

But to understand aging in mammals and in worms, obtaining a set of molecular markers of the aging process, even if only in organs that may have a subsidiary physiological role, will be very helpful. These markers constitute molecular definitions of the aging process and can be used in classifying how mutants and physiological perturbations act to "reset" some, or even all, of the features of aging. Thus, gene array comparisons of *C. elegans* aging from any tissues are important, even if the trigger is in the nervous system.

Besides the *daf-2* insulin-like signaling pathway, dauer arrest is also regulated by the *daf-7/daf-1/daf-4/daf-8/daf-14/daf-3* TGF-beta signaling pathway. The signals in these two pathways are neither redundant nor sequential: Animals missing either of these two signals shift their metabolism and arrest at the dauer stage. Only mutations in the insulin-like pathway genes cause dramatic increases in longevity. This suggests that the transcriptional program for longevity does not depend on TGF-beta transcriptional cofactors in the same manner that the metabolic switch to fat storage depends on both pathways.

daf-7 encodes a TGF-beta neuroendocrine signal that is produced by the ASI neuron (89). *daf-7* expression in this neuron is inhibited by dauer-inducing pheromone (89, 96). *daf-1* encodes a type I TGF-beta class ser/thr receptor kinase, *daf-4* encodes a type II TGF-beta class ser/thr receptor kinase, and *daf-8* and *daf-14* encode Smad proteins (34, 49), which couple TGF-beta signals from receptor kinases to the control of transcription. A *daf-3* null allele completely suppresses the dauer constitutive phenotype of mutations in *daf-1*, *daf-4*, *daf-7*, *daf-8*, and *daf-14* (110). Thus, mutations in *daf-3* bypass the need for any of the DAF-7 signal transduction pathway genes, suggesting that the major function of this signaling pathway is to antagonize DAF-3 gene activity. *daf-3* encodes a Smad protein that is most closely related to vertebrate DPC4, which is a cofactor for Smad1, Smad2, and Smad3 (86).

The DAF-16 Fork head and the DAF-3 Smad proteins are active in the absence of upstream signaling to induce arrest at the dauer stage and a shift to energy storage metabolism. It is interesting that the *daf-7* response pathway acts in neurons, similar to the *daf-2* response pathway (49). DAF-16 may interact with DAF-3 on the promoters of genes that regulate metabolism and reproductive vs. dauer development.

DROSOPHILA INSULIN SIGNALING

Drosophila orthologs of the mammalian and nematode insulin receptor, insulin receptor substrate, *age-1* PI3 kinase, AKT/PKB, PTEN, and S6 kinase regulate cell, organ, and total body size (83a). Mutations in all of these genes except for PTEN cause a dramatic decrease in cell size, whereas a mutation in PTEN causes the opposite phenotype (83a). It is interesting that mutations in *Drosophila* IRS also affect metabolism, causing a dramatic increase in fat storage, similar to mutations in the *C. elegans* insulin signaling pathway. IRS mutant females also live longer than wild type, although the males do not (14a). Many reductions of function mutants in the *Drosophila* insulin pathway do not show an extension of life span, but a rare insulin receptor allele combination does cause a modest increase in life span (107a). Unlike in *C. elegans* and perhaps mammals, there may be a narrow window of decreased insulin signaling in *Drosophila* that can prolong life (14a). One interesting feature of the fly insulin receptor mutant is that the low fertility is coupled with a decrease in juvenile hormone (107a). Thus, as in the case of worm insulin regulation of life span, neuroendocrine outputs of the insulin pathway may be key. A *Drosophila daf-16* ortholog is known but has not been implicated in either the fly insulin pathway or in longevity (E. Hafen, personal communication).

Drosophila has seven insulin-like genes (DILPs) (10). Overexpression of DILP2, which has a C peptide and is most closely related to insulin, causes larger cells, larger organs, and larger animals (10). A deficiency that removes five of the DILPs (plus other genes) suppresses the large eye phenotype caused by overexpression of the wild type insulin receptor in the eye (10). The different DILPs are expressed in seven bilaterally paired cells in the brain. Neurosecretory cells that project to the ring gland where insulin should be released into the hemolymph (10) probably exist. These insulin-producing neurons may couple to the juvenile hormone-releasing cells of the fly corpus allata, a neuroendocrine organ.

One of the mammalian insulins, IGF-I and its receptor, is implicated in the control of body and organ size of this type. It is interesting that tissue specific knockout mutations of the mouse insulin receptor gene in pancreas and liver cause a decrease in postnatal growth of those organs, suggesting that both the insulin and IGF-I also feed into the regulation of cell size in mammals (59, 73). Cell size regulation was shown by genetic mosaic analyses of IRS, AKT, and S6 kinase to be autonomous in *Drosophila*.

There is no indication that the *C. elegans* insulin signaling pathway controls cell size. However, a TGF-beta signaling system that is distinct from the DAF-7 metabolic control pathway regulates body size in *C. elegans* (57). Size regulation by insulin-like signaling appears to have maintained in the fly and vertebrate lineages, but in *C. elegans*, it is served by another pathway. Thus, there is a distinction between insulin signaling within the *C. elegans* nervous system to control aging and the autonomous regulation of cell size in *Drosophila*, and between the control of cell size by that pathway in vertebrates and one invertebrate, *Drosophila*, but not in another, *C. elegans*.

Aside from the insulin pathway, other mutations and inbred lines that increase or decrease *Drosophila* life span have been identified. For example, the *methuselah* mutation causes a 35% increase in fly life span as well as increased resistance to stress. *Methuselah* encodes a probable G protein–coupled (7 TM) receptor (67). Its ligand and mechanisms of coupling to stress resistance are unknown; an interesting possibility is that the Methuselah protein mediates the sensory inputs to the fly insulin pathway. Another locus that extends longevity is *Indy* (I'm not dead yet), which resembles a dicarboxylate transporter and is expressed in the gut and fat body (92). Moreover, *Indy* increases fecundity at later ages. Its implied role in intermediary metabolism suggests that the *Indy* mutation may induce a state similar to caloric restriction.

In one case, a gene identified by mutation recovered from a genetic screen in the laboratory, *methuselah*, may have variants in natural populations. In particular, the common ATATC haplotype has a sharp geographic (north-south) cline in U.S. populations, which, intriguingly, is associated with an 18% difference in life span (97). It would be interesting to examine these natural populations for differences in their reproductive schedule. Extensive studies show that life span can be rapidly selected as an indirect outcome of artificial selection for age at reproduction. Samples from natural populations of *Drosophila* contain genetic variants that can be rapidly selected, within 15 generations, for 50% or greater differences in life span on the basis of choosing individuals that are reproductive at early versus later ages (93). Selection was reversible, indicating that these life history variants depended on existing gene combinations not new mutations. Among the genes that differed in quantitative expression between young- and old-selected lines were heat shock proteins, e.g., *hsp 22* (60). An overarching conclusion from fly aging genetics is that stress resistance is coupled to longevity (94), as in *C. elegans*. Other gene candidates are being sought by QTL analysis and show complex interactions with gender and population density (17, 115).

SENSORY INPUT TO AGING IN *C. ELEGANS*

Dauer arrest in *C. elegans* is normally regulated by an uncharacterized dauer pheromone that is detected by sensory neurons (36). The dauer pheromone causes downregulation of *daf-7*, the TGF-beta ligand of the pathway that acts in parallel to the *daf-2* pathway (89, 96). It might also regulate the expression or release of any of the 37 worm insulins that have been detected, but this has not been established. The detection of the dauer pheromone depends on ciliated sensory endings; cGMP signaling in those sensory neurons has been implicated, suggesting that the pheromone receptor may be a 7 transmembrane receptor that couples to cGMP phosphodiesterase, as in mammalian odor sensation (7). Although no evidence currently exists of any coupling of sensory input to *daf-2* insulin-like signaling, mutations that uncouple sensory neurons from the environment cause an increase in longevity that is suppressed by *daf-16* (3). Thus, it is likely that sensory input is needed to activate the *daf-2* receptor, presumably via the activation of insulin agonists or inactivation of insulin antagonists by food or other sensory inputs. Particular sensory neurons produce a dauer inhibitory signal because laser ablation of those neurons causes dauer arrest (5). This cell ablation induced dauer arrest is dependent on *daf-16* gene activity, supportive of a coupling to the *C. elegans* insulin-like signaling pathway (5). Moreover, some neurons activate dauer arrest rather than reproductive development (96).

The insulin hormone may be produced by the sensory neurons or neurons connected to those sensory neurons. Serotonin may be involved in the signaling pathway. A mutation in *C. elegans* tryptophan hydroxlyase (TPH), which catalyzes the rate-limiting first step in serotonin synthesis, causes several behavioral defects that are associated with starvation: With abundant food, the mutant feeds more slowly, retains more eggs, accumulates larger stores of fat, and some animals arrest at the dauer stage (105). The metabolic phenotypes are a result of serotonin inputs to both the TGF-beta and the insulin-like pathways. The expression of a *daf-7:GFP* fusion gene is decreased in the *tph-1*(null) mutant, and dauer arrest of a temperature sensitive *daf-7*(mutant) at low temperature is enhanced. In addition, high gene dosage of *tph-1* suppresses dauer arrest of *daf-7*(*e1372*) at high temperature. This suggests that temperature may modulate serotonin levels to influence dauer arrest. There is also serotonergic input to the parallel insulin-like signaling pathway. A *daf-16* mutation suppresses the dauer arrest and fat accumulation *tph-1*(*mg280*) phenotypes (105). Moreover, like *daf-2* insulin-receptor mutants (but unlike *daf-7* pathway mutations), *tph-1*(mutant) hermaphrodites have an extended reproductive life span that is dependent on *daf-16* gene activity. Consistent with the effect on reproductive life span, *tph-1* animals have 25% longer life spans that is suppressed by *daf-16* (K. Ashrafi & G. Ruvkun, personal communication). Bacterial food and low temperature may normally upregulate serotonin signaling to in turn upregulate DAF-7 and insulin-like neuroendocrine signals. In the *tph-1* mutant, these hormones may be decoupled from such food signaling. Serotonin signaling has also been implicated in the control of mammalian feeding and metabolism.

Temperature is a potent regulator of dauer arrest (36). All mutations in the DAF-7 TGF-beta pathway, including *daf-7* null mutations, are temperature sensitive, whereas null mutations in the insulin pathway cause dauer arrest at all temperatures; they are not temperature sensitive (46). There is explicit temperature sensory input to this endocrine pathway. This pathway includes the AFD thermosensory neurons, which connect to the interneurons AIY and AIZ, which may, in turn, connect to secretory as well as motor control neurons. A mutation that decouples AIY from this thermoregulation of dauer arrest renders dauer arrest nontemperature sensitive. The antagonistic high and low temperature processing pathways of the *C. elegans* thermoregulatory pathway is similar to the organization of the vertebrate hypothalamus, which contains distinct warm and cold temperature processing units (46). The coupling of thermosensory input to the *daf-2* insulin-like signaling pathway may be homologous to the hypothalamic modulation of autonomic input to the pancreatic beta cells. Because the life span of *C. elegans* and other cold blooded animals is highly dependent on temperature, it is possible that these thermoregulatory circuits feed into life span regulation.

DIAPAUSE, HIBERNATION, SLEEP, AND LONGEVITY

The *C. elegans* insulin-like signaling pathway genes control longevity as part of a global endocrine system that controls whether the animals grow directly to reproductive adults or arrest at the dauer diapause stage. The connection between longevity and diapause control may not be parochial to *C. elegans*. Diapause arrest is an essential feature of many vertebrate and invertebrate life cycles, especially in regions with seasonal temperature and humidity extremes (109). Animals in diapause arrest slow their metabolism and their rates of aging and can survive for periods much longer than their reproductive life span.

Caloric restriction may cause a decline in mammalian insulin and insulin-like signaling, which, in turn, causes a repertoire of diapause-like responses, such as the expression of an ancient set of aging protective genes that confer decreased aging during the diapause period. Because these diapauses are pre-reproductive, they would be selected. Induction of diapause-like states postreproductively could also induce longevity, but the program would, in this case, have been selected for its prereproductive action.

From the connection between diapause and life span in worms and the similarity between insulin-like regulation of worm life span and caloric restriction, one may predict that humans who live much longer than normal may share some features of the diapause state. This could be reflected in a lower overall rate of metabolism, either in the whole animal or in key cells, such as particular neurons. It could also be reflected in a lower body temperature, either all of the time or during specific times, for example, sleep. It would be very interesting to analyze brain activities of the progeny of centenarians, some of whom are destined to live longer than normal. One might find a deeper sleep or altered pattern of brain electrical activities in those who inherited a longevity phenotype.

Reproductive senescence is delayed in weak *daf-2* mutants, and this is also dependent on *daf-16* gene activity (32a, 105). The coregulation of reproductive senescence and longevity of the entire animal is reminiscent of their co-regulation in humans as well: Female centenarians tend to continue to bear children much later (after 45 years old) than noncentenarian humans (87), as discussed above. There is a strong correlation between late fertility and longevity: (*a*) both serotonin and *daf-2* mutants cause late reproduction, (*b*) the correlation between late reproduction and longevity in many species, and (*c*) the regulation of fertility by insulin signaling in mammals.

Just as environmental extremes can select for variation in the genetic pathways that regulate *C. elegans* dauer formation, famines and droughts in human history may have selected for analogous variants in the human homologues of the *daf* genes. In fact, heterozygous mice carrying either the *db* or the *ob* recessive diabetes genes survive fasting ~20% longer than wild-type controls (87). The high frequency of type II diabetes in many human populations may be the legacy of such selections.

CLK MUTANTS AND METABOLISM

Mutations in the *C. elegans* coenzyme Q biosynthetic gene *clk-1* cause a modest increase in life span, in addition to a twofold slowing of the development rate (25). Because coenzyme Q is an essential component of the electron transport chain of mitochondria as well as bacteria, it is not surprising that the rate of living is affected in this mutant. What is surprising is that a null mutant in this gene is actually viable. It turns out that the *C. elegans* bearing this mutation is viable on wild-type *E. coli* but not on an *E. coli* mutant that produces no coenzyme Q (51). Thus, the *clk-1* null mutant uses coenzyme Q from *E. coli*. Presumably, the level or regulation of the *E. coli* derived coenzyme Q is not normal, leading to the modest increase in life span. These data weakly support the view that metabolic rate and aging are connected but do not implicate *clk-1* in any explicit regulation of aging.

FREE RADICALS AND AGING

The free radical theory of aging has been debated for years (103). It is supported by the general correlation between metabolic rate and longevity, by both comparing species as well as comparing life histories within a species. The increase in life span associated with declines in metabolic rates in insulin pathway mutants as well as *clk-1* coenzyme Q biosynthetic pathway mutants also support this model. Also supportive is the increase in life span that results from overexpression of free radical detoxifying enzymes, such as CuZn SOD, and free radical scavenging chemicals, such as vitamin E and others (103).

Recent genetic analysis suggests that free radical production can be increased by decoupling electron transport of ubiquinone to O_2. A missense mutation in *C. elegans* cytochrome b560 causes a decrease in life span and oxygen hypersensitivity (as well as radiation hypersensitivity, a reasonable pleiotrophy because radiation induces free radicals, which damage DNA). Paradoxically, this is the opposite phenotype from the *clk-1* defect in coenzyme Q biosynthesis. A model that explains this is that the cytochrome missense mutation causes a toxic build up of ubisemiquinone (a free radical that can generate superoxide) because the normal pathway for further reduction of singly reduced coenzyme Q is compromised (50).

The best evidence for the involvement of free radicals in longevity has come from the treatment of *C. elegans* with free radical detoxifying mimetics, which cause a 44% increase in life span of wild type and 60% increase in the life span of the above mutants that are expected to produce more free radicals than normal (71).

Similarly, increased expression of free radical scavenging enzymes, such as superoxide dismutase, increase invertebrate and vertebrate life span in some cases (85, 91) but not in others (48). Mouse knockout mutations in both mitochondrial Mn superoxide dismutase (64) and cytosolic CuZn SOD (100) show effects on whole animal and neural mortality. Why the conflicting results have been observed is not clear, but the effects of distinct genetic backgrounds with differences in the rate of production or detoxification are a likely variable. Some free radical generation may also be necessary to induce stress response pathways that are protective. These studies demonstrate that neurons are highly sensitive to free radical damage, consistent with the view from *C. elegans* that insulin signaling in the nervous system is the key to longevity control. In fact, expression of free radical scavenging enzymes in the nervous system is highly protective (58, 106, 116). Expression of catalase and SOD in the hypothalamic region, an ancient locus of the brain, is under GH/IGFI control, as expression of these genes is regulated by insulin signaling in the worm (43).

YEAST

Unlike metazoans, yeast do not have a segregated soma and germline. Despite this, an asymmetry in cell division does mark the parent cell, and the life span of those cells is subject to genetic control. This control of mother cell longevity, as measured by further cell division, is probably quite distinct from the postmitotic cell longevity of, for example, the animal brain, as recently reviewed by Guarente & Kenyon (39). In summary, even though the longevity control of a unicellular organism would be expected to be distinct from an organism with a separate mortal soma and immortal germline, points of tantalizing similarity emerge: A chromatin remodeling pathway has been implicated in yeast aging control that may also regulate the life span of *C. elegans* (111), and the activity of the control protein in this pathway Sir2 is regulated NAD, a surrogate measure of metabolism. Although yeast do not have an insulin signaling pathway, they appear to have a precursor to that metabolic control pathway, the yeast adenylate cyclase and SCH9, which function in glucose/nutrient signaling pathways (23a). SCH9 is homologous to the serine threonine kinase Akt/PKB of insulin signaling pathways in *Caenorhabditis*, *Drosophila*, and mammals. Mutations in the adenylate cyclase and SCH9 genes can extend the longevity of nondividing cells by up to threefold (23a).

CONCLUSIONS

Our purpose in this review is to outline the prospects of unifying mechanism in the genetics of aging. In case after case, from mice to worms to flies to yeast, genetic variants that modify metabolism also modify life span. These effects, collectively, are as general as that of caloric restriction, which also increases longevity and resistance to stress in many situations. The evolutionary theory of aging proposes that the life span is indirectly selected on the basis of the reproductive schedule. In turn, the reproductive schedule is coordinated by neural and endocrine mechanisms in multicellular organisms. Therefore, to consider that genes determining the life span could be expressed in neuronal and endocrine cells in diverse animals is no longer far-fetched. Consistent with this hypothesis are experiments in *Drosophila* and *C. elegans* in which life span was manipulated by the expression of genes in specific neurons. Genetic approaches may, thus, be able to identify a set of circuits that regulate longevity that were established in ancestral metazoans.

ACKNOWLEDGMENTS

CEF and GR are supported by grants from the NIH, and GR is also supported by the Ellison Medical Research Foundation. We thank Cathy Wolkow for allowing us to use some of the figures she has composed. CW, Valter Longo, and Richard Miller provided valuable comments on the manuscript.

Visit the Annual Reviews home page at www.AnnualReviews.org

LITERATURE CITED

1. Alessi DR, Andjelkovic M, Caudwell B, Cron P, Morrice N, et al. 1996. Mechanism of activation of protein kinase B by insulin and IGF-1. *EMBO J.* 15:6541–51

2. Alessi DR, James SR, Downes CP, Holmes AB, Gaffney PRJ, et al. 1997. Characterization of a 3-phosphoinositide-dependent protein kinase which phosphorylates and activates protein kinase B. *Curr. Biol.* 7:261–69

3. Apfeld J, Kenyon C. 1999. Regulation of life span by sensory perception in *Caenorhabditis elegans. Nature* 402:804–9

4. Bargmann CI. 1998. Neurobiology of the *Caenorhabditis elegans* genome. *Science* 282:2028–33

5. Bargmann CI, Horvitz HR. 1991. Control of larval development by chemosensory neurons in *Caenorhabditis elegans. Science* 251:1243–46

6. Bartke A, Brown-Borg HM, Bode AM, Carlson J, Hunter WS, Bronson RT. 1998. Does growth hormone prevent or accelerate aging? *Exp. Gerontol.* 33:675–87

7. Birnby DA, Link EM, Vowels JJ, Tian H, Colacurcio PL, Thomas JH. 2000. A transmembrane guanylyl cyclase (DAF-11) and Hsp90 (DAF-21) regulate a common set of chemosensory behaviors in *Caenorhabditis elegans. Genetics* 155:85–104

8. Blander G, Zalle N, Leal JF, Bar-Or RL, Yu CE, Oren M. 2000. The Werner syndrome protein contributes to induction of p53 by DNA damage. *FASEB J.* 14:2138–40

9. Brody JA, Grant MD, Frateschi LJ, Miller SC, Zhang H. 2000. Reproductive longevity and increased life expectancy. *Age Aging* 29:75–78

10. Brogiolo W, Stocker H, Ikeya T, Rintelen F, Fernandez R, Hafen E. 2001. *Curr. Biol.* 11:213–21

11. Brown-Borg HM, Borg KE, Meliska CJ, Bartke A. 1996. Dwarf mice and the ageing process. *Nature* 384:330

12. Bruning JC, Gautam D, Burks DJ, Gillette J, Schubert M, et al. 2000. Role of

brain insulin receptor in control of body weight and reproduction. *Science* 289: 2122–25

13. Carey JR, Judge DS. 2000. *Longevity Records: Life Spans of Mammals, Birds, Amphibians, Reptiles, and Fish*, Monographs on Population Aging, Vol. 8. Odense: Odense University Press. 241 pp.

14. Castro E, Ogburn CE, Hunt KE, Tilvis R, Louhija J, et al. 1999. Polymorphisms at the Werner locus: I. Newly identified polymorphisms, ethnic variability of 1367Cys/Arg, and its stability in a population of Finnish centenarians. *Am. J. Med. Genet.* 82:399–403

14a. Clancy DJ, Gems D, Harshman LG, Oldham S, Stocker H, et al. 2001. Extension of life span by loss of CHICO, a *Drosophila* insulin receptor substrate protein. *Science* 292:104–6

15. Colman PD, Kaplan BB, Osterburg HH, Finch CE. 1980. Brain poly(A)RNA during aging: stability of yield and sequence complexity in two rat strains. *J. Neurochem.* 34:335–45

16. Coschigano KT, Clemmons D, Bellush LL, Kopchick JJ. 2000. Assessment of growth parameters and life span of GHR/BP gene-disrupted mice. *Endocrinology* 141:2608–13

17. Curtsinger JW, Fukui HH, Resler AS, Kelly K, Khazaeli AA. 1998. Genetic analysis of extended life span in *Drosophila melanogaster*. I. RAPD screen for genetic divergence between selected and control lines. *Genetica* 104:21–32

18. De Haan G, Van Zant G. 1999. Genetic analysis of hemopoietic cell cycling in mice suggests its involvement in organismal life span. *FASEB J.* 13:707–13

19. Devi G, Ottman R, Tang M, Marder K, Stern Y, et al. 1999. Influence of APOE genotype on familial aggregation of AD in an urban population. *Neurology* 53:789–94

20. Do KA, Broom BM, Kuhert P, Duffy DL, Todorov AA, et al. 2000. Genetic analysis of the age at menopause by using estimating equations and Bayesian random effects models. *Stat. Med.* 19:1217–35

21. Dorman JB, Albinder B, Shroyer T, Kenyon C. 1995. The *age-1* and *daf-2* genes function in a common pathway to control the life span of *Caenorhabditis elegans*. *Genetics* 141:1399–406

22. Durham SK, Suwanichkul A, Scheimann AO, Yee D, Jackson JG, et al. 1999. FKHR binds the insulin response element in the insulin-like growth factor binding protein-1 promoter. *Endocrinology* 140:3140–46

22a. Eigenmann JE, Amador A, Patterson DF. 1988. Insulin-like growth factor I levels in proportionate dogs, chondrodystrophic dogs, and giant dogs. *Acta Endocrinol.* 106:448–53

23. Estevez M, Attisano L, Wrana JL, Albert PS, Massague J, Riddle DL. 1993. The *daf-4* gene encodes a bone morphogenetic protein receptor controlling *C. elegans* dauer larva development. *Nature* 365:644–49

23a. Fabrizio P, Pozza F, Pletcher S, Gendron CM, Longo VD. 2001. Regulation of longevity and stress resistance by Sch9 in yeast. *Science* 292:288–90

24. Falconer DS, Mackay TFC. 1996. *Introduction to Quantitative Genetics*. London: Longman. 464 pp. 4th ed.

25. Felkai S, Ewbank JJ, Lemieux J, Labbe JC, Brown GG, Hekimi S. 1999. CLK-1 controls respiration, behavior and aging in the nematode *Caenorhabditis elegans*. *EMBO J.* 18:1783–92

26. Finch CE. 1990. *Longevity, Senescence, and the Genome*. Chicago: Univ. Chicago Press. 922 pp.

27. Finch CE. 1994. Commentary on latent capacities for gametogenic cycling in the semelparous invertebrate *Nereis*. *Proc. Natl. Acad. Sci. USA* 91:11769–70

28. Finch CE, Austad SN. 2001. Organisms with slow aging. *Exp. Gerontol.* 36(Spec. issue)

29. Finch CE, Kirkwood TBL. 2000. *Chance, Development, and Aging.* New York: Oxford Univ. Press. 278 pp.

30. Finch CE, Morgan TE, Rozovsky I, Xie Z, Weindruch R, Prolla T. 2001. Microglia and aging in the brain. In *Microglia in the Degenerating and Regenerating CNS,* ed. WJ Streit. New York: Springer-Verlag. In press

31. Finch CE, Rose MR. 1995. Hormones and the physiological architecture of life history evolution. *Q. Rev. Biol.* 70:1–52

32. Finch CE, Tanzi RE. 1997. Genetics of aging. *Science* 278:407–11

32a. Gems D, Sutton AJ, Sundermeyer ML, Albert PS, King KV, et al. 1998. Two pleiotropic classes of daf-2 mutation affect larval arrest, adult behavior, reproduction and longevity in *Caenorhabditis elegans. Genetics* 150:129–55

33. George JC, Bada J, Zeh J, Brown SE, O'Hara T, Suydam R. 1999. Age and growth estimates of bowhead whales (*Balaena mysticetus*) via aspartic acid racemization. *Can. J. Zool.* 77:571–80

34. Georgi LL, Albert PS, Riddle DL. 1990. daf-1, a *C. elegans* gene controlling dauer larva development, encodes a novel receptor protein kinase. *Cell* 61:635–45

35. Gerdes LU, Jeune B, Ranberg KA, Nybo H, Vaupel JW. 2000. Estimation of apolipoprotein E genotype-specific relative mortality risks from the distribution of genotypes in centenarians and middle-aged men: Apolipoprotein E gene is a "frailty gene", not a "longevity gene". *Genet. Epidemiol.* 19:202–10

36. Golden JW, Riddle DL. 1984. A pheromone-induced developmental switch in *Caenorhabditis elegans*: Temperature-sensitive mutants reveal a wild-type temperature-dependent process. *Proc. Natl. Acad. Sci. USA* 81:819–23

37. Gottlieb S, Ruvkun G. 1994. daf-2, daf-16, and daf-23: genetically interacting genes controlling dauer formation in *Caenorhabditis elegans. Genetics* 137:107–20

38. Granner D, Andreone T, Sasaki K, Beale E. 1983. Inhibition of transcription of the phosphoenolpyruvate carboxykinase gene by insulin. *Nature* 305:549–51

39. Guarente L, Kenyon C. 2000. Genetic pathways that regulate aging in model organisms. *Nature* 408:255–62

40. Gudmundsson H, Gudbjartsson DF, Frigge M, Gulcher JR, Stefansson K. 2000. Inheritance of human longevity in Iceland. *Eur. J. Hum. Genet.* 8:743–49

41. Guo S, Rena G, Cichy S, He X, Cohen P, Uterman T. 1999. Phosphorylation of serine 256 by proteiun kinase B disrupts transactivation by FKHR and mediates effects of insulin on insulin-like growth factor–binding protein-1 promoter activity through a conserved insulin response sequence. *J. Biol. Chem.* 174:17184–92

42. Hall RK, Yamasaki T, Kucera T, Waltner-Law M, Obrien R, Granner DK. 2000. Regulation of phosphoenolpyruvate carboxykinase and insulin-like growth factor–binding protein-1 gene expression by insulin. The role of winged helix/forkhead proteins. *J. Biol. Chem.* 275:30169–75

43. Hauck SJ, Bartke A. 2000. Effects of growth hormone on hypothalamic catalase and Cu/Zn superoxide dismutase. *Free Radic. Biol. Med.* 28:970–78

44. Heijmans BT, Westendorp RG, Slagboom PE. 2000. Common gene variants, mortality and extreme longevity in humans. *Exp. Gerontol.* 35:865–77

45. Herskind AM, McGue M, Holm NV, Sorensen TI, Harvald B, Vaupel JW. 1996. The heritability of human longevity: a population-based study of 2872 Danish twin pairs born 1870–1900. *Hum. Genet.* 97:319–23

46. Hobert O, Mori I, Yamashita Y, Honda H, Ohshima Y, et al. 1997. Regulation of interneuron function in the *C. elegans* thermoregulatory pathway by the ttx-3 LIM homeobox gene. *Neuron* 19:345–57

47. Honda Y, Honda S. 1999. The daf-2 gene network for longevity regulates oxidative

stress resistance and Mn-superoxide dismutase gene expression in *Caenorhabditis elegans*. *FASEB J.* 13:1385–93

48. Huang TT, Carlson EJ, Gillespie AM, Shi Y, Epstein CJ. 2000. Ubiquitous overexpression of CuZn superoxide dismutase does not extend life span in mice. *J. Gerontol. A* 55:B5–B9

49. Inoue T, Thomas JH. 2000. Targets of TGF-beta signaling in *Caenorhabditis elegans* dauer formation. *Dev. Biol.* 217: 192–204

50. Ishii N, Fujii M, Hartman PS, Tsuda M, Yasuda K, et al. 1998. A mutation in succinate dehydrogenase cytochrome b causes oxidative stress and aging in nematodes. *Nature* 394:694–97

51. Jonassen T, Larsen PL, Clarke CF. 2001. A dietary source of coenzyme Q is essential for growth of long-lived *Caenorhabditis elegans* clk-1 mutants. *Proc. Natl. Acad. Sci. USA* 98:421–26

52. Kalantaridou SN, Davis SR, Nelson LM. 1998. Premature ovarian failure. *Endocrinol. Metab. Clin. N. Am.* 27:989–1006

53. Kenyon C, Chang J, Gensch E, Rudner A, Tabtlang R. 1993. A *C. elegans* mutant that lives twice as long as wild type. *Nature* 366:461–64

54. Kimura KH, Tissenbaum A, Liu Y, Ruvkun G. 1997. daf-2, an insulin receptor-like gene that regulates longevity and diapause in *Caenorhabditis elegans. Science* 277:942–46

55. Kirkwood TB, Austad SN. 2000. Why do we age? *Nature* 408:233–38

56. Klass M. 1983. A method for the isolation of longevity mutants in the nematode *C. elegans* and initial results. *Mech. Aging Dev.* 22:279–86

57. Krishna S, Maduzia LL, Padgett RW. 1999. Specificity of TGF-beta signaling is conferred by distinct type I receptors and their associated SMAD proteins in *Caenorhabditis elegans. Development* 126:251–60

58. Kubisch HM, Wang J, Bray TM, Phillips JP. 1997. Targeted overexpression of Cu/Zn superoxide dismutase protects pancreatic beta-cells against oxidative stress. *Diabetes* 46:1563–66

59. Kulkarni RN, Brüning JC, Winnay JN, Postic C, Magnuson MA, Kahn CR. 1999. Tissue-specific knockout of the insulin receptor in pancreatic beta cells creates an insulin secretory defect similar to that in type 2 diabetes. *Cell* 96:329–39

60. Kurapati R, Passananti HB, Rose MR, Tower J. 2000. Increased hsp22 RNA levels in *Drosophila* lines genetically selected for increased longevity. *J. Gerontol. A* 55:B552–59

61. Lakowski B, Hekimi S. 1998. The genetics of caloric restriction in *Caenorhabditis elegans. Proc. Natl. Acad. Sci. USA* 95:13091–96

62. Larsen P. 1993. Aging and resistance to oxidative damage in *Caenorhabditis elegans. Proc. Natl. Acad. Sci. USA* 90:8905–9

63. Larsen PL, Albert PS, Riddle DL. 1995. Genes that regulate both development and longevity in *Caenorhabditis elegans. Genetics* 139:1567–83

64. Lebovitz RM, Zhang H, Vogel H, Cartwright J Jr, Dionne L, et al. 1996. Neurodegeneration, myocardial injury, and perinatal death in mitochondrial superoxide dismutase–deficient mice. *Proc. Natl. Acad. Sci. USA* 93:9782–87

65. Lee CK, Klopp RG, Weindruch R, Prolla TA. 1999. Gene expression profile of aging and its retardation by caloric restriction. *Science* 285:1390–93

66. Lee CK, Weindruch R, Prolla TA. 2000. Gene-expression profile of the aging brain in mice. *Nat. Genet.* 25:294–97

67. Lin YJ, Seroude L, Benzer S. 1998. Extended life span and stress resistance in the Drosophila mutant *methuselah. Science* 282:943–46

67a. Lithgow GJ, White TM, Melov S, Johnson TE. 1995. Thermotolerance and extended life span conferred by single-gene mutations and induced by thermal stress. *Proc. Natl. Acad. Sci. USA* 92:7540–44

68. Ljungquist B, Berg S, Lanke J, McClearn GE, Pedersen NL. 1998. The effect of genetic factors for longevity: a comparison of identical and fraternal twins in the Swedish Twin Registry. *J. Gerontol. A* 53:M441–46

69. Lombard DB, Beard C, Johnson B, Marciniak RA, Dausman J, et al. 2000. Mutations in the WRN gene in mice accelerate mortality in a p53-null background. *Mol. Cell. Biol.* 20:3286–91

70. Martin GM, Oshima J. 2000. Lessons from human progeroid syndromes. *Nature* 408:263–66

71. Melov S, Ravenscroft J, Malik S, Gill MS, Walker DW, et al. 2000. Extension of life span with superoxide dismutase/catalase mimetics. *Science* 289:1567–69

72. Meyer MR, Tschanz JT, Norton MC, Welsh-Bohmer KA, Steffencs DC, et al. 1998. APOE genotype predicts when—not whether—one is predisposed to develop Alzheimer disease. *Nat. Genet.* 19:321–22

73. Michael MD, Kulkarni RN, Postic C, Previs SF, Shulman GI, et al. 2000. Loss of insulin signaling in hepatocytes leads to severe insulin resistance and progressive hepatic dysfunction. *Mol. Cell* 6:87–97

74. Migliaccio E, Giorgio M, Mele S, Pelicci G, Reboldi P, et al. 1999. The p66shc adaptor protein controls oxidative stress response and life span in mammals. *Nature* 402:309–13

75. Miller RA. 1999. Kleemeier Award Lecture: Are there genes for aging? *J. Gerontol. A* 54:B297–307

76. Miller RA, Chrisp C, Atchley W. 2000. Differential longevity in mouse stocks selected for early life growth trajectory. *J. Gerontol. A* 55:B455–61

77. Miller RM. 2001. Biomarkers and genetics of aging in mice. In *Cells and Surveys: Should Biomarkers Be Included in Social Science Research?*, ed. CE Finch, JW Vaupel, K Kinsella, 8:180–212. Washington, DC: Natl. Acad. 347 pp.

78. Morgan TE, Xie Z, Goldsmith S, Yoshida T, Lanzrein AS, et al. 1999. The mosaic of brain glial hyperactivity during normal aging and its attenuation by food restriction. *Neuroscience* 89:687–99

79. Morris JZ, Tissenbaum HA, Ruvkun G. 1996. A phosphatidylinositol-3-OH kinase family member regulating longevity and diapause in *Caenorhabditis elegans*. *Nature* 382:536–39

80. Nakae J, Park BC, Accili D. 1999. Insulin stimulates phosphorylation of the forkhead transcription factor FKHR on serine 253 through a Wortmannin-sensitive pathway. *J. Biol. Chem.* 274:15982–85

81. Nasrin N, Ogg S, Cahill C, Biggs W, Nui S, et al. 2000. DAF-16 recruits the CBP co-activator to the IGFBP-1 promoter in HepG2 cells. *Proc. Natl. Acad. Sci. USA* 97:10412–17

82. Ogg S, Paradis S, Gottlieb S, Patterson GI, Lee L, et al. 1997. The fork head transcription factor DAF-16 transduces insulin-like metabolic and longevity signals in *C. elegans*. *Nature* 389:994–99

83. Ogg S, Ruvkun G. 1998. The *C. elegans* PTEN homolog daf-18 acts in the insulin receptor-like metabolic signaling pathway. *Mol. Cell* 2:887–93

83a. Oldham S, Bohni R, Stocker H, Brogiolo W, Hafen E. 2000. Genetic control of size in *Drosophila*. *Philos. Trans. R. Soc. Lond. B* 355:945–52

84. Paradis S, Ruvkun G. 1998. *Caenorhabditis elegans* Akt/PKB transduces insulin receptor-like signals from AGE-1 PI3 kinase to the DAF-16 transcription factor 67. *Genes Dev.* 12:2488–98

84a. Paradis S, Ailion M, Toker A, Thomas JH, Ruvkun G. 1999. A PDK1 homolog is necessary and sufficient to transduce AGE-1 PI3 kinase signals that regulate diapause in *C. elegans*. *Genes Dev.* 13:1438–52

85. Parkes TL, Elia AJ, Dickinson D, Hilliker AJ, Phillips JP, Boulianne GL. 1998. Extension of *Drosophila* life span by overexpression of human SOD1 in motorneurons. *Nat. Genet.* 19:171–74

86. Patterson G, Koweek A, Wong A, Liu Y, Ruvkun G. 1997. The DAF-3 Smad protein antagonizes TGF-beta–related receptor signaling in the *Caenorhabditis elegans* dauer pathway. *Genes Dev.* 11:2679–90

87. Perls TT, Alpert L, Fretts RC. 1997. Middle-aged mothers live longer. *Nature* 389:133

88. Perls TT, Shea-Drinkwater M, Bowen-Flynn J, Ridge SB, Kang S, et al. 2000. Exceptional familial clustering for extreme longevity in humans. *J. Am. Geriatr. Soc.* 48:1483–85

88a. Perls TT, Silver MH, Lauerman JT. 2000. *Living to 100: Lessons in Living to Your Maximum Potential at Any Age.* New York: Basic Books. 272 pp.

88b. Pierce SB, Costa M, Wisotzkey R, Devadhar S, Homburger SA, et al. 2001. Regulation of DAF-2 receptor signaling by human insulin and ins-1, a member of the unusually large and diverse *C. elegans* insulin gene family. *Genes Dev.* 15:672–86

89. Ren P, Lim C-S, Johnsen R, Albert PS, Pilgrim D, Riddle DL. 1996. Control of *C. elegans* larval development by neuronal expression of a TGF-beta homolog. *Science* 274:1389–92

90. Rena G, Guo S, Cichy SC, Unterman TG, Cohen P. 1999. Phosphorylation of the transcription factor forkhead family member FKHR by protein kinase B. *J. Biol. Chem.* 274:17179–83

91. Reveillaud I, Niedzwiecki A, Bensch KG, Fleming JE. 1991. Expression of bovine superoxide dismutase in *Drosophila melanogaster* augments resistance of oxidative stress. *Mol. Cell. Biol.* 11:632–40

92. Rogina B, Reenan RA, Nilsen SP, Helfand S. 2000. Extended life span conferred by cotransporter gene mutations in *Drosophila. Science* 290:2137–40

93. Rose MR, Mueller LD. 2000. Ageing and immortality. *Philos. Trans. R. Soc. Lond. B* 355:1657–62

94. Rose MR, Vu LN, Park SU, Graves JL Jr. 1992. Selection on stress resistance increases longevity in *Drosophila melanogaster. Exp. Gerontol.* 27:241–50

95. Schachter F, Faure-Delanef L, Guenot F, Rouger H, Froguel P, et al. 1994. Genetic associations with human longevity at the APOE and ACE loci. *Nat. Genet.* 6:29–32

96. Schackwitz WS, Inoue T, Thomas JH. 1996. Chemosensory neurons function in parallel to mediate a pheromone response in *C. elegans. Neuron* 17:719–28

97. Schmidt PS, Duvernell DD, Eanes WF. 2000. Adaptive evolution of a candidate gene for aging in *Drosophila. Proc. Natl. Acad. Sci. USA* 97:10861–65

98. Schmoll D, Walker KS, Alessi DR, Grempler R, Burchell A, et al. 2000. Regulation of glucose-6-phosphatase gene expression by protein kinase B-alpha and the forkhead transcription factor FKHR: evidence for insulin response unit (IRU)-dependent and independent effects of insulin on promoter activity. *J. Biol. Chem.* 275:36324–33

99. Schwartz MW, Woods SC, Porte D, Seeley RJ, Baskin DG. 2000. Central nervous system control of food intake. *Nature* 404:661–71

100. Shefner JM, Reaume AG, Flood DG, Scott RW, Kowall NW, et al. 1999. Mice lacking cytosolic copper/zinc superoxide dismutase display a distinctive motor axonopathy. *Neurology* 53:1239–46

101. Smith JD. 2000. Apolipoprotein E4: an allele associated with many diseases. *Ann. Med.* 32:118–27

102. Sneider H, MacGregor AJ, Spector TD. 1998. Genes control the cessation of a woman's reproductive life: a twin study of hysterectomy and age at menopause. *J. Clin. Endocrinol. Metab.* 83:1875–80

103. Sohal RS, Weindruch R. 1996. Oxidative stress, caloric restriction, and aging. *Science* 273:59–63

104. Stengard JH, Weiss KM, Sing CF. 1998. An ecological study of association between coronary heart disease mortality rates in men and the relative frequencies

of common allelic variations in the gene coding for apolipoprotein E. *Hum. Genet.* 103:234–41

105. Sze JY, Victor M, Loer C, Shi Y, Ruvkun G. 2000. Food and metabolic signaling defects in a *C. elegans* serotonin null mutant. *Nature* 403:560–64

106. Takagi Y, Mitsui A, Nishiyama A, Nozaki K, Sono H, et al. 1999. Overexpression of thioredoxin in transgenic mice attenuates focal ischemic brain damage. *Proc. Natl. Acad. Sci. USA* 96:4131–36

107. Tang ED, Nunez G, Barr FG, Guan KL. 1999. Negative regulation of the forkhead transcription factor FKHR by Akt. *J. Biol. Chem.* 274:16741–46

107a. Tartar M, Kopelman A, Epstein D, Tu MP, Yin CM, Garofalo RS. 2001. A mutant *Drosophila* insulin receptor homolog that extends life span and impairs neuroendocrine function. *Science* 292:107–10

108. Taub J, Lau JF, Ma C, Hahn JH, Hoque R, et al. 1999. A cytosolic catalase is needed to extend adult lifespan in *C. elegans* dauer constitutive and clk-1 mutants. *Nature* 399:162–66

109. Tauber MJ, Tauber CA, Masaki S. 1986. *Seasonal Adaptation of Insects.* New York: Oxford Univ. Press. 411pp.

110. Thomas JH, Birnby DA, Vowels JJ. 1993. Evidence for parallel processing of sensory information controlling dauer formation in *Caenorhabditis elegans*. *Genetics* 134:1005–117

111. Tissenbaum HA, Guarente L. 2001. Increased dosage of a sir-2 gene extends life span in *Caenorhabditis elegans*. *Nature.* In press

111a. Tissenbaum HA, Ruvkun G. 1998. An insulin-like signaling pathway affects both longevity and reproduction in *Caenorhabditis elegans*. *Genetics* 148:703–17

112. Tomizawa M, Kumar A, Perrot V, Nakae J, Accili D, et al. 2000. Insulin inhibits the activation of transcription by a C-terminal fragment of the forkhead transcription factor FKHR. A mechanism for insulin inhibition of insulin-like growth factor–binding protein-1 transcription. *J. Biol. Chem.* 275:7289–95

113. Vanfleteren JR, De Vreese A. 1995. The gerontogenes age-1 and daf-2 determine metabolic rate potential in aging *Caenorhabditis elegans*. *FASEB J.* 9:1355–61

114. Van Voorhies WA, Ward S. 1999. Genetic and environmental conditions that increase longevity in *Caenorhabditis elegans* decrease metabolic rate. *Proc. Natl. Acad. Sci. USA* 96:11399–403

115. Vieira C, Pasyukova EG, Zeng ZB, Hackett JB, Lyman RF, Mackay TF. 2000. Genotype-environment interaction for quantitative trait loci affecting life span in *Drosophila melanogaster*. *Genetics* 154:213–27

116. Wang P, Chen H, Qin H, Sankarapandi S, Becher MW, et al. 1998. Overexpression of human copper, zinc-superoxide dismutase (SOD1) prevents postischemic injury. *Proc. Natl. Acad. Sci. USA* 95:4556–60

117. Wise PM, Smith MJ, Dubal DB, Wilson ME, Krajnak KM, Rosewell KL. 1999. Neuroendocrine influences and repercussions of the menopause. *Endocr. Rev.* 20:243–48

118. Wolkow CA, Kimura KD, Lee MS, Ruvkun G. 2000. Regulation of *C. elegans* life span by insulinlike signaling in the nervous system. *Science* 290:147–50

119. Wong PC, Rothstein JD, Price DL. 1998. The genetic and molecular mechanisms of motor neuron disease. *Curr. Opin. Neurobiol.* 8:791–99

Annu. Rev. Genomics Hum. Genet. 2001. 2:463–92

ENU MUTAGENESIS: Analyzing Gene Function in Mice

Rudi Balling

German Research Centre for Biotechnology, D-38124 Braunschweig, Germany;
e-mail: balling@gbf.de

Key Words mouse genetics, functional genomics, ethlynitrosourea, phenotyping, screening

■ **Abstract** With the completion of the human genome, sequence analysis of gene function will move into the center of future genome research. One of the key strategies for studying gene function involves the genetic dissection of biological processes in animal models. Mouse mutants are of particular importance for the analysis of disease pathogenesis and transgenic techniques, and gene targeting have become routine tools. Recently, phenotype-driven strategies using chemical mutagenesis have been the target of increasing interest. In this review, the current state of ENU mutagenesis and its application as a systematic tool of genome analysis are examined.

INTRODUCTION

We are at the beginning of a new era in mouse genetics. After the enormous success achieved via the use of the fruitfly *Drosophila melangogaster* as a model organism and tool for studying the mechanisms of embryonic development, we now see a tremendously growing interest in mouse genetics and genomics. The driving force is the conviction that mice are the best model organisms for the study of the pathogenesis of human disease. The completion of the human and mouse genome sequencing projects and the need to validate gene function support this trend.

The field of mouse genetics is almost 100 years old. Already by 1905, Cuenot had reported that the mouse mutant *Agouti yellow* is a Mendelian trait (13, 79). In those days, the only mouse mutants available were those maintained by mouse fanciers. For more than 50 years, mouse geneticists, led by the staff at the Jackson Laboratory, collected mouse mutants that occurred spontaneously in various mouse colonies around the world (www.jax.org). This changed quite a bit with the interest in the effects of radiation on mutation frequency. W. and E. Russel in Oakridge National Labs, Oak Ridge, Tennessee, and M. Lyon, B. Cattanach, and colleagues in Harwell, England, systematically used radiation to produce new mouse mutants and thereby laid the foundation for many of the large-scale mutagenesis approaches that we see today. Not only did they provide a wealth of new

1527-8204/01/0728-0463$14.00

mutants, they also produced the first in-depth region-specific chromosome maps of the mouse (8, 36, 37, 74).

Mouse genetics, nevertheless, remained a tedious and slow enterprise for many more years. Lack of polymorphic markers was the main reason that mapping of new mutants was a rate-limiting step. Microsatellite markers and polymerase chain reaction (PCR) changed this situation (15). Suddenly, many groups found themselves in the middle of competitive efforts to positionally clone the dozens of mouse mutants with an interesting phenotype. The field desperately needed more efficient ways to produce new mutants, in order to exploit the full potential of this model organism.

PHENOTYPE-DRIVEN VERSUS GENOTYPE-DRIVEN MUTAGENESIS

The production of transgenic mice and the development of embryonic stem cells and gene targeting opened a new avenue in the systematic production of mouse mutants. For the first time, a mutant could be produced for any gene that had been cloned (26, 75). The use of gene trapping has led to the construction of genome-wide libraries of embryonic stem (ES) cells in which every gene in the mouse genome is inactivated and can be recovered as a mutant through the generation of chimeras (28, 91, 93). "Give me a gene, and I'll give you a mutant." This scenario is now reality. With the availability of the complete mouse genome sequence, we can indeed envisage the possibility of producing a mutant for every gene in the genome.

However, what about the reverse? "Give me a phenotype, and I'll give you a mutant." Phenotype-based screens make no assumptions about the nature of the genes that cause a specific phenotype. In this case, the animal is exposed to a mutagen that acts in a random, shotgun-like approach on the entire genome. Consecutively, a large number of animals are screened in a systematic way to identify individuals that display the specific phenotype of interest. Following the isolation of such an animal, test breeding is used to confirm the genetic nature of the trait. Standard genetic and molecular tools of positional or candidate cloning are then used to isolate and identify the mutated gene. Whereas in gene targeting experiments the gene is known at the beginning of the experiment, phenotype-driven mutagenesis starts with a specific phenotype and only at the end is the responsible gene identified. Phenotype-driven approaches have been extremely important for *Drosophila* genetics. The dissection of *Drosophila* embryonic development was entirely phenotype-driven, with the number, identity, and shape of the fly segments and appendages as the screening endpoints (49). None of the developmental control genes had been cloned at the beginning of this important project. In fact, nobody would have been able to predict the molecular nature of the genes that might be involved in designing the *Drosophila* body plan.

We are now in a similar situation with many of the complex traits that underlie the pathogenesis of human diseases. It is very difficult to make predictions about the nature of the genes that are involved in these diseases. Knocking out one gene after the other and analyzing them one by one to reveal the loss-of-function phenotypes is impractical for dissecting the pathogenesis of a specific disease. In this context, genome-wide, systematic phenotype-driven mutagenesis screens have seen an explosion of interest in recent years. "Give me an assay, and I'll give you a mutant. Give me the mutant, and I'll give you the gene." This is now the motto for mouse genetics.

ENU (N-ETHYL-N-NITROSOUREA): THE MOST POWERFUL MUTAGEN IN MICE

The function of a gene is defined by its loss-of-function phenotype. This concept, developed by H. Mueller, is still an important guide to understanding gene function. The standard gene targeting experiment aims at the production of null mutations, through the insertion of DNA into the gene under investigation. In most cases, the production of a complete loss-of-function mutation is achieved. Only in some instances does differential splicing, or other circumstances, lead to residual gene activity. The production of specific mutations, i.e., designed point mutations, hypomorphs, or hypermorphs, through knockin experiments is much more difficult and time consuming. The main obstacle, however, is the lack of a priori knowledge on the molecular nature of the mutations that will result in specific effects. In addition, many mutations found in human genetic diseases are not the result of insertional mutations but are caused by point mutations. For this reason, the ideal mutagen for phenotype-driven mutagenesis experiments should lead to point mutations.

Mutagenic Action of ENU

ENU is such a mutagen. This synthetic compound was described as the "most potent mutagen in mice" (70). At optimal concentrations, mutation rates of 1 new mutation per gene in every 700 gametes, or even more, can be obtained (19, 21, 61, 63, 68, 69, 78). ENU is an alkylating agent and primarily induces point mutations (Figure 1). This is in contrast to radiation or mutagenic agents, such as chlorambucil, that induce deletions or inversions and often involve more than one gene (22, 67). ENU acts through the alkylation of nucleic acids, without the need for metabolic activation. The ethyl group of ENU can be transferred to nucleophilic sites, i.e., on the N1, N3, or N7 groups of adenine; the O6, N3, or N7 of guanine; the O2, O4, or N3 of thymine; and the O2 or N3 of cytosine. As a result of the formed DNA adduct, mispairing and base-pair substitution occur during the next round of DNA replication. The most common mutations induced by ENU are AT to TA transversions and AT to GC transitions (55). More than 82% of the mutations sequenced show these types of mutations, most likely the result of mispairing after the alkylation at O4 and O2 thymine (48).

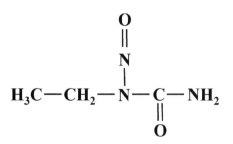

Figure 1   Ethylnitrosourea, the most potent mutagen in mice.

ENU is not only mutagenic but also toxic. For this reason, treatment protocols have to stay within a narrow range of dose and time of exposure in order to achieve the optimal balance between a high mutation frequency and low toxicity. In addition to its effect on nucleic acids, ENU can also lead to nonheritable alterations in proteins as a result of carbamoylation of amino acids. ENU is also a potent carcinogen, leading to various types of tumors in treated mice. However, the most important characteristic of ENU is its efficiency in mutagenizing male spermatogonial stem cells in mice. Postspermatogonial stem cells and female germ cells yield mutation rates that are one order of magnitude below that of male stem cell spermatogonia. Mature spermatozoa exhibit even lower mutation rates after ENU treatment. The mechanism of these cell-type specific differences is not known, although differences in the capacity to repair ENU-induced lesions are most likely involved.

The observation of a threshold, below which ENU lesions are usually not observed, points to the presence of a saturable DNA repair system. Above a certain ENU concentration, the repair systems may be saturated, resulting in a dose-dependent linear relationship between ENU dose and mutation rate. A number of different repair systems seem to be involved, including the activity of an O6-alkylguanine-DNA alkyltransferase (AGT) and nucleotide excision repair (NER). Genetic manipulation of the DNA-repair capabilities of mice may possibly increase the mutation frequency after mutagen exposure or may even shift the spectrum of induced mutations. An exhaustion of the DNA-repair capacity may also explain why the fractionated application of ENU results in much higher tolerated exposure doses compared to single exposures but, nevertheless, leads to a marked increase in mutation frequency (21).

Strain Differences

Different strains of mice vary markedly in their response to ENU (29). F1 hybrid animals are quite resistant, but many of the inbred strains cannot tolerate higher doses of ENU without marked effects in lethality and/or sterility. Concentrations between 200 and 400 mg/kg body weight administered intraperitoneally are the most commonly used dosage schemes. Instead of a single injection, a total dose of 400 mg/kg body weight given as a fractionated dose of four weekly injections of 100 mg/kg body weight is a very effective mutation regimen. Injection of ENU into male mice typically leads to a transient period of sterility. This sterility phase can last up to 14 weeks and is a result of the depletion of spermatogonial stem cells. The spermatocytes need to be repopulated from the remaining stem cells before treated males can be used for the production of F1 mice. However, for many mouse strains, the doses mentioned above are too high and lead either to reduced survival or to permanent sterility. In this case, careful titration of the ENU doses used needs to be done before large-scale experiments are initiated.

Mutation Rates

Mutation rates for a specific gene can vary significantly. Most of the mutation rates have been determined on the basis of specific-locus tests that use mutations in seven visible traits (14, 20). These specific locus tests were applied to very large numbers of mutagenesis treatments and, thereby, provided an accurate figure for the mutation frequencies of these genes in mice. The traits included the *albino*, *brown*, *pink-eye dilution*, *chinchilla*, *piebald spotting*, and *agouti* locus as well as the *short-ear* locus (64). However, in this limited number of loci, differences in the specific mutation rates of up to 10-fold have already been observed. One of the most important factors is, of course, the size of the gene sampled. Cold spots or hot spots for mutation induction can be observed, arguing that additional factors affecting mutation frequencies must play a role. These could include the presence of functional domains or regions that are critical for the protein structure and function or the requirement or nonrequirement of specific gene regions, such as an intact 3' end. Unfortunately, we do not know the mutation frequencies that we can expect for most genes. The number of genes for which specific locus tests are available is too small, and the number of alleles that have been recovered is, with the exception of some of the coat-color genes, too small to be statistically robust. For phenotype-driven mutagenesis, one has to keep in mind that the endpoints of these screens are measurements in physiological parameters that reflect the activity of many individual and highly integrated metabolic pathways. These pathways consist of many independent genes so that the hit rate for these pathways should be much higher than the hit rate for a specific gene.

The mutation frequencies observed in the specific-locus experiments suggest that in any individual ENU-treated mouse many different mutations are simultaneously induced. Assuming a total number of 25,000 to 40,000 genes in the mouse genome (46, 73) and a specific locus mutation frequency of 1 in 1000, a single mouse should have between 25 and 40 different mutagenized genes that should give rise to a recessive phenotype. A commonly raised concern is that this high frequency of multiple mutations in a single ENU-treated animal would lead to the manifestation of phenotypes that are the result of a multigenic rather than a monogenic trait and, therefore, would complicate the identification of the responsible gene(s). On both theoretical as well as practical grounds, this concern can be rejected. Even if an individual mouse carries between 20 and 40 mutations that can lead to a recessive phenotype, the chances are quite small that different mutations that produce a specific phenotype only when inherited together will segregate together and continue to produce this multigenic phenotype in the next generation. In each generation, 50% of the mutations are lost, and only the mutation for which the specific phenotype continues to be selected is maintained in the colony.

In summary, ENU fulfills many of the requirements for an ideal mutagen for phenotype-driven mutagenesis. It is very efficient, leading to a high mutation rate combined with low toxicity. It affects germ cells, particularly spermatogonial stem cells, and induces point mutations, leading to a wide range of missense, nonsense, and splice-site mutations.

GENETIC SCREENING STRATEGIES

Dominant Screens

A number of different strategies can be used in ENU-mutagenesis experiments to screen for phenotypically interesting mice. These include genome-wide screens for dominant or recessive mutations (Figure 2). Furthermore, one can carry out region-specific screens or sensitized screens. In a dominant screen, male mice are injected with ENU, and a large number of F1 offspring are produced by mating the treated males to unaffected females. Each of the F1 animals carries a unique set of altered alleles, which are screened for phenotypic abnormalities. Mice carrying mutant phenotypes are then test bred to confirm the genetic nature of the abnormality. Affected offspring can routinely be recovered by outcrossing, whereas homozygotes can be obtained by intercrossing heterozygous animals. Most of the ENU screens carried out in recent years were dominant screens, and only recently have the first large-scale recessive screens been initiated. The reason for the popularity of screening for dominant phenotypes lies in the simplicity of the logistics of the mutagenesis and breeding protocols and the enormous success in the recovery of mutants through this approach. In the two major large-scale screens carried out to date, about 2% of all F1 mice analyzed displayed a heritable phenotypic abnormality (27,47).

A convincing example for the efficiency and success of screening for a specific dominant phenotype is the identification of the circadian rhythm mutant *Clock* (87). Using a phenotyping assay that was based on the automatic recording of the activity pattern of mice with a running wheel, screening of <400 F1 mice was sufficient to isolate the *Clock* mouse, a mutant with a dominant abnormality in the circadian rythm pattern. Using biochemical or molecular tools to identify a gene involved in such a complex physiological trait would have been difficult. Whenever our knowledge about genes and pathways involved in specific biological phenomena is limited, a genetics approach may be the most straightforward way of getting a first glimpse.

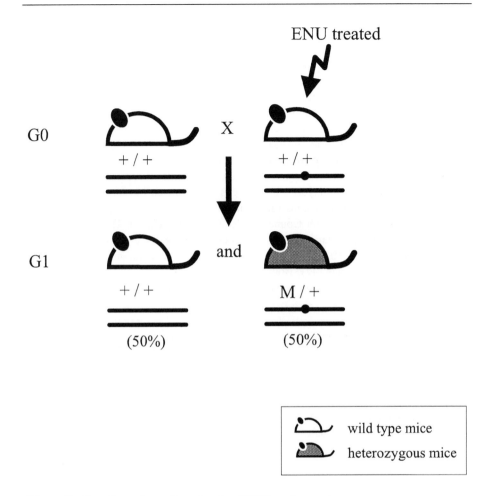

Figure 2   Breeding scheme for a dominant ENU-mutagenesis screen.

At the National Research Center for Environment and Health (GSF) in Neuherberg, Favor and colleagues screened more than 500,000 mice over the course of more than 15 years, with the goal of identifying dominant cataract mutants (18). Using a slit lamp to identify mice with problems in the transparency of the lens, close to 200 mutants, which constitute now the Neuherberg Cataract Collection, were isolated. The precise mutation frequency obtained is difficult to judge in this case because the experiments were part of risk assessment studies and, thus, were not optimized for efficient mutant recovery. Therefore, a number of different mutagens, concentrations, and conditions were used. The current estimate for the number of loci revealed by this collection is ~30 to 50. For some of the mutants recovered, e.g., for the crystalline loci, up to 10 different alleles were identified. However, many of the mutant loci were only recovered as a single allele.

Just as the cataract screen used a qualitative trait as an endpoint, a similar extensive screen was carried out by the group of Pretsch and colleagues, also in Neuherberg, for dominant mutants with deviations in the activity of 10 different metabolic enzymes in blood (i.e., LDH, TLPi, GP-6-DHG, etc.) (56). For each enzyme, ~10 loss-of-function mutants were isolated. These either displayed a complete or partial loss or an upregulation of enzyme activity. As the chromosomal map position for the mutants with increased enzyme activity does not correspond to the location of the structural gene(s), the mutations in these cases must affect regulatory loci. The fact that for all enzymes loss-of-function mutants were isolated shows that dominant screens are well suited to identify this class of mutations.

During the last four years, two groups have carried out large-scale and systematic genome-wide screens for dominant mutations (27, 47; also see Table 1). At the Medical Research Council (MRC) Genetics Unit in Harwell and at the Institute of Mammalian Genetics at the GSF Research Center in Neuherberg, a total of more than 50,000 mice were screened for a wide range of dominant phenotypic abnormalities (47). Both studies were directed toward the identification of mouse mutants that can serve as models of human disease. The GSF screen

TABLE 1   Major ENU screens: centers, websites, and contacts

ENU screen center and location	Website	Contact
Australia National University, Medical Genome Centre, Canberra	www.jcsmr.anu.edu.au/group_pages/mgc/MedGenCen.html	Christopher Goodnow
Baylor College of Medicine Mouse Genome Centre Houston, Texas	www.mouse-genome.bcm.tmc.edu	Monica Justice
Case Western Reserve University, Genetics Dept., Cleveland, Ohio, USA	jhn4@po.cwru.edu	Joe Nadeau
GSF, Neuherberg, Germany	www.gsf.de/isg/groups/enu-mouse.html	Martin Hrabé de Angelis
Jackson Laboratory, Bar Harbor, Maine, USA	www.jax.org	John Schimenti, Wayne Frankel
Mammalian Genetics Unit, Medical Research Council, Harwell, UK	www.mgu.har.mrc.ac.uk/mutabase/	Steve Brown
Oak Ridge National Laboratory Oak Ridge, Tennessee, USA	www.bio.ornl.gov/htmouse	Darla Miller, Gene Rinchik
RIKEN Genomic Sciences Centre Yokohama, Japan	tshirois@lab.nig.ac.jp	Shiroishi Toshihiko
University of Pennsylvania Philadelphia, USA	bucan@pobox.upenn.edu	Maja Bucan
University of Toronto, Mouse Models Initiative Toronto, Canada	www.mshri.on.ca/develop/rossant/enu_project/enu_homepage.htm	Janet Rossant

focused on specific abnormalities, e.g., congenital malformations, on blood-based screens for biochemical alteration of clinical relevance, on immunological defects, and on more complex traits, such as behavior or predisposition to allergies. The Harwell screen also carried out a basic blood-based screen but set its focus on the identification of neurobehavioral mutants. Each mouse was checked for more than 100 different traits. As previously mentioned, the total recovery rate for dominant mutants was in the range of 2%.

Recessive Screens

Despite the success of the large-scale dominant ENU screens, it is clear that dominant mutants cannot be recovered for all genes. Many of the human diseases display a recessive mode of inheritance and can be diagnosed only in affected homozygotes, not in heterzyogotic carriers. For this reason, screens for recessive mouse mutants are essential despite the increased logistical demands and the additional effort involved (Figure 3).

In mice, systematic and large-scale genome-wide recessive screens have not yet been carried out. Such screens also involve the production of F1 offspring from mutagenized G0 male mice, similar to the strategy for dominant screens. The resulting F1 progeny (G1), which are heterozygous for many different newly induced mutations distributed throughout the genome, are again mated to wild-type mice to establish families of G2 siblings. These G2 animals share the same set of mutations derived from their father, and each G2 animal has a 50% chance of inheriting any single mutation carried by the F1 male. If the G2 animal does carry a particular mutation, then 25% of the offspring from a cross between the G2 female and her father will be homozygous for the mutation.

As an alternative to the use of a backcross of the G2 females with the G1 male founders, random intercrossing between G2 siblings can be done—a breeding scheme that goes back to Haldane and was originally designed to recover lethal mutations (25). The intercross strategy has been successfully used in the two large-scale zebrafish ENU screens carried out in Tübingen and in Boston in recent years. By crossing 8–10 pairs of G2 heterozygotes, one can expect to recover on average of two sets of heterozygotes. The resulting homozygous phenotypes arise more than once and, therefore, can be compared to each other, which helps to rule out nongenetic alterations. All of the current mouse recessive screens, however, employ the backcross strategy described above.

Region Specific Screens: The Use of Deletions and Inversions

Genome-wide recessive screens are very labor intensive and time consuming. In order to combine the advantages of a recessive screen with that of a one-generation screen, one possibility is to look for mutations that fail to complement a particular chromosomal deletion (62, 63) (Figure 4). F1 offspring from ENU-treated males are mated to mice that carry a deletion covering a certain defined chromosomal

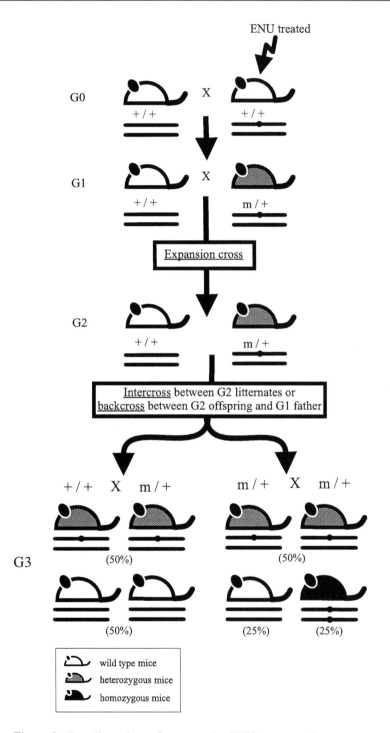

Figure 3 Breeding scheme for a recessive ENU-mutagenesis screen.

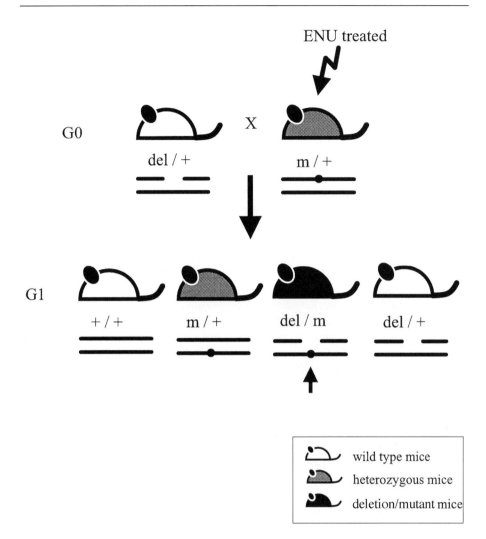

Figure 4   Breeding scheme for F1-deletion ENU-mutagenesis screen; del, deleted.

region. Based on the principle of allelic noncomplementation, the offspring that inherit the deletion from one parent and an ENU-induced mutation from the other parent will display a mutant phenotype. Using this breeding scheme allows for the screening of recessive phenotypic abnormalities within a large number of mutagenized mice. This strategy requires much less effort compared to the three-generation breeding scheme previously described. Furthermore, one of the main advantages of a region-specific screen is that any mutation that is identified is immediately localized to the specific chromosomal region, circumventing the need for mapping the mutation. Region-specific saturation screens have been extensively used by *Drosophila* geneticists (2) and in principle allow researchers to obtain saturation mutagenesis for a specific chromosomal region. Thus, a mutant could be retrieved for each gene in the deletion interval.

G. Rinchik has applied this approach systematically for the characterization of an 11 cM deletion complex surrounding the albino locus on mouse chromosome 7 (59, 63). Testing more than 4500 gametes, researchers identified 24 ENU-induced lethals. One of these mutants is *eed*, a gene with a sequence similar to the *Drosophila* polycomb gene family (72). Three alleles of *eed* were identified, two of which are null alleles and lead to an arrest of embryonic development around gastrulation. One of the alleles, however, is a hypomorph, and its classification allowed for the identification of a later role of *eed*, namely, its function in the development of the axial skeleton and the establishment of positional information along the anterior-posterior axis. A standard gene targeting allele would not have been sufficient to provide the information about the second function of *eed*. This example shows the importance of producing a number of different alleles for the same gene; only then will we have a chance to reveal the full spectra of functions that any single gene can have.

At the present time, the number of chromosomal deletions that can be used for region-specific screens is very limited. The existing ones mainly cover the regions linked to coat-color markers or other loci that were part of the specific locus test. In order to make the fullest use of the potential of deletions, expanding the collection of deletion mutants dramatically will be necessary. Two recent developments address this point. Bradley and colleagues have used homologous recombination in ES cells to specifically target and introduce new deletions into any chromosomal region of interest (35, 57, 94). For this purpose, LoxP sites are introduced by genetic engineering into an anchor locus in ES cells. Transient expression of CRE recombinase then results in the deletion of either the "floxed" DNA-region or, depending on the orientation, the two LoxP sites in an inversion. Even more sophisticated is the targeted introduction of a coat-color marker, such as tyosinase or agouty, directly into the ES-cell targeting vectors (30, 95). This leads to the marking of new deletion breakpoints and adds further to the usefulness and versatility of these deletions as a genetic resource. The second technology for the induction of deletions in defined regions of a chromosome involves the irradiation of targeted ES cells (85, 92). This approach generates in one experiment a whole set of ES-cell clones with deletion complexes around the targeted locus. The difference between the LoxP and CRE technology is that the deletion breakpoints induced by irradiation are not exactly known and have to be characterized later at the molecular level.

Marked Inversions, Balancer Screens

One of the limitations of deletion-based mutagenesis experiments is that larger deletions very often lead to a reduction in the fitness and fecundity of the mice. For some regions, it is even impossible to produce deletion heterozygotes because of the haploid insufficiency of specific genes in a region. A more sophisticated and versatile tool for screens of whole chromosomal regions is an inversion that carries both dominant and recessive markers. Inversions lead to suppression of recombination in a chromosomal region, which is of tremendous importance for the maintenance of a mutant stock. The ideal marked balancer inversion is lethal in its homozygous state and carries a dominant visible marker. In crosses of offspring from chemical mutagenesis experiments, the marked balancer inversion then tags the wild-type nonmutagenized chromosome. Homozygous wild-type mice are automatically eliminated from crosses because of the lethality of the inversion. Heterozygote carriers of new mutations can be easily recognized because of the presence of the dominant visible marker. The failure to recover a G3 class of mice that are wild type for the dominant marker even allows for the isolation of recessive lethals.

A screen using a dominantly marked, recessively lethal inversion of this kind is currently being carried out by Rinchik and colleagues at Oak Ridge National Laboratories (60). In this case, the *Eh* (*hairy ears*) inversion is used to mark the inversion on the distal part of chromosome 15. The *Caraculoid* mutation, located in the inversion region, is used as a dominant visible marker. G3 mice homozygous for the *Eh* inversion and, therefore, carrying two nonmutagenized chromosomes die, whereas mice with wild-type ears represent the test class. G3 mice with hairy ears mark the heterozygous carriers and can be easily propagated with any molecular genotyping.

Sensitized Screens

Obtaining mutants will not be possible for all genes. This is a result of a redundancy of gene function and a wide range of compensatory mechanisms that exist in cellular and metabolic networks. Phenotype-based screens allow for the identification of those gene functions that are not backed up by redundancy. A common observation is that phenotypic effects of mutations are dependent on the genetic background. Both penetrance and expressivity can be highly influenced by other genes. One of the best-known examples of this kind of modifier is *Mom1*, the modifier of Min-1 (16). *Min-1* mice carry a dominant mutation in the homologue of the human familial polyposis gene, which is involved in the development of multiple adenomas throughout the intestines (41). *Mom-1* severely affects the multiplicity of tumor formation by acting as a strong modifier. *Mom-1* was mapped to a 4 cM region on mouse chromosome 4 and was recently identified as a secretory phospholipase.

In these cases, the modifiers occurred as natural variants in different laboratory mouse inbred strains. However, ENU mutagenesis can be used to systematically produce and screen for modifiers. The availability of one mutant allows for the systematic identification of modifying loci. In such a case, F1 offspring from ENU-treated animals are mated to mice that already carry a mutation and display a specific phenotype of interest. The resulting offspring are then analyzed for an increase or a decrease in the severity of this phenotype. Genetic modifier mapping and cloning has not been extensively used in mice. With the availability of the complete mouse genomic sequence and the dramatically increasing number of ENU-mouse mutants that display strong genetic background effects, this will certainly change.

Allele-Specific Screens

To fully exploit the power of mouse genetics and genomics, producing multiple alleles of specific genes will be necessary. In many cases, gene targeting will produce the basic null allele. However, point mutations that lead to hypomorphic, hypermorphic, or neomorphic alterations in the protein can be extremely valuable for the analysis of gene function. With one mutant at hand, using ENU mutagenesis for the production of additional alleles is not difficult. Noncomplementation in compound heterozygotes is used as the screening endpoint. The cloning of the *Kreisler* gene is one of the examples where it was essential to have available a second, ENU-induced allele (12). ENU-induced alleles that are located in clusters will also play an important role in the molecular analysis of gene function. It is becoming increasingly clear that insertions like those that often accompany standard gene targeting procedures can have long distance effects on neighboring genes or genes that are even much further away from the targeted gene. The introduction of point mutations by ENU mutagenesis can help to unravel the role of the neighboring members of a gene cluster.

The Need for Genetic Reagents

There is no doubt that within the coming years hundreds, if not thousands, of new ENU-induced mouse mutants will be generated. The maintenance of mutants, however, has not received enough attention. The breeding of mutants on a daily basis is another equally important issue. For dominant traits, particularly those with visible phenotypes and normal penetrance, this is fairly straightforward. But many of the mutants will have a recessive phenotype where the homozygotes are either lethal or less viable and where wild-type and heterozygous carriers cannot be easily discriminated from each other. Maintaining these mutants as heterozygotes without having to carry out test breeding in every case will be a major challenge. One possibile solution is to use molecular genotyping. However, this method is not only expensive, but prone to error. The use of visibly marked balancer inversions would be an enormous help with this problem. Rinchik clearly pointed out this problem and strongly argued for a parallel investment in the development of appropriately marked chromosomal rearrangements as genetic reagents in mutation recovery, analysis, and maintenance of mutants (60). This up-front investment in visibly marked balancers would definitely pay off in the long run by reducing the genotyping of carries to a quick visible inspection. So far, the multiple rounds of gene targeting, required for the engineering of specific inversion mutants, was prohibitive. It is probably just a question of time before mouse geneticists realize the severity of the situation and joint efforts are directed toward the generation of this kind of genetic reagent.

Large Scale—Small Scale

During the last five years, life science research has seen a major change. Driven by improvements in technology, systematic, large-scale, and high-throughput approaches have become increasingly popular in certain research fields. Genomics, particularly genome sequencing, has been at the forefront of this change, but now this trend is also moving into other areas of biology, e.g., cell biology and structural biology. Mouse mutagenesis has seen a similar development. Although ENU mutagenesis was pioneered more than 20 years ago (70) and proof of principle was already perfectly established by that time, only during recent years have major centers decided to direct entire programs toward the purpose of generating hundreds of new mouse mutants by ENU (27, 47).

It is important to point out that, despite the catalytic role that the set up of these centers has played and continues to play, chemical mutagenesis is equally suited as a research tool in smaller labs. Even in labs with a mouse facility of only one or two mouse rooms, it is feasible to set up a functional ENU-mutagenesis screen that is able to genetically dissect a specific pathway or phenotype of interest (5, 6, 40, 76, 77). The successful search for circadian rhythm mutants, such as *Clock*, has already been mentioned (33, 53, 87). Similarly, Anderson and colleagues initiated a phenotype-based screen to identify genes that are involved in early mouse development (1, 31).

PHENOTYPIC SCREENING: YOU FIND WHAT YOU LOOK FOR

The driving force for many activities in mouse genetics is the hope of obtaining animal models of human diseases. The demonstration that we can obtain mutant phenotypes in mice that very closely resemble disease phenotypes in humans is still not appreciated widely in biomedical research. One of the reasons is that, so far, the identification of new mouse mutants was more or less dependent on serendipity and was primarily applied to phenotypes that are externally visible. Deviations in coat color, circling behavior, kinky tails, or small eyes are just a few examples. The power of a phenotype-driven mutagenesis, however, lies in the possibility of screening for any phenotype for which an assay can be developed. This can include blood-based abnormalities, such as high glucose levels, hypercholesteremia, or increased levels of urea; or they may involve very sophisticated screening procedures, such as context-based learning paradigms, diet-induced alterations in certain metabolic parameters, or age-dependent onset of degenerative diseases. The combined experience from the *Drosophila*, zebrafish, yeast, and bacterial genetics community tells us, "You find what you look for." The efficiency of obtaining mouse mutants for a wide range of different phenotypes has been demonstrated in two large-scale ENU screens that were carried out in Harwell and at the GSF in Munich.

Dysmorphological Mutants

Externally visible phenotypes have been at the heart of mouse genetics for almost a hundred years and were also included in the large-scale ENU screens mentioned above. In the meantime, more than 50,000 mice were screened in toto by these labs, and hundreds of new dysmorphological mutants were isolated (27, 47). Although the clinical relevance of a coat-color mutation, for example, might not be immediately apparent, many of these mutants are of key importance for the analysis of disease-related aspects. Examples of this are mouse mutants, which have an altered coat color, e.g., agouti or mahogony. The changes in coat color not only reflect underlying problems in the migration and aggregation of neural crest cells but also point to additional problems in the hematopoetic system. Similarly, mice with no hairs, e.g., the nude mouse mutant, often have additional problems with their immune system.

Clinical-Chemical Mutants

The screens in Harwell and Munich used basic blood-based screens that focused on the identification of hematologial changes, defects of various organ systems, and changes in metabolic pathways and electrolyte homeostasis. The guiding principle was to use routine procedures and parameters that have been used for decades in human clinical diagnosis. This allowed for the screening of large numbers of mice for a broad range of clinical-chemical and hematological parameters.

More than 120 different parameters were screened for each animal. For the hematological investigations, a blood sample was taken from three-month-old mice that were fasted overnight. The sample was then divided into several fractions for a number of different screens. 50 μl were used to measure basic hematolocial parameters, such as the mean corpuscular hemoglobin (MCH), the mean corpuscular hemoglobin concentration (MCHC), or the mean corpuscular volume (MCV). The parameters allow for the identification of mice with anemia and the distinction of microcytosis, macrocytosis, or hypochromic anemia (in thalassemia, iron deficiency), normochromic anemia (e.g., chronic diseases, aplastic anaemia, acute bleeding), and hyperchromic anemia (e.g., vitamin B12 deficiency, folic acid deficiency) as well as cholesterol levels or electrolyte concentrations. 130-μl plasma was used for the determination of a number of different parameters that include various enzyme activities (e.g., alkaline phosphatase, a-amylase, creatine kinase, aspartate-aminotransferase), specific substrates (e.g., glucose, cholesterol, triglycerides), or electrolytes (e.g., potassium, sodium serum chloride, calcium).

Although these assays already allow a fairly broad diagnostic survey for many diseases, a number of diseases are still not covered. These include, for example, renal diseases, which might best be detected by analyzing urine. Therefore, efforts are underway to adapt diagnostic tools from human clinical analysis for the analysis of mouse urine. Because mice physiologically excrete large amounts of low-molecular weight proteins in their urine, measurement of total urinary protein is not suitable for the detection of kidney lesions. However, employing SDS-polyacrylamide-gel-electrophoresis can be used to circumvent this problem and is currently being tested for its suitability to detect early and late glomerular and tubular lesions (58). One of the largest screens, to date, for the activity of erythrocyte enzymes was carried out by W. Pretsch and colleagues at the GSF during a period of almost 20 years (56). A total of more than 13,000 animals were screened for 10 different enzymes. Thirty-six different mutants with either reduced or enhanced enzyme activity were recovered, i.e., for glyceraldehyde-3-phosphate dehydrogenase (GAPDH), glucose phosphate isomerase (GPI), gluthathione reductase (GR), glucose-6-phosphate dehydrogenase (G6PD), lactate dehydrogenase A (LDH), malate dehydrogenase (MDH), phosphoglycerate mutase (PGAM), and triosephosphate isomerase (TPI).

The inclusion of further markers that might also be of interest is still limited by the amount of blood that is needed for the hematological analysis. Miniaturization of assays to allow for the analysis of very small amounts of blood with high speed, cost effectiveness, and accuracy is, therefore, a major goal in the development of new phenotyping assays for mice. The technology of high-throughput electrospray-tandem-mass-spectrometry (ESI-MSMS) bears great potential for solving some of the problems related to the limitations in blood volume and sample size. Rolinski et al. (66) are using ESI-MSMS to screen ENU-treated mice for metabolite profiles of amino acids and acylcarnitines. More than 80 defined disorders are known in humans. This method can potentially also be adapted to screen for many other parameters, which are, so far, not included in standard clinical-chemical diagnostic routines.

Immunologically Relevant Phenotypes

The immune system is a particularly approachable system for phenotype analysis. Because many immunologically relevant cell populations can be analyzed directly from peripheral blood, abnormalities in the quantity of these cell populations, such as B or T cells, or the quantification of the expression levels of selected cell-surface proteins on these cells can be easily detected. Fluorescence activated cell sorting (FACS) analysis or enzyme linked immunosorbent assay (ELISA)–based screens have been developed and used to recover more than 50 mutants from the GSF screen in Munich (23). These include a broad range of immunological disorders, such as common variable immunodeficiency (CVID), autoimmunity, allergy, MHC defects, and SCID as well as many other subtle changes in the expression of selected molecules of immunological interest. A similar large-scale screen, specifically designed to recover immunological mutants, is currently being carried out by C. Goodnow and colleagues in Australia.

Immunology only makes sense within the context of immunity. Mice with abnormal parameters in cell populations of the immune system are not necessarily affected in their response to an infection by pathogens. A primary screen, i.e., a challenge assay with bacteria, viruses, or parasites, is theoretically possible, but it would be too labor and cost intensive. For this reason the German Research Center for Biotechnology in Braunschweig, the GSF, and the Technical University in Munich are establishing a collaboration in which secondary screens in the form of infection challenges with bacteria and viruses are applied. Only those mice that have been pulled out of a primary immunological screen are analyzed in the much more demanding secondary infection screens. The hope is that in the primary screen an enrichment for mutants with an increased or decreased susceptibility to infection has occurred (34). Additional assays are also set up to identify mutants with an impaired immune response to vaccination (23).

Behavior and Sensory Mutants

For a long time, the analysis of behavioral phenotypes in mice was viewed with scepticism because of doubts that the observed traits would be of clinical relevance. Although the clinical picture is often only observed or measurable in humans, many neurobiological and physiological parameters can be detected in both mice and humans. These are sometimes called endophenotypes and can be used in genetic experiments, i.e., as a more reliable marker in genetic linkage analysis (84). Screens for behavioral abnormalities in mice need to take a number of points into consideration. The first of these is the importance of remembering that the identification of a single mutant animal among many normal mice is necessary. Many of the behavioral assays, however, have been developed to analyze populations of individuals and determine deviations from the mean. Behavioral parameters often have large individual variation, most likely because confounding environmental factors do exist but are unknown to the investigator. As a consequence, paying close attention to the baselines of the phenotype employed is very important. Furthermore, the genetic background can have a very strong influence on the penetrance or expressivity of phenotype. This can complicate or completely prevent the mapping of behavioral traits.

Tarantino et al. (84) describe the use of five behavioral assays that can be performed at a fairly high-throughput level: rotarod, zero maze, acoustic startle response, prepulse inhibition, and habituation of the acoustic startle response. These assays can be used to identify mutants with defects in neuromuscular function, anxiety-related behavior, and sensorimotor reactivity and gating. The range of assays and a discussion of their pitfalls and merits are not within the scope of this review. It is clear, however, that the genetic dissection of behavioral traits will attract increasing attention in the coming years. One of the most convincing arguments for the feasibility of screens for complex biological phenotypes was the recovery of the circadian rhythm mutant *Clock* (discussed above) (33, 87). A comprehensive screen called the SHIRPA screen, encompassing a battery of up to 40 simple tests, is used routinely in the mutagenesis screen in Harwell (65). This test allows the analysis of a detailed profile for deficits in muscle and lower motorneuron, spinocerebellar, sensory, neuropsychiatric, and autonomic function. The procedure only requires a simple testing arena and takes ~10 min per mouse.

Mouse mutants are very useful for the genetic dissection of pathways involved in sensory perception, such as vision, hearing, or nociception. The large collection of mouse cataract mutants produced by Favor and colleagues in Neuherberg (mentioned above) consists of more than 170 dominant cataract mutants (18). Complementation tests and chromosomal mapping experiments suggest that at least 50 different genes are involved in this phenotype. Pinto et al. describe a whole range of tests that are suitable for identifying mouse mutants with defects in the visual system (54). These tests involve some that do not depend on the behavior of a conscious animal, such as the electroretinogram and the visually evoked potential of the cortex. They also describe a number of other behavioral tests, e.g., maze-based tests, cued fear conditioning, as well as tests based on conditioned suppression, visual placing, optokinetik nystagmus, pupillary reflex, and light-induced shifts in circadian phase. Similar to what was previously mentioned for behavioral tests in general, it is important to carefully control for the genetic background of the different mouse strains under investigation. C3H/HeJ mice are blind because they carry the retinal degeneration (*rd*)-mutation. This prevents assays that test for a response to light in these animals.

A large number of mutants that have problems in hearing have been obtained from the large-scale ENU-mutagenesis screens. In a Europe-wide collaboration that involved the ENU screens in Harwell and in Munich, more than 40 deaf mouse mutants were identified. A very simple "click box," testing for an acoustically induced Preyer reflex, was used to screen for mutants with hearing problems. About half of the recovered mutants were classified as nonsyndromic deafness. Many others have additional problems, e.g., circling behavior or other congenital abnormalities. The first genes have since been cloned from these screens (32). The Europe-wide deafness screen nicely demonstrates that one of the most powerful strategies for the exploitation of mouse genetics lies in a collaboration between geneticists and physiologists. In addition to vision and hearing, other aspects of sensory perception can be analyzed through the use of ENU-mouse mutants. Zimmer and colleagues have started to isolate mouse mutants with abnormalities in nociception (27). A variety of assays have been or can be developed for taste, touch, smell, etc., opening useful entry points into the genetic and molecular basis of complex sensory traits.

Age-Related and Chronic Disease Phenotyping

Many diseases are age related and only develop at later stages in life. The current ENU screens are not designed to identify such late onset diseases because the majority of phenotyping is done at the time of weaning and at three months of age. The reason for this is that the logistics of keeping mice on the shelf for phenotyping, i.e., at the age of nine months or later, not only has a tremendous influence on the costs of such a screen but also leads to a very longtime frame for carrying out such an experiment. Nevertheless, a number of labs have started to include screens for degenerative, age-dependent diseases, i.e., by setting aside about 10% of the animals for screening at a later stage. One of the diseases that shows a strong age component is cancer. Currently, none of the screens addresses this problem, with the exception of the Japanese ENU-mutagenesis program, which specifically tries to address late onset diseases, including cancer (S. Toshihiko, personal communication). The problems associated with the extended time frame for late onset disease screens could be alleviated by the use of surrogate markers that are early indicators for a developing disease. A comet assay is used in the GSF mutagenesis screen to identify mutants with an increased genetic susceptibility to radiation (71).

Phenotyping for Pharmacogenomic Traits, i.e., Drug Metabolism

Other promising areas that can be approached by genetic means are those of pharmacogenomics and toxicogenomics. Individual differences in drug metabolism or adverse side reactions to drugs or toxic compounds have a strong genetic component. Although many of the cytochrome P450 enzymes have species-specific properties, there is no doubt that many of the principles involved in pharmacokinetics and pharmacodynamics can be studied in mice. Furthermore, the wide range of inbred mouse strains and their response to drugs have not been investigated well. Screening for mouse mutants that show alterations in their capacity for drug metabolism would be a very straightforward way to dissect the molecular basis of drug metabolism. One can imagine the design of an assay that utilizes a whole battery of chemicals combined with the subsequent monitoring of the metabolites produced as part of a mutagenesis screen.

Chromosomal Mapping and Cloning: Speeding It Up

After the successful identification of mouse mutants induced by ENU, the next step is to map the chromosomal location of the mutant locus. The principle strategy for this employs the generation of a backcross panel in which the mutant mouse strain is crossed with another strain that is genetically divergent from the genetic background of the mutant. Offspring from the F1 generation are either backcrossed or intercrossed in order to generate an F2 generation of mice in which the mutant phenotype segregates and can be chromosomally mapped by linkage analysis. After phenotyping the backcross panel into affected and nonaffected mice, DNA from the panel is produced and genotyped. For this purpose, a set of genome-wide spaced markers is used that are polymorphic between the genetic background of the mutant and the tester stock.

The larger the number of backcross progeny is, the smaller the minimal interval in which the gene responsible for the mutant phenotype has to be located. With

~50 backcross mice, researchers can map a mutation to a specific chromosome and locate the affected gene to an interval of 10 to 20 cM. However, with backcross panels of ~500–1000 mice, they are able to narrow down the critical region to less than 1 cM. With about 2000 kb in 1 cM and 1 gene every 50 to 100 kb, this will reduce the number of candidate genes to less than 100, still a formidable task for the endgame of mutation detection.

A significant enhancement to backcrossing experiments can be obtained through the use of so called speed backcrosses (86). Instead of using live male mice and natural matings for the production of backcross progeny, sperm is obtained from the epididymides of mice and then used for in vitro fertilization (24, 38, 39). From an individual male, sufficient sperm can be obtained and used as either frozen or fresh sperm to produce hundreds of offspring in a single in vitro fertilization experiment. This allows a very fast production of backcross panels from mutant mice, alleviating the pressures of time and animal space. For phenotypes that can be scored before birth (e.g., polydactyly, prenatal lethality, etc.), the transfer of almost the entire genetic analysis, involving genotyping and phenotyping from the animal house to the laboratory bench, is possible. Knowledge about the chromosomal location of a mutation is also important for a quick assessment of whether different mutants might be allelic. Although mapping mutants to a different chromosome or chromosomal interval excludes allelism, mapping a mutant to the same chromosomal region is not proof that the same gene is affected. Nevertheless, this is important information and helps to design future experiments to test the hypothesis.

Identification of the gene that is mutated in a mouse mutant can still be a tedious process. Until recently, positional cloning included very labor intensive and technically demanding methods, e.g., chromosomal walking or jumping. With the availability of the complete mouse genome sequence, this has dramatically changed. Now, positional candidate cloning will be the guiding principle for moving from a mutant to a gene. Once the minimal interval of a mutation is established, one can immediately proceed to scan the available mouse genome sequence and other databases for any genes located in this interval. On the basis of flanking information, such as expression data, human conserved synteny data, and other information, a list of candidate genes can be drawn up. Although any gene in the minimal interval initially qualifies as a candidate, inclusion of additional genetic, genomic, or biochemical data can be very useful to the prioritization of potential candidate genes for sequence alterations.

With the introduction of efficient means to produce and map mouse mutants, the technology for genotyping and for mutation detection will present the next bottleneck. A number of new methods may soon solve this problem (88). One of the most promising is the development of chips, or high-density oligo arrays, which allow simultaneous screening of large numbers of polymorphic markers. However, in principle, the technology for genotyping and mutation detection in mice is the same as that for single nucleotide polymorphism analysis in humans. The main difference is that in mice, owing to the availability of inbred strains, we are usually dealing with a diallelic system. This allows us to quickly compare the mutant DNA sequence to that of DNA from unaffected mice derived from the same genetic background. Such a comparison is particularly important for deciding which sequence changes might be causally related to a mutant phenotype instead of being a polymorphism without functional significance.

The era of mouse phenotyping has just begun. ENU mutagenesis is only one of many ways to produce mouse mutants. The need for systematic, efficient, and competent phenotyping is also there for knockout, gene-trap, or spontaneous mutants. For many traits, i.e., in the field of cardiovascular or neuroendocrine physiology, mouse phenotyping still lags behind those available for human clinical analysis as well as rat phenotyping. For many disease areas, mouse phenotyping does not exist at all. In order to fully exploit the genetic tools of the mouse, there is an urgent need to develop new phenotyping technologies, e.g., molecular imaging, which is one of the most promising areas. Techniques such as PET or SPECT use radio-labeled molecules to image molecular interactions of biological processes in vivo (52). PET uses positron-labeled molecules in very low mass amounts in order to image and measure with minimal disturbance the function of biological processes. After identifying a specific biological process to be studied, a positron-labeled molecule is synthesized through which an assay for the process can be performed. A PET scanner analyzes the changing tissue concentration of the labeled molecule and its labeled product over time or their accumulated concentration at a given time. An example for such a molecule is 2-deoxy-D-glucose (2DG), which is used for quantitative PET studies and has become one of the most commonly used molecular imaging probes for PET studies of cancer. Recently, micro-PET systems have been developed for rats and mice that allow the analysis of sample sizes that are 2000-fold smaller than those created by the standard PET imaging systems used for humans (9, 11).

Sequence-Based Screens

Whereas the entire approach described in this review rests on the notion of a phenotype-driven mutagenesis screen, it is important to realize that ENU mutagenesis can also be used to screen offspring from mutagenized mice directly for sequence changes in specific genes (4, 23). Given an average mutation frequency per locus of about 1:1000, an average size of a human/mouse gene of $\sim$1–2 Kb, and an estimated likelihood that $\sim$10% of sequence changes lead to a phenotypic effect, one would expect about 0.5 to 1 sequence changes per 100,000 bp (4). Although this sounds like a large number, current developments in mutation technology make this a likely scenario and a promising strategy for future functional genomic studies. In particular, approaches that integrate parallel processing of many genes would provide very fast access to mutations in any sequence of interest and the creation of allelic series of mutants for further study. One can imagine the establishment of large sperm archives with corresponding tissue samples for mutation detection derived from ENU-treated mice. Such an archive could become a worldwide resource for the identification and isolation of mutants. So far, the logistics, the efficiency of mutation detection, and the cost have prevented such an approach. The time will soon arrive when these limitations will be overcome.

ENU Mutagenesis in Embryonic Stem Cells

Instead of exposing live mice to a chemical mutagen such as ENU, an attractive alternative is to treat embryonic stem cells with the mutagen and then select in vitro those ES-cell clones in which mutations have been induced. The question of whether such mutagenized ES cells would maintain their germline competence presents a critical concern. Recently, this was answered with a clear yes (10, 42). These authors reported the successful mutagenesis and germline transmission of ENU-treated ES cells, using HPRT and HSV-*tk* as markers for piloting the experiment. The mutation frequencies were similar to those observed in live mice. This frequency could even be significantly increased by inhibiting the DNA repair enzyme, O6-alkylguanine-alkyltransferase (AGT), using O6-benzyl-guanine (O6-BG). Mutagenized ES cells can be used for genome-based as well as for phenotype-based mutagenesis and screening strategies. One of the most exciting potential applications, however, lies in the possibility of carrying out phenotypic screens directly in ES cells. Knowledge about the factors and conditions required for a controlled development and differentiation of stem cells is obtained rapidly. In theory, ES cells can differentiate in any cell type so that the combination of mutagenesis and phenotype analysis in vitro could become a very powerful strategy.

ARCHIVING AND DISTRIBUTION

The number of genes within the human and mouse genome is in the range of 26,000 to 35,000 (46, 73). In the long run, we will need at least one mouse mutant for every gene in the genome. In practice, we need more than one mutant per gene because different mutations in different regions of a gene will unravel different functions. This is the reason for the importance of allelic series, such as null alleles, hypo-, hyper-, and antimorphs. It would be impossible to keep all the mutants that are currently generated and those that will be produced in the future on the shelf as live animals. Although mouse embryo freezing was developed many years ago (90), it did not solve the problem of cryoconservation adequately. In order to freeze down a mouse strain as preimplantation embryos, between 200 and 400 embryos had to be collected. Just for this purpose, the size of the mutant mouse colony often had to be transiently increased rather than decreased. A breakthrough came with the development of sperm freezing combined with in vitro fertilization (44, 50, 83). Unfortunately, another 10 years passed before this technique was developed into a routine and robust method that works in most labs that work with mouse embryos (81, 86). Now for the first time, one male mouse is sufficient to very quickly derive by in vitro fertilization a large colony of offspring within a short time. This was a decisive factor for the success of large-scale ENU-mutagenesis programs. Unfortunately, there is still a big difference in the success rate of sperm freezing and in vitro fertilization between varying inbred mouse strains. Although this method works very well for many standard lab mouse strains and for most hybrids, e.g., BALB/c and C3H/HeJ, one of the major concerns is still the low recovery rate for C57BL16 mice (45). A number of protocols are currently being developed to overcome this problem. In parallel to sperm cryoconservation, methods are available to freeze mouse oocytes (43, 89) and even entire mouse ovaries (7, 82). The Jackson Laboratory in Bar Harbor, Maine, had been the main resource for storage and distribution of mouse strains, until recent years when additional mouse resource centers, i.e. Harwell, the GSF, and RIKEN, joined efforts to serve the "mouse community" by providing information, infrastructure, and resources. EMMA, the European Mouse Mutant Archive, will be an additional important component to meet the needs of the research community in the postgenomic era.

INFORMATICS AND DIGITAL MANAGEMENT

With the establishment of the first large-scale ENU-mutagenesis programs, the need for efficient documentation became immediately apparent. The tradition of paper-based records for maintaining and tracking animal breeding records was not sufficient to cope with the large amount of data that resulted from these mutagenesis projects. In the meantime, sophisticated electronic informatics systems for animal husbandry and mutagenesis have been developed and are continuously being improved (51, 80). These systems have been implemented on a client-server basis, where an important component is that data entry, i.e., weaning or phenotyping data on individual mice, are at "point of generation." This includes a number of personal computer workstations that are connected via a local area network to a central database and World Wide Web server.

The amount of information that will be collected on many of the mutants generated in the future will be enormous. There is an urgent need to develop tools for tracking, searching, and analyzing genetic, molecular, and phenotypic data. This includes the necessity to develop structured phenotypic vocabularies. A Gene Ontology (GO) project has been started as a collaboration among the Saccharomyces Genome Database (SGD), FlyBase, Mouse Genome Database (MGD), and the Mouse Gene Expression Database (GXD), with the goal of developing independent ontologies for describing functions, biological processes, and cellular locations of gene products (3, 17) (http://www.geneontology.org). In addition, MGD is working on an effort to integrate data from the various mutagenesis and phenotyping centers in order to provide a platform for efficient searching and description of mouse phenotype data. The implementation and deployment of computer-based data recording systems for mouse mutagenesis and phenotyping data do need a major commitment with respect to resources and personnel. This part of mouse genetic programs is often underappreciated, although there is no doubt that investments made early pay off in the long run. Sharing data and resources worldwide was always a major component of mouse genetics and only easy access and availability will allow this field to flourish as it has in the past.

SUMMARY AND PERSPECTIVES

Systematic mutagenesis screens are a hallmark of experimental genetics. Starting in the fly room of T.H. Morgan and the discovery by H.J. Muller that X rays can dramatically increase the mutation frequency, the search for interesting phenotypes that deviated from the norm were at the heart of this approach. The same strategy was applied in bacterial genetics to dissect the basic mechanisms of replication and gene regulation. The revolution induced by the insight obtained from the systematic phenotype screens in the fruitfly *Drosophila melanogaster* concerning the segmentation of insects speaks for itself. The extension of this approach to the yeast *Sacharomyces cerevisiae*, the worm *C. elegans*, and the fish *Brachydanio rerio* further demonstrated the power of genetic dissection of complex biological phenomena. Why did it take so long before systematic mouse mutagenesis became an undisputed tool in the toolbox of mouse genetics?

Part of the answer lies in the logistics and costs that are required to operate a mouse facility. Particularly in the United States, where cage charges can be prohibitive, mutagenesis experiments have been considered too long-term oriented, and quicker returns were expected from alternative approaches, such as gene targeting or working with other species. Although ENU mutagenesis was pioneered in the United States, it was not possible for the National Institutes of Health to finance a number of mutagenesis centers until two European centers made a strong commitment to embark on large-scale ENU-mouse screens. However, one also has to keep in mind that just a few years ago the prospects of cloning a mouse mutant through positional or candidate cloning were very much different from what they are today. The genetic map was sparse. Microsatellite markers had not yet been developed, and the availability of the complete sequence of the mouse genome was not even considered. All this has changed.

What can we expect from the future of mouse genetics? Producing mutants is no longer the bottleneck. The discovery steps of the future will be in the analysis of the phenotype. One can already see the mutagenesis centers transforming into phenotyping centers. Most of them will specialize in one or a few areas and try to become centers of excellence in a given field. The most fruitful application of mutagenesis, however, will emerge when this study is combined with the latest technology of genomics and proteomics. Systematic differential analysis by comparing the biochemistry, cell biology, homeostasis of metabolic pathways, or gene regulatory phenomena of a mutant vs. a wild-type animal will be the key to understanding complex biological systems.

Although ENU-induced mouse mutants initially give us information about monogenic traits, these mutants will be essential tools for the future dissection of multigenic and multifactorial traits. The search for modifiers and quantitative trait loci has just begun. From this year on, we will have the complete human genome sequence at our hand. The mouse sequence will follow very soon. This will have a dramatic influence on mouse genetics. Positional cloning will be exchanged for candidate evaluation. Using comparative information from other species will be only a "mouse" click away. As previously mentioned, one of the keys to the success of mouse genetics in the past was the accessibility and availability of the mice. In the age of sperm freezing and in vitro fertilization, this might become even easier. The rate-limiting step is probably not the availability of mice but the availability of well-trained people, particulary mouse pathophysiologists, that are able to detect and to make sense of the phenotypic alterations in mouse mutants. We need at least one mutant for every gene in the mouse genome. No doubt, the future of mouse genetics is bright.

ACKNOWLEDGMENT

I thank Christopher Schippers for his help.

Visit the Annual Reviews home page at www.AnnualReviews.org

LITERATURE CITED

1. Anderson KV. 2000. Finding the genes that direct mammalian development—ENU mutagenesis in the mouse. *Trends Genet.* 16:99–102
2. Ashburner M. 1989. *Drosophila: A Laboratory Handbook*, pp. 343–403. New York: Cold Spring Harbor Lab. Press
3. Ashburner M, Blake J, Botstein D, Cherry JM, Eppig JT, et al. 1999. On the representation of "gene function and process" in genome databases. *Proc. Genome Sequencing Biol. Conf.* New York: Cold Spring Harbor Lab. Press
4. Beier DR. 2000. Sequence-based analysis of mutagenized mice. *Mamm. Genome* 11:594–97
5. Bode VC. 1984. Ethylnitrosourea mutagenesis and the isolation of mutant alleles for specific genes located in the t-region of mouse chromosome. *Genetics* 17:457–70
6. Bode VC, McDonald JD, Guenet JL, Simon D. 1988. Hph-1: a mouse mutant with hereditary hyperphenylalaninemia induced by ethylnitrosourea mutagenesis. *Genetics* 118:299–305

7. Candy CJ, Wood MJ, Whittingham DG. 1997. Long-term fertility of recipients of cryopreserved mouse ovaries. *J. Reprod. Fertil. Abstr. Ser.* 19:66
8. Cattanach BM, Burtenshaw MD, Raspberry C, Evans EP. 1993. Large deletions and other gross forms of chromosome imbalance compatible with viability and fertility in the mouse. *Nat. Genet.* 3:56–61
9. Chatziioannou AF, Cherry SR, Shao Y, Silverman RW, Meadors K, et al. 1999. Performance evaluation of microPET: a high resolution lutetium oxyorthosilicate PET scanner for animal imaging. *J. Nucl. Med.* 40:1164–75
10. Chen Y, Yee D, Dains K, Chatterjee A, Cavalcoli J, et al. 2000. Genotype-based screen for ENU-induced mutations in mouse embryonic stem cells. *Nat. Genet.* 24:314–17
11. Cherry SR, Shao Y, Silverman RW. 1997. MicroPET: a high resolution PET scanner for imaging small animals. *IEEE Trans. Nucl. Sci.* 44:1109–13
12. Cordes SP, Barsh GS. 1994. The mouse segmentation gene *kr* encodes a novel basic

domain-leucine zipper transcription factor. *Cell* 79:1025–34

13. Cuénot L. 1905. Les races pures et les combinaisons chez les souris. *Arch. Zool. Exp. Gén.* 4:123–32

14. Davis AP, Woychik RP, Justice MJ. 1998. Effective chemical mutagenesis in FVB/N mice requires low doses of ethylnitrosourea. *Mamm. Genome* 10:308–10

15. Dietrich W, Katz H, Lincoln S, Shin H-S, Friedman J, et al. 1992. A genetic map of the mouse suitable for typing intraspecific crosses. *Genetics* 131:423–47

16. Dietrich WF, Lander ES, Smith JS, Moser AR, Gould KA, et al. 1993. Genetic identification of *Mom1*, a major modifier locus affecting *Min*-induced intestinal neoplasia in the mouse. *Cell* 75:631–39

17. Eppig JT. 2000. Algorithms for mutant sorting: the need for phenotype vocabularies. *Mamm. Genome* 11:584–89

18. Favor J, Neuhäuser-Klaus A. 2000. Saturation mutagenesis for dominant eye morphological defects in the mouse *Mus musculus*. *Mamm. Genome* 11:520–25

19. Favor J, Neuhäuser-Klaus A, Ehling UH. 1988. The effect of dose fractionation on the frequency of ethylnitrosourea-induced dominant cataract and recessive specific locus mutations in germ cells of the mouse. *Mutat. Res.* 198:269–75

20. Favor J, Neuhäuser-Klaus A, Ehling UH. 1991. The induction of forward and reverse specific-locus mutations and dominant cataract mutations in spermatogonia of treated strain DBA/2 mice by ethylnitrosourea. *Mutat. Res.* 249:293–300

21. Favor J, Neuhäuser-Klaus A, Ehling UH, Wulff A, Zeeland AA. 1997. The effect of the interval between dose applications on the observed specific-locus mutation rate in the mouse following fractionated treatments of spermatogonia with ethylnitrosourea. *Mutat. Res.* 374:193–99

22. Flaherty L, Messer A, Russell LB, Rinchik EM. 1992. Chlorambucil-induced mutations in mice recovered in homozygotes. *Proc. Natl. Acad. Sci. USA* 89:2859–63

23. Flaswinkel H, Alessandrini F, Rathkolb B, Decker T, Kremmer E, et al. 2000. Identification of immunological relevant phenotypes in ENU mutagenized mice. *Mamm. Genome* 11:526–27

24. Glenister PH, Thornton CE. 2000. Cryoconservation—archiving for the future. *Mamm. Genome* 11:565–71

25. Haldane JBS. 1956. The detection of autosomal lethals in mice induced by mutagenic agents. *J. Genet.* 54:327–42

26. Hogan B, Beddington R, Costantini F, Lacy E. 1994. *Manipulating the Mouse Embryo: A Laboratory Manual.* New York: Cold Spring Harbor Lab. Press

27. Hrabé de Angelis M, Flaswinkel H, Fuchs H, Rathkolb B, Soewarto D, et al. 2000. Genome-wide, large-scale production of mutant mice by ENU mutagenesis. *Nat. Genet.* 25:444–47

28. Joyner AL, Auerbach A, Skarnes WC. 1992. The gene trap approach in embryonic stem cells: the potential for genetcis screens in mice. *Ciba Found. Symp.* 165:277–88

29. Justice MJ, Carpenter DA, Favor J, Neuhäuser-Klaus A, Hrabé de Angelis M, et al. 2000. Effects of ENU dosage on mouse strains. *Mamm. Genome* 11:484–88

30. Justice MJ, Noveroske JK, Weber JS, Zheng B, Bradley A. 1999. Mouse ENU mutagenesis. *Hum. Mol. Genet.* 8:1955–63

31. Kasarskis A, Manova K, Anderson KV. 1998. A phenotype-based screen for embryonic lethal mutations in the mouse. *Proc. Natl. Acad. Sci. USA* 95:7485–90

32. Kiernan AE, Ahituv N, Fuchs H, Balling R, Avraham KB, et al. 2001. The Notch ligand Jagged1 is required for inner ear sensory development. *Proc. Natl. Acad. Sci. USA* 98:3873–78

33. King DP, Zhao Y, Sangoram AM, Wilsbacher LD, Tanaka M, et al. 1997. Positional cloning of the mouse circadian *Clock* gene. *Cell* 89:641–53

34. Lengeling A, Pfeffer K, Balling R. 2001. The battle of two genomes: genetics of bacterial host/pathogen interactions in mice. *Mamm. Genome* 12:261–71

35. Liu P, Zhang H, McLellan A, Vogel H, Bradley A. 1998. Embryonic lethality and tumorigenesis caused by segmental aneuploidy on mouse chromosome 11. *Genetics* 150:1155–68

36. Lyon MF, King TR, Gondo Y, Gardner JM, Nakatsu Y, et al. 1992. Genetic and molecular analysis of recessive alleles at the pink-eyed dilution (p) locus of the mouse. *Proc. Natl. Acad. Sci. USA* 89:6968–72

37. Lyon MF, Morris T. 1966. Mutation rates at a new set of specific loci in the mouse. *Genet. Res.* 7:12–17

38. Marschall S, Hrabé de Angelis M. 1999. Cryopreservation of mouse spermatozoa doubles your mouse space. *Trends Genet.* 15:128–31

39. Marschall S, Huffstadt U, Balling R, Hrabé de Angelis M. 1999. Reliable recovery of inbred mouse lines using cryopreserved spermatozoa. *Mamm. Genome* 10:773–76

40. McDonald JD, Bode V, Dove VC, Shedlovsky A. 1990. Pahhph-5: a mouse mutant deficient in phenylalanine hydroxylase. *Proc. Natl. Acad. Sci. USA* 87:1965–67

41. Moser AR, Pitot HC, Dove WF. 1990. A dominant mutation that predisposes to multiple intestinal neoplasia in the mouse. *Science* 247:322–24

42. Munroe RJ, Bergstrom RA, Zheng QY, Smith R, John SW, et al. 2000. Mouse mutants from chemically mutagenized embryonic stem cells. *Nat. Genet.* 24:318–20

43. Nakagata N. 1989. High survival rate of unfertilized mouse oocytes after vitrification. *J. Reprod. Fertil.* 73:479–83

44. Nakagata N. 1993. Production of normal young following transfer of mouse embryos obtained by in vitro fertilization between cryopreserved. *Reprod. Fertil.* 99:77–80

45. Nakagata N. 2000. Cryopreservation of mouse spermatozoa. *Mamm. Genome* 11:572–76

46. Nature. 2001. *The Human Genome.* 409 pp.

47. Nolan PM, Peters J, Strivens M, Rogers D, Hagan J, et al. 2000. A systemic, genome-wide, phenotype-driven mutagenesis programme for gene function studies in the mouse. *Nat. Genet.* 25:440–47

48. Noveroske JK, Weber JS, Justice MJ. 2000. The mutagenic action of N-ethyl-N-nitrosourea in the mouse. *Mamm. Genome* 11:478–83

49. Nüsslein-Volhard C, Wieschaus E. 1980. Mutations affecting segment number and polarity in *Drosophila. Nature* 287:795–801

50. Okuyama M, Isogai S, Saga M, Hamada H, Ogawa S. 1990. In vitro fertilization (IVF) and artificial insemination (AI) by cryopreserved spermatozoa in mouse. *J. Fertil. Implant.* 7:116–19

51. Pargent W, Heffner S, Schäble KF, Soewarto D, Fuchs H, Hrabé de Angelis M. 2000. MouseNet^a database: digital management of a large-scale mutagenesis project. *Mamm. Genome* 11:590–93

52. Phelps ME. 2000. PET: the merging of biology and imaging into molecular imaging. *J. Nucl. Med.* 41:661–81

53. Pickard GE, Sollars PJ, Rinchik EM, Nolan PM, Bucan M. 1995. Mutagenesis and behavioral screening for altered circadian activity identifies the mouse mutant, *Wheels. Brain Res.* 705:255–66

54. Pinto LH, Enroth-Cugell C. 2000. Tests of the mouse visual system. *Mamm. Genome* 11:531–36

55. Popp RA, Bailiff EG, Skow LC, Johnson FM, Lewis SE. 1983. Analysis of a mouse alpha-globin gene mutation induced by ethylnitrosurea. *Genetics* 105:157–67

56. Pretsch W. 2000. Enzyme-activity mutants in *Mus musculus.* I. Phenotypic description and genetic characterization of ethylnitrosourea-induced mutations. *Mamm. Genome* 11:537–42

57. Ramirez-Solis R, Liu P, Bradley A. 1995. Chromosome engineering in mice. *Nature* 14:720–24

58. Rathkolb B, Decker T, Fuchs E, Soewarto D, Fella C, et al. 2000. The clinical-chemical screen in the Munich ENU Mouse

Mutagenesis Project: screening for clinically relevant phenotypes. *Mamm. Genome* 11:543–46

59. Rinchik EM. 1991. Chemical mutagenesis and fine-structure functional analysis of the mouse genome. *Trends Genet.* 7:15–21

60. Rinchik EM. 2000. Developing genetic reagents to facilitate recovery, analysis, and maintenance of mouse mutations. *Mamm. Genome* 11:489–99

61. Rinchik EM, Carpenter DA. 1999. N-ethyl-N-nitrosourea mutagenesis of 6- to 11-cM subregion of the *Fah-Hbb* interval of mouse chromosome 7: completed testing of 4557 gametes and deletion mapping and complementation analysis of 31 mutations. *Genetics* 152:373–83

62. Rinchik EM, Carpenter DA, Long CL. 1993. Deletion mapping of four loci defined by N-ethyl-N-nitrosourea–induced postimplantation—lethal mutations within the *pid-Hbb* region of mouse chromosome 7. *Genetics* 135:1117–23

63. Rinchik EM, Carpenter DA, Selby PB. 1990. A strategy for fine-structure functional analysis of a 6- to 11-cM region of mouse chromosome 7 by high-efficiency mutagenesis. *Proc. Natl. Acad. Sci. USA* 87:896–900

64. Rinchik EM, Russell LB. 1990. Germ-line deletion mutations in the mouse: tools for intensive functional and physical mapping of regions for the mammalian genome. In *Genome Analysis*, ed. K Davies, S Tilghman, pp. 121–58. New York: Cold Spring Harbor Lab. Press

65. Rogers DC, Fisher EM, Brown SD, Peters J, Hunter AJ, Martin JE. 1997. Behavioral and functional analysis of mouse phenotype: SHIRPA, a proposed protocol for comprehensive phenotype assessment. *Mamm. Genome* 8:711–13

66. Rolinski B, Arnecke R, Dame T, Kreischer J, Olgemöller B, et al. 2000. The biochemical metabolite screen in the Munich ENU Mouse Mutagenesis Project: determination of amino acids and acylcarnitines by tandem mass spectrometry. *Mamm. Genome* 11:547–51

67. Russell LB, Hunsicker PR, Cacheiro NLA, Bangham JW, Russell WL, et al. 1989. Chlomambucil effectively induces deletion mutations in mouse germ cells. *Proc. Natl. Acad. Sci. USA* 86:3704–8

68. Russell WL, Hunsicker RR, Carpenter DA, Cornett CV, Guinn GM. 1982. Effect of dose fractionation on the ethylnitrosourea induction of specific-locus mutations in mouse spermatogonia. *Proc. Natl. Acad. Sci. USA* 79:3592–93

69. Russell WL, Hunsicker PR, Raymer GD, Steele MH, Stelzner KF, et al. 1982. Dose-response curve for ethylnitrosourea-induced specific-locus mutations in mouse spermatogonia. *Proc. Natl. Acad. Sci. USA* 79:3589–91

70. Russell WL, Kelly PR, Hunsicker PR, Bangham JW, Maddux SC, Phipps EL. 1979. Specific-locus test shows ethylnitrosourea to be the most potent mutagen in the mouse. *Proc. Natl. Acad. Sci. USA* 76:5918–22

71. Schindewolf C, Lobenwein K, Trinczek K, Gomolka M, Soewarto D, et al. 2000. Comet assay as a tool to screen for mouse models with inherited radiation sensitivity. *Mamm. Genome* 11:552–54

72. Schumacher A, Faust C, Magnuson T. 1996. Positional cloning of a global regulator of anterior-posterior patterning in mice. *Nature* 383:250–53

73. Science. 2001. *The Human Genome.* 291 pp.

74. Searle AG. 1974. Mutation induction in mice. *Adv. Radiat. Biol.* 4:131–207

75. Sedivy JM, Joyner AL. 1992. *Gene Targeting.* New York: Oxford Univ. Press

76. Shedlovsky A, Guenet JL, Johnson LL, Dove WF. 1986. Induction of recessive lethal mutations in the T/t-H2 region of the mouse genome by a point mutagen. *Genet. Res.* 47:135–42

77. Shedlovsky A, King TR, Dove WF. 1988. Saturation germ line mutagenesis of the murine t region including a lethal allele at

the quaking locus. *Proc. Natl. Acad. Sci. USA* 85:180–84

78. Shedlovsky A, MacDonald JD, Symula D, Dove WF. 1993. Mouse models of human phenylketonuria. *Genetics* 134:1205–10

79. Silver L. 1995. *Mouse Genetics*. New York: Oxford Univ. Press

80. Strivens MA, Selley RL, Greenaway SJ, Hewitt M, Liu X, et al. 2000. Informatics for mutagenesis: the design of *Mutabase*— a distributed data recording system for animal husbandry, mutagenesis, and phenotypic analysis. *Mamm. Genome* 11:577–83

81. Sztein J, Farley JS, Young AF, Mobraaten LE. 1997. Motility of cryopreserved mouse spermatozoa affected by temperature of collection and rate of thawing. *Cryobiology* 35:46–52

82. Sztein JM, Sweet H, Farley JS, Mobraaten LE. 1998. Cryopreservation and orthotopic transplantation of mouse ovaries: new apparoach in gamete banking. *Biol. Reprod.* 58:1071–74

83. Tada N, Sato M, Tamanoi J, Mizorogi T, Kasai K, et al. 1990. Cryopreservation of mouse spermatozoa in the presence of raffinose and glycerol. *J. Reprod. Fertil.* 89:511–16

84. Tarantino LM, Gould TJ, Druhan JP, Bucan M. 2000. Behavior and mutagenesis screens: the importance of baseline analysis of inbred strains. *Mamm. Genome* 11:555–64

85. Thomas JC, LaMantia C, Magnuson T. 1998. X-ray–induced mutations in mouse embryonic stem cells. *Proc. Natl. Acad. Sci. USA* 95:1114–19

86. Thornton CE, Brown SDM, Glenister PH. 1999. Large numbers of mice established by in vitro fertilization with cryopreserved spermatozoa: implications and

applications for genetic resource banks, mutagenesis screens, and mouse backcrosses. *Mamm. Genome* 10:987–92

87. Vitaterna MH, King D-P, Chang AM, Kornhauser JM, Lowrey PL, et al. 1994. Mutagenesis and mapping of a mouse gene, *Clock*, essential for circadian behavior. *Science* 264:719–25

88. Wells C, Brown SDM. 2000. Genomics meets genetics: towards a mutant map of the mouse. *Mamm. Genome* 11:472–77

89. Whittingham DG. 1977. Fertilization in vitro and development to term of unfertilized mouse oocytes previously stored at −196°C. *J. Reprod. Fertil.* 49:89–94

90. Whittingham DG, Leibo SP, Mazur P. 1972. Survival of mouse embryos frozen to −196°C and −269°C. *Science* 178:411–14

91. Wiles MV, Vauti F, Otte J, Fuchtbauer EM, Ruiz P, et al. 2000. Establishment of a gene-trap sequence tag library to generate mutant mice from embryonic stem cells. *Nat. Genet.* 24:13–14

92. You Y, Bergstrom R, Klemm M, Lederman B, Nelson H, et al. 1997. Chromosomal deletion complexes in mice by radiation of embyonic stem cells. *Nat. Genet.* 15:285–88

93. Zambrowicz BP, Friedrich GA, Buxton EC, Lilleberg SL, Person C, et al. 1998. Disruption and sequence identification of 2000 genes in mouse embryonic stem cells. *Nature* 392:608–11

94. Zheng B, Mills AA, Bradley A. 1999. A system for rapid generation of coat color-tagged knockouts and defined chromosomal rearrangements in mice. *Nucleic Acids Res.* 27:2354–60

95. Zheng B, Sage M, Cai W-W, Thompson DM, Tavsanli BC, et al. 1999. Engineering a mouse balance chromosome. *Nat. Genet.* 22:375–78

Annu. Rev. Genomics Hum. Genet. 2001. 2:493–510

THE HUMAN REPERTOIRE OF ODORANT RECEPTOR GENES AND PSEUDOGENES

Peter Mombaerts

The Rockefeller University, New York, New York 10021; e-mail: peter@rockefeller.edu

Key Words olfaction, smell, odor, olfactory, nose

■ **Abstract** The nose of *Homo sapiens* is a sophisticated chemical sensor. It is able to smell almost any type of volatile molecule, often at extraordinarily low concentrations, and can make fine perceptual discriminations between structurally related molecules. The diversity of odor recognition is mediated by odorant receptor (OR) genes, discovered in 1991 by Buck & Axel. OR genes form the largest gene families in mammalian genomes. A decade after their discovery, advances in the sequencing of the human genome have provided a first draft of the human OR repertoire: It consists of ~1000 sequences, residing in multiple clusters spread throughout the genome, with more than half being pseudogenes. Allelic variants are beginning to be recognized and may provide an opportunity for genotype-phenotype correlations. Here, I review the current knowledge of the human OR repertoire and summarize the limited information available regarding putative pheromone and taste receptors in humans.

INTRODUCTION

Often underestimated and underappreciated, the human nose is an advanced device of molecular recognition, superior to any electronic nose that has ever been manufactured. A common misconception is that humans can smell 10,000 odors; this view stems from a confusion between detection, discrimination, and identification of odors. Of the millions of volatile molecular species that have been catalogued by chemists, a significant fraction can be detected by the human nose. It follows that we can detect hundreds of thousands, if not millions, of distinct odors. We are less proficient at discriminating between odors than we are at simply detecting them. Our capacity to identify or name single molecular species by their odor quality is also extremely restricted: The best paid "noses" in the fragrance industry can identify by name a few thousand compounds at most.

Terrestrial animals, such as humans, smell air-borne odors, in contrast to aquatic or amphibian animals, which can also smell water-soluble molecules with low volatility, such as amino acids. Odorants entering the human nose stimulate olfactory sensory neurons (OSNs), located within the olfactory epithelium. These neurons are generated in situ from progenitor cells at the base of the epithelium,

1527-8204/01/0728-0493$14.00

in a process of neurogenesis that continues throughout adult life at low rate. OSNs are bipolar neurons containing two processes: a single dendrite and a single axon. The receptive end of an OSN is the dendrite, which terminates in cilia (specialized subcellular compartments that protrude into the nasal cavity). The axon of an OSN penetrates the skull through holes of the cribriform plate and synapses with second-order neurons and interneurons within glomeruli (globose structures located in the outer part of the olfactory bulb). Sensory information is processed within the bulb and relayed to the olfactory cortex and several other centers in the brain, where it gives rise to the olfactory percept of, for instance, a lily. The olfactory pathway is unique among the senses in that the cell detecting the stimulus, the OSN, is a neuron that projects its axon directly to the brain.

OSNs derive their functional heterogeneity from the expression of any of a large number of odorant receptors (ORs). How odorants interact with their receptors on the surface of OSN cilia is poorly understood in chemical terms. This conversion of chemical information into electrical impulses involves signal amplification via G-protein activation, elevation of adenylyl cyclase activity, and opening of cyclic-nucleotide-gated channels by cAMP. The influx of cations through these channels depolarizes the cell membrane, ultimately resulting in an increase in the frequency of action potentials that travel down the axons to the glomeruli.

This review is concerned with the molecules responsible for the initial step of olfactory signal transduction: the repertoire of human ORs. Other reviews in the field of olfaction are available (6, 7, 16, 37, 45–47). I focus on genomics and sketch the first draft of the human OR repertoire that has emerged a decade after their discovery, in the wake of recent major advances in the human genome project. The human- or primate-specific high proportion of pseudogenes in the OR repertoire is discussed. Areas of relative ignorance, such as OR expression in human OSNs and functional characterization, are outlined. Attention is drawn to the emerging realization of the highly polymorphic nature of the human OR repertoire: No two individuals may have an identical OR repertoire and, perhaps, an identical sense of smell. The discovery of candidate receptors for other chemosensory stimuli—pheromones and tastants—is briefly reviewed.

THE PRE-GENOMIC PHASE

Discovery and Characterization of OR Genes

Genes for odorant receptors were identified in rat by Buck & Axel (8) in 1991, based on evidence for a G-protein coupled pathway in vertebrate olfactory signal transduction (53, 67) and relying on the assumption that ORs are encoded by a large family of genes specifically expressed in OSNs. The elegant design of their discovery was one of the first applications of degenerate polymerase chain reaction (PCR). It made use of the presence of motifs within seven-transmembrane proteins, a type of receptor to which all known G-protein coupled receptors belong. OR genes could be grouped in small families of genes according to their cross-hybridization properties (8).

A minimum estimate of the complexity of the OR repertoire was originally 100–200 genes in rat (8), followed by a more realistic estimate of 500–1000 OR genes in rat and mouse (5). The estimate of 1000 OR genes in mammalian genomes has become a standard phrase in introductions to research articles, reviews, and textbooks, although the experimental evidence supporting it is limited and generally not available for any species other than human, mouse, and rat. These genes are interchangeably referred to as olfactory receptor, odorant receptor, or odor receptor genes, and they are abbreviated as OR genes. I prefer the term odorant receptor because olfactory receptor is ambiguous: It can be confused with the olfactory sensory neuron (OSN), which physiologists often refer to as olfactory receptor by analogy with photoreceptor in the retina. The term odor receptor is rather colloquial and may also convey the misleading notion that individually these receptors recognize odors or smells as we experience them in daily life (for instance, the odor of a lily).

For several years, the products of OR genes were candidate or putative receptors for odorants, as none of the ∼1000 ORs was functionally matched with any of the millions of possible odorous ligands. The failure of heterologous expression systems explains most of the poor progress. Functional proof that ORs can mediate cellular responses to odorants (36, 81) was not delivered until 1998. The first evidence consisted of the demonstration that responsiveness to short-chain aliphatic aldehydes, such as octanal, is conferred upon rat OSNs by adenovirally mediated gene transfer of a specific OR, i.e., the rat *I7* gene (81). Ligands for a human OR were described a year later (78); it is surprising to note that, to date, this OR, named OR17-40, is the only human OR that has been functionally characterized.

Following the identification of OR genes in rat, counterparts in other vertebrate species, including human, proved easy to obtain, mainly by PCR amplification with degenerate oligonucleotide primers for conserved motifs (45, 47). Because the coding region of OR genes is intronless, cDNA is not required and genomic DNA is a suitable and convenient template for obtaining PCR fragments of OR genes: Genomic DNA of a koala is easier to obtain than cDNA prepared from its olfactory mucosa. Full-length ORF sequence information requires the isolation of cDNA or genomic clones. With these methods, thousands of partial or full OR open reading frame (ORF) sequences have been reported from a least two dozen vertebrate species (28, 45, 47). A specialized database archives this wealth of information (68).

OR genes have ORFs of ∼1 kb that are intronless. The predicted amino acid sequence features seven putative transmembrane domains, which is characteristic of G-protein coupled receptors. Various consensus amino acid motifs can be discerned. The transmembrane domains 3, 4, and 5 are hypervariable regions: These domains could together form the odorant-binding pocket, in analogy with other seven-transmembrane proteins. Seventeen hypervariable residues have been identified within these domains (56) and deserve particular attention for model building and functional characterization of the odorant-binding pocket.

Expression Patterns of Vertebrate OR Genes

A vertebrate OR gene is expressed in a minute subset of OSNs, although the percentage of positive cells is difficult to count or even to estimate. These observations, combined with limited single-cell RT-PCR data (40), have fueled the increasingly popular one receptor–one neuron, or singular expression, hypothesis. By analogy with lymphocytes, which express a single immunoglobulin or T-cell receptor, it is attractive to think of the olfactory system as composed of sensory neurons that each express a single OR type. Only a single piece of evidence against this hypothesis exists: Some OSNs in rat may express a specific combination of two ORs (58). Because a given OSN is broadly responsive to odorants and a given odorant, in turn, activates numerous glomeruli (6), there is no one-to-one correlation between ligands and receptors. Olfactory coding is combinatorial in nature, but the relevant stimulus features are thus far elusive.

Rat and mouse OR genes are expressed in OSNs within one of four, approximately even-sized zones of the olfactory epithelium (59, 75). The zonal organization of the olfactory epithelium was not anticipated by anatomical and physiological data, and its biological significance remains unclear. Within a zone, cell bodies of OSNs expressing a given OR are distributed in a punctate, apparently random fashion. Axons of OSNs that express the same OR gene converge with extremely high precision on a small number of glomeruli that reside in recognizable locations within the olfactory bulb (14, 44, 49, 50, 60, 70, 74, 77, 82). This anatomical convergence provides a substrate for the integration of sensory information relayed by a single OR type: It suffices that a fraction of OSNs expressing a given OR are stimulated for the OR-specific glomeruli to be activated. With 1800 glomeruli in the mouse olfactory bulb (64) and ~1000 OR genes, the principle of convergence yields on average two glomeruli per OR. As no reliable data exist about the numbers of glomeruli in the human olfactory bulb, the number of intact human ORs cannot be predicted by applying this rule.

Class I Versus Class II OR Genes

There have been various attempts to classify ORs, with the ultimate objective to define structure-function relationships and to establish scenarios of evolutionary history. An unambiguous division emerged after ORs were identified in catfish, which appear to harbor an OR repertoire that is ~10% that of mammalian species (52). Interestingly, fish-like ORs were later identified in the amphibious animal *Xenopus laevis*, along with ORs that are more homologous to those of terrestrial vertebrates, such as rat and mouse (23). The argument was made that the fish-like, Class I ORs may recognize water-soluble ligands, whereas the tetrapod-specific, Class II ORs may detect airborne ligands (23, 24). This hypothesis was supported by the segregated expression of Class I ORs in the part of the nasal cavity of *X. laevis* that is accessible by water and of Class II ORs in a distinct compartment accessibly by air but not by water (23). According to this view, Class I ORs in terrestrial vertebrates would be evolutionary relics.

Class I ORs deserve special attention, in light of the surprising finding of a single extended cluster on human chromosome 11 when the first draft of the human genome sequence became available for database mining (see below). Several Class I ORs have been described in mouse. Five mouse Class I ORs (*S6*, *S18*, *S19*, *S46*, and *S50*) were identified by single-cell RT-PCR in OSNs that respond to aliphatic odorants with a transient calcium increase (40); if this is accepted as evidence that these ORs recognize those odorants, then Class I ORs do not recognize specifically water-soluble odorants, as these aliphatic odorants are volatile. Other Class I ORs were discovered accidentally by researchers sequencing regions flanking the β-globin cluster in mouse; related genes reside at similar positions in the genomes of human (11, 22) and chicken (11). A rat Class I OR gene, *RA1c*, is expressed not only in OSNs, but also in distinct regions of the cerebral cortex and brainstem (57). Such ectopic expression has been confirmed for a highly homologous mouse OR gene, *MOL2.3*, which is expressed in OSNs that project their axons to distinct glomeruli (14). Interestingly, in all cases, mouse and rat Class I OR genes are expressed in the most dorsal zone of the olfactory mucosa (11, 14, 40, 57).

The Human OR Repertoire

During the first decade, characterization of the human OR repertoire was fragmentary and nonsystematic. I have previously reviewed the literature on this subject (46). Three efforts deserve special attention.

Lancet and colleagues have extensively characterized a cluster of human OR genes on 17p13 (4, 26, 29, 30, 65, 69). A region of 412 kb comprises 17 OR sequences, 10 of which are genes with an uninterrupted ORF and 6 are pseudogenes in all individuals examined; one gene is a pseudogene in some subjects but has an intact ORF in others. Interestingly, six OR coding regions (30) were not detected in earlier experiments in which degenerate PCR was applied on cloned DNA of this region, indicating that genomic sequencing of vertebrate species will provide much more, and possibly surprising, information compared to the traditional approach of cloning OR fragments by degenerate PCR. This study revealed the first allelic variant of a human OR gene, differing in 2/648 nucleotides. The OR gene cluster has a very complex organization, with genes in both transcriptional orientations (30). A cluster of 15 OR sequences on chimpanzee 19p15 is orthologous to the OR cluster on human 17p13 (66). The overall layout of the chimp cluster is conserved. Four of the OR-like sequences are pseudogenes, two of which have different deleterious mutations than those in human. Sequence comparison with orthologous OR genes in other African apes (66) suggests that all 15 OR genes from the genomic cluster were intact at the time of the orangutan-African ape divergence, which occurred ∼9 million years ago. Attempts to correlate OR sequences on human 17p13 with their counterparts in the mouse genome were less clear (39): A pairwise association could be established in some but not all cases, although the 13 orthologous mouse OR genes (4 of which are pseudogenes) are contained within a single cluster on mouse chromosome 11B3-11B5.

Rouquier, Trask, Giorgi, and colleagues undertook a genome-wide characterization of the human OR repertoire (63, 72), thus providing a framework for contemporary research. Mapping by fluorescence in situ hybridization (FISH) revealed that OR genes reside on all human chromosomes except chromosome 20 and the Y chromosome. A total of 28 loci, perhaps as many as 53 sites, were labeled by FISH. Surprisingly, many OR clones hybridize by FISH to multiple sites in the genome: Forty-four locations, spread over 13 chromosomes, were identified, 33 of which contain sequences that cross-hybridize to one or more other genomic locations (72). This was the first indication that large-scale duplications may have shaped the present human OR gene repertoire. OR gene clusters may be responsible for chromosomal microrearrangements through unequal homologous recombination (25a). This group provided the first report of a high incidence of pseudogenes in the human OR repertoire: 63 of 87 (72%) distinct OR PCR products representing clusters in different regions of the genome are from pseudogenes (63). The ORF of the pseudogenes is typically interrupted by more than one deleterious mutation, such as nonsense mutations or frameshifts resulting from insertion or deletion events. An unusual poly-T expansion of 5 to 12 nucleotides within transmembrane domain 4 was found in 18 of the 63 pseudogenes. By contrast, pseudogenes are few and far between in rat and mouse; only seven pseudogenes have thus far been described in mouse (39, 80). A sampling of the OR repertoire of various species suggests that primates, except New World monkeys, are peculiar among mammals, having accumulated a high proportion of pseudogenes in their OR repertoires (62). The pseudogenization of the primate OR repertoire may reflect their decreased dependence on the sense of smell for survival and reproduction.

A third effort, by Ziegler and colleagues (21, 79, 83), focused on OR genes linked to the major histocompatibility complex (MHC) in human and mouse. A total of 34 OR genes and pseudogenes were described in a major and a minor cluster on human 6p21. The gene density of one OR sequence per 23 kb in the major cluster is similar to that of the 17p13 cluster (one OR sequence per 24 kb). This characterization resulted in the identification of numerous OR alleles (see below).

THE GENOMIC PHASE

In early 2001, a first draft of the human genome sequence was reported by Celera Genomics (76) and the International Human Genome Project Consortium (34). Two groups have capitalized on the wealth of information publicly available from the latter project and have each produced a first draft of the human OR repertoire (31, 84). The overall conclusions are in accord with the information accumulated in the pre-genomic phase: There are ∼1000 OR sequences in the human genome; the coding regions are intronless; the genes are spread over dozens of clusters; and more than half of OR sequences are pseudogenes. A major new finding is that chromosome 11 contains nearly half of the OR repertoire (9, 31, 84), including a single cluster of ∼100 OR Class I sequences.

Weizmann Institute Analysis

A first and thorough characterization was carried out by Glusman, Lancet, and colleagues (25, 28, 31) by combing the available draft sequence of 85% of the human genome, complemented with mining of additional sequence databases and in-house cloning of OR fragments by PCR. A total of 906 OR genes and pseudogenes were identified, corresponding to a repertoire of $\sim$1000 OR sequences when the remainder of the human genome sequence is deposited. Of these sequences, two thirds had not been identified previously. Sixty-three percent of OR sequences appear to represent pseudogenes, but when only full-length ORF sequences are included, this fraction drops to 53%. A tentative repertoire of 322 intact human ORs is thus derived (31), representing 1% of the $\sim$30,000 genes of which the human genome is now thought to consist. Similarly, the total length of the human genome occupied by the olfactory subgenome is estimated at 30 Mb, or 1% of genomic DNA. OR sequences reside on all chromosomes except 20 and Y, as was documented earlier. They are not evenly distributed: Chromosomes 1, 6, 9, 11, 14, and 19 account for three quarters of the OR repertoire, and chromosome 22, the smallest autosome, contains a single OR gene. Eighty percent of ORs are embedded within 24 clusters of more than 6 ORs. Most clusters range in size from 100 kb to 1 Mb. At least 64 clusters of more than two ORs can be identified.

Following their classification into families ($>$40% amino acid identity) and subfamilies ($>$60% amino acid identity) (28, 38), 13 families of Class II ORs and 4 families of Class I ORs were discerned within the human repertoire (31). The most important new findings to result from this first draft are that chromosome 11 harbors 42% of all ORs and that it contains the two largest OR clusters, each comprising more than 100 OR sequences. One cluster is on 11p15 and contains 102 Class I OR sequences, half of which are pseudogenes. Members of this class had been described earlier in human, mouse, and rat (see above). Their preponderance in the human genome renders it less plausible that Class I ORs are relics in genomes of terrestrial species and that they are specialized in the detection of water-soluble odorants. Interestingly, the β-globin locus resides in the middle of the cluster of Class I OR sequences (T. Olender & D. Lancet, personal communication), and this may be evolutionarily conserved in mouse and chicken.

Comparison of the 24 clusters with more than 6 ORs suggests that the 2 large OR clusters on chromosome 11 are the ancestral gene clusters. The cluster of Class II (tetrapod-specific) sequences may have given rise to all other clusters by way of sequential cluster duplication, whereas the cluster of Class I (fish-like) sequences appears to not have developed further. Clustering is likely to result from the traditional concept of gene duplication followed by divergence. It will be interesting to determine to what extent the human OR repertoire has been shaped by whole-cluster duplication versus intracluster duplication of individual OR genes. A definitive evolutionary reconstruction of the human OR repertoire is now within reach but will be dependent on completion of the human genome sequence and accurate chromosome assignments of OR sequences.

The authors conclude (31) that, "The observed OR clusters lack a clear internal structure as observed for homeobox, beta globin and immunoglobulin genes. Rather, OR genes appear to be disposed in haphazard arrangements that appear to correlate only with their phylogenetic classification. This, together with the large number of singleton OR genes scattered throughout the human genome, suggests that there may be no functional importance to clustering."

Senomyx Analysis

Zozulya and colleagues (84) have also reported a first draft of the human OR repertoire, but their focus was specifically on ORs with uninterrupted ORFs and the chromosomal mapping was not a high priority. This analysis is mostly in agreement with that of the Weizmann group (31). They identified 347 candidate human ORs; the minor discrepancy with the Weizmann group may be attributed to potential misclassification of certain OR sequences as pseudogenes by the Weizmann group. Importantly, these 347 intact coding regions were recloned by genomic PCR using the mixed DNA of a pool of 10 individuals, confirming the correct identification of intact OR ORFs. Chromosome 1 harbors 155 intact ORs (45%), followed by 42 ORs on chromosome 1, 26 ORs on chromosome 9, and 24 ORs on chromosome 6. The average amino acid identity for a pair of ORs is in the 35%–40% range, with values as low as 20%. The average human OR is 315 amino acids long. Multiple sequence alignment revealed more than a dozen amino acid sequence motifs of various lengths, which, combined with particular conserved single residues, define a sequence signature of an OR.

Nomenclature

No nomenclature scheme has been embraced by the community of olfactory researchers. Lancet and colleagues have proposed for several years to name OR genes according to family (>40% amino acid identity) and subfamily (>60%) assignment, with sequences from all vertebrate species contributing to this classification (28, 38). They have argued that an independent and retrospective analysis of the human OR repertoire by principal component analysis validates these cut-offs (31). Zozulya and colleagues have introduced a slightly different naming scheme that combines phylogenetic relationships with genomic location (84). In the near future, a pragmatic, convenient, and generally accepted nomenclature should be adopted, which must also deal effectively with the enormous potential for variability in the human OR repertoire (see below). The *Drosophila* olfactory community has already come to an agreement (18) for the naming of the ~60 odorant receptor genes, effectively eliminating the four parallel nomenclatures.

Caveats

Much of this database mining was performed on high-throughput, unannotated raw sequences—a so-called draft sequence (34, 76). Moreover, ~5%–15% of the human genome sequence was not included in the first draft of the public consortium. Some of the chromosome assignments of OR sequences and even entire sequence contigs are uncertain; there are discrepancies between various maps. This may be a particular problem for ORs because of the extensive sequence similarity and the large duplications (72). Because this draft is a composite of a few individuals, the difficulties in distinguishing between alleles and highly related OR genes and the occurrence of allelic pseudogenes of intact ORs may confuse attempts to obtain a global view of "the" human OR repertoire.

THE POST-GENOMIC PHASE

Variability in the Human OR Repertoire

A priority will be to describe variations in the OR repertoire within the human population and to correlate them to perceptual differences and preferences in smell. Not only does the OR repertoire constitute the largest gene family in the human genome, but it may also display a substantial degree of variability—the task ahead is thus very challenging. There are several isolated indications of a great degree of variability in the human OR repertoire.

A first type of variation was uncovered in the form of differences in the number and chromosomal location of a 36-kb sequence block that contains three OR sequences, two of which are pseudogenes. This block is repeated subtelomerically such that humans can have between 7 and 11 copies of this block on any of several chromosomes (71).

A second source of variation is that a member of the OR cluster on 17p13 has an uninterrupted ORF in some individuals but is a pseudogene in others, a result of a nonsense mutation (30). A similar situation has been encountered for three OR genes in an independent study (21, 79).

A third factor of variation is represented by conventional polymorphic alleles that result in amino acid substitutions. Two studies merit discussion. Lancet and colleagues identified numerous single nucleotide polymorphisms in the OR gene cluster on 17p13, some of which affect the coding region of intact ORs (26, 65). The average nucleotide diversity for the ∼1-kb coding region of intact ORs was 0.03%, and the average number of haplotypes was 2.3 for intact ORs. A second study (21, 79) focused on two clusters comprised of ∼34 OR sequences (15 of which are intact ORs) adjacent to the MHC on 6p21. Using immortalized cell lines from different ethnic origins and with different MHC haplotypes, investigators screened these ORs, along with three others on different chromosomes, for polymorphisms among the human population. Every OR gene examined has multiple alleles: The average number of alleles was 3.4, and one OR gene could encode seven different amino acid sequences. Three of the MHC-linked OR sequences exhibited both intact ORFs and pseudo-ORFs. It is conceivable that some of these polymorphic alleles are not functional (and are thus pseudogenes) because certain structural features, for instance, residues or motifs critical for G-protein activation, are compromised.

Both studies may be viewed as early signs of the fluidity of the human OR repertoire. Only the tip of the proverbial iceberg is in sight. Perhaps all ~1000 OR sequences appear in an intact, potentially functional form within the human population at large. The distribution of pseudogenes may be such that in any given member of the human species only one third of sequences may encode an intact OR and the remaining two thirds may code for defective ORs. The subset of intact ORs may be nonoverlapping between individuals: It is conceivable that no two human beings have exactly the same OR repertoire and, by inference, an identical sense of smell. One is tempted to speculate that this variability has been selected for in evolution, much like the numerous MHC haplotypes within the human population that provide a safeguard against significant "holes" in the immune defense mechanisms and against emerging threats of infectious agents.

A Human OR Repertoire May Be Double the Size

In mouse, an OR gene is expressed monoallelically in a given OSN (13). Approximately 50% of OSNs express the gene from the maternal allele, the other 50% from the paternal allele, and biallelic expression is not observed (70). The olfactory system of mice is thus a mosaic of 2×1000 distinct populations of OSNs. In inbred strains of mice, both alleles are identical, so their olfactory system is in fact composed of 1000 distinct populations of OSNs, each expressing a different OR type. The mechanisms of monoallelic expression remain enigmatic.

Let us assume that monoallelic expression can be extrapolated to human OR genes. It may very well be the case that humans are heterozygous at most OR loci. With ~330 intact ORs each having multiple alleles, an individual would thus have 2×330 types of ORs. Each of these OR types would be expressed in a distinct population of OSNs, and many of them could be regarded as separate channels conveying distinct sensory information. The common occurrence of OR allelism may explain the need for monoallelic expression. This mode of expression would avoid the possibility of dual, ambiguous response patterns resulting from coexpression of different OR types from the same locus and would thus increase the signal-to-noise ratio in the olfactory apparatus. In short, the functional OR repertoire available to a given human being may be double the size than the number of functional OR genes.

Are All Human OR Genes Receptors for Odorants?

A minimal criterion for a gene to encode a receptor for odorants is that it ought to be expressed within OSNs. This is an obvious requirement, but a decade after their discovery, expression of OR genes within human OSNs has yet to be documented by a method that offers single-cell resolution, such as in situ hybridization, single-cell RT-PCR, or immunohistochemistry (46). RT-PCR expression in samples of whole olfactory mucosa has been reported for ORs from the 17p13 cluster; 9 of

11 intact OR genes of that cluster were expressed in the form of spliced RNA, but 5'RACE experiments were not successful (69). RT-PCR is not quantitative and does not permit the demonstration of neuron-specific expression within the tissues samples, which are heterogeneous in their cell-type composition. The absence of satisfactory expression data in humans can be attributed to obstacles in obtaining large and nondegraded samples of human olfactory mucosa from either biopsies or necropsies. Moreover, in contrast to other species, the human olfactory mucosa does not form a contiguous sheet but is fractionated as islands or pockets surrounded by nonolfactory mucosa. In addition, there are no antibodies for ORs from the human or other species.

This paucity of expression data may be mitigated by evidence from other, experimentally tractable vertebrate species that a randomly chosen OR gene is generally expressed at the mRNA level in OSNs. In other words, genes considered as OR genes by virtue of their homology to other OR genes most likely encode receptors for odorants; this is the null hypothesis. But several, as yet isolated and unexplained, observations have been reported over the years that vertebrate OR genes are occasionally expressed in nonolfactory tissues. For instance, more than 50 ORs have been cloned by PCR from human testis cDNA (54, 73); interestingly, only a single pseudogene has been described among these testicular ORs. These genes might be OR-like genes with a sperm cell–specific expression and function, rather than OR genes that are expressed both in OSNs and in male germ cells. Further sporadic reports of ectopic expression of OR genes include the notochord of chick (51), the heart of rat (19), the brainstem of rat (57) and mouse (14), and erythroid cells of both human and mouse (22). Likewise, numerous expressed sequence tags (ESTs) for ORs have been recovered from nonolfactory human organs (17), although genomic contamination may explain the presence of these intronless coding regions among EST libraries. A hypothesis that must be ruled out is that a fraction of the human OR repertoire, as with any other vertebrate OR repertoire, may be expressed exclusively in nonolfactory tissues and is, therefore, not involved in olfaction. The extraordinary diversity of ligand specificities of ORs may have been exploited in evolution to "recruit" OR members for specific chemosensory roles in nonolfactory cell types. A bolder hypothesis proposed on speculative grounds is that ORs may function as "area code molecules" and shape the embryonic development of the vertebrate body plan (17).

Thus, the possibility that some OR genes do not encode receptors for odorants should not be forgotten and suppressed. All human ORs should thus be regarded as candidate ORs.

Function of Human ORs in the Sense of Smell

A persistent problem in the field of vertebrate olfaction is that there are very few robust assays to assign odorous ligands to a given, cloned OR or to identify ORs for a given odorous ligand (36, 81). Typically, vertebrate ORs cannot be expressed on the surface of heterologous cells, precluding a rapid characterization of ligand-receptor relationships. Additionally, the few assays that have been reported may not be generally applicable to all ORs. In 1999, expression in the human embryonic kidney 293 cell line and *X. laevis* oocytes permitted the identification of helional as an odorous ligand for the first, and thus far only, human OR, OR17-40 (78). Commercial applications of the near-complete characterization of the human OR repertoire will be critically dependent on robust assays that permit high-throughput screening and are invariably applicable to human ORs.

Recent progress in the molecular basis of taste perception was derived from psychophysical genetics: Variations in the detection and preference of bitter or sweet compounds were linked genetically to seven-transmembrane proteins, which became candidate taste receptors (1, 12, 41, 48). By contrast, olfactory psychophysics has not been instrumental in suggesting ORs for particular odorous stimuli, with one possible exception in mouse (32). There are numerous specific anosmias (2)—increases in detection thresholds for specific compounds—but their genetic basis has remained obscure, presumably owing to the complexity of the human OR repertoire. Identification of polymorphisms and application of microarray technology may provide a solution to the problem of relating variations in olfactory perception to specific alleles or haplotypes within the human OR repertoire. Massively parallel, whole-genome approaches could thus provide candidate high-affinity ORs for particular odorous ligands.

OTHER CHEMOSENSORY RECEPTORS IN HUMAN

A Single Putative Pheromone Receptor is Intact in Human

Many vertebrate species, including rodents, possess a specialized, or accessory, olfactory system, which may mediate the detection of pheromones (35). These are chemical signals that provide information about gender, dominance, and reproductive status between individuals of the same species. They elicit in the recipients innate and stereotyped reproductive and social behaviors, along with profound neuroendocrine and physiological changes. Sensory neurons located in a specialized structure, the vomeronasal organ (VNO), detect these pheromones, although they may also detect other types of chemosensory stimuli and pheromones may also interact with sensory neurons of the main olfactory epithelium. The chemosensory receptors of the VNO are termed vomeronasal receptors (VRs). Two families of VR genes, also encoding putative seven-transmembrane proteins, have been proposed to encode pheromone receptors, but the evidence is circumstantial. The first family of VR genes discovered consists of ∼100 members in rat (20). Whereas the mouse V1R repertoire may contain as many as one third of pseudogenes (15), the human repertoire appears to be riddled with pseudogenes; seven distinct human V1R pseudogenes reside on various chromosomes (27). We (61) and others (55) have identified a single human V1R-like gene displaying an intact ORF that encodes a putative protein with distinct homology to mouse and rat V1Rs. Sequence determination of this gene, termed *V1RL1*, in 11 individuals revealed 3 functional alleles within the human population (61). We further demonstrated that spliced messages are present in the main olfactory mucosa of humans (61); neuron-specific expression has yet to be demonstrated. The *V1RL1* gene may very well be the sole remaining functional member of the human V1R repertoire, which may thus consist of 99% of pseudogenes.

An obvious explanation for this massive decay of an entire gene family is our much decreased dependence on pheromones for regulating social and reproductive behavior. There is only one generally accepted pheromone-like effect in humans, the synchronization of the menstrual cycle between women who live in close proximity with each other (43). In humans, a VNO-like organ develops prenatally and becomes vestigial in adults; detection of pheromones may be taken over by the main olfactory system (35).

Candidate Human Taste Receptors

A family of seven-transmembrane receptors that may represent bitter taste receptors has recently been discovered (1, 12, 41, 48). Based on the genetic linkage of a locus at 5p15 that controls variations within the human population in the detection of the bitter compound 6-n-propyl-2-thiouracil, a family of a few dozen human T2R genes was identified that are expressed in taste receptor cells. One of these receptors, hT2R-4, responds to denatonium and 6-n-propyl-2-thiouracil upon gene transfer into the human embryonic kidney cell line 293 (12). A related mouse receptor, mT2R-5, responds to cycloheximide, and polymorphic alleles of mT2R-5 in various inbred strains correlate with their ability to detect this compound, which tastes bitter to humans (12). It is not clear if all T2Rs are bitter receptors. A given taste receptor cell appears to express a broad fraction of the T2R repertoire, in sharp contrast to OSNs, which may express a single OR from an even larger repertoire. Interestingly, the T2Rs have distant sequence homology to rodent V1Rs.

Two other seven-transmembrane proteins with taste receptor cell-specific expression, T1R1 and T1R2, were identified earlier (33), but functional or genetic evidence that these molecules are also taste receptors is not yet present. Very recently, a third member of this family (T1R3) with $\sim$30% amino acid homology to T1R1 and T1R2 was proposed as a sweet receptor (3, 35a, 42, 50a, 64a). The evidence is largely based on fine-mapping of the mouse T1R3 ortholog to the *Sac* locus, which controls variations in the detection of sweet-tasting compounds in mice. The T1Rs have distant sequence homology to rodent V2Rs.

CONCLUSION

Analysis of the chemical composition of the external environment is the most ancient sensory modality: Even single-cell organisms, such as bacteria, must have chemosensory receptors to sample the outside world. In all species, including humans, it has become a recurrent theme that this function requires large gene families. A first draft of the human OR repertoire has become available as a result of recent advances in the human genome project. There are indications of a potentially enormous fluidity within the human OR repertoire. Although the complexity of the variations in the human population may present a formidable challenge in the near future, the OR repertoire may offer an extraordinary opportunity to define the genetic basis of our individuality in terms of how we smell.

ACKNOWLEDGMENT

I thank Mona Khan for her support and helpful suggestions.

Visit the Annual Reviews home page at www.AnnualReviews.org

LITERATURE CITED

1. Adler E, Hoon MA, Mueller KL, Chandrashekar J, Ryba NJ, et al. 2000. A novel family of mammalian taste receptors. *Cell* 100:693–702
2. Amoore JE. 1967. Specific anosmia: a clue to the olfactory code. *Nature* 214:1095–98
3. Bachmanov AA, Li X, Reed DR, Li S, Chen Z, et al. 2001. Positional cloning of the mouse saccharin preference (Sac) locus. *Chem. Senses.* In press
4. Ben-Arie N, Lancet D, Taylor C, Khen M, Walker N, et al. 1994. Olfactory receptor gene cluster on human chromosome 17: possible duplication of an ancestral receptor repertoire. *Hum. Mol. Gen.* 3:229–35
5. Buck LB. 1992. The olfactory multigene family. *Curr. Biol.* 2:467–73
6. Buck LB. 1996. Information coding in the vertebrate olfactory system. *Annu. Rev. Neurosci.* 19:517–44
7. Buck LB. 2000. The molecular architecture of odor and pheromone sensing in mammals. *Cell* 100:611–18
8. Buck L, Axel R. 1991. A novel multigene family may encode odorant receptors: a molecular basis for odor recognition. *Cell* 65:175–87
9. Buettner JA, Glusman G, Ben-Arie N, Ramos P, Lancet D, Evans GA. 1998. Organization and evolution of olfactory receptor genes on human chromosome 11. *Genomics* 53:56–68
10. Bulger M, Bender MA, van Doorninck JH, Wertman B, Farrell CM, et al. 2000. Comparative structural and functional analysis of olfactory receptor genes flanking the human and mouse β-globin clusters. *Proc. Natl. Acad. Sci. USA* 97:14560–65
11. Bulger M, von Doorninck JH, Saitoh N, Telling A, Farrell C, et al. 1999. Conservation of sequence and structure flanking the mouse and human β-globin loci: The β-globin genes are embedded within an array of odorant receptor genes. *Proc. Natl. Acad. Sci. USA* 96:5129–34
12. Chandrashekar J, Mueller KL, Hoon MA, Adler E, Feng L, et al. 2000. T2Rs function as bitter taste receptors. *Cell* 100:703–11
13. Chess A, Simon I, Cedar H, Axel R. 1994. Allelic inactivation regulates olfactory receptor gene expression. *Cell* 78:823–34
14. Conzelmann S, Levai O, Bode B, Eisel U, Raming K, et al. 2000. A novel brain receptor is expressed in a distinct population

of olfactory sensory neurons. *Eur. J. Neurosci.* 12:3926–34

15. Del Punta K, Rothman A, Rodriguez I, Mombaerts P. 2000. Sequence diversity and genomic organization of vomeronasal receptor genes in the mouse. *Genome Res.* 10:1958–67

16. Doty RL. 2001. Olfaction. *Annu. Rev. Psychol.* 52:423–52

17. Dreyer WJ. 1998. The area code hypothesis revisited: Olfactory receptors and other related transmembrane receptors may function as the last digits in a cell surface code for assembling embryos. *Proc. Natl. Acad. Sci. USA* 95:9072–77

18. Drosophila Odorant Recept. Nomencl. Comm. 2000. A unified nomenclature system for the *Drosophila* odorant receptors. *Cell* 102:145–46

19. Drutel G, Arrang JM, Diaz J, Wisnewsky C, Schwartz K, et al. 1995. Cloning of OL1, a putative olfactory receptor and its expression in the developing rat heart. *Recept. Channels* 3:33–40

20. Dulac C, Axel R. 1995. A novel family of genes encoding putative pheromone receptors in mammals. *Cell* 83:195–206

21. Ehlers A, Beck S, Forbes SA, Trowsdale J, Volz A, et al. 2000. MHC-linked olfactory receptor loci exhibit polymorphism and contribute to extended HLA/OR-haplotypes. *Genome Res.* 10:1968–78

22. Feingold EA, Penny LA, Nienhuis WA, Forget BG. 1999. An olfactory receptor gene is located in the extended human ß-globin gene cluster and is expressed in erythroid cells. *Genomics* 61:15–23

23. Freitag J, Krieger J, Strotmann J, Breer H. 1995. Two classes of olfactory receptors in *Xenopus laevis*. *Neuron* 15:1383–92

24. Freitag J, Ludwig G, Andreini I, Rössler P, Breer H. 1998. Olfactory receptors in aquatic and terrestrial vertebrates. *J. Comp. Physiol. A* 183:635–50

25. Fuchs T, Glusman G, Horn-Saban S, Lancet D, Pilpel Y. 2001. The human olfactory subgenome: from sequence to structure and evolution. *Hum. Genet.* 108:1–13

25a. Giglio S, Borman KW, Matsumoto N, Calvar V, Gimelli G, et al. 2001. Olfactory receptor-gene clusters, genomic inversion polymorphisms, and common chromosome rearrangements. *Am. J. Hum. Genet.* 68:874–83

26. Gilad Y, Segré D, Skorecki K, Nachman MW, Lancet D, et al. 2000. Dichotomy of single nucleotide polymorphism haplotypes in olfactory receptor genes and pseudogenes. *Nat. Genet.* 26:221–24

27. Giorgi D, Friedman C, Trask BJ, Rouquier S. 2000. Characterization of nonfunctional V1R-like pheromone receptor sequences in human. *Genome Res.* 10:1979–85

28. Glusman G, Bahar A, Sharon D, Pilpel Y, White J, et al. 2000. The olfactory receptor gene superfamily: data mining, classification, and nomenclature. *Mamm. Genome* 11:1016–23

29. Glusman G, Clifton S, Roe B, Lancet D. 1996. Sequence analysis in the olfactory receptor gene cluster on human chromosome 17: recombinatorial events affecting receptor diversity. *Genomics* 37:147–60

30. Glusman G, Sosinsky A, Ben-Asher E, Avidan N, Sonkin D, et al. 2000. Sequence, structure, and evolution of a complete human olfactory receptor gene cluster. *Genomics* 63:227–45

31. Glusman G, Yanai I, Rubin I, Lancet D. 2001. The complete human olfactory subgenome. *Genome Res.* 11:685–702

32. Griff IC, Reed RR. 1995. The genetic basis for specific anosmia to isovaleric acid in the mouse. *Cell* 83:407–14

33. Hoon MA, Adler E, Lindemeier J, Battey JF, Ryba NJ, et al. 1999. Putative mammalian taste receptors: a class of taste-specific GPCRs with distinct topographic selectivity. *Cell* 96:541–51

34. Int. Hum. Genome Seq. Consort. 2001. Initial sequencing and analysis of the human genome. *Nature* 409:860–921

35. Keverne EB. 1999. The vomeronasal organ. *Science* 286:716–20

35a. Kitagawa M, Kusakabe Y, Miura H, Ninomiya Y, Hino A. 2001. Molecular genetic identification of a candidate receptor gene for sweet taste. *Biochem. Biophys. Res. Commun.* 283:236–42

36. Krautwurst D, Yau KW, Reed RR. 1998. Identification of ligands for olfactory receptors by functional expression of a receptor library. *Cell* 95:917–26

37. Lancet D. 1996. Vertebrate olfactory reception. *Annu. Rev. Neurosci.* 9:329–55

38. Lancet D, Ben-Arie N. 1993. Olfactory receptors. *Curr. Biol.* 3:668–74

39. Lapidot M, Pilpel Y, Gilad Y, Falcovitz, Sharon D, et al. Mouse-human orthology relationships in an olfactory receptor gene cluster. *Genomics* 71:296–306

40. Malnic B, Hirono J, Sato T, Buck LB. 1999. Combinatorial receptor codes for odors. *Cell* 96:713–23

41. Matsunami H, Montmayeur JP, Buck LB. 2000. A family of candidate taste receptors in human and mouse. *Nature* 404:601–4

42. Max M, Shanker YG, Huang L, Rong M, Liu Z, et al. 2001. *Tas1r3*, encoding a new candidate taste receptor, is allelic to the sweet responsiveness locus *Sac. Nat. Genet.* 28:58–63

43. McClintock MK. 1971. Menstrual synchrony and suppression. *Nature* 229:244–45

44. Mombaerts P. 1996. Targeting olfaction. *Curr. Opin. Neurobiol.* 6:481–86

45. Mombaerts P. 1999. Molecular biology of odorant receptors in vertebrates. *Annu. Rev. Neurosci.* 22:487–509

46. Mombaerts P. 1999. Odorant receptor genes in humans. *Curr. Opin. Genet. Dev.* 9:315–20

47. Mombaerts P. 1999. Seven-transmembrane proteins as odorant and chemosensory receptors. *Science* 286:707–11

48. Mombaerts P. 2000. Better taste through chemistry. *Nat. Genet.* 25:130–32

49. Mombaerts P, Wang F, Dulac C, Chao SK, Nemes A, et al. 1996. Visualizing an olfactory sensory map. *Cell* 87:675–86

50. Mombaerts P, Wang F, Dulac C, Vassar R, Chao SK, et al. 1996. The molecular biology of olfactory perception. *Cold Spring Harbor Symp. Quant. Biol.* 61:135–45

50a. Montmayeur JP, Liberles SD, Matsunami H, Buck LB. 2001. A candidate taste receptor gene near a sweet taste locus. *Nat. Neurosci.* 4:492–98

51. Nef S, Nef P. 1997. Olfaction: transient expression of a putative odorant receptor in the avian notochord. *Proc. Natl. Acad. Sci. USA* 94:4766–71

52. Ngai J, Dowling MM, Buck L, Axel R, Chess A. 1993. The family of genes encoding odorant receptors in the channel catfish. *Cell* 72:657–66

53. Pace U, Hanski E, Salomon Y, Lancet D. 1985. Odorant-sensitive adenylate cyclase may mediate olfactory reception. *Nature* 316:255–58

54. Parmentier M, Libert F, Schurmans S, Schiffmann S, Lefort A, et al. 1992. Expression of members of the putative olfactory receptor gene family in mammalian germ cells. *Nature* 355:453–55

55. Pentages E, Dulac C. 2000. A novel family of candidate pheromone receptors in mammals. *Neuron* 28:835–45

56. Pilpel Y, Lancet D. 1999. The variable and conserved interfaces of modeled olfactory receptor proteins. *Protein Sci.* 8:969–77

57. Raming K, Konzelmann S, Breer H. 1998. Identification of a novel G-protein coupled receptor expressed in distinct brain regions and a defined olfactory zone. *Recept. Channels* 6:141–51

58. Rawson NE, Eberwine J, Dotson R, Jackson J, Ulrich P, et al. 2000. Expression of mRNAs encoding for two different olfactory receptors in a subset of olfactory receptor neurons. *J. Neurochem.* 75:185–95

59. Ressler KJ, Sullivan SL, Buck LB. 1993. A zonal organization of odorant receptor

gene expression in the olfactory epithelium. *Cell* 73:597–609

60. Ressler KJ, Sullivan SL, Buck LB. 1994. Information coding in the olfactory system: evidence for a stereotyped and highly organized epitope map in the olfactory bulb. *Cell* 79:1245–55

61. Rodriguez I, Greer CA, Mok MY, Mombaerts P. 2000. A putative pheromone receptor gene expressed in human olfactory mucosa. *Nat. Genet.* 26:18–19

62. Rouquier S, Blancher A, Giorgi D. 2000. The olfactory receptor gene repertoire in primates and mouse: evidence for reduction of the functional fraction in primates. *Proc. Natl. Acad. Sci. USA* 97:2870–74

63. Rouquier S, Taviaux S, Trask BJ, Brand-Arpon V, van den Engh G, et al. 1998. Distribution of olfactory receptor genes in the human genome. *Nat. Genet.* 18:243–50

64. Royet JP, Souchier C, Jourdan F, Ploye H. 1988. Morphometric study of the glomerular population in the mouse olfactory bulb: numerical density and size distribution along the rostrocaudal axis. *J. Comp. Neurol.* 270:559–68

64a. Sainz E, Korley JN, Battey JF, Sullivan SL. 2001. Identification of a novel member of the T1R family of putative taste receptors. *J. Neurochem.* 77:896–903

65. Sharon D, Gilad Y, Glusman G, Khen M, Lancet D, et al. 2000. Identification and characterization of coding single nucleotide polymorphisms within a human olfactory receptor gene cluster. *Gene* 260:87–94

66. Sharon D, Glusman G, Pilpel Y, Khen M, Gruetzner F, et al. 1999. Primate evolution of an olfactory receptor cluster: diversification by gene conversion and recent emergence of pseudogenes. *Genomics* 61:24–36

67. Sklar PB, Anholt RRH, Snyder SH. 1986. The odorant-sensitive adenylate cyclase of olfactory receptor cells: differential stimulation by distinct classes of odorants. *J. Biol. Chem.* 261:15538–43

68. Skoufos E, Marenco L, Nadkarni PM, Miller PL, Shepherd GM. 2000. Olfactory receptor database: a sensory chemoreceptor resource. *Nucleic Acids Res.* 28:341–43

69. Sosinsky A, Glusman G, Lancet D. 2000. The genomic structure of human olfactory receptor genes. *Genomics* 70:49–61

70. Strotmann J, Conzelmann S, Beck A, Feinstein P, Breer H, et al. 2000. Local permutations in the glomerular array of the mouse olfactory bulb. *J. Neurosci.* 20:6927–38

71. Trask BJ, Friedman C, Martin-Gallardo A, Rowen L, Akinbami C, et al. 1998. Members of the olfactory receptor gene family are contained in large blocks of DNA duplicated polymorphically near the ends of human chromosomes. *Hum. Mol. Genet.* 7:13–26

72. Trask BJ, Massa H, Brand-Arpon V, Chan K, Friedman C, et al. 1998. Large multi-chromosomal duplications encompass many members of the olfactory receptor gene family in the human genome. *Hum. Mol. Genet.* 7:2007–20

73. Vanderhaeghen P, Schurmans S, Vassart G, Parmentier M. 1997. Specific repertoire of olfactory receptor genes in the male germ cells of several mammalian species. *Genomics* 39:239–46

74. Vassar R, Chao SK, Sitcheran R, Nuñez JM, Vosshall LB, et al. 1994. Topographic organization of sensory projections to the olfactory bulb. *Cell* 79:981–91

75. Vassar R, Ngai J, Axel R. 1993. Spatial segregation of odorant receptor expression in the mammalian olfactory epithelium. *Cell* 74:309–18

76. Venter JC, Adams MD, Myers EW, Li PW, Mural RJ, et al. 2001. The sequence of the human genome. *Science* 291:1304–51

77. Wang F, Nemes A, Mendelsohn M, Axel R. 1998. Odorant receptors govern the formation of a precise topographic map. *Cell* 93:47–60

78. Wetzel CH, Oles M, Wellerdieck C,

Kuczkowiak M, Gisselmann G, et al. 1999. Specificity and sensitivity of a human olfactory receptor functionally expressed in human embryonic kidney 293 cells and *Xenopus laevis* oocytes. *J. Neurosci.* 19:7426–33

79. Younger RM, Amadou C, Bethel G, Ehlers A, Fischer Lindahl F, et al. 2001. Characterization of clustered MHC-linked olfactory receptor genes in human and mouse. *Genome Res.* 11:519–30

80. Xie S, Feinstein P, Mombaerts P. 2000. Characterization of a cluster comprising approximately 100 odorant receptor genes in mouse. *Mamm. Genome* 11:1070–78

81. Zhao H, Ivic L, Otaki JM, Hashimoto M, Mikoshiba K, et al. 1998. Functional expression of a mammalian odorant receptor. *Science* 279:237–42

82. Zheng C, Feinstein P, Bozza T, Rodriguez I, Mombaerts P. 2000. Peripheral olfactory projections are differentially affected in mice deficient in a cyclic nucleotide-gated channel subunit. *Neuron* 26:81–91

83. Ziegler A, Ehlers A, Forbes S, Trowsdale J, Volz A, et al. Polymorphisms in olfactory receptor genes: a cautionary note. *Hum. Immunol.* 61:1281–84

84. Zozulya S, Echeverri F, Nguyen T. 2001. The human olfactory receptor repertoire. *Genome Biol.* 2:0018.1–12

SUBJECT INDEX

science, 344–45
challenges arising from,
364–66
computation for, 349–53
as discovery science,
343–44
examples of, 355–64
framework for, 354–55
future of, 364–66
genomic sequences in
humans and model
organisms, 344
new approach to decoding
life, 343–72
perturbation of biological
systems, 345–47
quantitative
high-throughput
biological tools, 347–49

T

Tandem mass spectrometry,
359
TaqMan assay, 247–48
Targets
antimicrobial, through
understanding
pathogenesis, 263–64
gene expression profiles
and drug, 264
Tattered (TD) mouse, 324–25
Tay-Sachs disease (TSD),
85–86
TCR
See Transcription-coupled
repair
Teratology
biochemical, 313–15
Tetracycline (tet)-regulable
system, 185
Therapeutic index
relative width of, 10
Thiopurine
S-methyltransferase
(TPMT), 15, 19
Time-resolved fluorescence
detection, 244

Timing
of replication, 163
TNF polymorphism, 380
TPMT
See Thiopurine
S-methyltransferase
Transcription
regulable systems for,
185–86
reverse, 181
Transcription-coupled repair
(TCR), 42
Transcriptional initiation,
182
Transcriptional silencing,
161, 163–64
Transcriptome
aging, 445–46
Transducers
needed by checkpoints, 45
Transducing human CD34$^+$
cells, 200–1
TRANSFAC database, 349
Transferrin
isoelectrofocusing, 132,
138
Transgene expression
regulation of, 183–86
Transgenesis, 3
Translocations
balanced, 155
Transmembrane attractant
receptors, 362
*trans*population models
of diversity, 85–86
HBB alleles, 86
HEXA alleles, 85–86
Transporters
drug, genetic
polymorphisms in, 20–21
Treatment efficacy, 9
Trichothiodystrophy (TTD),
48–49
TSD
See Tay-Sachs disease
TTD
See Trichothiodystrophy

Tuberculosis, 376–77
Tumor suppresser gene
inactivations, 57

U

UBE3A gene, 155–56,
164–65
Understanding Gene Testing
(website), 222
U.S. Department of Defense,
414
U.S. Department of Health
and Human Services,
406, 429
U.S. government agencies
on the Web, 229
U.S. Multicenter AIDS
Cohort Study, 383
USH genes, 274–80
and isolated deafness or
retinitis pigmentosa,
285–88
USH2A gene, 279–80
and isolated retinitis
pigmentosa, 287–88
Usher 2A, 285
Usher 1B, 280–83
cochlear phenotype, a
primary sensory cell
defect, 280–82
retinal phenotype, 282–83
Usher 1C, 283–84
Usher 1D, 284–85
Usher syndrome (USH)
from genetics to
pathogenesis, 271–97
loci identification, 274
three clinical subtypes,
272–74
toward pathogenesis,
280–85
the *USH* genes, 274–80,
285–88
Usherin, 280
USH1B gene, 274–77
and isolated deafness,
286–87